T0350883

Practical Pharmaceutical Engineering

Practical Pharmaceutical Engineering

Gary Prager
Livingston, NJ, US

This edition first published 2019
© 2019 John Wiley & Sons, Inc.

All rights reserved. No part of this publication may be reproduced, stored in a retrieval system, or transmitted, in any form or by any means, electronic, mechanical, photocopying, recording or otherwise, except as permitted by law. Advice on how to obtain permission to reuse material from this title is available at http://www.wiley.com/go/permissions.

The right of Gary Prager to be identified as the author of this work has been asserted in accordance with law.

Registered Office
John Wiley & Sons, Inc., 111 River Street, Hoboken, NJ 07030, USA

Editorial Office
111 River Street, Hoboken, NJ 07030, USA

For details of our global editorial offices, customer services, and more information about Wiley products visit us at www.wiley.com.

Wiley also publishes its books in a variety of electronic formats and by print-on-demand. Some content that appears in standard print versions of this book may not be available in other formats.

Limit of Liability/Disclaimer of Warranty
In view of ongoing research, equipment modifications, changes in governmental regulations, and the constant flow of information relating to the use of experimental reagents, equipment, and devices, the reader is urged to review and evaluate the information provided in the package insert or instructions for each chemical, piece of equipment, reagent, or device for, among other things, any changes in the instructions or indication of usage and for added warnings and precautions. While the publisher and authors have used their best efforts in preparing this work, they make no representations or warranties with respect to the accuracy or completeness of the contents of this work and specifically disclaim all warranties, including without limitation any implied warranties of merchantability or fitness for a particular purpose. No warranty may be created or extended by sales representatives, written sales materials or promotional statements for this work. The fact that an organization, website, or product is referred to in this work as a citation and/or potential source of further information does not mean that the publisher and authors endorse the information or services the organization, website, or product may provide or recommendations it may make. This work is sold with the understanding that the publisher is not engaged in rendering professional services. The advice and strategies contained herein may not be suitable for your situation. You should consult with a specialist where appropriate. Further, readers should be aware that websites listed in this work may have changed or disappeared between when this work was written and when it is read. Neither the publisher nor authors shall be liable for any loss of profit or any other commercial damages, including but not limited to special, incidental, consequential, or other damages.

Library of Congress Cataloging-in-Publication Data

Names: Prager, Gary, 1943– author.
Title: Practical pharmaceutical engineering / Gary Prager.
Description: First edition. | Hoboken, NJ : John Wiley & Sons, 2019. | Includes index. |
Identifiers: LCCN 2017058784 (print) | LCCN 2018005908 (ebook) | ISBN 9781119418849 (pdf) |
    ISBN 9781119418719 (epub) | ISBN 9780470410325 (cloth)
Subjects: LCSH: Pharmaceutical technology. | Biochemical engineering. | Production engineering.
Classification: LCC RS192 (ebook) | LCC RS192 .P73 2018 (print) | DDC 615.1/9–dc23
LC record available at https://lccn.loc.gov/2017058784

Cover design by Wiley
Cover image: © hakkiarslan/Getty Images

Set in 10/12pt Warnock by SPi Global, Pondicherry, India

Printed in the United States of America

V10005993_112018

Contents

Preface

The purpose of this text is to serve several purposes. The main intent is to provide a working knowledge for personnel involved in various aspects of pharmaceutical operations with technical and engineering tools to address tasks not necessarily related to their particular expertise. For instance, a chemist dealing with quality control issues can, after using the information contained in parts of Chapter 3 (Heating, Ventilating, and Air Conditioning (HVAC)), have the basic tools to deal with germane HVAC issues, should he be selected to represent the quality control user on a team responsible for design, construction, and installation of an HVAC system for a quality control laboratory. At a minimum the information provided should allow one to have a basic understanding of issues that may arise during the contractor/user briefings.

As a practical source of material for dealing with common engineering issues that arise, the intent is to also provide a basis to deal with resolving common situations encountered in pharmaceutical operations; it is not intended to present esoteric problems and/or situations requiring detailed resolutions and solutions. Such complex technical problems are not often encountered in "real-life" situations. The majority of the examples are based on everyday working situations. The need to design and size a bioreactor is performed by vendors, and the installation, commissioning, and validation are generally the responsibility of the owner/operator. Consequently, details such as verifying motor sizing and performance are typically common concerns of the user; knowing how to quickly determine motor revolutions per minute (rpm) during validation is more of a common task than detailed design of a shell and tube heat exchanger, albeit a knowledge of both topics are important for successful results.

This approach is particularly relevant since the era of in-house process engineering and design is becoming more and more irrelevant in pharmaceutical operations, as evidenced by the diminishing need for corporate design specifications and design standards and outsourcing of detailed design projects.

As a result of this redirection of assets, the emphasis on pharmaceutical engineering is now more project oriented rather than process design and modifications performed by in-house engineering operations. For instance, preparation of P&IDs is, for the most part, the responsibility of the contractor, consultant, or system vendor, ergo, the emphasis on quick and practical results.

While intended as a practical tool for practical pharmaceutical operations, this text can be envisioned as an instructional tool for an undergraduate engineering course that can augment standard coursework such as unit operations, thermodynamics, organic chemistry, biochemistry, and reaction kinetics, when considering a pharmaceutical engineering introductory course. An instructor could devise related exercises that can conform to the existing curricula (creation of appropriate exercises could also be a one or two credit graduate problem).

Since many pharmaceutical engineering courses are evening classes, the instructor could have the proposed problems be compatible with the requirements and interests of the students.

This approach would "customize" the course content.

Of course, the intent of this text is to provide an informative, useful source for many existing and evolving areas of pharmaceutical and biotechnology operations. Hopefully, it can serve as a beginning point for projects and/or a refresher for others.

While this text is authored and not edited, a great deal of assistance by others was required for completion. My friend and longtime associate, Stuart Cooper, PE, is a major reason impetus for this task. His detailed commentary to my input and his IT skills are truly masterful. Stu, my heartfelt thanks for your contributions. Also, special thanks to William B. Jacobs, PhD, whose regulatory knowledge helped make a more cohesive product. The contribution by Robert Bracco of Pfizer helped make his tableting operations input a more complete presentation.

Professor Angelo Perna, Professor Deran Hanesian, and Ed May of NJIT also assisted with their content advice and encouragement.

A special "thanks" is in order for my Wiley editors and staff, Bob Esposito, Michael Leventhal, Beryl Mesiadhas, and Vishnu Narayanan for their input and monitoring of this text.

Often, there is one person who can influence your life significantly as both a teacher and mentor. In my life this individual was the late Dr. T.T. Castonguay of the University of New Mexico Chemical Engineering Department. In addition to forcing us to master the basics, his "life lessons," integrated with class work, served as a foundation for many alums; thanks, Doc., and, of course, a bittersweet thank you to my late wife Robin. Despite her condition, she always managed to keep me focused on this task; I'll always miss you.

1

US Regulations for the Pharmaceutical Industries

1.1   Introduction

In brief, the Food and Drug Administration (FDA) is tasked with protecting the public health of residents of the United States. It is not the only agency within the government that can identify with that goal, but it *is* the agency that is responsible for ensuring citizens safe, efficacious access to an array of products that include food, drugs, and medical devices. The scope of this chapter is to concentrate on the pharmaceutical aspects of the FDA's mission; however, it is important to understand the structure of the agency, its history, and its role in the regulatory arena.

In an ideal world, there would be no need for oversight, as all actions would be for the general good of society as a whole, as opposed to individual gain at the unfair expense, be it monetary, health, or some other metric, of others. That is not a political statement, but rather leads to an understanding that most regulations, and certainly the establishment of most of *regulatory agencies*, come about as the result of egregious acts that call for remedy. That is not to say that organizations have not been created as advisory advocates for industries, independent of scandal, as in the creation of the US Pharmacopeia (USP [1]) in 1820 and the Association of Official Agricultural Chemists (now AOAC International [2]) in 1897; however, the establishment of regulatory agencies historically has been reactive rather than proactive.

Practical Pharmaceutical Engineering, First Edition. Gary Prager.
© 2019 John Wiley & Sons, Inc. Published 2019 by John Wiley & Sons, Inc.

It would be naïve, however, to suggest that regulatory agencies, including the FDA, are independent of political influence; they are not, nor can they be, given the structure of our legal system. The centerpiece of our legal system is the US Constitution, which establishes the structure of our country and also defines how we self-regulate. The legislative branch, working within the framework of the Constitution, establishes federal statutes (or legislations) that reinforce the principles of the Constitution and establish control of our society. The rules and proposed rules, as well as notices of federal agencies and organizations, executive orders, and documents are published daily in the Federal Register.

The Code of Federal Regulations (CFR) is the codification of the rules posted in the Federal Register. It is updated once each calendar year and issued quarterly. There are currently 50 titles in the CFR, with 21 CFR covering Food and Drugs. This codification is meant to clarify regulations, denoting the intent of the legislation passed. However, as might be expected, the regulations are subject to interpretation. Ultimately, disputes about the interpretation of legislation, as well as its constitutionality, are clarified by the Judicial Branch, which reviews specific complaints or disputes and can elect to apply its opinion narrowly to the specific dispute or as an overarching opinion having much broader impact. At the time of publication, the CFR can be accessed online at http://www.ecfr.gov/cgi-bin/ECFR?page=browse. This, as well as any other online address in this text, is subject to change.

The crafting of statutes, the codification of the legislation, and the interpretation of both the intent and the scope of regulations are all subject to the vagaries of human judgment and influence; hence the previous statement that regulatory agencies are subject to political influence. Reviewing the timeline of the formation of the FDA as provided on its own website (http://www.fda.gov/AboutFDA/WhatWeDo/History/Milestones/ucm128305.htm) illustrates the difficulty of establishing regulation in the face of competing influences. Be that as it may, once the regulations and regulatory agencies are established, there has historically been remarkable resistance to the politicization of the agencies themselves. The "greater good" prevails.

1.2   The FDA: Formation of a Regulatory Agency

The seminal event that led to the formation of the precursor to the FDA was the discovery of adulterated antimalarial drugs (quinine) being imported into the United States at time when malaria was a major health concern. In 1848 Congress required US Customs Service inspectors to stop the importation of these drugs when it passed the Drug Importation Act, effectively sealing off the United States from unscrupulous overseas manufacturers. Almost 50 years later, it was again the US Customs Service that was tasked, at importers expense, with the inspection of all tea entering the United States when the Tea Importation Act of 1897 was implemented.

In 1862, President Abraham Lincoln appointed Charles M. Wetherill, a chemist, to serve in the newly created US Department of Agriculture (USDA). The USDA housed the Bureau of Chemistry, a precursor to the FDA, where Wetherill began investigating the adulteration of agricultural products. Succeeding USDA Chief Chemists Peter Collier (1880) and Dr. Harvey W. Wiley (1883) expanded the food adulteration studies and campaigned for a federal law regulating foods. For his efforts, Dr. Wiley is regarded as the "Father of the Pure Food and Drugs Act," having vigorously crusaded for its eventual passage.

In 1902 the Biologics Control Act was passed to ensure purity and safety of serums, vaccines, and similar products used to prevent or treat diseases in humans by licensing biologics manufacturers and regulating the interstate commerce of biologics.

The first major legislation was passed in response to growing outrage, fanned by muckraking writers, over the unsanitary conditions in meat-packing plants and the presence of poisonous preservatives and dyes in foods. The original Food and Drug Act was passed in 1906 prohibiting interstate commerce of misbranded or adulterated foods, drinks, and drugs. The Federal Meat Inspection Act was passed the same day. The next year, the Certified Color Regulations listed seven color additives that were considered safe in food. Poisonous, colorful coal-tar dyes were banned from foods.

From 1912 to 1933, a series of minor back-and-forth legislative and judicial rulings effectively increased the regulations against misleading therapeutic statements, mislabeling of contents, and other deceptive practices. Also imposed were more stringent requirements for the dispensing of narcotic substances and the qualitative and quantitative labeling of package contents. Still under the auspices of the USDA, the precursor to the FDA began to be separated from nonregulatory research, which was placed under the aegis of the Bureau of Chemistry and Soils in 1927. The beginning of the separation of regulation of meat and dairy products from FDA control began in 1930, the same year the name was officially changed to the FDA.

This new agency recommended a complete revision of the obsolete 1906 Food and Drugs Act, launching a 5-year legislative battle. The second major regulatory revision, the 1938 Federal Food, Drug, and Cosmetic Act (FD&C) was largely passed as a result of a 1937 incident in which 107 persons were killed by consuming Elixir Sulfanilamide containing the poisonous solvent diethylene glycol. As a result, new provisions were added:

- Control was extended to cosmetics and therapeutic devices.
- New drugs were required to be shown to be safe *prior* to marketing.
- Eliminated the need to prove intent to defraud in misbranding cases.
- Provided safe levels of poisonous components that were unavoidable.
- Authorized standards of identity, quality, and fill weights for foods.
- Authorized inspections of manufacturing facilities.
- Added court injunctions to the previously authorized penalties of seizures and prosecutions.

That same year, however, regulation of advertising of all FDA-regulated products with the exception of prescription drugs was transferred to the Federal Trade Commission (FTC).

In 1940, the FDA was transferred from the USDA to the Federal Security Agency, precursor to the Department of Health, Education, and Welfare (HEW). In the 1940s a Supreme Court decision extended liability for violations by companies to officials responsible within the company regardless of their knowledge of the violations. Two particular amendments were passed requiring the FDA to test and certify the purity and potency of the drugs insulin and penicillin. Other legislation extended the reach of government and the maintenance of public health and confirmed the agency's regulatory control over interstate commerce. At the end of the decade, the FDA published for the first time guidance to the industry and procedures for appraisal toxicity of chemicals in food.

In the 1950s, there was an increased oversight of both food and drug products, including their labeling. Drugs that required medical supervision were restricted in their sale, requiring a licensed practitioner to authorize purchases. The purpose for which a drug is offered was required to be on the label as part of the directions for use of that product. The factory inspection was found to be too vague and therefore was reinforced by a further amendment in 1953. The FDA increased its oversight of the safety of foods with the Miller pesticide amendment, the food additives amendment, and the color additives amendment.

In the 1960s the United States was spared of the tragedy suffered by Western European families because the drug thalidomide was kept off the US market, preventing birth defects affecting potentially thousands of babies. This success, by the FDA medical officer Frances Kelsey, aroused strong public support for stronger drug regulation. As a result the Kefauver–Harris drug amendments were passed to ensure drug efficacy and greater drug safety. These amendments required that drug manufacturers prove to the FDA the effectiveness of their products before placing them on the market. The FDA contracted with the National Academy of Sciences and National Research Council to evaluate the effectiveness of 4000 drugs that had been approved on the basis of safety alone between 1938 and 1962. Other legislation enacted in the 1960s included Drug Abuse Control Amendments, to combat abuse of stimulants, depressants, and hallucinogens, and a Consumer Bill of Rights.

In the 1970s further consumer protections were put into place with the first patient package insert for oral contraceptives that delineated the risks and benefits of taking the drug. The Comprehensive Drug Abuse Prevention and Control Act replaced previous laws and categorized drugs based on abuse and addiction potential versus their therapeutic value. Some responsibility shifted among government agencies with the Environmental Protection Agency (EPA) taking over the FDA program for setting pesticide tolerances. Regulation of biologics – including serums, vaccines, and blood products – was transferred from the National Institute of Health (NIH) to the FDA. Over-the-counter drug reviews began to enhance the safety, effectiveness, and labeling of drugs sold over-the-counter. The Bureau of Radiological Health was transferred to the FDA to protect humans against unnecessary exposure to radiation from products in the home, in industry, and in healthcare professions.

The 1980s saw the FDA revise regulations on drug testing, greatly increasing protections for subjects upon whom new drugs were tested. In reaction to deaths caused by cyanide placed in Tylenol bottles, packaging regulations requiring tamper-resistant closures was enacted. The FDA also promoted research and marketing of drugs needed for treating rare diseases with the Orphan Drug Act. To promote competition and lessen costs, the FDA allowed the marketing of generic versions of brand-name drugs without requiring repeating the research necessary to prove them to be safe and effective. At the same time, they gave brand-name companies the right to apply for up to 5 years of additional patent protection for the new medicines they had developed to make up for the time lost, while the products were going through the FDA's approval process.

Acquired immune deficiency syndrome (AIDS) tests for blood were approved by the FDA to prevent the transmission of the causative agent to recipients of blood donations. The marketing of prescription drugs was limited to legitimate commercial channels in order to prevent the distribution of mislabeled, adulterated, subpotent, and/or counterfeit drugs to the public.

Investigational drug regulations were revised, expanding access to investigational drugs for patients with serious diseases with no alternative therapies. This trend was continued in the early 1990s as regulations were established to accelerate a review of drugs for life-threatening diseases.

In 1994 the Dietary Supplement Health and Education Act established specific labeling requirements, a regulatory framework, and authorized the FDA to promulgate good manufacturing practice (GMP) regulations for dietary supplements. Dietary supplements and dietary ingredients were classified as food, and a commission was established to recommend how to regulate any claims appearing on the labels. As a result of this, 21 CFR part 111 Current Good Manufacturing Practice (cGMP) in manufacturing, packaging, labeling, or holding operations for dietary supplements was established.

Also in the 1990s was a relaxation of some regulations on pharmaceutical manufacturers including an expansion of allowable promotional material on the approved use of drugs. It was during this period that the FDA attempted to extend its reach to the tobacco industry, defining nicotine as a drug and smoking or smokeless tobacco products to be combination of drug delivery systems, restricting the sale of such materials to minors. The FDA was forced to rescind its rule in 2000 when the Supreme Court upheld a lower court ruling supporting a lawsuit by a tobacco company against the FDA.

In the 1990s there was increased focus on the effectiveness of drugs as influenced by gender and, in 2002, in children. This was a reaction to the discovery that drugs commonly tested on male subjects left unresolved the question of how female subjects responded to exposure to these drugs. Similarly, the safety and efficacy of drugs prescribed for children was required.

In the 2000s there was again a response to the current events. The Public Health Security and Bioterrorism Preparedness and Response Act of 2002 was designed to improve the country's ability to prevent and to respond to public health emergencies. In response to questions about the jurisdiction of various departments within the FDA, the Office of Combination Products was formed to oversee products that fall into multiple jurisdictions, for example, medical devices that contain a drug component.

The cGMP initiative focused on the greatest risks to public health in manufacturing procedures applying a consistent approach across FDA. It also ensured that process and product quality standards did not impede innovation of new products.

In general, in this new century the FDA has continued to respond and grow in three main areas:

1) Responding to specific external forces, as in COX-2 selective agents and dietary supplements containing ephedrine alkaloids as health risks. The Drug Quality and Security Act (DQSA) of 2013 in response to an epidemic of fungal meningitis linked to a compounded steroid, among other provisions, outlined steps for an electronic and interoperable system to identify and trace certain drugs throughout the United States.
2) The FDA increased its influence on product development (for both human and nonhuman species) by encouraging specific remedies and also by expanding how the FDA can collaborate in the process of developing therapeutic products from laboratory to production to end use. Establishment of user fees for drugs, medical devices, and biosimilar biologic agents that are targeted to fund expedited reviews.
3) Has promoted a continuation of improved dissemination of information to both physicians and patients.

In summary, the FDA was created out of necessity in response to events that threatened the health and safety of citizens with regard to their food and medical supplies. It has continued to oversee our food and drug supply for both humans and animals as it has evolved. Perhaps the most influential pieces of legislation were the Food and Drugs Act of 1906, the Food Drug and Cosmetic Act of 1938, the Kefauver–Harris Amendments of 1962, and a Medical Device Amendments of 1976. Until 1990 all US laws and regulations relating to medical products were in reaction to medical catastrophes. A proactive stance, with new laws and regulations written to avoid medical calamities began in the 1990s.

There are corresponding agencies around the world that operate independently according to their individual mandates from their legislative bodies. In some cases the relations in the United States are more restrictive than those agencies of other countries; in other cases the United States is less restrictive in its oversight. Given the ever-increasing interrelationships of multinational companies and their markets, there is great impetus to align the regulatory requirements of individual countries into harmonized code. International agencies are working toward that end at this time. However, the trend in regulation, while vacillating, has been toward the more restrictive, including more detailed accountability and traceability of all products. This is likely to continue.

With the trend toward greater regulation, greater international harmonization and acceptance of the FDA as a partner in producing safe, efficacious, high-quality products, and learning to work with this development will be most beneficial not only for the consumers but also to the manufacturers. The FDA focuses on ensuring public safety within the scope of their mandate, and it is in the best interest of all. Rather than view the FDA as an adversary to be controlled, the FDA should be viewed as a partner in product development.

1.3 FDA's Seven Program Centers and Their Responsibility

1.3.1 Center for Biologics Evaluation and Research

This is the center within the FDA that regulates biological products for human use including blood, vaccines, tissues, allergenics, and cellular and gene therapies. Biologics are derived from living sources and many are manufactured using biotechnology. They often review cutting-edge biomedical research, evaluating scientific and clinical data submitted to determine whether or not the products meet the Center for Biologics Evaluation and Research (CBER)'s standards for approval. The approvals may be for newly submitted biologicals or for new indications for products already approved for a different purpose.

1.3.2 Center for Drug Evaluation and Research

The Center for Drug Evaluation and Research (CDER) oversees over-the-counter and prescription drugs including biological therapeutics and generic drugs. For regulatory purposes, products such as fluoride toothpaste, antiperspirants and dandruff shampoos, and sunscreens are all considered to be drugs.

1.3.3 Center for Devices and Radiological Health

FDA's Center for Devices and Radiological Health (CDRH) is tasked with eliminating unnecessary human exposure to man-made radiation from medical, occupational, or consumer products in addition to ensuring the safety and effectiveness of devices containing radiological materials. The CDRH is particularly concerned about the lifecycle of the product from conception to ultimate disposal in a safe manner.

1.3.4 Center for Food Safety and Applied Nutrition

Center for Food Safety and Applied Nutrition (CFSAN) is responsible for ensuring a safe, sanitary, wholesome, and properly labeled food supply. It is also responsible for dietary supplements and safe, properly labeled cosmetic products. As needed, it may work in conjunction with other centers as, for example, with CDER or enforcement of the FD&C Act or products that purport to be cosmetics but meet the statutory definitions of a drug.

1.3.5 Center for Veterinary Medicine

The Center for Veterinary Medicine (CVM) regulates the manufacture and distribution of food additives, drugs, and medical devices that will be given to animals. The animals may be either for human consumption or companion animals. One growing area of interest is that of genetically modified or genetically engineered animals. The FDA has expressed an interest in regulating these animals; however, depending upon the animal species and its intended use, the FDA will regulate these animals in combination with other federal departments and agencies such as the USDA and the EPA.

1.3.6 Office of Combinational Products

Combination products are defined in 21 CFR 3.2(e) as:

1) A product composed of two or more regulated components, i.e. drug/device, biologic/device, drug/biologic, and drug/device/biologic, that are physically or chemically combined or mixed and produced as a single entity.

2) Two or more separate products packaged together in a single package or as a unit and composed of drug and device products, device and biological products, or biological and drug products.

3) A drug, device, or biological product packaged separately that according to its investigational plan or proposed labeling is intended for use only with an approved individually specified drug, device, or biological product where both are required to achieve the intended use, indication, or effect and where upon approval of the proposed product the labeling of the approved product would need to be changed, e.g. to reflect a change in intended use, dosage form, strength, route of administration, or significant change in dose.

4) Any investigational drug, device, or biological product packaged separately that according to its proposed labeling is for use only with another individually specified investigational drug, device, or biological product where both are required to achieve the intended use, indication, or effect.

1.3.7   Office of Regulatory Affairs

The Office of Regulatory Affairs (ORA) oversees the field activities of local FDA field operations. It also provides FDA leadership on imports inspections and enforcement policy, inspects regulated products and manufacturers, conducts sample analyses of regulated products, and reviews imported products offered for entry into the United States. The ORA also advises the commissioner and other officials on regulations and compliance-oriented matters and develops FDA-wide policy on compliance and enforcement. The ORA develops and/or recommends policy programs and plans activities between the FDA and state and local agencies.

1.4   New Drug Development

While the overall focus of this book is on the manufacture of pharmaceuticals, it is useful to understand how drugs are developed. Every new formulation must undergo a series of tests to prove it is both safe and efficacious to the consumer. The FDA estimates that it takes over 8 years, from concept to approval for public consumption of a new drug. At any stage in the investigation, or during postmarket evaluations, the drug may be deemed unsafe and restricted from market. The FDA does not actually test the drug itself for safety and efficacy, but rather reviews data submitted by the drug company sponsor.

1.4.1   Discovery

A typical drug development pathway involves the generation of large numbers of molecules of similar structures with the intention of identifying the most promising candidates for further development. The rationale behind this is that slight variations on a known structure may attenuate the behavior of the known molecule in a desirable fashion. That is to say, substitution on a well-characterized structure may be expected to increase beneficial properties of the chemical or alternately decrease detrimental characteristics.

The discovery of a new drug involves more than formulation development. On the lab scale, research and development will determine the potential drug stability and active ingredients, as well as any other requirements. A formal protocol for nonclinical studies must be designed to establish exactly how the preclinical study will be performed, including the types of animals to be tested, the duration and frequency of the test, and how the data will be handled. Finally chemistry, manufacturing, and controls (CMC) will be established to allow larger scale production of the drug under GMP.

Scale up from bench to manufacture requires consideration of the following:

Active ingredients: identity, purity, and stability.
Raw materials specifications and identification.
Intermediate products.
Filtration and/or purification process.
Solubility, particulate size, disintegration, dissolution (for pills and capsules).
Sterility requirements.
Final drug specifications.
Dose uniformity.
Required QC tests.
Methodologies for QC assays.
Validations: QC assay method
  Equipment
  Cleaning
Record keeping and documentation

The pertinent area of the CFR regarding investigation into the potential of a new drug for human use is 21 CFR 312, Investigational New Drug Application (IND or INDA). In this part of the regulations, procedure requirements governing use of investigational new drugs including stipulations for the submission for review to the FDA are found.

1.4.2 Investigational New Drug Application

It is illegal to transport unapproved drugs across state lines for any purpose. Thus there exists the necessity to request an exemption from this federal statute in order to conduct clinical trials. In order to transport a new unapproved drug, an IND or INDA must be filed to get an exemption from the statute. Form 1571 can be obtained from the FDA website (http://www.fda.gov/opacom/morechoices/fdaforms/cder.html).

Required information for the submission includes the data collected from the preclinical animal pharmacology and toxicology studies, showing the safety of the proposed drug. It must be demonstrated that the manufacturer can reliably reproduce and supply consistent batches of the said drug, so information about the composition, manufacture stability, and controls for manufacture must be supplied. Finally the detailed protocols for the proposed clinical studies, including the qualifications of the clinical investigators, and commitments to obtain informed consent from the research subjects, commitments to review the study by an institutional review board (IRB), and a firm commitment to adhere to investigational new drug regulations must be submitted.

The investigation of a drug for potential human applications is initiated and overseen by a sponsor committed to properly conduct a study, be they an institution or organization, a company, or even an individual. They are responsible for the management, from start to finish, of a clinical trial. Alternately, they may provide financing for the study by investigators who will actually initiate and complete the study. The sponsor does not, however, relinquish responsibility simply by financing a project proposed by an individual investigator.

Once the required NDA is submitted to the FDA, it is assigned an IND number that is to be used in all correspondence with the FDA regarding the application. The FDA or more specifically the CDER will review the IND. The IND is reviewed on medical, chemistry, pharmacology/toxicology, and statistical bases to review the safety of the proposed study. If the review is complete and acceptable with no deficiencies, the study may proceed. If not, a clinical hold is placed on the study and the sponsor is notified, affording him the opportunity to submit new data.

INDs are not approved by the FDA. An IND becomes effective 30 days after receipt by the FDA unless a clinical hold is imposed. The clinical hold can be placed at any time and is an order by the FDA to suspend or delay a proposed or ongoing clinical investigation. The clinical hold is commonly placed upon the study for deficient study design, unreasonable risk to subjects, inclusion of an unqualified investigator, misleading investigator brochure submission, or insufficient information to assess the risk to test subjects.

Once the IND is in effect, it must be maintained so that current information is submitted to the FDA. Toward this end, amendments are made to the original protocol. These may be either protocol amendments or information amendments. Three types of protocol amendments may be submitted: for a new protocol, a change in protocol or a new investigator carrying out a previously submitted protocol. Informational amendments fall outside the scope of the protocol amendments. An information amendment is any amendment to an IND application with information essential to the investigational product that is not within the scope of protocol amendments, safety reports, or annual reports. This may include new technical information or discontinuation of the clinical trial.

A written safety report that transmits information about any adverse drug experience or adverse events associated with the use of the drug is to be submitted to the FDA and all participating investigators along with Form 3500A as soon as possible, but no more than 15 calendar days after initial notification to the sponsor. In the case of serious adverse events, the report must be submitted no later than 7 days after the receipt of information by the sponsor. The sponsor will follow up and investigate all safety and relevant information and report to the FDA as soon as possible.

An annual report is to be sent to the FDA to update the IND about the progress of the investigation and all changes not reported in amendments or other reports. It should be submitted within 60 days of the calendar date that the IND went into effect.

Additionally, meetings may be scheduled with the FDA at various stages of investigation. Meetings may be held pre-IND to discuss, for example, CMC issues. Meetings may also be held at the end of Phase I, Phase II, or pre-new drug application (NDA).

The IND can be withdrawn by the sponsor at any time without prejudice. The FDA and all pertinent IRBs will be notified. Any remaining drugs will be disposed of by the sponsor or returned to the sponsor.

An IND may go on inactive status at the request of the applicant or the FDA if, for example, no human subjects entered the study within a period of 2 years, or if the IND remains under a clinical hold for 1 year or more. An inactive application may be reactivated if activities under the IND have recommenced. An IND that remains on inactive status for 5 years or more may be terminated.

The IND may also be terminated for cause by the FDA. Such cause may be determination that test subjects may be exposed to significant or unreasonable risk or if methods, facilities, and controls used for the manufacturing are inadequate to maintain appropriate standards for quality and purity of the proposed drug as needed for subject safety. Additional grounds for termination may be found in 21 CFR 312.44.

1.4.3 Preclinical Studies (Animal)

Before a drug can be tested on a human being, it must be shown to be safe. This can be established by compiling data from previous nonclinical studies on the drug, by compiling data from previous clinical testing or data from markets in which the drug has previously been sold, if relevant, or new preclinical studies may be undertaken. Both in vivo and in vitro laboratory animal studies are used.

These preclinical studies must be able to show any potential toxic effects under the conditions of the proposed clinical trial. The toxicity studies should include single and repeated dose studies, reproductive studies, genotoxicity, local tolerance studies, and the potential for carcinogenicity or mutagenicity. Additionally pharmacology studies to establish safety and pharmacokinetic studies to determine how the drug reacts in the body (absorption, distribution, metabolism, or excretion) may be performed.

At this stage the FDA will generally ask for a pharmacological profile of the drug, a determination of the acute toxicity in at least two species of animals, and a short-term toxicity study. Under 21 CFR 312.23(a)(8) the basic safety tests are most often performed in rats and dogs. Selection of a safe starting dose for humans, suggestion of the target organs subject to toxic reactions, and a margin of safety between therapeutic doses of a toxic substance will be established.

Good laboratory practice (GLP) covers several different aspects of preclinical studies. An organizational chart delineating responsibilities and reporting relationships is essential. A quality assurance unit (QAU) is required to ensure that the study takes place under GLP standards. The testing facility must be of the proper size and condition to allow proper conduct of the studies. Feed, bedding supplies, and equipment must be stored separately and protected from contamination. A separate space must be maintained for the storage of test and control items. Laboratory space for routine and specialized procedures must be separated and data reports and specimens must have a separate, limited access area.

Any equipment used for data collection or assessment must be maintained, calibrated, and kept clean. Written standard operating procedures (SOPs) must be maintained for all aspects of specimen or data handling. All prepared solutions and reagents must be properly labeled with the name of the contents, the concentration, the preparer, the expiration date, the date of preparation, and the required storage conditions.

There must be a written protocol that clearly indicates the objectives and methods for the study. The study must be conducted in accordance with the approved study protocol. Proper forms will be used for the collection of data. If data is collected manually, the data must be recorded legibly and in ink, at the time it is observed or determined, with the dated signature of the person collecting the data.

1.4.4 Clinical Studies

Once the IND is in effect, clinical trials may begin. These are conducted in at least three phases under good clinical practices (GCP).

1.4.4.1 Phase I Studies

Traditional Phase I studies are the first exposure of humans to the drug and are designed to evaluate how the drug acts in the body and how well it is tolerated. The human pharmacological studies evaluate the pharmacokinetic parameters, generally in healthy volunteers who are not the target market for the drugs, although some patients may be included in Phase I studies. These studies generally start out with single dose, followed by escalated dosage and short-term repeated dose studies. These trials are very closely monitored. Well-designed Phase I experiments will greatly aid the design of Phase II studies.

The FDA will periodically issue guidance to industry, outlining its then current thinking on pertinent topics. Such guidance does not establish legally enforceable responsibilities but rather should be viewed as recommendations. One such guidance was issued in June 2016, jointly by the CDER and the CBER providing information for industry, researchers, physicians, IRBs, and patients about the implementation of FDA's regulation on charging for investigational drugs under an IND for the purpose of either clinical trials or expanded access for treatment use (21 CFR 312.8), which went into effect on 13 October 2009.

Another guidance was developed by the Office of New Drugs in the CDER in 2006 was for exploratory IND studies. There exists a great deal of flexibility in existing regulations regarding the amount of data that needs to be submitted with an IND application. This guidance suggests that industry as a whole has been submitting more information for an IND than is required by regulations. The guidance sought to clarify the manufacturing controls preclinical testing and clinical approaches that should be considered when planning limited early exploratory IND studies in humans. Within the guidance the phrase "exploratory IND study" is

"intended to describe the clinical trial that:

is conducted early in Phase 1
involves very limited human exposure, and
has no therapeutic or diagnostic intent (e.g., screening studies, micro dose studies).

These exploratory IND studies precede traditional Phase I dose escalation, safety, and tolerance studies of investigational new drug and biological products.

In vitro testing models may examine binding sites, the effect on enzymatic activities, toxic effects, and other pharmacologic markers. These initial screening tests often require only small quantities of the drug of interest; any in vitro testing may eliminate unlikely candidates. Those candidates that provide the expected pharmacologic response will then be produced in larger quantities for in vivo testing in small animals to determine the efficacy and safety of the drug. In vitro testing is generally cheaper and less restrictive than in vivo testing, and the screening at this level is quite important.

The expense of conducting human trials is formidable; therefore the agency observed that "new tools are needed to distinguish earlier in the process those candidates that hold promise from those that do not." Traditionally, an IND is filed for one chemical entity that proved most promising during in vitro testing and subsequently showed promise in supporting toxicological data during studies of the investigational drug in animals.

The guidance suggests that exploratory IND studies having no therapeutic or diagnostic intent, be used in very limited population studies of short duration to limit human exposure, but further refine the efficacy and safety of the potential drug. For example, they can be used to determine if the method of action or response in humans is the same as that in the test animals (e.g. a binding property or enzyme inhibition). This further refinement can help select the most promising candidate from a group of products designed for a particular therapeutic effect in humans.

In-depth description of the exploratory filing as opposed to the traditional IND filing is beyond the scope of this chapter. Information for the candidate product in an exploratory IND application is similar to that of the traditional IND application including physical, chemical, and/or biological characteristics as well as the source (animal, plant, biotechnology, or synthetic derivation), the therapeutic class, doses, and administration routes intended for human trial.

Analytical characterization of the candidate product may be offered under two scenarios within the IND application. In the first case the chemicals used will be the same batch as those used in in vitro animal testing. Their use is to qualify the potential drug. It is recommended that the impurity profile of the drug be established to the extent possible; however at this stage in product development, not all impurities need be fully characterized. If issue arises during toxicological studies, it can be addressed at that time using appropriate agency guidance even when the sponsor files a traditional IND for further clinical investigation.

The second case is where the candidate drug to be used in clinical studies may not be from the same batch as that used in the preclinical studies. The focus in this situation is to demonstrate that the batch to be used is representative of the batch used in nonclinical toxicology studies, and this must be supported by relevant analytical comparisons.

Safety is, of course, paramount and the preclinical safety programs may be tailored to the exploratory study design, for example, micro-dose studies that are designed to evaluate pharmacokinetics or imaging of specific targets, such as binding affinity, and are not designed to induce pharmacologic effects. The single exposure to micro quantities is comparable with routine environmental exposures; therefore routine safety pharmacology studies are not needed. All preclinical safety studies supporting the application will be consistent with GLP.

1.4.4.2 Phase II Studies

Phase II studies are exploratory to determine the safety and efficacy of the drugs. These are generally referred to as therapeutic exploratory studies. The population is larger than that of Phase I. These studies are designed to demonstrate the therapeutic activity of the treatment and to assess the short-term safety of exposure to the drug. Dose response studies of this Phase will help to refine the appropriate dose ranges or regimens, thereby optimizing the design of the extensive Phase III studies.

1.4.4.3 Phase III Studies

Phase III studies are done in larger populations of patients to confirm the results of the Phase II studies. These are generally called confirmatory clinical trials. The purpose of this study is to determine the short- and long-term risk–benefit balance of the active ingredient and to assess its overall therapeutic value. The data gathered from Phase III trials can be extrapolated to the general population. At the completion of Phase III studies, the data are submitted to the FDA as part of an NDA, with the intention of marketing of introducing the drug to market.

1.4.4.4 Phase IV Studies

Phase IV studies are generally referred to as postmarketing studies, with the attendant implication that the drug has proven safe. It is, however, important to realize that critical information can be gained from postmarket studies. The finest designed Phase I, II, and III studies can have only a finite number of subjects taking part in the studies. The population that comprises the studies may not be large enough to statistically show an adverse effect that is limited in occurrence to a small segment of the general population. Once the drug is introduced to the marketplace, a much larger, more diverse population will, essentially, become test subjects in a Phase IV study. Careful analysis of the data may reveal adverse reactions that were not apparent in prior Phase studies.

1.4.4.5 Institutional Review Board

The studies should be conducted under GCP, ensuring that the reports from the clinical trials in the data gathered are credible and accurate and that the rights, integrity, and confidentiality of trial subjects are protected. These principles are in accordance with the Declaration of Helsinki regarding participation in medical experiments. One key provision of that declaration is the right to self-determination and informed consent by any participants in a study.

Informed consent forms should be obtained from the IRB and provided to the test subject along with any other written material that will explain exactly what his or her participation in the trial entails. The participant in this trial is being put at risk for the purpose of helping his fellow man and or financial gain. That money changes hands is insufficient reason to withhold information about the study. Informed consent is just that a full and complete disclosure of all the risks known to the best knowledge of the sponsor or investigator to which the subject will be exposed. Each participating test subject must sign a consent form prior to any initiation of the clinical trial.

An IRB is composed of a minimum of five experts with different backgrounds including scientific and nonscientific areas and at least one who is independent of the institutional trial site. The IRB is designated to review and monitor medical and biomedical research using human subjects. It is their purview to review the protocols and to maintain the safety of the test subjects at all times.

The object of clinical trials is to demonstrate the safety and efficacy of drugs for use in the human population. Although the drugs have passed initial stages of testing in preclinical (animal) populations, demonstrating that their injection or ingestion will leave the animals unharmed, how the drug interacts with the human body is still an unknown quantity. The trial of any unknown drug must contain some unknown risk to the test subject that must be addressed prior to the commencement of any human testing. The welfare of the individual test subject cannot be overlooked when qualifying drugs that may be beneficial to the greater population. The risk/benefit ratio to the test subject must be determined, balancing any perceivable risks or discomforts to which the test subject may be exposed against the anticipated benefits the drug may provide. Clinical testing of the drug can only be performed if that ratio is justifiable. Paramount above all is the test subject safety.

The IRB is the ultimate arbiter and will determine if the clinical test will go forward. Any clinical trials will be performed in compliance with instructions and/or approvals provided by the IRB. Some of the factors that the IRB will consider include what types of people may join a population of test subjects, the scheduled treatments, medications and dosages, procedures and tests, and the overall length of the designed study.

The IRB will also determine that the investigators and all support staff are qualified by training, education, and experience to perform the studies. Staffing should be adequate to perform all of the necessary duties as outlined in the protocol. Any medical decisions concerning treatment of the patient must be made by a physician or other qualified medical person.

1.4.4.6 Clinical Data Monitoring Committees

The collection and handling of data is of critical importance during these trials. A properly designed trial will yield much information about the effects of the drug in the human body. It is imperative that the data is properly collected and is reviewed in a timely fashion, and the evaluation of the data is acted upon as necessary. Proper handling of the data will allow the sponsor to assess the progress of the clinical trial to determine whether to continue the study, modify, or terminate it if the safety of the subjects becomes an issue when the expected efficacy of the drug is not presented.

Sponsors of studies evaluating new drugs or devices are required to monitor these studies (21 CFR 312.50 and 312.56 for drugs and biologics) on an ongoing basis and may find it advisable to establish data monitoring committees (DMCs) to evaluate the accumulating data in clinical trials. The DMC may advise the sponsor of discovered adverse effects that may compromise the safety of the trial subjects as well as the continuing evaluation of the validity and scientific merit of the trial.

1.4.4.7 Quality Assurance

The materials to be studied must be produced under GMPs (alternately called cGMPs: current good manufacturing practices); that is to say, they must be produced in accordance with all the standard practices and procedures normally associated with producing pharmaceutical materials. The manufacture, handling, storage, dispensation, and ultimate dissemination of the investigational drug must be performed in adherence to the procedures outlined in the investigational protocol. If the clinical investigations conducted under the IND are terminated prior to the completion of the experiments as outlined in the investigational protocol, all stocks of the investigational drug should be returned to the sponsor or otherwise disposed of as a sponsor dictates.

There must be in place adequate quality assurance systems not only to ensure proper collection, tabulation, and reporting of data but also to maintain conformance with the procedures outlined in the investigation of protocol. Quality assurance must extend from the manufacture of the investigational drug, through the selection of both the individual and the collective test subjects, and through the administration of investigational drug to the subjects as well as the subsequent analytical procedures used to gather data and the ultimate compilation of that data into a report. Any laboratory analyses must use validated procedures and be appropriate for the data that they are intended to provide.

The clinical trial protocol is a formal document that is submitted and accepted as a template for the study. Information provided in the protocol include descriptors of the study, the date, name, and address of the sponsor and or investigators authorized to initiate the protocol, the medical expert, trial sites, and clinical laboratories where the investigation will take place.

1.4.4.8 Investigator's Brochure

The drug to be investigated will be described in detail including its manufacture and a summary of the procedures and results of the nonclinical (animal) studies that serve as a basis for determining the dosage and application schedule of the investigational drug to the human subjects. The vector for conveying this information is in the investigator's brochure (IB). This is a compilation of all data, clinical and nonclinical, relevant to the study of the investigational drug and human test subjects. IB should begin with a summary, highlighting pharmaceutical, pharmacological, pharmacokinetic, toxicological, physical, and chemical information that has been gathered and is relevant to the development of a clinical study. Included in more detail is that the summary will be the physical, chemical, and pharmaceutical properties and formulation of the drug. The result includes any nonclinical studies including pharmacokinetics, drug metabolism in preclinical subjects including toxicology, the effects of any studies that were conducted on humans including safety and efficacy, and the drug's interaction with the human body. For studies researching new indications for existing drug, results of previous investigational studies should be included here, including postmarket investigations if any were conducted.

Here clearly defined description of the objective of the study as well as the experimental design of the study will be detailed.

The basis for the selection of the individuals that will partake in the study as test subjects as well as any reasons for exclusion of potential test subjects will be defined. For example, test subjects may be required to have a particular condition that is potentially responsive to the investigational drug. The inclusion of "normal" subjects may be sufficient to demonstrate the safety of the drug, but may be unable to support any findings of efficacy of the drug, as they would not have the physiological condition targeted by the proposed drug. Alternately, it may have been determined in the preclinical studies that the investigational drug is potentially dangerous to a limited number of people with specific indications and that risk may be minimized by excluding the defined subset from the investigational group. As described earlier, the risk–benefit analysis may be such that the potential therapeutic value of the investigational drug is outweighed by the overall risk to these potential test subjects, excluding this subset of the human population. The risk cannot be ignored, and the only safe way to proceed with a study of this type is to exclude from the potential test population those who would be harmed by the drug.

The treatment of the test subjects must be described in detail including how the drug will be administered and how the health of the subjects will subsequently be monitored, as well as the methods used to determine the safety and efficacy of the drug in human usage. Any methodologies for obtaining samples for analyses from the subjects must be detailed, and the handling of the data (recording, analyzing, etc.) included in the ultimate reports must be detailed in the protocol.

1.4.4.9 Informed Consent

Second only to the safety of the test subject is the maintenance and respect of the privacy of the individual. It is important to acknowledge that abuse of research patients by investigators has occurred in relatively recent times, perhaps most egregiously in the infamous Tuskegee syphilis study in 1928. Initially started with the best of intentions, the Great Depression caused the financial sponsor of the study to withdraw funding to the US Public Health Service (PHS). The study sought to treat the occurrence of syphilis in black men living in various counties in Mississippi, Virginia, Georgia, Alabama, North Carolina, and Tennessee, in a test population of over 2000 men, 25% of whom had tested positive for syphilis. With restricted finances the PHS was unable to treat the infected men and the focus of the project changed.

There was a question at the time, of whether or not the progress of the disease within the black population was different from that in the white population. It was therefore decided to track the progression of the disease in the infected men without informing them that they were infected. The subjects received routine examinations but with either no treatment or substandard ineffectual treatment for their underlying condition, syphilis. Not only did the PHS not treat the subject for their disease, but also they prevented other government agencies from treating the "patients." When, in 1943, the PHS routinely began to use penicillin to treat patients under its purview, it specifically excluded those subjects of the Tuskegee syphilis study. Further it tracked the test subjects through the end of the study in the early 1970s, preventing them from receiving treatment. Even then the study was not ended by the PHS voluntarily, but rather only after being exposed by a reporter.

While we would like to think that nothing this outrageous could occur again, safeguards have been put in place to ensure that it does not. Indeed, the safeguards, including investigational protocol, endeavor to prevent any harm from occurring to any patient, mentally, physically, or as a violation of their rights to privacy. It should also be noted that while animal test subjects may be harmed and, indeed, sacrificed, the treatment and care of animals used in preclinical studies must be designed to minimize pain and suffering of the animals.

Informed consent granted by the research subject means that they or their legal representatives have been fully informed of all pertinent aspects of the proposed drug trial. This information is to be presented to the potential test subject so that they can decide whether or not to participate in the study, based upon their evaluation of the risks to which they themselves would be subjected. Neither the investigator nor any of his representatives is to exert any influence upon the potential subject, nor may any unreasonable time constraints to be placed upon the decision-making process. The information required for consent must be presented in a clear, unambiguous manner and understood by the potential test subject. Any and all questions about the trial must be answered completely to the satisfaction of the potential subject or his legal representative. Transmission of this information should be written as well as oral, and the explanations should be made in the presence of a witness who is uninvolved with the study. The witness will sign the informed consent confirming that information was properly transmitted to the potential subject or their legal representative and that all questions were properly asked and answered. Once all of these conditions have been met, the subject is to sign and personally date the informed consent form, attesting that he has been presented with complete information about the test to which he is committing himself and that he willingly agrees to partake in the investigation. Any new information discovered during the trial that affects the informed consent will be transmitted to the subject or the legal representative, and a modified informed consent will be signed at that time. A copy of the original informed consent and copies of any amendments or changes to the informed consent should be provided to the subject and/or his legal representative as such changes are made.

The participation of a test subject in a research trial is completely voluntary, and the subject may decide to end his or her participation in the research trial at any time, without penalty. As part of informed consent, the subject will be informed of trial treatments and procedures including invasive procedures and be made aware that he or she may not receive the experimental treatment but rather be part of a control group. The subject will be fully informed of his responsibilities, as a member of the research population, of any expectations required of him during the study period. He must be informed of any anticipated risks or inconveniences as well as any expected benefits of the treatment that he may receive.

The duration, as well as the scope of the trial, is to be made known to the potential subject for his or her information. The potential research subject must also be informed of any foreseeable reasons for termination of his or her participation in the trial by the research team. The subject must be informed of any payment or expenses accruable to him or her as a result of participation in the trial.

Access to any data that may identify the subject, such as original medical records, shall be limited; however the subject must be made aware of that access to the said records by authorized researchers, and monitors or the IRB will be made. The subject also must be aware that records enabling the identification of individual research participants will not be released to the general public and that every effort will be made to conform to the legal requirements that the test subject's identity remain confidential.

As mentioned earlier, the handling of the data is critical. Therefore, the methods of statistical analysis to be used on experimental data, including the structure of the design, shall be provided. Further, the limitations placed upon access to the data must be specified in the investigational protocol. This is important for maintaining the privacy of individuals as well as assuring the integrity of the data used to form the reports, thereby enabling independent assessment of the study's results by third parties. It is axiomatic in any regulatory industry that if data is not recorded, it did not occur. The data must be properly acquired and recorded and the records maintained in an accessible and safe location for reasonable period of time.

1.5 Commercializing the New Drug

The ultimate goal for a new drug is commercialization. The IND is simply an investigational permit allowing transportation of an unapproved drug to a test site where it can undergo testing in human subjects to determine if it should be approved for sale to human populations. Application must be made to the FDA to market a new drug, assuming that the pharmaceutical company, after evaluation of the clinical studies, decides to proceed to market. At this point the pharmaceutical company needs to submit to the FDA an NDA seeking permission to market the new drug. The applicable Regulations are found under 21 CFR 314.

As noted earlier, the US FDA is the primary regulatory body for the dissemination and sale of drug products within the United States, with corresponding agencies serving a similar function in their respective countries. Clearance by the US FDA to introduce a new drug into the domestic market does not guarantee access to foreign markets. It is therefore in the best interest of domestic drug manufacturers to fulfill international requirements simultaneously with domestic requirements, thereby gaining entrance to markets worldwide. Harmonization of worldwide requirements is an ongoing process with which the US FDA is committed.

The FDA has standardized the format for submission of NDAs. There exists a structure for the submission of INDs and evolving harmonized standards; however outlining only one such standardized format in depth is sufficient to illustrate the level of detail required. For illustrative purposes, the US application process and forms, as extracted from the applicable CFRs, will be described basically as issued. Following that will be a description of harmonized submission format with notations of where corresponding sections of US submission requirements are applicable. Finally, a more descriptive explanation of the sections of the US submission requirements will expand the understanding of what is required to submit a new drug for market approval.

1.5.1 New Drug Application

An NDA has been required since the revision to the FD&C Act of 1938. The initial NDA required by the 1938 Act required only the establishment of the safety of the drug under study. The 1962 Kefauver–Harris amendment to the FD&C Act additionally required demonstration of the efficacy of the drug for its intended use. Further in addressing the safety aspects of the potential drug, it must be shown that the benefits of the drug outweigh the risks associated with the drug.

Three copies of the NDA are submitted, providing an archival copy, a review copy, and a field copy.

An NDA for a new chemical entity will generally contain, under 21 CFR 314.5, the information as follows, although different groupings have been submitted. The description of each section below is not comprehensive.

1.5.1.1 NDA Application Form 356h

Each NDA or supplement to approved NDAs must have an application form (Form 356h): Application to Market a New or Abbreviated New Drug or Biologic for Human Use (21 CFR 314 & 601) available online (www.fda.gov/downloads/AboutFDA/ReportsManualsForms/Forms/UCM082348.pdf) signed and dated by an authorized agent or official of the applicant representing the submission to the FDA. It will have the name, address, telephone number, and e-mail address of the applicant; the date of the application; the application number if previously issued; the names (established, proprietary, code, and chemical) of the drug product; dosage form and strength; route of administration; identification numbers of all INDAs referenced; the identification numbers of all drug Master Files and other applications referenced in the application; and the potential drug product's proposed indications for use.

Also included will be a statement of the submission classification (new, resubmission, etc.), a statement of the potential market for the drug product as either prescription or over-the-counter product, and a checklist that will identify what enclosures are required under the section the applicant is submitting. This form will be used for contact by the FDA regarding the submission.

1.5.1.2 Index

The NDA Application should include a comprehensive index by volume number and page number to the summary, the technical sections, and the supporting information for the archival copy of the NDA.

1.5.1.3 Summary

Writing at the level of submission to a peer-reviewed journal, a summary of the various parts of the submission, using tabular or graphical data where possible, should contain enough detail to impart a good general understanding of data (including quantitative aspects) and information included in the NDA.

The summary must contain the following information:

Proposed labeling text (referring to support sections annotating inclusion of each statement in the labeling) including any medication guide required.

The pharmacologic class and rationale for intended use.

Any prior or pending foreign marketing history including any countries in which applications for marketing are pending or in which the drug has been withdrawn for safety or effectiveness issues.

A summary of the following sections of the NDA:

- Chemistry, manufacturing, and controls
- Nonclinical pharmacology and toxicology
- Human pharmacokinetics and bioavailability
- Microbiology section (if applicable)
- Clinical data, including statistical results

The summary will conclude with a discussion of the risk–benefit analysis and proposed additional studies or Phase IV.

1.5.1.4 Technical Sections

1.5.1.4.1 *Chemistry, Manufacturing, and Controls*

A full description of the manufacturer(s), the components, and the specifications of the drug substance includes:

- A full description of the *drug substance* including its physical and chemical characteristics and stability; the manufacturer's name and address; the method of synthesis, isolation, and purification of the drug; and all process controls and specifications as well as analytical methods used to ensure identity, strength, quality, and purity of the drug substance and bioavailability of the drug product.
- A list of all components used in manufacture of the *drug product* and their specifications (regardless of whether they appear in the drug product) and a statement of the composition of the drug product and any manufacturer (including address) and stability data with proposed expiration dating.

  Any drug product batch history records of drug products used for bioavailability or bioequivalence studies, including manufacturer, specifications, and other criteria as above.

  Proposed or actual master production record including equipment and production process to be used for commercial manufacture of the drug product.
- Environmental impact claim or categorical exclusion under 21 CFR 25.30 or 25.31 or an environmental assessment under 21 CFR 25.40.
- The applicant may, at its option, submit a complete chemistry, manufacturing, and controls section 90–120 days before the anticipated submission of the remainder of the application.
- A statement certifying delivery of the field copy of the NDA to applicable FDA district office.

1.5.1.4.2 *Nonclinical Pharmacology and Toxicology*

Descriptions of the in vitro and in vivo studies, preferably presented in graphical or tabular format, includes:

- The pharmacological actions of the drug in relation to its proposed therapeutic indication and studies that otherwise define the pharmacologic properties of the drug or are pertinent to any adverse effects.
- Studies of the toxicological effects of the drug as they relate to the drug's intended clinical uses, including, as appropriate, studies assessing the drug's acute, subacute, and chronic toxicity and studies of toxicities related to the drug's mode of administration or conditions of use.
- As appropriate, studies of the effects of the drug on reproduction and fetal development.
- Also included should be any studies of the absorption, distribution, metabolism, and excretion of the drug in animals.
- For nonclinical laboratory study, a statement should also be made regarding GLP being used throughout the studies. Alternately, if the studies were not conducted under GLP, an explanation should be provided for each incident of noncompliance.

1.5.1.4.3 Human Pharmacokinetics and Bioavailability

Description of the human pharmacokinetic data and human bioavailability data or information supporting a waiver of the submission of in vivo bioavailability data includes:

- Description of each bioavailability and pharmacokinetic study of the drug in humans including analytical and statistical methods used.
- A statement about the rationale for establishing the tests, analytical procedures, and acceptance criteria including data and information supporting that rationale.
- A summarizing discussion and analysis of the pharmacokinetics and metabolism of the active ingredients and the bioavailability and/or bioequivalence of the drug product.

1.5.1.4.4 Microbiology

This section should detail, for anti-infective drugs only:

- This should include a description of the biochemical basis of the drug's action. Antimicrobial spectra of the drug, including preclinical studies, to establish effective use concentrations.
- Any known resistance factors or studies thereof.
- A description of clinical microbiological laboratory procedures.

1.5.1.4.5 Clinical Data

A description of the clinical investigations of the drug includes:

- Each clinical pharmacology study of the drug with a brief comparison of the results of the human studies with the animal pharmacology and toxicology data.
- A description of each controlled clinical study pertinent to the proposed use of the drug, including the protocol and a description of the statistical analyses used to evaluate the study.
- A description of each uncontrolled clinical study, a summary of the results, and a brief statement why the study is classified as uncontrolled.
- A description and analysis of any other data or information relevant to evaluation of the safety and efficacy of the drug product received by the applicant derived from any source, foreign or domestic, including controlled and uncontrolled studies of uses of the drug other than those proposed in the application.
- An integrated summary of the data demonstrating substantial evidence of effectiveness for the claimed indications. Included will be evidence required to support the dosage and administration section of the labeling.

- Safety summary and update including an integrated summary of all available information about the safety of the drug product, including pertinent animal data, demonstrated or potential adverse effects of the drug, clinically significant drug/drug interactions, and other safety considerations. The safety data shall be presented by gender, age, and racial subgroups. When appropriate, safety data from other subgroups of the population of patients treated also shall be presented.

  The applicant shall also update periodically its pending application with new safety information learned about the drug that may reasonably affect the statement of contraindications, warnings, precautions, and adverse reactions in the draft labeling or medication guide.
- If the drug has the potential for abuse, a description and analysis of studies or information related to the abuse of the drug, including a proposal for scheduling under the Controlled Substances Act, is required. Studies related to overdosage including information on dialysis, antidotes, or other treatments if known shall be provided.
- A summary of the risks and benefits of the drug including a discussion of why the benefits exceed the risks under the conditions stated in the labeling.
- A statement with respect to each clinical study involving human subjects that either was conducted in compliance with the IRB regulations or were not subject to them and that it was conducted in compliance with the form consent regulations.
- If a sponsor transferred any obligations for the conduct of any clinical study to a contract research organization, the name and address of said organization, the clinical study transferred, and a listing of each obligation transferred or if all obligations transferred a general statement to that effect.
- Any audit or review by the sponsor of original records to verify the accuracy of case reports in the course of monitoring the study should be listed.

1.5.1.4.6 *Statistical*
This section describes the statistical evaluation of clinical data including:

- Description and analysis of each controlled clinical study with supporting documentation and statistical analysis.
- A summary of information about the safety of the drug product and documentation and supporting statistical analyses using evaluating the safety information.

1.5.1.4.7 *Pediatric Use*
This section will include a description of the investigation of the drug for use in pediatric populations, including an integrated summary of information that is relevant to the safety and effectiveness as well as the risk–benefit determinations in pediatric populations.

1.5.1.5 Samples and Labeling
The FDA may request samples be sent to, generally, two or more agency laboratories that will perform all necessary tests and validate the analytical procedures. Such samples will be:

- Four representative samples in quantities sufficient to perform the required tests in triplicate to determine if the drug substance and drug product meet the NDA specifications of the following:
  - The proposed drug product.
  - The drug substance used in the drug product above.
  - Reference standards and blanks (standards recognized by an official compendium excluded).
- Samples of the finished market package, if requested.

The following must be submitted in the archival copy of the NDA:

- Three copies of the analytical procedures and related descriptive information contained in the CMC section that are necessary for the FDA's laboratories to perform all tests on the drug substance and the drug product.
  This includes any supporting data for accuracy, specificity, precision, and ruggedness and complete results of the applicant's tests on each sample.
- Four copies of the draft or 12 copies of the final printed labeling for the drug product including, if applicable, any medical guide required.

1.5.1.6 Case Report Forums and Tabulations

For the archival copy of the NDA:

- Case report tabulations for each adequate and well-controlled Phase I and Phase II studies, from the earliest Phase I studies and for safety data from other clinical studies. The tabulations must include the data on all patients in each study unless the FDA agrees in advance subject was not pertinent to a review of the drug's safety or efficacy.
- Case report forms for each patient who suffered an adverse event and had to leave the study or died must be included.
- Additional data to be provided by the applicant include additional case report forms and tabulations needed to conduct a proper review of the NDA.
- Prior to submitting an NDA to the FDA, applicants may meet with the agency to discuss the presentation and format of supporting information. Alternate formats must be agreed upon by both parties.

1.5.1.7 Other

The following general requirements apply to the submission of information within the summaries and within the technical sections:

- Information previously submitted may be incorporated by reference to the file by name, reference number, volume, and page number in the agency's records where the information can be found. Resubmission is not required.
- A complete and accurate translation of each part of the NDA that is not in English and a copy of each original literature publication translated into English will be provided.
- If the NDA is submitted with an obtained "right of reference or use," a written statement signed by the owner of the data must be included, and access to the underlying data must be granted to the FDA.

1.5.1.8 Patent Information

The information pertaining to the drug should be submitted for drug substance, drug product, and method of use. This section will include patent number and expiration date, type of patent, name of patent owner, name of US representative of a foreign patent owner, and declaration if patent covers the drug submitted. A more complete description of types of patents that must and must not be submitted is described in 21 CFR 314.53.

1.5.1.9 Patent Certification

With regard to patents claiming drug, drug product, or method of use, a 505(b) submission, for each such patent the applicant shall provide the patent number and certify, in its opinion and to the best of its knowledge, one of the following circumstances:

1) Paragraph I Certification – that the patent information has not been submitted to the FDA.
2) Paragraph II Certification – that the patent has expired.

3) Paragraph III Certification – that the date on which the patent will expire.
4) Paragraph IV Certification – that the patent is invalid, unenforceable, will not be infringed by the manufacture, use, or sale of the drug product for which the application is submitted.

For more information regarding patents, refer to 21 CFR 314 as it goes into far greater detail about patents than above.

1.5.1.10 Claimed Exclusivity

A new drug product, upon approval, may be entitled to a period of marketing exclusivity. To claim exclusivity, it must submit with the NDA prior to approval the following:

- A statement claiming exclusivity.
- A reference to the appropriate paragraph under 21 CFR 314.108 that supports the claim.
- Claims under 21 CFR 314.108(b)(2) must provide information to show that, to the best knowledge or belief, a drug has not previously been approved under Section 505(b) containing any active moiety in the drug for which approval is sought.
- An NDA claiming exclusivity under 21 CFR 314.108(b)(4) or (b)(5) must show that the NDA contains "new clinical investigations" that are "essential to approval of the NDA or supplement" and were "conducted or sponsored by the applicant."
  - "New Clinical Investigations" requires a certification that to the best of the applicant's knowledge, each of the clinical investigations meet the definition set forth in 21 CFR 314.108(a).
  - "Essential to approval" is a list of all published studies or publicly available reports known to the applicant that to the best of the applicant's knowledge is complete and accurate and finds that the publications do not provide a sufficient basis for approval.
  - "Conducted or sponsored by" requires a certified accountant's statement that, if not the sponsor, the applicant provided 50% or more of the cost of the investigation.

1.5.1.11 Financial Certification or Disclosure

The application shall contain a financial certification or disclosure statement or both as required. The applicable forms FDA 3454 (no financial interest) and FDA 3455 (disclosure of financial interest) shall be signed by the investigator and sponsor(s) CFO and submitted in this section.

1.5.1.12 Format of an Original NDA

A complete archival copy of the NDA that will be maintained by the FDA during the review of the NDA to permit individual reviewers to refer to information that is not contained in their particular technical sections of the NDA, to give other agency personnel access to the NDA for official business, and to maintain in one place a complete copy of the NDA.

A review copy of the NDA shall be provided with technical sections individually bound together with the application form and a copy of the summary.

A field copy contains the technical section, a copy of the application, a copy of the summary, and a certification that the field copy is a true copy of the technical section contained in the archival and review copies of the NDA.

Sufficient binding folders may be obtained from the FDA to bind the archival, the review, and the field copies of the NDA.

Electronic format submissions must be in a form that the FDA can process, review, and archive. Electronic submission is evolving, and the FDA periodically issues guidance as to file formats, media, and organization.

The above is the traditional format for submitting an NDA in the United States. The pharmaceutical business is, however, composed of worldwide enterprises. From a financial viewpoint, it makes sense for companies to move toward worldwide standardization of submission processes. From a government viewpoint standardization of submission and approval formats would allow faster access to worldwide markets for domestic companies and would also speed in the importation of new pharmaceuticals developed in other countries. Toward that end experts from around the world have been working to harmonize regulations.

1.6   Harmonization

For pharmaceutical products the International Conference on Harmonization of Technical Requirements for Registration of Pharmaceuticals for Human Use (ICH [3]) is the organization whose unique mission is to reduce or eliminate redundant testing during the research and development phase of drug development. It brings together the regulatory authorities and experts from the pharmaceutical industries of the United States, Europe, and Japan to discuss scientific and technical aspects of data submission required for product registration (acceptance for sale).

1.6.1   Common Technical Document

One outcome of these meetings is called the "Common Technical Document" (CTD), a harmonized format for submitting new product applications. This format was agreed upon in November 2000, in San Diego, USA, with the agreed-upon implementation date in the three regions of July 2003. Currently the United States is accepting applications and submissions in both formats.

An FDA Draft Guidance for Industry Submitting Marketing Applications According to the ICH-CTD Format – General Considerations was issued August 2001.

The CTD is composed of five modules in the following format:

1.6.1.1   Module 1. Administrative Information and Prescribing Information

1.1 FDA Form 356h
1.2 Comprehensive Table of Contents of the Submission including Module 1
The location of each document should be identified by referring to the volume numbers that
    contain the relevant documents and any tab identifiers.
1.3 Administrative Documents Specific to Each Region (for example, application forms, pre-
    scribing information)

This is a region-specific module containing, for example, application forms for use in the region. The content in the format can be adjusted for the particular regulatory agency to which the forms are being submitted. It is not technically a part of the CTD.

The corresponding sections of the traditional submission to the FDA that should be included in Module 1 of the CTD are:

Index and FDA Form 356h
Labeling
Patent information on any patent that claims the drug
Patent certifications (not for Biologics License Application [BLA])
Debarment certification
Establishment description
Field copy certification (not for BLA)
Financial certification including User Fee cover sheet
Other information

1.6.1.2 Module 2. Common Technical Document Summaries

2.1 Overall Technical Document Table of Contents (Modules 2–5)
2.2 CTD Introduction to the Summary Documents
2.3 Overviews and Summaries
  Quality overall summary
  Nonclinical overview
  Clinical overview
  Nonclinical written and tabulated summaries
    Pharmacology
    Pharmacokinetics
    Toxicology
  Clinical summary
    Biopharmaceuticals studies and associated analytical methods
    Clinical pharmacology studies
    Clinical efficacy
    Clinical safety
    Literature references
    Synopses of individual studies

This should be a single page and begin with a general introduction to the pharmaceutical, including its pharmacologic class, mode of action, and propose clinical use.

Information on quality should be presented in a structured format as described in FDA Guidelines M4Q, M4S, and M4E.

Module 2 is equivalent of a summary of all technical sections of the traditional NDA.

1.6.1.3 Module 3. Quality

3.1 Module 3 Table of Contents
3.2 Body of Data
3.3 Literature References

Information on quality should be presented in a structured format as described in Guideline M4Q.

The information here is similar to what is included in CMC in the traditional NDA.

1.6.1.4 Module 4. Nonclinical Study Reports

4.1 Module 4 Table of Contents
4.2 Study Reports and Related Information
4.3 Literature References

Nonclinical study reports should be presented in the order described in Guideline M4S.

The information here is similar to what is included in Nonclinical Pharmacology and Toxicology in the traditional NDA.

1.6.1.5 Module 5. Clinical Study Reports

5.1 Module 5 Table of Contents
5.2 Tabular Listing of Clinical Studies
5.3 Clinical Study Reports
5.4 Literature References

The human study reports and related information are presented here in the order described in Guideline M4E.

The corresponding sections of the traditional submission to the FDA that should be included in Module 5 of the CTD are:

Human Pharmacokinetics and Bioavailability
Clinical Microbiology
Clinical Data
Safety Updates
Statistical Information
Case Report Forms and Tabulations

The Guidelines M4Q, M4S, and M4E give far greater detail about the exact order in which the data is presented. For example, M4S specifies that where multiple studies of the same type are summarized within pharmacokinetics and toxicology sections, the studies should be ordered by species, by route, and then by duration (shortest duration first).

What is important to note here is that, essentially, the information contained in the traditional FDA submission is the same as in the CTD. The difference is in how the presentation is organized. Standardized submissions will ensure that each area will receive the information that it requires to evaluate the new drug in the same format while still satisfying any particular area requirements of formats or forms. Obviously the advantage of harmonization is that once Modules 2–5 are prepared for one country, the same four modules can be submitted to other members of the harmonization project unchanged, with only the region-specific forms of Module 1 addressed to the country in who's market the drug company would like to enter.

1.7   Review Process of US NDA

Three copies of the application are required.

- Archival copy.   The complete archival copy of the application contains all the sections in the application. FDA retains the archival copy to permit individual reviewers to refer to information that is not contained in their particular technical sections. The archival copy can be submitted in either electronic format in accordance with 21 CFR 11 or hard copy.
- Review copy.   This copy contains the title sections. It is required to be separately bound to the top of the Application Form 356h and a copy of the summary section, Modules 1 and 2 in CTD.
  Review copies that may be necessary include:
  - Quality (Module 3)
  - Nonclinical (Module 4)
  - Clinical (Module 5) – safety and efficacy documents for clinical reviewer
  - Clinical (Module 5) – safety and efficacy documents for the statistical reviewer
  - Clinical (Module 5) – clinical pharmacology and pharmacokinetics documents (or bioequivalence documents) for Abbreviated New Drug Applications (ANDA's)
  - Clinical (Module 5) – clinical microbiology documents
- Field.   Only the CMC section, or Module 4, Quality, in CTD format. The field copy should be submitted to the local district office of the FDA and a signed statement of submission attached.

The archival and a review copy of the drug marketing application are submitted to the CDER in Beltsville, Maryland. The first step is a review for completeness to ensure that sufficient data and information have been submitted in each area "filing" the application. Incomplete NDAs result in a formal "refuse-to-file" action, in which case the applicant receives a letter detailing the decision and the deficiencies noted. This decision must be made within 60 calendar days after CDER initially receives the NDA.

If the complete NDA is accepted, it undergoes technical reviews with each reviewer submitting a written evaluation of the NDA to the FDA division or office director who then evaluates the reviews and recommendations of the reviewers and decides the action to be taken on the application. A letter will be generated that provides either an approval, approvable, or non-approvable decision as well as a basis for that decision.

The technical reviewers each focus on a specific area of their expertise.

Medical reviewers evaluate the clinical sections of the application including the results of clinical trials and all toxicology and human pharmacology.

Biopharmaceutical review is performed by pharmacokineticists who evaluate the rate at which and the extent to which the drug's active ingredient is made available to the body as well as the ways it is metabolized, distributed, and eliminated from the human body.

Statistical reviewers are statisticians who validate the statistical relevance of the data in the NDA primarily by evaluating the methods used to conduct the studies and the methodology used to analyze the data.

Pharmacology/toxicology review team members evaluate the results of animal testing and the relationship of the animal drug effects to potential effects in humans.

Chemists perform the chemical review of the chemistry and manufacturing control sections of the NDA. The identification, manufacturing control, and analytical procedures are reviewed for suitability and accuracy. They confirm the ability to reproduce the drug reliably as well as its stability.

As part of the NDA review process, the FDA conducts preapproval inspections (PAIs) to verify the accuracy and completeness of the manufacturer-related information in the NDA. They will evaluate the manufacturing controls used to produce the pharmaceuticals that were studied in the preclinical and clinical trials. They will evaluate the manufacturers' compliance with cGMPs and collect a variety of drug samples for analysis to confirm methods validation, methods verification, and forensic screening for substitution. If the manufacturing facility is found wanting during the PAI process, approval of the NDA will be withheld until the deficiencies are addressed and corrected.

The timeframe for reviewing NDA is that within 180 days of receipt of application for new drug under Section 505(b) or an abbreviated application for new drug under Section 505(j). FDA will review it and send the applicant either an approval letter under 314.105 or a Complete Response Letter under 314.110. This 180-day period is called the "initial review cycle." The applicant may, anytime before the NDA is approved, withdraw an application or an abbreviated application, and it may later be submitted again for consideration. The initial review cycle may be adjusted by mutual agreement between FDA and applicant.

Repeated meetings may take place between the FDA and the applicant. A meeting prior to submission of the NDA is helpful to discuss the presentation of the data supporting the application. This pre-NDA meeting also helps reviewers become familiarized with the data to be submitted to facilitate its review. At this meeting a summary of the clinical studies would be discussed as well as a proposed format for organizing the submission.

About 90 days after the initial submission of the NDA, another meeting may be held in order to discuss any deficiencies or issues that are discovered on the initial review. Alternately the FDA may communicate with the applicant by telephone, letters, faxes, e-mails, or meetings.

"Correct the deficiencies" notifications are probably communicated by the FDA to the applicant. Major scientific issues are usually reserved for discussion and noted in a complete response letter after the initial review process is completed.

At the end of the review, a complete response letter will be sent if the FDA determines that the NDA submission will not be approved in its present form for one or more reasons. A complete response letter usually will describe all of the specific deficiencies that the agency has identified in an application.

An end of review conference provides an opportunity for the applicant to meet with the FDA to discuss the deficiencies. The purpose of this meeting is to address what further steps are necessary for the application to be approved.

Once the decision on action recommendation is made by the reviewers and their supervisors, the decision must ultimately be evaluated and agreed to by the division director. Once the director signs the approval action letter, the product can be legally marketed in the United States.

1.8 Current Good Manufacturing Practice in Manufacturing, Processing, Packing, or Holding of Drugs

cGMP regulations for drug and biological products are geared toward commercial manufacturers for all types of pharmaceutical products for administration to humans or animals. 21 CFR 210 and 211 are the relevant regulations, with §210 being general applicability and definitions and §211 covering the 10 main categories addressed by the cGMPs.

1.8.1 Organization and Personnel

1.8.1.1 Quality Control Unit

Quality is the responsibility of everybody in the pharmaceutical manufacturing organization. It is the responsibility of each manufacturer to establish, document, and implement a system for managing quality throughout the manufacturing operation.

The quality control unit shall have the responsibility and authority to approve or reject all components, drug product containers, closures, in-process materials, packaging material, labeling, and drug products, and the authority to review production records to assure that no errors have occurred or, if errors have occurred, that they have been fully investigated and corrective action taken if unexpected results occur during production.

The quality control unit shall be responsible for approving or rejecting drug products manufactured, processed, packed, or held under contract by another company. They shall have the responsibility for approving or rejecting all procedures or specifications impacting on the identity, strength, quality, and purity of the drug product. In order for the quality control unit to function properly, adequate laboratory facilities for testing and evaluation of all components of the pharmaceutical including raw materials and packaging materials as well as finished goods must be made available to the quality control unit.

The responsibilities and procedures of the quality control unit shall be in writing and the written procedures will be followed.

1.8.1.2 Personnel

The pharmaceutical manufacturer is expected to employ personnel whose training and experience qualify them to perform their jobs. Additionally there should be an adequate number of people to perform all of the activities required and to supervise the production or processing of pharmaceutical products. The particular job functions should be specified in writing.

Training of operators should be conducted by qualified individuals on continuing basis. At a minimum, training should address the immediate responsibilities the employee performs. However, conveying an understanding of larger scope of the production process empowers the employees to recognize threats or concerns that may occur outside of their immediate prevue. The effectiveness of training should be assessed periodically and records of such training and assessment should be maintained.

1.8.1.3   Personnel Hygiene

Maintenance of good personnel hygiene is required of all personnel, maintaining good sanitation and health habits. They should be provided with clean clothing and protective apparel such as head, face, and hand coverings, where appropriate to avoid contaminating drug products. Personnel with illnesses such as communicable diseases or open lesions should not be allowed potential contact with and subsequent contamination of the drug product.

1.8.1.4   Consultants

Consultants who advise on the manufacture, processing, packing, or holding of pharmaceuticals should also have proper education, training, and experience to enable them to provide competent advice within their area of expertise. Records should be kept of all pertinent contact information for the consultant, their qualifications, and what services they provided to the company.

1.8.2   Building and Facilities

1.8.2.1   Design and Construction Features

The design and construction features of the buildings wherein pharmaceutical manufacturing will occur should be such to suit the type of the drug manufactured, processed, packaged, and held in that facility. The building should be capable of being maintained as a clean environment wherein proper processing may occur.

There should be adequate room to receive, identify, and store and quarantine drug product containers, closures, labeling, process components, in-process intermediates, and drug products that securely safeguard against contamination or premature release pending quality control sampling, testing, and release for manufacture.

The facility should be designed so that workflow minimizes the chance for contamination or adulteration of product.

The building must have adequate lighting, ventilation, air filtration, air heating, and cooling. Plumbing must be such that water is available in sufficient quantities to supply all needs. Likewise source facilities and refuse containers must be sized accordingly.

Sewage, trash, and other refuse in and from the building and immediate surrounding area shall be disposed of in a safe and sanitary manner.

Adequate washing within the facilities must be made available for personnel to maintain personal hygiene and sanitary conditions on the factory floor. Maintenance of facilities must be an ongoing continuous process.

1.8.3   Equipment

The equipment used to manufacture pharmaceuticals must be designed to the proper size and in the proper location to produce the desired pharmaceutical. In general the construction equipment must be sturdy, easily cleanable, sanitizable, and easily maintained. The equipment should be regularly and reliably calibrated and there should be in place on location SOPs explaining how to use the equipment as well as how to maintain, clean, and sanitize it. Records shall be maintained recording any and all calibrations, cleaning or sanitization, and/or maintenance operations for each piece of equipment used in the manufacture of drug products.

Automated systems, both mechanical and electronic, may be used provided that they are regularly and routinely inspected, checked, and calibrated according to established SOPs. Such systems will be put in place that prevent unauthorized alteration of computer controls, records, master batch records, or other records either intentionally or by error. Input and output checks of such systems shall be performed with sufficient frequency to assure the integrity of the records, calculations, and process controls.

1.8.3.1 Filters

Filters that come in contact with liquids that are components of injectable drug products shall be of such construction as to not release fibers into said liquid.

1.8.4 Control of Components and Drug Product Containers and Closures

The containers and closures should be clean and kept under clean conditions and protected from contamination. Each individual lot of untested components, drug containers, and closures shall be received and isolated, pending quality control unit testing and release for use. The testing methodology shall be in accordance with SOPs specifying the appropriate test procedure and sampling size. The component should either be accepted and released for production or rejected and returned to the manufacturer. Rejected materials should be prevented from entering production stream while awaiting transport out of the facility.

1.8.5 Production and Process Controls

Each drug product shall have a written procedure describing in detail the manufacturing control process required to produce the pharmaceutical of the required identity, strength, quality, and purity. The components for the drug manufacturer should be weighed, measured, or subdivided as appropriate. Weighing, measuring, or subdividing operations for components shall be adequately supervised. Each component dispensed to manufacturing shall be examined by a second person to assure that the component was released by the quality control unit; the weight or measure is correct as stated in batch production records, and the container is properly identified. The components are then added to the batch by one person and their addition verified by a second person. If the weighing and measuring, subdividing, and/or adding to the batch is done by automated equipment, only one person is necessary to assure proper proportions.

All such containers shall be identified at all times to indicate their contents and phase of production, if applicable.

Written records must be kept as part of the master production records or master batch records, for all in-process and final drug product testing. The quality control unit will review all records. Any deviation from the written procedure must be recorded, investigated, and justified. The product produced is suspect until the review takes place. The final disposition of the suspect product must be recorded. The master production record in all production records will be maintained by document control. No products are released until a complete review of the entire production record by the corporate authorities.

Retention samples will be kept from each batch of finished product released for 1 year after the expiration date. These retention samples should be in their production packaging unless overly large, in which case smaller samples might be stored in appropriate containers with corresponding appropriate labeling.

Actual yield and percentages of theoretical yield shall be determined as appropriate at the conclusion of each phase of the manufacturing procedure.

When appropriate, time limits for each manufacturing phase shall be established to ensure product quality.

Appropriate SOPs to prevent contamination by objectionable microorganisms in both sterile and non-sterile drug products must be established and followed to eliminate microbial contamination of said products.

Written procedures shall be established and followed to reprocess, as appropriate, nonconforming batches of drug product or drug product intermediaries. Said reprocessing shall be supervised by the quality control unit.

1.8.6 Packaging and Labeling Control

Written procedures shall describe in sufficient detail the maintenance of strict control over labeling issued for use in drug packaging operations. Each batch of labels will be stringently examined for correctness and appropriate for each batch of pharmaceutical. The label reflects the proper identity and conformity to the labeling specified in the master or batch production records. All excess labeling bearing a lot with control number shall be destroyed. A methodology will be maintained to reconcile issues of labeling issued, use, and returned.

Written procedures for receiving and evaluating raw materials shall be in place including specific instructions for receiving, reviewing, releasing, and distributing all components. Any labeling or packaging materials meeting appropriate written specifications may be approved and released for use. Any labeling or packaging materials that do not meet appropriate written specifications will be rejected to prevent their use in operations for which they are unsuitable.

Records will be maintained for each shipment of its different labeling and packaging material indicating receipt, examination, and testing and whether accepted or rejected including quantities.

The labels for the final product shall be at minimum the name of the product, the active ingredients in quantities thereof, quantization of the contents, batch or control number, expiration date, storage and handling conditions, directions for use, warnings and precautions, and then the name and address of the manufacturer or marketer.

Obsolete and outdated labels and packaging will be destroyed and their disposition recorded.

Use of gang-printed labeling for different drug products or different strands or net contents of the same drug product is prohibited unless there is adequate differentiation by size, shape, or color to prevent misapplication.

If cut labeling is used packaging and labeling operation must include one of the following:

- Dedication of labeling and packaging lines for each different strength of each different drug product.
- 100% verification of correct labeling by use of appropriate electronic or electromechanical equipment.
- 100% visual inspection for correct labeling during or after completion of finishing operations for hand applied labeling. The examination will be conducted by one person and independently verified by second person.

On package printing the operation shall be confirmed to conform to the drug production record.

Issuance of labeling shall be under strict control. Labeling issued for a batch shall be inspected for conformity to the batch or master control record. Unused labels shall be returned to secured storage and any discrepancies among issued, used in production, and returned labels shall be resolved. All excess labeling bearing lot or control numbers shall be destroyed.

Unless specified by 21 CFR 211, over-the-counter drugs shall be produced in tamper-evident packaging, and all such features shall be prominently printed on the package and not obscured should any of the features be compromised.

Packaged and labeled drug shall be inspected during finishing operations to confirm that packages and containers have the proper labels. A representative sample shall be obtained at the conclusion of finishing operations and shall be visually inspected for correct labeling.

Appropriate expiration dating determined by stability testing shall be imprinted clearly on each package. The dating will be appropriate for the storage conditions the package may be exposed to. If the drug product is to be reconstituted, both the intact package and the reconstituted drug product expiration data shall be imprinted on the package.

1.8.7 Holding and Distribution

Each manufacturer will have written procedures describing the warehousing of drug products that will include the quarantine of drug products before release by the quality control unit. The drug products will be stored under appropriate conditions of temperature, humidity, and light assuring that the integrity of the drug product is not compromised by environmental conditions.

Similarly there shall be written procedures describing the distribution of drug product including a procedure whereby the oldest approved stock of drug product is distributed first. Temporary and appropriate deviations from this requirement are permitted.

There must exist a system in which each lot of drugs can be tracked, thereby facilitating the recall if necessary. The written instructions for implementing a recall must clearly define how the recalled lot(s) shall be implemented, including who will bear responsibility for each phase of the recall and how the recalled drug products shall be handled. A method to determine the effectiveness of the recall shall be established.

1.8.8 Laboratory Controls

The appropriate organizational unit shall determine any specification, standards, statistically based sampling plans, and test procedures to be used. These shall be reviewed and approved by the quality control unit for appropriate applicability and adequacy. Compliance with the written procedures used to assure the proper purity, identity, and strength of the final drug product, as well as proper labeling shall be documented at the time that the appropriate checks are made. Any deviation from established criteria is reason to quarantine and restrict from distribution any drug product pending review and release by the QAU, whose rationale for action will also be documented.

Scientifically sound and appropriate specifications, standards, sampling plans, and analytical procedures shall be established to assure that components, drug product containers, closures, in-process materials, labeling, and drug products conform to appropriate standards of identity, strength, quality, and purity. Testing procedures published by established independent organizations such as the Association of Analytical Communities (AOAC) may be cited, or independently derived testing procedures may be developed; however, in either case, the rationale and efficacy of specific methodology must be established and documented.

Written specifications for the receipt, quarantine, sampling, testing, and acceptance or rejection of raw materials will be followed. Instruments, apparatus, gauges, and recording devices shall be calibrated at suitable intervals in accordance with established written programs to maintain their appropriate levels of accuracy and precision. Instruments, apparatus, gauges, and recording devices that do not meet established specifications shall not be used. A record of calibrations shall be maintained and instruments that do not meet established specifications shall be identified and securely isolated.

The satisfactory performance of each batch should be confirmed by established, written testing procedures, and no batch shall be released or distributed prior to confirmation by the QAU that it is in compliance with established criteria.

There should be written testing program designed to assess the stability characteristics of the drug products under various conditions. The results of such testing should be used to determine appropriate storage conditions expiration dates.

Reserves sample shall be maintained for periods appropriate relative to their expiration dates, typically 1 year past the expiration date of the last lot of the drug containing the active ingredient. The reserves sample shall consist of at least twice the quantity necessary for all tests required to determine whether the active ingredient meets its established specifications.

1.8.9   Records and Reports

If you don't write it down, you didn't do it. Good documentation ensures availability of data needed for validation, review, and statistical analysis. The data should be accessible in a format that lends itself to review and analysis such that the data may lead to modifications of established procedures.

Any record of production, control, or distribution is required to be maintained and that specifically associated with a batch of the drug product shall be retained for least 1 year after the expiration date of the batch or, in the case of certain OTC drug products lacking expiration dating, 3 years after the distribution of the batch.

All required records, or copies thereof, shall be readily available for authorized inspection during the retention period at the establishment where the activities occurred. The records may be photocopied or reproduced in other fashion as part of the inspection. Records can be immediately retrieved from an alternate location by computer or other electronic means should be deemed in compliance.

Written records shall be maintained such that the data can be used for evaluating, at least annually, the quality standards of each drug product to determine the need for any changes in the drug product specifications or of the manufacturing or the control procedures. Written procedures must be in place to review a representative number of batches either approved or rejected and, where applicable, all records associated with said batches. A review procedure of complaints, recalls, returned or salvage drug products, and investigations for each product shall be established. Procedures shall also be established to assure that responsible officials are notified in writing of any investigations, any recalls, reports of special observations issued by the FDA, or any regulatory actions relating to cGMPs brought by the FDA.

Records must show the capital equipment cleaning and use, except for routine maintenance such as lubrication and adjustments. The persons performing and checking the cleaning and maintenance performed shall date and initial the log, indicating that the work was performed at the time such observations are made. Entries in the log should be in chronological order.

Records shall be kept of all component, drug product container, closure, and labeling materials. These records shall include the results of a test or examination performed and the disposition of rejected components, drug product containers, closure, and labeling.

Master production and control records for each drug product including each batch size shall be prepared, dated, and signed (full signature) by one person and independently checked, dated, and signed by a second person. Written procedures shall be established, describing the preparation of master production and control records and said procedures shall be followed. Master batch control records shall include the name and strength of the product and a description of the dosage form, the name and quantity of each active ingredient per unit of the drug product and a statement of the total weight or measure of any dosage unit, a complete list of components sufficiently specific to indicate any special quality characteristic, manufacturing statement weight or measure of each component, any calculated excess of component, theoretical weight, and a theoretical yield including the maximum and minimum percentages of theoretical yield beyond which investigation is required according to 21 CFR 211.192.

Batch production and control records shall be prepared for each batch of drug product produced and shall include complete information relating to the production and control of each batch, with an accurate representation of the appropriate master production or control record and recordings of lot identification for each component used. The records shall be checked for accuracy, dated, and signed. Documentation shall be maintained of each significant step in the manufacture, processing, packaging, or holding of the batch. The information recorded shall include dates, identity of major equipment and lines, weights, measures and lot numbers of components in process and laboratory control results.

It shall be recorded that the packaging and labeling area is inspected before and after use, including complete labeling control records and specimens of same identification of persons performing and directly supervising and checking each significant step in the packaging and labeling operation.

All production records shall be reviewed and approved by the quality control unit to determine compliance with all established approved written procedures before any batch is released or distributed.

Laboratory records shall include complete data derived from all tests necessary to ensure compliance with establish specifications and standards.

Distribution records shall contain the name and strength of the product and description of the dosage form, name and address of the consignee, date and quantity shipped, and lot or control number of the drug product.

Complaint files shall be maintained. There shall be written procedures describing the handling of all written and oral complaints regarding the drug product. The procedure will include provisions for review by the quality control unit and for review to determine whether the complaint represents a serious and unexpected adverse drug experience that is required to be reported to the FDA. A written record of each complaint shall be maintained in a file designated for drug product complaints. Access to this record will most likely be the first request of an FDA representative on a visit to the manufacturing facility.

1.8.10 Returned and Salvaged Drug Products

Drug products returned from the market shall be identified and quarantined pending determination of their disposition. If there is any doubt of their safety, identity, strength, quality, or purity, return drug product shall be destroyed unless subsequent testing proved the drug product meets appropriate standards of safety, identity, strength, quality, and purity. Records of returned drug product shall be maintained and shall include the name and labeled potency of the drug product dosage form, lot number, reason for the return, quantity return, date of disposition, and ultimate disposition of the returned drug product.

1.8.11 Other

While description of the cGMP in manufacturing, processing, packaging, and holding of drugs is generally limited to the 10 categories above, it is important to understand that the regulatory arena is a dynamic, changing environment. Toward that end, it is beneficial to adopt a broader view to anticipate possible changes. Two categories are gaining wider attention in the processing world in areas of quality issues, and a brief description of change control and validation here is in order.

1.8.11.1 Change Control

A formal change control system that evaluates and documents all changes that could impact the intermediate and final products should be established. The system shall include approval of changes in specifications, analytical methods, facilities, raw materials, support system, processing steps, labeling, and packaging. Any proposals for GMP relevant changes should be drafted, reviewed, and approved by the appropriate organizational units and reviewed and approved by the quality unit(s).

1.8.11.2 Validation

There are two ways to ensure the complete integrity of the final drug product: verify or validate. Verifying would involve the destructive testing of each and every unit. Validating a process ensures that by following the validated, written procedure, the ultimate drug product will meet all required specifications. The approach to validation, including the validation of production processes, cleaning procedures, analytical methods, in-process control test procedures, computerized systems, and persons responsible for design, review, approval, and documentation of each validation phase, should be documented.

The critical parameters/attributes should normally be identified during the development stage or from historical data, and the necessary ranges for the reproducible operation should be defined. This should include:

1) Process parameters that could affect the critical quality attributes must be identified.
2) The range for each critical process parameter expected to be used during routine manufacturing and process control must be determined.

Validation should extend to those operations determined to be critical to the quality and purity of the drug product.

1.9 Compliance

The FDA conducts regular inspections for a variety of reasons under the general authority granted by FD&C Act Sections 702 and 704, which allows it to conduct investigations and collect samples of suspected drugs. The goal of the FDA's inspections is to minimize consumer exposure to adulterated drug products. Inspections may be routine GMP reviews, in response to a specific complaint, re-inspection after a warning letter, a check of recall effectiveness, or PAI for an NDA.

FDA conducts inspections of establishments that manufacture drug products for use inside and outside the United States and foreign establishments that intend to conduct clinical studies on their new drug products or market their products inside the United States.

Some of the administrative tools available to the FDA to ensure compliance with cGMPs include notices of observations (Form 483), warning letters, recalls, product withdrawal, drug license suspension or revocation, debarment, penalties, and disqualifications.

The FDA can invoke both civil and criminal judicial enforcement. In general, criminal sanctions against persons are only used when a prior warning or other type of notice was issued and failure to take corrective action exists.

FDA enforcement options include seizure, wherein it may order a halt to production of a drug manufacturing facility when the product is held in an unacceptable environment. Quarantining a warehouse is considered a "mass seizure" as it may include products that are not subject to contamination. A seizure may be specific for a product if, for example, labeling is noncompliant. At the same time a seizure is made, further injunctive action may be taken. These recommendations are made by the FDA compliance officer in district management. They are promptly acted upon by the Division of Compliance Management and Operations (DCMO).

An injunction is initiated to stop or prevent violation of the law; it is not necessary to show that the law has been violated, only to show that there is a likelihood that it may be violated if an injunction is not entered. An injunction does not preclude additional or concurrent action such as recall or seizure. Inspection warrants may be requested when inspection has been refused completely or when faced with refusals in limited areas.

Civil penalties are provided for and may be brought in any US District Court within whose jurisdiction, any act or omission constituting a violation, may have occurred. Such action may be taken for, among other infractions, failure to give notification or to take corrective action as required, the introduction or delivery of a noncompliant product into interstate commerce, the failure to properly maintain records or to permit inspections, nonresponse to prior warning/notice, failure to report, or failure ahead of obtaining product certification before distribution of the product.

The FDA is authorized to conduct inspections on factories, warehouses, establishments, vehicles, and all pertinent equipment, finished and unfinished materials, containers, and labeling where food, drugs, devices, or cosmetics are manufactured or held. The FDA authority extends to inspection of clinical laboratories and clinical study facilities as well as contract test laboratories, clinical study monitors, clinical study sponsors, and IRBs. The FDA is constrained to "reasonable" inspections, that is, inspection at reasonable times, within reasonable limits and in a reasonable manner.

Announced inspections are generally expected by the company and may include PAI requested by the company or international inspections. Notification is generally a few weeks before the inspection.

Unannounced inspections are conducted for specific reasons such as a product recall or a recall effectiveness check. It may be as a response to some complaint or as part of an FDA routine compliance inspection program.

During an inspection, the FDA inspector can review and inspect any and all documents related to the product under question. The inspector however cannot inspect and review financial or pricing data, personal data, sales data, and research data (unless related to the product safety).

At present the FDA uses a systems approach program for compliance inspections as opposed to the top-down approach and bottom-up approach previously employed. System inspections fall into six categories.

1.9.1 Quality System

The quality system is inspected to ensure compliance with cGMPs, internal procedures, and established specifications. The system includes the quality control unit, all product defect evaluations, and evaluation of return and salvaged drug products.

1.9.2 Facilities and Equipment System

Similarly the inspection activities of facilities and equipment systems and resources used in production of drugs or drug products are expected to ensure compliance with approved internal procedures and cGMP regulations.

Typical targets would be maintenance documents and records, equipment qualifications, calibration, preventative maintenance, cleaning validation records, and utility validations and calibrations.

1.9.3   Materials System

Review of material systems includes the control of finished products, incoming raw materials, containers, and closures.

1.9.4   Production System

When examining the manufacture of drugs and drug products, inspection of batch compounding, dosage form, production and process sampling and testing, and process validation may be performed.

1.9.5   Packaging and Labeling System

Records of the packaging and labeling process, printed labels and packaging materials, receiving examination, and uses of labels and packaging materials may be inspected.

1.9.6   Laboratory Control System

This inspection may include the examination of personnel to establish education and training qualifying them for their work assignments. Production, control, or distribution records that should be maintained by cGMP regulation should include inspection audit.

1.9.6.1   Inspection Strategies

The FDA inspects drug manufacturers in three ways:

1) Full system inspection – includes a minimum of four of the six systems, one of which must be quality, to ensure that all systems are under control and comply with cGMP regulations.
2) Abbreviated inspection – includes the quality system and one other of the six systems.
3) Compliance inspections – conducted to evaluate or verify that corrective actions have been taken after regulatory action as a result of some aspect being found to be noncompliant. Specifically, areas that have been found deficient and subjected to corrective actions and systems are used to determine the overall compliance status of the firm after the corrective actions are taken. Manufacturers are expected to bring all deficiencies into compliance, not just those cited by FDA Form 483. The compliance inspection also includes cause inspections that investigate specific problems that come to the FDA's attention. The problems may be indicated in the Field Alert Report (FAR).

1.9.6.2   Inspection Process

In any inspection, the inspector is required to present two pieces of picture ID to the firm. A responsible individual should be designated to meet with the inspector and he can examine and record the credential information but cannot make a copy of the credentials. The FDA expects senior management and responsible officials to attend the opening and closing (exit interview) meetings.

At the opening meeting the FDA inspectors will present Notice of Inspection, Form 482, and explain the reasons for their visit. The inspectors should be provided with a quiet space and all requested documents and records to review. Inspectors may request a tour of the facility, and the company should ensure that all required personnel are present to answer questions.

The FDA may also take photographs and obtain samples; if the company resists an inspection, warrant can be sought from FDA headquarters.

At the conclusion of the inspection Form 483, a Notice of Observations will be issued that include significant observations made during their visit. At this meeting the inspector will go through each observation in finding and giving the company the opportunity to respond. Corrections made during the FDA visit will be noted on Form 483; the observations will remain on the list. Findings are listed in order of their significance on Form 483. Upon completing the review, the inspector will issue a copy of the observations to the highest-ranking officer in the firm. The firm is not required to respond to the 483, but it is industry practice to send the FDA letter within 15 calendar days detailing corrections or correction plans.

1.9.6.3   Warning Letters
A warning letter is a written communication from FDA to an individual or firm indicating that one or more products, practices, processes, or other activities are in violation of the FD&C Act. The FDA will issue a warning letter to the firm if the deficiencies are serious or if the product can cause a serious public health risk and if the firm continues violative conduct. The firm must respond to the warning letter as soon as possible.

1.9.6.4   Establishment Inspection Report
The EIR is essentially a diary of the inspection and is prepared after the conclusion of the inspection. It will include the reason for the inspection, the date of inspection, the scope of the inspection, what type of inspection was performed as well as the findings and observations, and a brief description of the product and processes. The conclusion of the inspection is included in the EIR and will be one of three categories:

1) No Action Indicated (NAI): no significant or no cGMP deviations.
2) Voluntary Action Indicated (VAI): cGMP deviations the firm can correct that do not compromise public health or safety.
3) Official Action Indicated (OAI): this is requested by the FDA due to the serious nature of the findings indicating that deviations may affect safety.

Appropriate enforcement action for the recommended official action indicated will be assigned by the agency. Further violation of the law by the company may result in the FDA asking the court to hold company in civil or criminal action of the decree. Under consent decree conditions, the FDA may order the firm to cease operation if it is still not in compliance with cGMP regulations.

1.10   Electronic Records and Electronic Signatures

Under 21 CFR 11, the criteria under which the agency considers electronic records, electronic signatures, and handwritten signatures executed to electronic records to be trustworthy, reliable, and generally equivalent to paper records and handwritten signatures executed on paper are established. This does not, however, apply to electronic transmissions of paper records.

1.10.1   Electronic Records

1.10.1.1   Closed Systems
Closed systems used to create, modify, maintain, or transmit electronic records shall safeguard the system to ensure the authenticity, integrity, and the confidentiality of electronic records by

employing procedures and controls to ensure that the records cannot readily be altered. Such procedures and controls shall include the following:

- Validation of systems to ensure accuracy, reliability, consistent intended performance, and the ability to discern invalid or altered records.
- Generation and protection of records to enable their accurate retrieval throughout the records retention period.
- Limit system access to authorized individuals and use secure, computer-generated, time-stamped audit trails to document actions that create, modify, or delete electronic records.
- Verify that persons who develop, maintain, or use electronic record/electronic signature systems have the education, training, and experience to perform their assigned tasks. Establish controls over the access to and distribution of system operation and maintenance documentation. Implement change control procedures to maintain an audit trail that documents modification of systems documentation.

1.10.1.2 Open Systems

Use of open systems to create, modify, maintain, or transmit electronic records shall employ procedures and controls designed to ensure the authenticity, the integrity, and the confidentiality of electronic records from the point of their creation to the point of their receipt. Such procedures may include document encryption and use of appropriate digital signature standards.

1.10.2 Electronic Signatures

A unique electronic signature shall be to one individual and shall not be reused by, or reassigned to, anyone else, and that individual's identity must be verified prior to sanction of its use. Biometrics such as fingerprint or scanned iris patterns may be used. Electronic signatures that are not based upon biometrics shall employ at least two separate components to establish identity, such as an identification code and a password. Such two component systems must be used only by their owners. The integrity of the generation of passwords or identification must be confirmed by periodic testing.

1.11 Employee Safety

While the FDA is the lead regulatory agency with regard to pharmaceuticals, its focus is on the safety and efficacy of the drug product, not on the hazardous conditions that employees in the industry may encounter. The agency primarily responsible for employee safety is the Occupational Safety and Health Administration (OSHA), housed within the US Department of Labor. The agency has overarching responsibility for employee safety in all industries, issuing regulations in response to diverse threats to health as varied as requiring respirators for manned entrance to confined chambers, hard hats and steel-toed shoes for construction sites, and ergonomic adjustments to prevent repetitive injuries in offices and assembly lines. While each of these (among many other conditions) may at some time apply to the pharmaceutical industry, we shall illustratively concentrate on one: process safety management of highly hazardous chemicals 29 CFR 1910.

Drug products and their in-process antecedents are chemicals to which the human body may react, not always in a favorable manner. While it may not be possible to eliminate exposure to known or potential hazardous chemicals in the workplace, it is in everyone's best interest to limit exposure to levels below which the body can resist the deleterious effects, that is, below the threshold level.

The most obvious chemical hazards seem quite straight forward, if extreme exposure to a chemical results in near instantaneous death, acute toxicity. That may have sufficed in the not too distant past, but we have better understanding of how chemicals may affect our bodies and realize that the harmful effect may not appear long after exposure, like asbestosis, or cumulatively, like cigarette smoking, a chronic toxicity. Some chemicals, like aflatoxin, are both chronic and acute, killing at low dosages, but also carcinogenic. Death is not the only harmful outcome. Methanol may not kill at low doses, but neurologic manifestations including seizures, symptoms of amyotrophic lateral sclerosis, and blindness may present shortly after ingestion.

We now understand that there are chemicals whose damage may not present for a generation, or present without an obvious link to chemical exposure. Genotoxic chemicals are destructive to our genetic material, DNA (deoxyribonucleic acid) and RNA (ribonucleic acid), which may cause mutations. Fertility may be impaired. Teratogenic toxins may not affect those directly exposed, but cause birth defects. The human body has the ability to tolerate many of these chemicals at low levels so it is important to know the threshold levels above which harmful effects may be experienced.

For many, if not most, of the chemicals to which we are exposed in the workplace, critical data has been compiled in a safety data sheet (SDS). This technical document lists many of the chemical properties of the chemical and the known effects on human health. It provides detailed information about the health effects of exposure; a hazard evaluation relating to handling, storage, or use; measures to protect workers at risk of exposure to the chemical; and emergency procedures to follow in the event of exposure. An SDS may be considered the starting place when evaluating hazard risks of chemicals.

OSHA, through 29 CFR 1910.119, addresses risks to employees by catastrophic release of hazardous chemicals from:

- Concentrations of chemicals above the threshold levels.
- Flammable liquid or gas in excess of 10 000 lb in one location.
- Hydrocarbon fuels used for workplace consumption.
- Flammable liquids stored or transferred below their boiling points in atmospheric tanks without refrigeration.

1.11.1 Process Safety Information

The employer shall complete a compilation of written process safety information before conducting any process hazard analysis to facilitate, identify, and understand the hazards posed by those processes involving highly hazardous chemicals. This process safety information shall include information pertaining to:

- Hazards of chemicals (suitable SDSs may be used) used or produced by the process, which shall include:
  - Toxicity information.
  - Permissible exposure (threshold limits).
  - Physical data (boiling point, flash point, etc.).
  - Reactivity data.
  - Corrosivity data.
  - Thermal and chemical stability data.

- Hazardous effects of inadvertent mixing of different materials that could foreseeably occur.
- Process technology information
  - Flow diagram of the process.
  - Process chemistry (reactions that take place).
  - Maximum intended inventory (bulk storage).
  - Safe upper/lower limits for temperatures, pressures, flows, or compositions.
  - An evaluation of the consequences of deviations, including those affecting the safety and health of employees.
- Process equipment information
  - Construction material.
  - Piping and instrument diagrams.
  - Electrical classification.
  - Relief system design and design basis.
  - Design codes and standards.
  - Material and energy balances for processes (built post 26 May 1992).
  - Safety systems (interlocks, detection, or suppression).

For existing equipment designed and constructed per superseded codes, standards or practices, the employer will certify that the equipment is designed, maintained, and operating in a safe manner.

1.11.2 Process Hazard Analysis

The employer shall perform an initial process hazard analysis (hazard evaluation) that is appropriate to the complexity of the process and shall identify, evaluate, and control the hazards involved in the process. The priority order for conducting process hazard analyses should be based on a rationale that includes such considerations as the extent of the process hazards, the number of potentially affected employees, and the operating history of the process.

One or more of the following methodologies shall be used: What-If, Checklist, What-If/Checklist, Hazard and Operability Study (HAZOP), Failure Mode and Effects Analysis (FMEA), Fault Tree Analysis, and/or an appropriate equivalent methodology.

The process hazard analysis shall address:

- The hazards of the process.
- The identification of any previous incident that had a likely potential for catastrophic consequences in the workplace.
- Engineering and administrative controls applicable to the hazards and their interrelationships (e.g. process monitoring and control instrumentation with alarms, hydrocarbon sensors).
- Failure of engineering and administrative controls consequences.
- Facility sitting.
- Human factors.
- A qualitative evaluation the possible safety and health effects of control failure on employees in the workplace.

A team with expertise in engineering and process operations, with at least one employee experienced and knowledgeable about the process being evaluated and one member knowledgeable in the specific process hazard analysis methodology used.

1.11.3   Operating Procedures

Written operating procedures will be developed and implemented by the employer to provide clear instructions for safely conducting processes that shall address:

- Steps for each operating phase:
  - Initial startup.
  - Normal operations.
  - Temporary operations.
  - Emergency shutdown including the conditions under which emergency shutdown is required, and the assignment of shutdown responsibility to qualified operators to ensure that emergency shutdown is executed in a safe and timely manner.
  - Emergency operations
  - Normal shutdown
  - Startup following a turnaround, or after an emergency shutdown
- Operating limits:
  - Consequences of deviation.
  - Steps required to correct or avoid deviation.
- Health and safety and considerations:
  - Properties of, and hazards presented by, the chemicals used in the process.
  - Precautions necessary to prevent exposure, including engineering controls, administrative controls, and personal protective equipment.
  - Control measures to be taken if physical contact or airborne exposure occurs.
  - Quality control for raw materials and control of hazardous chemical inventory levels.
  - Any special or unique hazards.
- Safety systems and their functions.

Operating procedures shall be readily accessible to employees. They shall be reviewed as needed to incorporate any changes to the operating processes.

The employer shall develop, document, and implement safe workplace practices for employees and contractors to control hazards during operations. These precautions may include lockout/tagout to prevent processes from proceeding while in an unsafe state or requiring a respirator and banning solo entry for confined space.

1.11.4   Training

Employees involved in or newly assigned to operate a process shall be trained in an overview of the process and in the operating procedures emphasizing the specific safety and health hazards, emergency operations, and applicable safe work practices. Refresher training shall be provided at least every 3 years. All training shall be documented stating the name of the employee, the date of training, and the means of verifying comprehension of the material.

1.11.5   New Facility Startup

1.11.5.1   Pre-startup Safety Review

A safety review shall be performed when starting up a new facility or when modifications to an existing plant are extensive enough to require changes to the process safety information. It will ensure that prior to operation:

- Construction and equipment is in accordance with design specifications.
- Safety, operating, maintenance, and emergency procedures are adequate and in place.

- Process hazard analysis has been performed and recommendations have been resolved or implemented before startup.
- Training of each involved employee has been completed.

1.11.6 Mechanical Integrity

Mechanical integrity shall be checked for:

- Pressure vessels and storage tanks.
- Piping systems.
- Relief and vent systems.
- Emergency shutdown systems.
- Controls (including monitoring devices and sensors, alarms, and interlocks).
- Pumps.

Written Procedures
    The employer shall establish and implement written procedures to maintain the ongoing integrity of process equipment.
Training for Process Maintenance Activities
    The employer shall train each employee involved in maintaining the ongoing integrity of process equipment in an overview of that process and its hazards and in the procedures applicable to the employee's job tasks.
Inspection and Testing
    Inspections and tests shall be performed on all process equipment and shall follow recognized and generally accepted good engineering practices. The frequency of inspections and process equipment tests shall be consistent with manufacturer's recommendations. Each inspection shall be documented.
Quality Assurance
    Qualification that equipment is suitable for the process design specifications and is properly installed and consistent with manufacturer's instructions. Appropriate maintenance materials and spare parts are suitable for the process application.

1.11.7 Hot Work Permit

A permit shall be issued for hot work performed on or near a covered process. It shall document that prior to initiating work, proper fire prevention and protection procedures have been implemented, the dates the work is permitted, and the object upon which the hot work performed.

1.11.8 Management of Change

Written procedures shall be put in place to manage changes to all aspects of the covered process operation. Prior to any change the following considerations are to be addressed:

- Technical basis for change.
- Impact of change on safety and health.
- Modifications to the operating procedures.
- Time necessary for the change.
- Authorization requirements for the proposed change.

Employees affected by the change shall be informed and trained as necessary prior to startup of the process. If necessary, process safety information shall be updated.

1.11.9 Incident Investigation

Any incident that resulted in or could have resulted in a catastrophic release of highly hazardous chemicals in the workplace shall be investigated as promptly as possible but within 48 h of the incident. The incident investigation team shall contain at least one employee knowledgeable in the process involved and other persons with sufficient knowledge and experience to investigate and analyze the incident. An incident report shall include:

- Date of incident.
- Date investigation commenced.
- Description of the incident.
- Factors contributing to the incident.
- Recommendations resulting from the investigation.

The employer shall promptly address and resolve the incident report findings and recommendations documenting any corrective actions taken. The report shall be reviewed with all personnel whose tasks are relevant to the incident findings. The report shall be retained for 5 years.

1.11.10 Emergency Planning and Response

An emergency action plan shall be established for the entire plant. The plan shall include procedures for handling small releases. If applicable, hazardous waste and emergency response provisions shall apply 29 CFR 1910.120(a),(p), and (q).

1.11.11 Compliance Audits

Certification of compliance with regulatory provisions will be done at least every 3 years to verify that procedures and practices are adequate and implemented. The audit shall be conducted and reported by at least one person knowledgeable of the process. Any deficiencies in the report shall be corrected and documented.

1.12 US EPA

As manufacturing entities, pharmaceutical manufacturing facilities are subject to regulation by a variety of regulatory agencies not directly responsible for the production of safe, efficacious drug products. The EPA, specifically the Office of Compliance, Chemical Industry Branch, is tasked with enforcement of a number of statutes impacting drug manufacturing facilities. The EPA has, in the past, relied on a command and control approach to regulate industrial facilities, but now is combining its traditional method with innovative techniques such as self-assessments and facility management systems.

There is little correlation among the complex web of requirements that results from regulations originating independently that target the same medium or activity. Many industrial facilities have found that using a complete facility, the environmental management system (EMS) approach yields cost effective solutions for tackling all of the requirements as a complete facility solution instead of as individual components.

There are five major statutes that impact pharmaceutical manufacturing under the auspices of the EPA:

- Clean Air Act (CAA).
- Safe Drinking Water Act (SDWA).

- Resource Conservation and Recovery Act (RCRA).
- Emergency Planning and Community Right-to-Know Act (EPCRA).
- Clean Water Act (CWA).

A brief summary of the meaning and impact of these statutes will serve as an introduction to the understanding the scope of the Federal environmental regulations. State and local jurisdictions may also impact the facility, and all should be monitored to keep abreast of pending changes.

1.12.1 Clean Air Act

Clean Air Act Regulatory Requirements

- Title I: National Primary and Secondary Ambient Air Quality Standards (NAAQS)
  - The Air Quality Act of 1967 requires the designation of air quality control regions (AQCRs) based on "jurisdictional boundaries, urban-industrial concentrations, and other factors including atmospheric areas necessary to provide adequate implementation of air quality standards" [Section 107(a)(1967)].
  - Recognizing the deleterious effects of poor air quality on human health, NAAQS have been set for ozone, carbon monoxide, particulate matter <10 μm, sulfur dioxide, nitrogen dioxide, and lead. Of these, all but lead may have a significant impact on the pharmaceuticals industry.
  - Given the varying levels of pollutants among the regions, conformation to the regulations is site specific. Undergoing a new source review, permitting for "attainment areas" (where desirable air quality standards exist) would require installation of the best available control technology (BACT), whereas permitting for nonattainment areas (NAAQS are exceeded) would require installation of more stringent lowest achievable emission rate (LAER) technology.
  - Major pharmaceutical industry sources must, irrespective of location, comply with performance standards set by the EPA, referred to as new source performance standards (NSPS) applicable to new sources or modified facilities. Requirements include monitoring, recordkeeping, and reporting.
- Title III: National Emissions Standards for Hazardous Air Pollutants (NESHAP) and Maximum Achievable Control Technology (MACT) Standards
  - NESHAP refers to standards for a select group or hazardous air pollutants for which additional risk-based standards were developed prior to 1990s CAA amendments. Monitoring, recordkeeping, and reporting are required for these pollutants. 1990 CAA identified 189 hazardous air pollutants (HAPs) for which standards of performance were to be developed based upon maximum achievable technology, not risk.
  - Existing NESHAPs for HAPs on the 1990 list are still applicable, and the EPA set MACT standards applicable to specific industries for so-called hazardous organic NESHAPS (HON).
- Title V: Permitting Program
  - Defined the minimum standards and procedures for state-operating permit programs. This consolidates all of a source requirement into one permit. Any major source is required to obtain a permit.
  - Major sources emit or has potential to emit.
    - 10 tons per year (TPY) or more of any hazardous air pollutant.
    - 25 TPY or more of any combination of HAPs.
    - 100 TPY of any air pollutant.

- Major sources in ozone nonattainment areas are defined as sources with the potential to emit.
  - 100 TPY or more of volatile organic compounds (VOCs), or nitrogen oxides (NO_x) in areas defined as moderate or marginal.
  - 50 TPY or more of VOCs or NO_x in areas classified as serious.
  - 25 TPY or more of VOCs or NO_x in areas classified as severe.
  - 10 TPY or more of VOCs or NO_x in areas classified as extreme.
- Other sources requiring permits regardless of source size include:
  - NSPS.
  - NESHAP.
  - PSD (Prevention of Significant Air Quality Deterioration)/NSR (New Source Review).
  - Acid rain.
- Title VI: Stratospheric Ozone Protection.
  - Provides for a phaseout of the production and consumption of chlorofluorocarbons (CFCs) and other chemicals that deteriorate the ozone layer.
- CAA Assessment Considerations
  - Many CAA requirements have been summarized into a comprehensive permit. The compliance assessor should review data derived from previous facility self-assessments.
- CAA regulatory requirements that may apply to the pharmaceutical industry.
  - 40 CFR Part 60
  - Subparts D_a, D_b, D_c, K_b, GG
  - 40 CFR Part 61
  - Subparts J, M, V, Y
  - 40 CFR Part 63
  - Subparts H, I, Q
  - 40 CFR Part 68
  - 40 CFR Part 82

1.12.2 Safe Drinking Water Act

Safe Drinking Water Act Regulatory Requirements

- SDWA mandated that EPA regulate to protect human health from contaminants in drinking water.
  - EPA developed drinking water standards.
  - Joint federal/state system to ensure compliance.
  - EPA to protect underground source of drinking water through control of underground injection of waste.
- Underground Injection Control Program
  - Permit program with five classes of injection wells.
    - Class I – Large volumes of hazardous and nonhazardous waste into deep, isolated rock formation separated from drinking water by impermeable clay.
    - Class II – Inject fluids, mostly brine, associated with oil and gas extraction. About 10 barrels of brine are required to yield 1 barrel of crude oil.
    - Class III – Inject super-hot steam or water into mineral formations that are then pumped to the surface and extracted. The fluid is treated and reinjected into the same formation. More than 50% of the salt and 80% of the uranium produced in the United States is obtained this way.

- o Class IV – hazardous or radioactive waste is injected into or above underground sources of drinking water. These wells are banned under UIC (Underground Injection Control) program.
  - o Class V – other injection methods, some quite technologically advanced, but some low-tech holes in the ground. Generally shallow and dependent upon gravity to drain into the ground.
- Public Water System Program
  - – Primary and secondary drinking water regulations.
  - o Primary have adverse effects on human health.
  - o Secondary affect aesthetic quality of water.
- Not EPA enforceable, but states may enforce.
  - – Established testing procedures, monitoring requirements, notifications, and reporting.
- SDWA Assessment Considerations
  - – If the facility has its own source of potable water and provides water to 25 unique individuals for 6 months, it is subject to national drinking water standards, in which case it must be monitored for required contaminants, is required frequency using an approved laboratory and tests, and is maintaining records.
- SDWA Regulatory Requirements
  - – 40 CFR Part 141
    - o Subparts B, G, F, C, H, I, D
  - – 40 CFR Part 143
  - – 40 CFR Part 144

1.12.3 Resource Conservation and Recovery Act

Resource Conservation and Recovery Act Regulatory Requirements:

- Amendment of the Waste Disposal Act of 1965 addresses hazardous and solid waste management. The Hazardous and Solid Waste Amendment (HSWA) of 1984 strengthened RCRA's provisions and added governance of Underground Storage Tanks. The objective is to protect human health and to conserve energy resources and valuable materials. A cradle-to-grave system was implemented, and some states are implementing more stringent regulations.
- Hazardous Waste Generation
  - – Determination of what waste is hazardous is the initial step in determining compliance and is detailed in 40 CFR 261. A waste may not be on the federal list, but exist on a state list of hazardous materials.
  - – Secondary materials generated by the drug industry may be classified as solid wastes or potentially hazardous waste. Potentially hazardous waste that is to be recycled is subject to rules about accumulation and disposal.
  - – Generators are classified by the amount of waste generated.
    - o Large quantity generator (LQG).
    - o Small quantity generator (SQG).
    - o Conditionally exempt small quantity generator (CESQG).
  - – Generators may accumulate hazardous waste for up to 90 days (180 for small quantity generators).
- Hazardous Waste Transportation Regulations
  - – Under 40 CFR 262, transporter must obtain an EPA identification number and specific manifesting and recordkeeping requirements.
- Hazardous Waste Treatment, Storage, and Disposal Regulations

- Any facility that treats, stores, or disposes (TSDF) of hazardous waste is subject to requirements under 40 CFR 264 and 265.
- Must obtain an operating permit and abide by TSD regulations, far more extensive than those for generators and transporters and include technical and administrative requirements.
- Land Disposal Restrictions (LDRs)
  - Hazardous waste is largely prohibited from land disposal.
    o Comply with a specified treatment standard.
    o Dispose in a "no migration unit."
- Underground Storage Tank (UST) Regulations
  - USTs containing hazardous materials and petroleum are regulated under 40 CFR 280. States have discretion to develop their own UST regulatory program.
  - In 1998 it was required that all existing USTs must add:
    o Spill, overfill, and corrosion protection
    o Close the existing UST
    o Replace with a new UST
- RCRA Assessment Considerations
  - Key Components
    o Knowledge of facility
    o Document review
      ▪ Facility maps, organization charts, manuals, photos, etc.
    o Assessment plan
      ▪ Trace Material Flow Through the Plant
- RCRA Regulatory Requirements
  - 40 CFR 261.5 and 262.34
  - 40 CFR 262
  - 40 CFR 263
  - 40 CFR 264 and 265
  - 40 CFR 268
  - 40 CFR 280

1.12.4 Emergency Planning and Community Right-to-Know Act

Emergency Planning and Community Right-to-Know Act Regulatory Requirements, also known as Superfund Amendments and Reauthorization Act (SARA) Title III, is intended to inform the general public emergency planning and emergency response personnel about potential hazards in the community.

- Hazardous Substance Notification
  - A release means any spilling, leaking, pumping, emitting, emptying, discharging, injecting, escaping, leaching dumping, or disposing of into the environment, excluding exposure only within a workplace.
  - Facilities releasing a hazardous substance equal to or exceeding the reportable quantity (RQ) must immediately notify the National Response Center at 800/424-8802 or 202/426-2675.
  - RQ ranges from 1 to 5000 lb.
- Emergency Planning and Notification
  - Release of a quantity equal to or exceeding its threshold planning quantity of an extremely hazardous substance shall notify the state emergency response commission (SERC) or governor (if no commission) and the local emergency planning commission (LEPC).

- Hazardous Chemical Reporting: Community Right-to-Know
  - Pharmaceutical facilities must submit a Safety Data Sheet (SDS) or a list of hazardous chemicals for which SDSs are required to the SERC, LEPC, and the fire department.
  - They must also submit a Tier I (Aggregate Information by Hazard Type) or Tier II (Specific Information by Chemical) form for all hazardous chemicals (above a threshold of 500 lb), indicating the aggregate amount at the facilities classified by hazard category.
  - If any agency requests a Tier II report, it must be submitted within 30 days of the request. A Tier II form may be submitted in lieu of a Tier I form.
- Toxic Chemical Release Inventory
  - A submission of the Toxic Chemical Release Inventory (TRI) reporting form (Form R) is required. This is a compilation of release information of toxic compounds into the community.
  - A complete Form R is required annually for each toxic chemical manufactured, processed, or otherwise used at each covered facility.
- EPCRA Assessment Considerations
  - Activities will focus primarily on reporting and recordkeeping.
  - Form R is highest profile reporting requirement.
- EPCRA Regulatory Requirements
  - 40 CFR 302
  - 40 CFR 355
  - 40 CFR 370
  - 40 CFR 372

1.12.5 Clean Water Act

1.12.5.1 Clean Water Act Regulatory Requirements

The intent is to restore and maintain the physical and biological integrity of the nation's water. Both direct and indirect discharges of waters are regulated.

- Effluents Limitations Guidelines and Categorical Pretreatment Standards
  - Establishes limitations for direct and indirect discharges.
  - Biological oxygen demand (BOD).
  - Chemical oxygen demand (COD).
  - Total suspended solids (TSS).
  - pH.
  - Priority and nonconventional pollutants.
- National Pollutant Discharge Elimination System (NPDES)
  - Controls indirect discharges by permits.
    - o Individual – specific facility.
    - o General – category of similar discharges within an area.
- Pretreatment program.
  - Controls direct discharges.
  - Goals
    - o Prevent damage to municipal wastewater treatment plants.
    - o Prevent pollutants from passing through the treatment plant untreated.
    - o Encourage reuse and recycling of municipal and industrial sludge.
- Policy on Effluent Trading in Watersheds
  - Water quality standards must be met and technology-based requirement remain in place.
  - Effluent trading potentially offers a number of economic, environmental, and social benefits.
- Spills of Oil and Hazardous Substances

- Prohibits oil discharges
- Oil pollution prevention
  - Establishes procedures to prevent discharge of oil.
- Reportable Quantities for Hazardous Substances
  - Designates hazardous substances and reportable quantities.
- CWA Assessment Considerations
  - Verify facility's operations are properly regulated by the permit and that monitoring results are representative of the facility's operations.
- CWA Regulatory Requirements
  - 40 CFR 439
  - 40 CFR 110
  - 40 CFR 112
  - 40 CFR 116

1.13   Process Analytical Technology

Process Analytical Technology (PAT) is not strictly a regulatory issue. It is an FDA Guidance establishing a risk-based framework that is intended to support innovation and efficiency in pharmaceutical product development, manufacturing, and quality assurance, working within existing regulations. It is founded on process understanding to facilitate innovation and risk-based regulatory decisions by industry and the agency. The framework has two components:

1) A set of scientific principles and tools supporting innovation.
2) A strategy for regulatory implementation that will accommodate innovation.

The regulatory implementation strategy includes creation of a PAT Team approach to CMC review and cGMP inspections as well as joint training and certification of PAT review and inspection staff.

The FDA considers PAT to be a system for designing, analyzing, and controlling manufacturing through timely measurements (i.e. during processing) of critical quality and performance attributes of raw and in-process materials and processes, with the goal of ensuring final product quality. A concise summation might be: know the product/process better and all will be well, that is, by better understanding the drug product and process, more effective tools can be brought to bear on the (newly discovered) critical control points with the ultimate result of delivering a better quality product more efficiently and at lower cost. The tighter controls would also (de facto) meet regulatory requirements.

This approach has not been widely adopted by the industry, not so much because of resistance, but rather because of the complexity, a lack of knowledgeable personnel, and uncertainty of regulatory implementation and acceptance. A broader view of reading the guidance may find intimations of the iterative quality approach of the International Organization for Standardization (ISO [4]). If so, that may ease the industry into broader implementation.

1.13.1   Process Understanding

By FDA definition, a process is generally considered well understood when

- All critical sources of variability are identified and explained.
- Variability is managed by the process.
- Product quality attributes can be accurately and reliably predicted.

In the FDA guidance, variability (e.g. in raw materials), caused by insufficiently understood chemical and mechanical attributes, is "adjusted" by experienced formulators and benchtop analyses. Better understanding of the unknowns would lead to real-time detection during production of these variations and allow instantaneous adjustments to process conditions, thus ensuring quality drug products that meet all specifications, thereby clearing regulatory hurdles.

By implementing during product development phase an experimental design to examine the effects of varied physical characteristics in real time on a small scale, transfer from bench to pilot plant and/or production would be optimized.

1.13.2   Principles and Tools

Currently, most unit operations are performed on a time basis that does not take into account physical, chemical, or biological variability. It is promulgated that proper implementation of PAT tools would lend greater understanding and therefore enable optimization. The PAT toolkit includes:

- Multivariate Tools for Design, Data Acquisition, and Analysis
  - Pharmaceutical processes and products are complex multi-factorial systems, therefore, to understand the impact of varied moieties (parts or functional groups of organic molecules), a multivariate mathematical analysis of the effect of variability on process/product characteristics would be key.
- Process Analyzers
  - Much advancement in the development of process analyzers, including the ability, in some analyzers, to nondestructively determine biological, chemical, and physical attributes of the materials being processed.
  - Measurements may be made:
    o At-line, in close proximity to the process stream.
    o Online, where sample is diverted (and may be returned).
    o Inline, where sample is not removed from process stream.
  - Much more rapid results vs. sample removal to a laboratory for analysis and, indeed, much more data, the volume of which would be controlled by computer-based knowledge systems.
  - Proper design of process equipment and placement of analyzer is critical to ensure that the data collected is relevant.
- Process Control Tools
  - Once the critical attributes have been determined and process analyzers are properly in place and detecting critical parameters, process controls must be designed that will in real time provide adjustments to the operation(s) controlling all of the critical parameters.
  - Use the cumulative data to construct algorithms that enable the process controllers to adjust operations to achieve an endpoint of product quality attribute(s) rather than an endpoint based on time.
  - Implementation across the manufacturing platform would yield vast quantities of real-time data that could be used to fine-tune and optimize the final drug product attributes.
- Continuous Improvement and Knowledge Management Tools
  - In short, knowledge is power.
    o The more relevant data collected, the better quality final drug product.
    o Collection of data over the lifecycle of the product, in addition to improving quality, would facilitate regulatory evaluation of postapproval changes.

Risk-Based Approach
Given an inverse relationship between the level of process understanding and the risk of producing a poor product for a defined system, it should be possible to develop less stringent regulatory approaches to manage change requests. Risk-based analysis and management may form a separate system.

Integrated Systems Approach
Advances in information gathering and dissemination drive a push for an integrated approach to drug product development, as opposed to a handoff from one department to another. Development, manufacturing, quality assurance, and information management should work together as a team in as early a stage as possible. Toward that end, the FDA has developed a new regulatory strategy that includes a PAT team approach to joint training, certification, CMC review, and cGMP inspections.

Real-Time Release
The FDA has indicated that it is willing to validate real-time release based on a combination of in-process controls and assessed material attributes. In real-time release, material attributes as well as process parameters are measured and controlled.

1.13.3 Strategy for Implementation

The FDA believes that current regulations are broad enough t allow implementation of PAT, but understands that flexibility, coordination, and communication is critical.

The FDA strategy includes:

- A PAT team approach for CMC review and cGMP inspections.
- Joint training and certification of PAT review, inspection, and compliance staff.
- Scientific and technical support for the PAT review, inspection, and compliance staff.
- The recommendations provided in their guidance.

1.13.3.1 PAT Regulatory Approach

One goal of the guidance is to tailor the FDA's usual regulatory scrutiny to meet the needs of PAT-based innovations that:

- Improve the scientific basis for establishing regulatory specifications.
- Promote continuous improvement.
- Improve manufacturing while maintaining or improving quality.

1.14   Conclusion

This chapter is intended to be a broad overview of the regulatory aspects of pharmaceutical manufacturing. As such, I have focused on the regulatory agencies of the United States, embodied in the CFR.

There are independent organizations that contribute to, and in fact are indispensable to, the regulatory environment. Listed below are a few broken into rough categories though they often overlap with other designations.

References

1 United States Pharmacopeia (USP) – annual compendium of drug information. www.usp.org (accessed 12 December 2017).
2 AOAC International – publishes Official Methods of Analysis of the AOAC. www.aoac.org (accessed 12 December 2017).

3 International Conference on Harmonization (ICH). www.ich.org (accessed 12 December 2017).

4 International Organization for Standardization (ISO) – also harmonization. https://www.iso.org/home.html (accessed 12 December 2017).

Further Reading

American Society for Quality (ASQ). https://asq.org.in/ (accessed 12 December 2017).

ASTM International – wide ranging international standards organization. www.astm.org (accessed 12 December 2017).

International Society for Pharmaceutical Engineering (ISPE) – Guides. www.ispe.org (accessed 12 December 2017).

National Formulary (NF) – publishes in conjunction as the USP-NF. This pharmacopeia has a role in U.S. Law for drugs that do not conform to USP-NF standards. www.uspnf.com (accessed 12 December 2017).

2

Pharmaceutical Water Systems

CHAPTER MENU

In addition to being the most prevalent liquid on the planet, water is widely used in virtually every facet of pharmaceutical and biotechnology operations. As will be seen, because of its widespread use, water is a convenient standard for manufacturing and other pharmaceutical-related operations. Also, because of water's use in other industries, a large database exists to simplify operations employing water in its various phases (i.e. liquid, gas– vapor/steam, and solid ice).

2.1 Pharmaceutical Water Systems Basics

The criteria for water that is used for pharmaceutical, biotech, and device manufacturing are regulated by specific procedures found in the US Pharmacopoeia/National Formulary (USP/ NF) and 21 CFR. However, prior to delving into these regulations and procedures, a basic knowledge of the workings of these water systems is desirable. This is particularly relevant for pharmaceutical engineering since often other elements of the drug and device organization (e.g. quality assurance, quality control, and purchasing) may lack the capability to comprehensively and analytically determine adequacy of water system design, installation, and operation. This is due in large part to the outsourcing of pharmaceutical activities that have traditionally been performed by in-house operations. The role of many activities relating to pharmaceutical water production has been rapidly changing to a point where design and installation is being replaced by operation and installation oversight. That is, the role of design and installation is

Practical Pharmaceutical Engineering, First Edition. Gary Prager.
© 2019 John Wiley & Sons, Inc. Published 2019 by John Wiley & Sons, Inc.

becoming an outsourced activity with a diminishing in-house capability. Consequently, an overview of the fundamentals of water systems and related engineering aspects are described herein. This chapter is intended for both experienced engineering personnel with pharmaceutical backgrounds but not necessarily focused on water systems and those individuals with detailed design and project experience. Similarly, those with minimal technical background should find the material of value. Also, engineers with water system knowledge may find some elements, such as the design fundamentals and associated chemistry of ion exchange units, reverse osmosis (RO), and evaporation/distillation systems of value.

While the material covered in this chapter is sufficient to understand the example problem solutions, some solutions are not straightforward and require some analysis. In reality, this is how "real-life" engineering problems are resolved.

With the decline of in-house process design activities in many pharmaceutical manufacturing operations and the recent emphasis on as-built supplied purified water manufacturing operations, it is becoming more difficult to verify the adequacy of many as-built purified water operations. In many cases, the final cost varies significantly from the approved bid cost. This variation is often the result of a multitude of contractor-initiated change orders required before, during, and subsequent to the installation phase. Hopefully, Chapter 2 will assist both the user and the contractor to obviate or reduce the need for extensive change orders for a purified water system.

A significant and evolving dynamic in pharmaceutical manufacturing, particularly in the area of biotechnology manufacturing, is the trend toward relying on outside procurement of pharmaceutical waters. Many operations, in an effort to minimize costs, are using this technique. There is also a parallel effort to minimize or eliminate the need for facility steam by employing single-use or disposable equipment in development (i.e. pilot plant operations) and manufacturing operations. By incorporating single-use equipment, the cost of high purity water and steam for cleaning operations is significantly reduced. Nevertheless, purified water and steam production are still very important unit operations in existing facilities involved in pharmaceutical manufacturing, as opposed to biotechnology products.

When referring to pharmaceutical water, the term actually identifies eight types of water used in the manufacture of drug products. The eight types of water are:

1) Non-potable
2) Potable (drinkable) water
3) USP purified water
4) USP water for injection (WFI)
5) USP sterile WFI
6) USP sterile water for inhalation
7) USP bacteriostatic WFI

The content of this chapter, although applicable to the eight types of pharmaceutical water identified, deals primarily with USP purified water and the most stringent type of produced water, WFI. For this chapter, these eight water classifications will all be considered as pharmaceutical-grade water, indicating that they have been manufactured in accordance with the regulations required in 21 CFR 210 and 21 CFR 211.

While pharmaceutical water operations are regulated by cGMP regulations (the definition of cGMP is identified later in the Definitions), actual deionized water quality standards are specified in the current version of the USP.

Table 2.1 identifies the standards for the two general categories of pharmaceutical quality water, specifically USP Grade Water and WFI Grade Water.

Table 2.1 Pharmaceutical water quality standards.

Water type	Parameter	Units	Value	Comments
USP	Conductivity at 25°C	µS/cm	1.3	µS/cm = microsiemens/cm
	Total organic carbon (TOC)	ppb	<500	ppb = parts per billion
	Bacteria	cfu/100 ml	500	cfu = colony-forming units per ml
	Nitrates	ppm	N/A	Numerical value not defined in USP; numerical limits are for Japan and Europe
	Ammonium (NH_4)	ppm	N/A	Numerical value not defined in USP; numerical limits are for Japan and Europe
WFI	Conductivity at 25 °C	µs/cm	1.3	
	Total organic carbon (TOC)	ppb	<500	
	Bacteria	cfu/100 ml	10	Recommended
	Nitrates	ppm	N/A	The same as USP criteria
	Ammonium	ppm	N/A	The same as USP criteria

When the product is WFI, the consensus is that this grade of water should be stored and circulated at a temperature at or above 65 °C (149 °F). Previous guidance recommended a WFI circulating temperature of 80 °C (176 °F). The benefits of the lower circulating temperature include the ability to replace standard 316 grades of stainless steel with high performance polymers such as polyvinylidene fluoride (PVDF). Another advantage is the lower energy costs required to cool water from 80 °C to 20–25 °C to 65 °C, the normal working temperature for pharmaceutical manufacturing operations involving WFI.

Pharmaceutical engineering is differentiated from other areas of engineering (e.g. chemical engineering, mechanical engineering) in large part because many terms used in chemical and mechanical and chemical engineering are identified by other words. Also, many concepts and terms used in pharmaceutical water operations are not common to the chemical process industry. Therefore, it would be helpful to become familiar with some of these terms as a reasonable starting point for Chapter 2 as well as other chapters.

The following terms are frequently encountered in purified water operations and brief definitions may be useful:

- Anion – A negatively charged ion.
- Bioburden – The quantity of microorganisms present in a sample or given surface area or volume. The term cfu is often employed to indicate the bioburden.
- Cation – A positively charged ion.
- Colony-forming unit (cfu) – Formation of microbes on a surface or volume. Quantitatively, the colony-forming unit content is given by cfu/ml.
- cGMP – Current good manufacturing practice. A compilation of Federal Regulations contained in Title 21 of the Code of Federal Regulations containing specific sections related to quality practices in pharmaceutical manufacturing. These regulations include Sections 210 and 211. Pharmaceutical water is addressed as a component used in the processing of drugs or drug components as defined in 21 CFR 210.1(a).
- DNA – Deoxyribonucleic acid. The salient component of chromosomes that contain the characteristics of organisms.

- Endotoxin – A poison in the form of a fat and sugar complex that is part of a cell wall in certain types of bacteria (gram negative). The endotoxin is released when the cell dies or when the cell is ruptured. In addition to a fever, endotoxins can cause septic shock.
- Flux – The rate of transfer of a fluid, particle, or energy across a given surface (e.g. for heat flux, units would be $Btu/ft^2 h$, and for fluids, units can be $gal/ft^2 min$).
- Fungus (fungi, plural) – Organisms that exist without chlorophyll and reproduce by spore production. Fungi survive by absorbing nutrients from organic (often live) species. Some fungal diseases affecting humans include ringworm, athlete's foot, and candidiasis.
- Heat capacity – Quantity of heat required to raise a unit of mass in the metric system (g) or a unit of force in English units (lbf) one temperature unit (°F or °C depending on the system employed). At constant pressure, the heat capacity is denoted as C_p, and at constant volume the notation is C_v. These terms, C_p and C_v, are derived from thermodynamics and briefly covered in this chapter.
- Latent heat – A thermodynamic term where energy (most often in the form of heat) is introduced into a system, resulting in no temperature change. If enough latent heat is applied at a specific temperature that is unique to the substance, the system will change phase. An example is heating water to 100 °C (212 °F). If a latent heat quantity of about 970 Btu/lb is added to the water, the water will begin vaporizing to steam, yet the temperature will remain at 100 °C (212 °F).
- Molality – The molality of a solute is the number of moles (see definition of the mole, defined later) of the solute divided by the mass of solvent in kilograms. It is important to note molality is not dependent on temperature since only the mass of the solute and the mass of the solvent enter into the definition.
- Molarity – Defined as the number of gram mole of solute per liter of solution. The dimensions for molarity are given in metric units (e.g. g mol/l of solution). It is most important to note that molarity is in metric units only. As an example, the molarity of a 1 l sodium chloride (NaCl) solution containing 117 g of sodium chloride is considered. The atomic weights for the constituents are:

Atomic weight of sodium, Na = 23

Atomic weight of chlorine, Cl = 35.5

The gram molecular weight of sodium chloride (NaCl) is 35.5 + 23 = 58.5 g NaCl/g mol NaCl. Since the measured amount of sodium chloride is 117 g, the total number of gram moles of sodium chloride (NaCl) is

$$\frac{117 \; \cancel{g \, NaCl}}{1} \times \frac{1 \, g \, mol \, NaCl}{58.5 \; \cancel{g \, NaCl}} = 2 \, g \, mol \, NaCl \, (solute)$$

Accordingly, a solution containing 2 g mol of sodium chloride solute dissolved in 1 l of pure water is a 2 M (or 2 molar) solution (refer to the definition of a mole). It is important to note that molarity is based on volume and is also a factor of temperature. The temperature dependence of molarity and molality should be noted.

- Mole – The atomic weight of a substance expressed in grams or pounds is called the gram molecular weight or pound molecular weight. Using potassium nitrate (KNO_3) as an example (for illustrative purposes, the atomic weights are rounded to the nearest whole number):

Atomic weight of potassium, K = 39

Atomic weight of nitrogen, N = 12

Atomic weight of oxygen, O = 16 (three oxygen molecules yields 16 × 3 = **48**)

Molecular weight of potassium nitrate, $KNO_3 = 39 + 12 + 48 = 99$. When dealing with the metric system, the basic dimensions are given as $99\,g\ KNO_3/g\,mol\ KNO_3$; in engineering units the dimensions are $99\,lb\ KNO_3/lb\,mol\ KNO_3$.

- Normality – A solution that contains a gram-equivalent weight of solute per liter of solution is identified as a normal solution. A gram-equivalent is the weight that reacts with a gram atom $(1.0\,g)$ of hydrogen ion or hydroxide ion (OH^{-1}). If a compound does not contain a hydrogen or hydroxide ion, then the valence of the cation (the positive valence) is used. Some equivalent weights are given by:

Compound	Molecular weight (g)	Equivalent weight (g)
HCl	36.5	36.5
H_2SO_4	98.1	49.05
NaOH	40.0	40.0
$Ca(OH)_2$	74.0	37.0
NaCl	58.5	58.5
$AlCl_3$	133.5	44.5

The normality of a solution is identified by the symbol N and preceded by a numerical value. For instance, $0.2\,N$ NaOH is defined as "two tenths normal NaOH." Also, 1 l of a $2\,N\ NaOH$ solution contains 2 gram-equivalent weights (g-eq wt) of solute (80 g NaOH) dissolved in 1 l of solution (usually water). In quantitative terms, normality can be expressed as

$$N = \frac{\text{no. of g-eq wt solute}}{\text{liters of solution}}$$

- pH – A logarithmic scale to determine the hydrogen ion concentration in aqueous solutions. The hydrogen ion concentration is identified by "pH" and the hydroxide ions "pOH." Mathematically, the definitions are [1]

$$pH = -\log\left[H^+\right], \text{where a pH below 7 is acidic}$$
$$pOH = -\log\left[OH^-\right], \text{where a pOH above 7 is base}$$

Further, recalling basic chemistry, $pH + pOH = 14$. Table 2.2 relates concentrations of strong acids and strong bases to the aqueous solution pH.

- Purified water – Water that typically has been processed by distillation, ion exchange, RO, or other acceptable process and employs US FDA Potable Water as a manufacturing base.

Table 2.2 Concentration of strong acid (HCl) and strong base (NaOH) and pH/pOH.

Concentration of HCl, molar	10^{-1}	10^{-3}	10^{-5}	0			
pH	1	3	5	7	9	11	13
pOH	13	11	9	7	5	3	1
Concentration of NaOH		10^{-13}	10^{-13}	10^{-9}	0		

Potable water is identified as water that is safe to drink and conforms to EPA criteria defined in 40 CFR 141.14, 141.21, specifically.

- Pyrogen – A fever-inducing substance with the biochemical structure composed of lipopoly-saccharides, carbohydrate with a general molecular structure $(C_6H_{10}O_5)_n$.
- Sanitization – A chemical and/or physical operation used to kill bacteria or reduce contamination. This is performed by reducing the number of microbial contaminants (cfu) to a safe level. Quantitatively, an effective sanitization level can be expressed as being less than or equal to 10^{-3}. Sanitization is less effective, and then sterilization.
- Sensible heat – A thermodynamic term where the addition of energy (most often in the form of heat) to a system results in an increase of the system temperature.
- Sterilization – An operation employing a chemical agent or physical process clause and a high level of sterile resistant microorganisms to be destroyed. Typically, the microbial level, subsequent to sterilization, is on the order of 10^{-6}. A mathematical relationship is available for calculating the sterility assurance level (SAL). Basically, the SAL is a term describing the likelihood a sterilization operation is at or below the sterilization level of 10^{-6} (i.e. one in a million surviving species). The formula for determining the SAL is given as

$$Log_{10} B = log_{10} A - \frac{F_0}{D_{121}}$$

where

B = probability of survival
A = bioburden or product
D_{121} = time required at 121 °C to reduce the population of organisms in the sample by 90%
F_0 = time in minutes at 121 °C.

2.1.1   Fundamentals of Fluid Mechanics for Pharmaceutical Water Systems

Virtually all operations involving the manufacture of pharmaceuticals, biotech products, cosmetics, and medical devices require water as a part of the operations. Unlike standard process designs encountered in the petrochemical, petroleum, and chemical process industries, pharmaceutical designs are all regulated by the provisions of 41 CFR 210 and 211(cGMP) as well as USP/NF criteria. However, the salient features in designing water systems are the same engineering procedures required to create a successful water operating system, regardless of the intended end use. Also, because water is such a commonly used chemical, regulatory organizations such as the FDA find water and water-related operations are very useful in performing facility/process evaluations. Water systems serve as a good baseline. This section will introduce, define, and in many cases review the design and regulatory criteria required for successful pharmaceutical operations and design.

In recent years, in an effort to become more cost effective, pharmaceutical and biotech firms have employed the concept of obtaining water systems on an as-built basis. This has resulted in the installation of water systems that are designed to accommodate a range of variables. Often, these designs may not be ideal for the facility they are intended to accommodate. Because of the growing paucity of in-house process engineering capability, many pharmaceutical firm employees are assigned the role of performing engineering like tasks regardless of their background.

As with all water systems, a basic knowledge of fluid mechanics is required. Because of practical nature of this book, many theoretical aspects have been deleted in order to introduce the most amount of relevant material in the fewest number of pages. Nevertheless, several excellent theoretical sources are located in the reference/further reading section of this book.

Water is known as a Newtonian fluid. Simply stated, when a force or pressure (force/area) is applied to the fluid, in this case water, the flow will be proportional to the applied force. The constant of proportionality is the viscosity denoted by the Greek letter μ. While viscosity of liquids can vary significantly, our concern is with the viscosity of water, which in most cases for engineering calculations is $1\,lb_m/(ft\,h) = 2.42$ in English units or $1\,cP$ (centipoises) in metric units at ambient conditions. Often, viscosity data and calculations may require seconds in lieu of hours. In this case, the conversion from English units would be

$$1lb_m /\left(ft\,h\right) = \frac{1}{1488}cP \tag{2.1}$$

The English units for viscosity are often very cumbersome for use in certain calculations. Consequently, there is a preference to employ centipoise units when viscosity is a process variable. Viscosity is also an intensive variable of a substance, in this case water. That is, for ambient water system calculations, the viscosity of water is one, whether the mass is $1\,g$, $1\,kg$, or $1000\,kg$. Variables that vary with a quantity are identified as extensive variables. A water flow is an extensive property. That is, by employing a larger pumping system, the flow can vary from 10 to $500\,gal/min$ by varying the pump size or valve setting. However, regardless of the flow, the viscosity of water at identical conditions will always be 1 centipoise ($1\,cP$), regardless of at what point in the flow system a measurement is determined. While intensive and extensive variables of a system are most commonly dealt with in thermodynamics, this differentiation of physical properties of systems, in this context, will be further dealt with when working with the Reynolds number, located further in this chapter.

Also, for this book, ambient conditions are defined as a temperature between 64.4 °F (18 °C) and 75.2 °F (24 °C) and an atmospheric pressure of $14.7\,lb/in^2$ (psi) at sea level. While the value $14.7\,psi$ is the standard value, for certain components such as pressure vessels and heat exchangers, the value of $15\,psi$ is accepted (this will be addressed in Chapter 4). While most water system designs are performed in English units in the United States, occasionally the atmospheric pressure at sea level is given in metric units as $760\,mm$ of mercury.

Another widely used pressure value is the bar. For engineering calculations, a bar can be considered equivalent to one atmosphere of pressure, the value of which is $14.7\,lb/in^2$.

In addition to atmospheric pressure, some purified water operations perform under a pressure in addition to atmospheric pressure. For instance, steam lines often contain steam that is well above $14.696\,lb/in^2$ during operation. For this situation, the term gage pressure is used. Gage pressure is the pressure a component such as steam is recorded. The gage pressure is the sum of the atmospheric pressure plus or minus the operating pressure above or below atmospheric pressure (a typical example of a below atmospheric pressure would be a process operating under a vacuum, and the reading would be a minus reading). For instance, steam generated at the boiling point of water, 212 °F (at sea level), is produced at a gage pressure of $0\,lb/in^2$ ($14.7\,lb/in^2$). However, if steam were generated at a temperature of 327.83 °F, the pressure would be $100\,psia$ (lb/in^2 absolute). This means the pressure reading would be $100\,lb/in^2$, but the calculated absolute pressure is actually $14.696\,lb/in^2$ (atmospheric pressure) plus the additional pressure generated at a higher temperature (327.83 °F), which is $85.304–85.3\,lb/in^2$.

When dealing with absolute and gage pressures, as is often the case in thermodynamics, compressor, and steam calculations and often encountered in pharmaceutical water operations, it is most important to understand and be capable of converting from gage units to

Table 2.3 Altitude and the boiling point of water.

Altitude (ft above sea level)	Boiling point (°F)
0	212.0
250	211.5
500	211.0
1000	210.1
2000	208.1
3500	205.3
4000	204.3
4500	203.4
5000	202.4
5250	202.0
6000	200.6
7000	198.7

absolute units. The conversion is easily remembered by adding the gage pressure to the atmospheric pressure, thereby obtaining the absolute temperature.

Just as atmospheric pressure varies with altitude, the boiling point of fluids also varies with altitude. Since this chapter is intended for pharmaceutical water manufacturing and related operations, the focus is on purified water production and related operations.

Table 2.3 compares various altitudes and their associated boiling points. This information will be applied in this chapter.

In addition to Table 2.3, for process and project calculations involving steam generation, a copy of steam tables with the values of saturated and superheated steam would be an important asset. An abbreviated steam table can be found in Table 2.13.

Temperature in degrees centigrade is important since many pharmaceutical operations require data be recorded using the Celsius scale. The following relationships can be used convert from one temperature scale to another by employing the following relationships:

$$\frac{°C}{°F-32} = \frac{100}{180} = \frac{5}{9} \tag{2.2}$$

$$°C = \frac{5}{9} \times \left(°F - 32\right) \tag{2.3}$$

$$°F = \frac{9}{5} \times \left(°C\right) + 32 \tag{2.4}$$

Note the values in parenthesis. Failure to perform conversions in this specified format can result in an erroneous final value.

While not typically employed in pharmaceutical water operations, there are two more temperature scales defined. These scales are used when dealing with extremely high and low temperatures and are identified as absolute temperatures. When dealing with Fahrenheit temperatures, the absolute scale is expressed in terms of degrees Rankine (°R) and Kelvin (K) in centigrade units. Absolute temperatures are most frequently employed when dealing with

Table 2.4 Comparison of temperature systems for pure water.

Scale	Freezing point of pure water	Boiling point of pure water
Centigrade (°C)	0	100
Fahrenheit (°F)	32	212
Absolute		
Kelvin (K)	273.16 K (373 K)	373.16 K (373 K)
Rankine (°R)	492 °R	672 °R

gas-phase operations. It is important to note °R and °F are often used interchangeably. If the units are employed interchangeably, care should be employed to verify the units are consistent with results. In this case, it is important to note that 1 °F = 1 °R; the same conversion holds when dealing with calculations, often thermodynamics in the centigrade scale. For pharmaceutical water manufacture, the gas phase is generally not employed except for producing high quality water such as WFI, which often employs evaporation- or distillation-type unit operations where steam is used. However, should the need arise to employ these particular absolute scales, Table 2.4 presents a comparison of important values for all temperature systems. The comparison fluid for Table 2.4 is pure water. If it is anticipated an assignment will require significant conversions of one system to another, it is suggested examples of converting to various temperature systems be performed.

A common terminology that can lead to common errors is the use of the symbol for temperature and time. Accepted convention uses the capital "T" when denoting temperature. For thermodynamic calculations, the temperatures used are most often given in absolute units, °R or K. When dealing with a formula using time as a variable, the lowercase letter "t" is used for this convention. This symbol and letter differentiation will become significant when the heat transfer aspects of purified water production are covered further in the chapter.

Two other constants used in pharmaceutical water calculations are the density of water (ρ), which is given as $62.3 \, \text{lb/ft}^2$ or 8.34 lb/gal at ambient conditions. Often, in fluid flow calculations, the unit of specific gravity will be employed in lieu of density. The specific gravity is the ratio of the density of a given fluid to the ratio of water. When equivalent units are divided by one another, the resulting value will be dimensionless. Consequently, the specific gravity of water, at ambient conditions, is 1.0 and not $62.3 \, \text{lb/ft}^3$. Also, when considering volumetric flow at ambient conditions, another common factor used in many water system calculations is $7.48 \, \text{gal/ft}^3$ for the volumetric density of water. Again, this is for water at ambient conditions.

In pharmaceutical water calculations, as is the case with all fluid mechanics calculations, it is important to use correct units associated for associated calculation variables. For instance, while the English units associated with viscosity are given in ft^2/s, many sources reference the viscosity of water at ambient conditions as 1 cP (metric units). Care must be used to verify proper dimensions are employed in performing required calculations.

In the science of fluid mechanics, we distinguish between compressible and incompressible fluids. Laws relating to the behavior of compressible fluids like air, steam, or gases fall within the domain of aerodynamics or gas dynamics. Because, for all pressures encountered in hydraulic applications, water in its liquid form retains a constant volume for a given amount of mass, it is considered to be incompressible. The assumption of incompressibility results in a considerable simplification of fundamental fluid mechanics principles. It allows us to measure the amounts of water in volumetric terms instead of a mass. Most pharmaceutical water systems installed in the United States use gallons per minute as a measurement parameter. As might be expected, water flowing into and through a pharmaceutical operation is most often flowing through a piping configuration whose main components are a series of pipes, fittings, and a variety of equipment such as pumps (when dealing with pharmaceutical water systems, usually only centrifugal pumps are employed), vessels, tanks, and other process components. Perhaps the simplest process to evaluate is a simple pipe containing water flowing through it.

While the flow rate is an important parameter, this value is also important for determining other order system variables. One such variable is the fluid velocity. The fluid velocity is given by the following relationship:

$$v = \frac{Q}{A}$$ (2.5)

where Q is the volumetric flow rate in (gal/min), v is the fluid velocity (ft/s), and A is the perpendicular cross section of the pipe (ft^2). (When performing process calculations, the cross-sectional flow area is more conveniently presented as

$$Q = \frac{v \pi D_i^2}{4}$$

Here, D_i is the inside pipe diameter given in ft.)

The following example employs some of the previous information to describe an elementary fluid flow configuration. The example also demonstrates the use of elementary dimensional analysis as a problem-solving technique. Dimensional analysis is a technique of using the units of quantities (i.e. ft, gal/min, ft^2, lb/ft^3) to assist in solution of engineering problems.

Another more direct method of determining the velocity of flowing water when the volumetric flow is given in gal/min is the relationship

$$v = \frac{0.4085 \left(\text{gal/min} \right)}{d^2}$$ (2.6)

where d is the inside pipe diameter given in inches.

Example 2.1   Water at 65 °F is flowing through a 2 in. 316L stainless steel pipe with a flow of 30 gal/min.

A. Calculate the volumetric flow rate (gal/min) flow of the water through the pipe after 1 min.
B. Calculate the velocity of the water (ft/s) through the pipe.

Solution
A. The inside diameter of a 2 in. pipe (Schedule 40) pipe is 2.067 in.

$$\left(2.067 \, \overline{\text{in}} \, \overset{1}{} \right) \times \frac{1 \text{ft}}{12 \, \underset{1}{\overline{\text{in}}}} = 0.172 \text{ft} \quad \left(\text{Note how the inches cancel to yield the diameter in feet} \right).$$

$$D^2 = \left(0.172 \text{ft} \right) \times \left(0.172 \text{ft} \right) = 0.0296 \text{ft}^2$$

$$\Pi = 3.142$$

$$A = \frac{\Pi D^2}{4}$$

$$A = \frac{\left(3.142 \right) \times \left(0.0296 \text{ft}^2 \right)}{4} = 0.0232 \text{ft}^2$$

A. Flow after 60 seconds

$$v = \frac{Q}{A}$$

$$v = \frac{30\,\text{gal}}{\text{min}} \times \frac{1\,\text{min}}{1} = 30\ \text{gal} \leftarrow \textbf{Solution}$$

B. Velocity of water through the pipe (ft/s)

$$\frac{30\,\text{gal}}{\text{min}} \times \frac{1\,\text{ft}^{3}}{7.48\,\text{gal}} \times \frac{1\,\text{min}}{60\,\text{s}} \times \frac{1}{0.0232\,\text{ft}^{2}} = 2.881\,\text{ft/s} \leftarrow \textbf{Solution}$$

(**Note:** In Part B of the solution, the units ft^3 are divided by ft^2 to yield ft.)
The velocity can be confirmed by using Equation (2.5) as a calculation check.

$$v = \frac{0.4085(\text{gal/min})}{d^{2}}$$

$$v = \frac{0.4085(30)}{(2.067)^{2}} = 2.87\,\text{ft/s} \leftarrow \textbf{Solution}$$

The check confirms Solution B is a reliable value.

The description of the system shown previously is very simplistic. An actual operating pharmaceutical water system is, as noted earlier, often a very complex operation composed of many components and equipments.

Note how the units in the previous example cancel out to display the desired units in the solution.

The solution to Example 2.1 uses term Schedule 40. Schedule refers to the wall thickness of a pipe. While Example 2.1 identifies the pipe as having a nominal 2 in. diameter, the actual inside diameter is 2.245 in. The term schedule is a standard word for all types of piping, regardless of composition (i.e. 304L[1]) (stainless steel, carbon steel, or a plastic such as PVDF). To maintain consistency in pipe sizing, a standard prepared by the American Standards Association has been prepared. This convention addresses steel and plastic piping ranging from 1/8 to 24 in. and higher. While there are 10 schedule numbers ranging from Schedule 10 to Schedule 160, most pharmaceutical water systems use Schedule 40 for most applications and occasionally Schedule 10 for non-process requirements. Similarly, the majority of piping arrangements are fabricated of 1, 1½, 2, and 2½ in. nominal diameters. Table 2.5 provides data for the most common types of pharmaceutical water system piping. Also included in Table 2.5 is the relevant data for Schedule 80 piping. It should be recalled that fittings such as tees and branches are fabricated of Schedule 80 wall thickness.

Piping lengths are generally supplied in standard 20 ft lengths. While the price for a specific piping type may be quoted on a per foot basis, unless otherwise specified, actual delivery will be in 20 ft lengths. Thus, when determining the length of piping required, it should be noted that, for estimating purposes, the total length required should be divided by 20 ft. The resulting value will

---

1 L denotes the type of steel identified is fabricated of a composition intended for welding applications. It differs from standard compositions (without the L) because of lower carbon content.

Table 2.5 Common pharmaceutical piping parameters.[a]

Nominal size (in.)	Outside diameter (in.)	Schedule no.	Inside diameter (in.)
1	1.315	10	1.097
1	1.315	40	1.049
1	1.315	80	0.957
1½	1.900	10	1.682
1½	1.900	40	1.610
1½	1.900	80	1.500
2	2.375	10	2.157
2	2.375	40	2.065
2	2.375	80	1.939

[a] ASA Standards B36.10-1950.

then provide the number of lengths required. While many water systems are supplied as "as-built" units with the vendor providing the piping, the user project engineering team should be aware of the approximate quantity of piping required for the project as well as the approximate cost, regardless of the piping material (e.g. 304L or 304 stainless steels or PVDF plastic piping).

As stated previously, pharmaceutical water is often transported through complex and serpentine piping networks requiring a great deal of energy to transport the water effectively. Additionally, to economically and efficiently operate a water system and a high velocity for water transport through the piping system is desired. The most effective technique for water for transport in piping systems is, as noted previously, using, in most designs, centrifugal pumps to move the water throughout the operating system. Determining the size of the centrifugal pump required is a necessary activity for water system design. Pump requirements begin with the Bernoulli theorem. This relationship is a means of expressing the application of the law of conservation of energy to flow of fluids, in this case piping arrangements. The total energy at any particular point above some arbitrary horizontal plane is equal to the sum of the elevation head, pressure head, and the velocity head as follows:

$$Z + 144\frac{P}{\rho_1} + \frac{v^2}{2g_c} = H \tag{2.7}$$

where Z is the elevation in ft, P is the pressure of the fluid at point Z (lb/in^2), ρ is the density of the fluid (62.3 lb/ft^3 for water), v is the flow velocity (ft/s), and g_c is the gravitational constant (32.17 ft/s^2). Head, H, is defined as the elevation of a fluid above the discharge point. This elevation, given in feet in English units, is the vertical distance from the highest origin elevation, point 2, to the discharge elevation, point 1. The term H is also referred to as the total available hydraulic energy at a point in the system and is referred to as the head. The numerical value for H will always be a constant value at a given point.

When dealing with fluid flow calculations and pumping problems in particular, an important conversion factor will be converting the head (H), which is normally given in feet, to pounds per square inch (psi). The required conversion factor is

$$\text{Fluid head}\left(\text{lb/in}^2\right) = \frac{\left(\text{Head in ft}\right) \times \left(\text{Specific gravity}\right)}{2.31} \tag{2.8}$$

Specific gravity (SG) is the ratio of the density of a fluid divided by the density of water. For pharmaceutical applications, the specific gravity of water is 1. It is worth noting that since this is a ratio, the dimensions for the fluids, lb/ft^3, are cancelled out and consequently the specific gravity is dimensionless.

Expressing the head in feet is also an important relationship. The head in feet will be of importance when dealing with centrifugal pumps, further in the chapter. By rearranging Equation (2.8), the head in feet is defined as

$$\text{Head (ft)} = \frac{(2.31)\left[\text{Fluid head}\left(lb/in^2\right)\right]}{\text{Specific gravity}} \tag{2.9}$$

If friction losses are neglected and no energy (heat) is added to, or taken from, a piping system (i.e. centrifugal pumps transporting water with no heat generation), the total head, H (given in ft), in the previous equation will be, as stated previously, a constant for any point in the fluid. It should be noted that a system that has no energy loss or gain across the system boundary is referred to as an adiabatic system in thermodynamic terms. Realistically, energy losses or energy increases or decreases do occur and must be included in the Bernoulli equation. Thus, an energy balance may be written for two points in a fluid; in this chapter, the fluid is water.

Note the friction loss from point 1 to point 2 (Z_1 and Z_2) is h_L foot-pounds per pound of flowing fluid; this is generally referred to as the head loss in feet of fluid (in this case, water). This loss is, predominately, in the form of heat generation and is most often manifested as a small temperature change. The equation may be written as

$$Z_1 + \frac{144P_1}{\rho_1} + \frac{v^2}{2g} = Z_2 + \frac{144P_2}{\rho_2} + \frac{v^2}{2g} + h_L \tag{2.10}$$

The term $v^2/2g_c$ is defined as the fluid kinetic energy or velocity head and will be used for pump system calculations and evaluations.

Equation (2.9) is the basic relationship from which virtually all useful relationships involving pharmaceutical water flow and virtually all fluids requiring flow information. The relationship, Equation (2.10), is commonly known as the Bernoulli equation.

Another relationship related to Equation (2.9) is an equation used for determining friction losses in the flowing system. The relationship is similar to the Darcy–Weisbach equation (Equation 2.11) and is given by

$$h_f = \frac{fLv^2}{2Dg_c} \tag{2.11}$$

where

h_f = head loss, ft
f = the friction factor, dimensionless (defined in greater detail shortly)
L = the equivalent piping length of the piping configuration, ft
v = fluid velocity, ft/s
D = inside pipe diameter, ft
g_c = gravitational constant, 31.17 ft/s^2 at sea level (for most engineering problems, the value employed is 32.2 ft/s^2, regard of geographical location)

It is important to note the units of h_f are given in ft and not lb/in^2 as is the case with the Darcy equation (Equation 2.11).

Example 2.2 Water is flowing from a vertical pipe at a velocity of 4 ft/s. The vertical distance from point 1 to point 2 is 50 ft. The water is flowing from an open tank and discharging into a collection pond.

(Point 1 is 0 ft and point 2 is 50 ft.)

Calculate the head loss (h_L) due to frictional losses.

Solution

$Z_1 = 0\,\text{ft}$

$Z_2 = 50\,\text{ft}$

Density, $\rho_1 = \rho_2 = 62.3\,\text{lb/ft}^3$

$P_1 = 14.7\,\text{lb/in}^2$

$P_2 = \dfrac{(50\text{ft})(1)}{2.31} = 21.64\,\text{lb/in}^2$ 　　(Inserting Equation 2.10)

$$0\,\text{ft} + \frac{144\,\text{in}^2 / \text{ft}^2\left(14.7\text{lb} / \text{in}^2\right)}{62.3\text{lb} / \text{ft}^3} + \frac{\left(16\text{ft}^2 / s^2\right)}{\left(64.4\text{ft} / s^2\right)} = 50\,\text{ft} + \frac{144\,\text{in}^2 / \text{ft}^2\left(21.64\text{lb} / \text{in}^2\right)}{62.3\text{lb} / \text{ft}^3}$$

$$+ \frac{\left(16\text{ft}^2 / s^2\right)}{\left(64.4\text{ft} / s^2\right)} + h_L$$

$h_L = 0 \leftarrow$ **Solution**

The head loss for this example is zero. This is an example of water flowing vertically from a discharge point, and there is little energy buildup or loss since the frictional loss due to water flow is virtually nonexistent for water flowing vertically downward over a small distance.

The value in the numerator, $144\,\text{in}^2/\text{ft}^2$, is the conversion factor required to convert square inches to square feet.

As can be seen in the aforementioned example, this technique for determining the friction loss can be quite cumbersome. One quick check to determine the correct approach to solving the approach to the Bernoulli equation correctly is to note that the value for h_L must always be greater than zero or positive. If a negative answer results, most probably, the values used for Z_1 and Z_2 are either reversed or incorrectly recorded.

A more useful technique in obtaining solutions to head loss and friction contributions would be a relationship that could directly yield results for the pressure difference between points 1 and 2, a more descriptive friction expression, and a more simplified calculation technique. Such a relationship exists in the form of the Darcy–Weisbach equation. This modified relationship is expressed quantitatively as

$$\Delta P = \frac{f\rho Lv^2}{144 \cdot 2gD} \tag{2.12}$$

For those interested in the derivation of the Darcy relationship, several texts on fluid mechanics are available, and several excellent references are located in the references at the end of this chapter.

The L term is the equivalent (ft) of the piping configuration of interest (i.e. 50 ft for Example 2.2). The letter f is referred to as the friction factor and will be referred to shortly. A brief examination of Equation (2.5) will reveal the friction is a dimensionless number.

As noted earlier, pharmaceutical water piping systems typically consist of piping and other components. These "other" components are typically elbows, reducers, tees, spool pieces (small lengths of piping installed to connect two piping systems or to accommodate for possible thermal expansion), valves, and process equipment. In determining the total length of a piping configuration, a variety of estimates can be employed to approximate the equivalent length of piping required for various valves and fittings commonly used in pharmaceutical water systems. Table 2.6 presents estimates for several of these components. This table is useful for confirming vendor supplied information such as a piping and instrumentation diagram (P&ID) are reasonably accurate. More precise techniques are described in this section in more detail.

It is most important to note that all fittings such as elbows, tees, and branches are Schedule 80 components, regardless of pipe diameter. This information will be of value when calculating equivalent lengths of piping networks.

Depending on the level of detail required to verify or confirm vendor designs or to prepare process calculations, Table 2.7 presents a more detailed data presentation than the information

Table 2.6 Equivalent pipe diameters.

Fittings and valves	Equivalent length, pipe diameters[a]
45 degree elbows	15
90 degree elbows (standard radius)	32
90 degree square elbows	60
Tees used as elbow entering run	60
Tees used as elbow entering branch	90
180 close return bends	75
Couplings	Negligible
Unions	Negligible
Gate valves (open)	7
Globe valves (open)	300
Angle valves (open)	170

[a] Add to length of pipe to obtain equivalent length.

Table 2.7 Friction loss in pipe fittings and valves in terms of equivalent feet of straight pipe.

Nominal pipe size (in.)[a]	1½	2	2½	3
45 degree elbows	1.9	2.4	2.9	3.6
90 degree elbows (standard radius)	2.7	3.5	4.2	5.2
90 degree square elbows	4.1	5.2	6.2	7.7
Tees used as elbow entering run	31.0	40.0	51.0	61.0
Gate valve (open)	3.2	4.3	5.3	6.4

[a] Cast fittings.

presented in Table 2.6. Both tables are acceptable for most vendor design confirmations. Note the modified format of Table 2.6 and how it differs from Table 2.7 in style. Observe also that Table 2.7 uses a nominal diameter of 1½ in. in lieu of 1 in.

For pharmaceutical water systems that are designed by an outside vendor, it is usually not required to accommodate consider fittings in the equivalent length calculation. For most water systems, the fittings represent a small fraction of the total equivalent length calculation. In many cases, the pressure drop due to fittings is less than 3–4 psig.

Example 2.3   A 30 ft length of 2 in. pipe (nominal diameter) begins from a tee and ends at a normally open (NO) globe valve. Estimate the equivalent length of the pipe run.

Solution
Referring to Table 2.6, the tee described is equivalent to 60 pipe diameters. Thus,

$$\frac{2\,\text{in}}{1} \times \frac{1\,\text{ft}}{12\,\text{in}} \times 60 \left(\text{pipe diameters} - \text{equivalent length}\right) = 9.99\,\text{ft} = \mathbf{10\,ft}$$

Similarly,

$$\frac{\overset{1}{2\,\text{in}}}{1} \times \frac{1\,\text{ft}}{\underset{1}{12\,\text{in}}} \times 300 \left(\text{pipe diameters}\right) = 49.99\,\text{ft} = \mathbf{50\,ft}$$

The total equivalent length for the configuration described is $30\,\text{ft} + 10\,\text{ft} + 50\,\text{ft} = \mathbf{90\,ft} \leftarrow \mathbf{Solution}$

As Example 2.3 describes, components of piping systems often have a distinct effect on piping systems. This is most relevant when employing Equation (2.5) for fluid flow and pressure drop determinations or contractor design verification. While the relationships in Table 2.7 are useful, a more accurate method of determining the effects of friction on fluid flow, water in this case, would be desirable. This more accurate technique is available through use of the friction factor, f, which was identified in the Darcy relationship (Equation 2.11). The friction factor is the term used to correct for resistance of liquid flow, which, for our purposes, is water flowing through piping arrangements.

For pharmaceutical water applications, two specific friction factors are most often used. The first friction is identified as the Moody friction factor. The second commonly used factor is the Fanning friction factor. These two parameters perform the same function; that is, both the Moody factor and the Fanning factor provide a technique to account, quantitatively, for flow losses in flowing water systems using a more accurate method than described in Table 2.7 and Example 2.3. Since both friction factors are used for the same purpose, it is reasonable to expect them to be similar; and, as expected, they are. The difference between the two friction factors is a factor of 4. The Fanning friction factor is 4 times greater than Moody friction factor. In mathematical form, the relationship can be defined as

$$\text{Fanning factor} = 4 \times \text{Moody factor} \tag{2.13}$$

The use of these parameters in pharmaceutical water systems will be detailed shortly.

A term noted on the abscissa (the x-coordinate) is the Reynolds number with the parenthetical notation

$$N_{\text{Re}} = \frac{Dv\rho}{\mu} \tag{2.14}$$

where

D = inside pipe diameter (while the diameter is given, the term actually refers to a characteristic diameter and not necessarily a circular value). The geometry can be a rectangular duct, a V-shaped conduit, or another geometry. The important detail is that the geometry selected uses a consistent parameter for measurement. If something other than a circular geometry is selected, the diameter should be the hydraulic radius, which is given as $R_h = A/P$ where A is the cross-sectional area of the flow geometry such as an open top rectangular duct or a semicircular duct and P is the wetted wall perimeter of the duct.
v = the linear velocity of the flowing fluid
ρ = the density of the flowing fluid, water in this case
μ = the fluid viscosity.

The Reynolds number is actually a measure of the ratio of inertial forces to viscous forces. A cursory analysis of Equation (2.14) will note that as the numerator increases, the Reynolds number increases. For fluid mechanics, a larger Reynolds number is desirable. This is also the case when dealing with heat transfer, a topic also covered in this chapter. The relationship between the Reynolds number and other dimensionless numbers, fluid mechanics, and heat transfer will be addressed with detail in Section 2.4.

While the units for the Reynolds number may vary, it is most important to, again, reiterate that units are consistent and when values are inserted the calculated result will be a dimensionless number.

The Reynolds number is a very relevant in several areas of fluid mechanics including scale-up of processes, safety relief valve sizing for pressure vessels, and, as noted heat transfer topics. For pharmaceutical water system applications, the Reynolds number is used for determining resistance flow factors identified as friction factors. As will be seen, the Reynolds number is also valuable for determining proper water flow regions or regimes.

Equation (2.8) presents the Darcy equation and the associated friction factor, f, in terms of the Moody friction factor. If presented in terms of the Fanning friction factor, Equation (2.8) can be shown as

$$\Delta P = \frac{4\rho f L v^2}{2(144 g D)} \tag{2.15}$$

The primary difference between the two friction factors is preference on the part of the system designer. Civil and mechanical engineers who employ fluid mechanics usually have a preference for the Moody equation. Similarly, chemical engineers generally employ the Fanning friction factor. Mechanical engineers use both relationships, often depending on what method is specified in corporate procedures or specifications. Within the past few years, with the availability of free online friction factor determination, there has been a trend toward using the Moody factor in calculations in lieu of the Fanning friction factor. As will been seen shortly, each factor will have an associated "chart" for flow calculations (i.e. the Moody friction factor will use the Moody chart and the Fanning friction factor will obtain data from the Fanning chart). The two charts differ only in the value of the decimal value of the friction identified on the ordinate values (left vertical scale). It is critical that the same chart, regardless of which is used, be used throughout the design. Failure to be consistent will, obviously, yield incorrect results.

While most tables indicate which type of friction factor is employed, Fanning or Moody (Darcy–Weisbach), the type of friction factor employed can be determined by observation if for some reason the type of friction factor is not identified. The type of friction factor employed can be discerned by:

- Observing the value of the friction factor for laminar flow (Reynolds number <2100) at a Reynolds number of 1000.
- If the value of the friction is 0.064, the Darcy friction factor is plotted (Moody diagram). For the laminar range, the friction factor, f, can be described as $f = 64/Re$.
- If the value of the friction factor is 0.016, the Fanning friction is plotted (recall the Fanning friction factor is one fourth the value of the Moody factor). For the laminar region, $f = 16/Re$.

Figure 2.1 is a diagram of a Moody friction factor chart. Again, it is important to note the Fanning factors are 0.25 the value of a Moody friction factor.

The Moody chart is based on the Colebrook–White equation, which is given as

$$\frac{1}{f^{0.5}} = \frac{-2\log_{10}(\varepsilon/D)}{3.7} + \frac{2.51}{Re\sqrt{f}} \tag{2.16}$$

It should be noted the term D is the diameter given in ft. The roughness factor, ε, is also represented in ft. The term $\log_{10}$ is the logarithmic value in the base 10 system.

As can be seen, Equation (2.16) requires an iterative or "trial-and-error" solution, starting with $f = 0.05$, to calculate a required friction factor.

Other effective mathematical models that eliminate the need to perform iterative calculations are the relationships given by Moody (Equation 2.17) and Churchill (Equation 2.18):

$$f = 0.0055\left[1 + \frac{\left(20000(\varepsilon/D) + 10^6\right)^{1/3}}{Re}\right] \tag{2.17}$$

$$f = 8\left(\left(\frac{8}{Re}\right)^{12} + \frac{1}{(A+B)^{1.5}}\right)^{1/12} \tag{2.18}$$

$$A = \left(-2.457\ln\left(\left(\frac{7}{Re}\right)^{0.9} + 0.27\frac{\varepsilon}{D}\right)\right)^{16}$$

$$B = \left(\frac{3730}{Re}\right)^{16}$$

Employing a computer or programmable calculator would, no doubt, shorten the solution time required. However, unless a very detailed value for the friction factor is required, the Churchill equation would normally not be employed for a friction factor determination. For most process and project applications, the Moody or Fanning relationships are quite adequate.

The decision to employ a Moody chart or a mathematical relationship is dependent upon the preciseness of the value required. If the friction factor is needed for a process calculation, Equations (2.10), (2.11), or (2.12) may be required. If a confirmation of a vendor design is sought, then the Moody chart may be preferred. However, for most requirements, the Moody chart is quite adequate, both for process calculations or a less detailed application. Another much quicker technique to find a friction factor will be described further in the text.

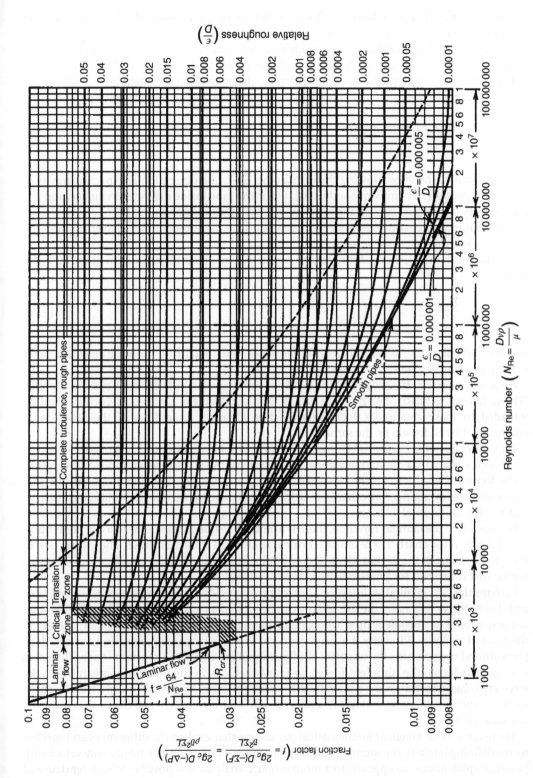

Figure 2.1 Moody friction factor chart. *Source*: Courtesy of American Society of Mechanical Engineers.

There are several approaches that allows for differentiating between the Fanning chart and the Moody chart. The most straightforward technique is by observing the diagonal line in the upper left-hand corner of the chart (Moody chart or Fanning chart). The line is identified as laminar flow and has a value of 64/Re for the Moody chart and 16/Re for the Fanning chart.

The symbol Re refers to the Reynolds number, alternatively abbreviated as N_{Re}. This value can be expressed in several forms. One very useful relationship for the Reynolds number is

$$N_{Re} = \frac{3162(\text{gal}/\text{min})(\text{SG})}{(\text{I.D.})(\mu)} \tag{2.19}$$

where gal/min is the water flowing through a pipe, SG is the dimensionless specific gravity of the flowing water (1 at ambient conditions), I.D. is the pipe inside diameter in inches, and μ is the viscosity in centipoises (1 cP for water at relatively ambient conditions).

This relationship is convenient since it eliminates the requirement for converting units to dimensionless values.

Another useful relationship for determining the Reynolds number, predicated on the mass flow is

$$N_{Re} = \frac{50.6Q\rho}{d\mu} \tag{2.20}$$

where Q is the flow rate in gal/min, ρ is the density of water (62.3 lb/ft^3), d is the inside diameter of the pipe in inches, and μ is the viscosity of water (1 cP).

Also, as can be seen the ordinate of a Moody or Fanning chart consists the Reynolds number, $DV\rho/\mu$. Equations (2.10) and (2.11) are merely modified forms of $Dv\rho/\mu$ presented in a format intended to make required calculations quicker. However, regardless of the relationship employed, the results will be the same.

This term, as noted earlier, is referred to as the Reynolds number and is most important in the design and analysis of pharmaceutical water systems. As is the case with the friction factor, f, the Reynolds number is a dimensionless value. However, the Reynolds number has very important uses in engineering other than being an independent variable for obtaining a friction factor value for flow calculations.

In addition to the laminar flow line, there is a laminar region that extends to a Reynolds number of 2100 on the chart. There is also a critical zone and a transition zone. The last section of the Moody chart is the region identified as complete turbulence (for rough pipes). It is this region that is relevant for pharmaceutical water systems.

Perhaps the best qualitative interpretation of the Reynolds number is to view the numerator and denominator as a ratio; the numerator can be viewed as a description of the variables that are responsible for turbulent flow, whereas the denominator represents molecular transfer in the form of viscosity. Viscosity is a fixed property of a substance. At given conditions such as pressure and temperature, the viscosity of a flowing substance will be constant. As the Reynolds number gets larger, the turbulent forces, primarily velocity in the numerator, overpowers the viscosity value, which remains constant, in the denominator. As identified earlier in the chapter, viscosity is an intensive variable of a pharmaceutical water system and thus will remain at a constant value.

There are several excellent mathematical paradigms that explain the differences and mechanisms defining laminar, transitional, and turbulent flow. However, since we are concerned with practical applications, as opposed to a more esoteric analysis, it is possible to differentiate and define the various flow regions as follows.

Pharmaceutical water systems are concerned with only fluid (water) flows in the turbulent zone. For pharmaceutical water, this translates to water that is flowing in a region where the Reynolds number is greater than 10 000.

It is generally desirable to have pharmaceutical water flow in a turbulent range and a friction factor of about 0.02 for Schedule 40 pipe with a 2 in. nominal diameter. One important reason is that with water flowing at a higher velocity (as can be seen in Figure 2.3, velocity is proportional to the Reynolds number, though not linearly), there is less time for formation of a biofilm layer (organic matter such as yeast, mold, bacteria, and endotoxins) on the inside wall of the piping. Also, higher flow rates diminish the chances that yeast, mold, and bacteria form and multiply in the piping. This is primarily due to the fact that there is constant circulation at all locales in the piping and minimum time is permitted for bacterial, fungus, or yeast growth. Another reason is at a higher Reynolds number, at a fixed pipe diameter; the volumetric flow is higher and can result in a shorter processing time for batch processing.

While higher flow is desirable in pharmaceutical water systems, there are design criteria that limit the maximum desirable flow through piped systems. It is generally not good practice to have water flow in piping to exceed 15 ft/s. Higher linear velocities than that tend to cause erosion of the inside wall of the piping. As a result, impurities in the form of solids and other constituents present (e.g. non-soluble salts such as calcium oxalate and some sulfates) in the water are not desired for effective pharmaceutical water system maintenance.

Just as flow at a higher Reynolds number is advantageous for movement of fluids, high Reynolds flow is a key constituent for efficient heat transfer, both of which are very relevant for effective pharmaceutical water system design and operation.

Upon examination of the Moody chart, a fractional term, ε/D, is also observed. This term is referred to as the relative roughness. The relative roughness is defined as the ratio of the roughness of a pipe's inside wall to the pipe inside diameter. Two important details should be remembered when using this ratio, ε/D, for fluid flow calculations. The two details are:

1) The letter in the numerator is the Greek letter, epsilon ε, and should not be mistaken for a lowercase letter, e, which is the symbol for the natural logarithm.
2) The ratio ε/D is a dimensionless value. Thus, if a value for ε is given in ft, the associated value for D should also be identified in ft.

On occasion, there may be a need to determine the value of a friction factor without having access to a Moody or a Fanning chart. Fortunately, there are reasonably accurate approximations when both the Moody and the Fanning charts are not available and values for the Reynolds number are known. Using a Moody chart, the following relationship can be used as reasonable estimates for the friction factor, f:

$$ f = \frac{0.235}{Re^{0.2}} \quad \left(\text{applicable for} \, Re > 2100 \right) \tag{2.21} $$

While there are other methods for determining the fraction factor in the absence of a Moody or Fanning chart (Equations 2.12 and 2.13), these other techniques generally require somewhat detailed computations to obtain reasonable solutions. Equation (2.21) is accurate for most friction factor determinations where a quick value is required.

While the aforementioned relationship is reasonably accurate, it is predicated on the assumption that smooth pipe is being evaluated.

While the Reynolds number is very important for fluid mechanics, it is also widely used in scale-up projects and, as will be shown in the next section, most important in heat transfer operations. In fact, a great deal of standard heat transfer calculations involves employing the

Reynolds number and two additional dimensionless groupings, the Nusselt number and the Prandtl number, both of which will be addressed also in the next section.

Example 2.4   Water is flowing at ambient conditions through 50 ft of 2 in. Schedule 40 PVDF (plastic) pipe. At a rate of 40 gal/min, calculate and determine the Reynolds number and Moody friction factor using two methods.

Solution
The PVDF piping is considered as smooth wall pipe.

A. Reynolds number calculation:
From Table 2.5, the inside diameter (i.d.) of Schedule 40 2 in. pipe is 2.065 in.
Referring to Equation (2.10),

$$Re = \frac{3162(40)(1)}{(2.065)(1)} = 61249 \leftarrow \text{Solution}$$

B. Friction factor determination:
For a Reynolds number of 61 249, Figure 2.1, by observation, indicates that the Moody friction factor is **0.021.** ← **Solution**
C. Friction factor calculation:

Referring to Equation (2.21),

$$f = \frac{0.235}{Re^{0.2}} = \frac{0.235}{61249^2} = 0.026 \leftarrow \text{Solution}$$

With a variance of less than 20% from the Moody chart value, the estimate given in Equation (2.21) is acceptable as an estimate when a Fanning chart is not available. It is also a reasonable method to employ when requiring a friction factor for purposes other than detailed process design calculations. A typical application of Equation (2.21) would be to quickly verify vendor data.

So that one does not lose sight of the purpose of the information presented earlier, the reader is reminded that the intent of the material previously presented is to determine the pressure drop required to transport quantities of water through a piping configuration. To achieve this task, centrifugal pumps, as stated previously, are generally employed to perform this task. The required pump capacity is determined by the calculated pressure drop, which is determined through the use of the Darcy–Weisbach equation.

A. Is the 1½ in. piping adequately sized for the specified flow rate?
B. Is the Reynolds number turbulent enough to confirm turbulent flow through the pipes?

Solution: It is observed in the Darcy equation that the variable for water velocity is in units of ft/s. Pharmaceutical water systems most often specify water flow in gal/min rather than ft/s. The units of gal/min are used because pump sizing is predicated on flow in gal/min.
While this value, 67 757, is about 7% larger than a rigorous calculation predicated on the
The decision on determining the volumetric flow rate is, hopefully, determined by an analysis of current and future pharmaceutical water requirements for a particular application. Once the flow rates are determined, it is the task of the process engineer to specify the pipe size (i.e. the pipe inside diameter). Most often, this task is accomplished via published data found in several texts. Often, if such data is available, it is usually found in the engineering department.

This may present a difficulty for the non-engineer who is assigned as the project representative of other organizations, such as the quality assurance or validation operations lacking such data. Also, obtaining the data from texts is rather expensive for a once or twice use. Consequently, it would be desirable to have access to a practical inexpensive method of correlating flow rate with diameter. Fortunately, there are two relationships that can be applied to determine these parameters.

One relationship that is applicable to pipes that are greater than 2 in. I.D. (>2 in.) is

$$Q = 1.2(d+2)^3 \tag{2.22}$$

where Q is flow rate in gal/min and d is the internal pipe diameter expressed in inches.

For piping that is 2 in. in diameter, or smaller (2 in $\leq$), the following relationship is applicable:

$$Q = \frac{d}{2}(d+2)^3 \tag{2.23}$$

The previous equations are specific to pharmaceutical water with a viscosity of 1 cP and a density of 62.3 lb/ft^3. These two relationships are specific to centrifugal pumping. They are not applicable to gravity flow (a system where no pump is employed) or suction configurations.

Another useful relationship for pipe sizing is given by

$$d = \left(\frac{Q}{10}\right)^{0.5} \tag{2.24}$$

As is the case with Equations (2.17) and (2.18), d represents the inside diameter in inches. Equation (2.18) generally provides greater results for diameter calculations.

Example 2.5   A pharmaceutical manufacturing operation has determined it will require a new 32 gal/min USP water purification system to support a new tablet manufacturing operation. The rate of 32 gal/min is to remain constant with no need to enlarge the system. There are three vendors submitting proposals and price quotes for the project. Two vendors both submit proposals using 2 in. Schedule 40 PVDF piping. The third bidder submitted a proposal employing 1½ in. Schedule 40 PVDF piping. The bids are to be evaluated within the next hour.

A. Calculated flow:
    For Schedule 40 1½ in. pipe (Table 2.5), the inside diameter of the pipe is 1.610 in. At ambient manufacturing conditions, the viscosity of water is 1 cP and the specific gravity is 1. By substituting the I.D. of 1.610 in. as the I.D. term, Equation (2.23) becomes

$$Q = \frac{1.610}{2}(1.610+2)^3$$

Q = 37.9 gal / min ← Solution

The result can be checked by using Equation (2.24):

$$d = \left(\frac{Q}{10}\right)^{0.5}$$

$$d = \left(\frac{32}{10}\right)^{0.5} = 1.79 \text{ in.}$$

While, as noted, Equation (2.23) yields a somewhat larger result, the value is only larger by about 10% and confirms the validity of selecting the smaller pipe diameter.

The selection of the 1½ in. PVDF piping for the water system would be adequate to accommodate a 32 gal/min flow rate.

Were these calculations to be required for an actual process design calculation, and not as a design adequacy determination, actual water flow data can be obtained by several sources. One reliable source of information is a text published by Ingersoll-Rand titled *Cameron Hydraulic Data*. For a 1½ in. Schedule 40 pipe, the published value is a 32 gal/min flow rate. Thus, Equation (2.16), as well as Equation (2.17), is highly accurate and suitable for rapid design verification calculations.

B. Reynolds number calculation:

Recalling the density of water at ambient manufacturing conditions is 62.3 lb/ft^3, by employing Equation (2.20), the relationship has the form

$$N_{Re} = \frac{50.6 Q \rho}{d \mu}$$

$$N_{Re} = \frac{50.6(32)(62.3)}{1.610(1)} = 67757 \leftarrow \textbf{Solution}$$

The definition in Equation (2.14) (i.e. $D v \rho / \mu$) a value of 67 757 is large enough to assure the flow is turbulent enough for the intended application.

A comparison of the calculated values and published values is found in Table 2.8. The published values are taken from *Cameron Hydraulic Data* published by Ingersoll-Rand° (16th Edition, 1979) [2].

It is observed the published data for 1½ in. pipe can vary from 4 to 180 gpm. While there is a significant variation of flow rates, the allowable flow velocity should not exceed 15 ft/s. Higher velocities offer the potential of pipe wear (i.e. erosion) of the inside wall at higher flows. In terms of volumetric flow, this translates to a maximum flow rate of 180 gal/min. For pharmaceutical water, a flow rate between 6 and 15 ft/s is most effective.

For calculations intended to be a check on vendor or in-house flow sizing calculations, using the nominal pipe diameter is sufficient since the results are within the margin of error (usually ±10%).

Pipe wear is important to both operator and designer of a pharmaceutical water piping and outbound system for both initial capital cost and projected life. Any section of piping must be replaced when wall thickness is not sufficient to withstand pipe stresses. At the outset, and

Table 2.8 Comparisons of published and calculated flow rates (Schedule 40 piping).

Nominal ID (in.)	Actual ID (in.)	Calculated flow (gal/min)	Published (gal/min)	Flow velocity (ft/s)	Variation (%)
1	1.049	14.87	16	4.00	7.06
1½	1.610	37.9	38	5.99	0.26
2	2.067	67.27	70	8.82	3.9
2½	2.469	107.11	110	7.37	2.63
3	3.000^a	150.0	150	6.81	0.0

^{a} Asphalt dipped new steel pipe.

therefore, the designer must provide adequate wall thickness for the anticipated pressures and an additional allowance for wear.

Pipeline wear occurs most often from mechanical erosion and chemical corrosion depending on the selection of pipe composition. While resistant to mechanical erosion in stainless steel configurations, the probability of specific types of corrosion is enhanced when stainless alloys are employed. Chemical corrosion occurs because of oxygen levels in the water and acidic pH of the water. Conversely, employing polymeric materials such as PVDF may introduce greater mechanical erosion than encountered in stainless steel piping. Corrosion by oxygen attack leads to formation of ferrous or ferric scale, which can be eroded by mechanical action such as high water velocity in the piping.

2.2 Pharmaceutical Water Equipment

One factor that permits a comprehensive coverage of pharmaceutical water production is the fact that virtually all operations are, for the most part, identical or modifications of equipment commonly used in process industries. Pumps, compressors, evaporators, and mixers are used widely in industries such as petroleum refining, food processing, and pulp and paper manufacturing. The biggest identifiable difference is materials selection (Chapters 8 and 9) and the need for a validation procedure for pharmaceuticals (Chapter 10). Consequently, many elements of this chapter may be considered a review of previously covered material, while for others, the information presented should be of value to individuals with little or no background in pharmaceutical water manufacturing operations and related equipment and systems.

2.2.1 Centrifugal Pumps

Movement of fluids via pumping is a major component of most process plants, and pharmaceutical manufacturing is no exception. With emphasis on energy consumption in manufacturing operations, designing, specifying, and maintaining pumps is a major element of efficient facility operation. The Hydraulic Institute estimates that pumps consume 25–50% of the energy consumption in process plant operations, including pharmaceutical operations exclusive of filling and packaging operations. Hence, it would be very helpful to have a reasonable knowledge of pump operations.

Figure 2.2 presents a diagram of a typical centrifugal pump.

Figure 2.2 Centrifugal pump.

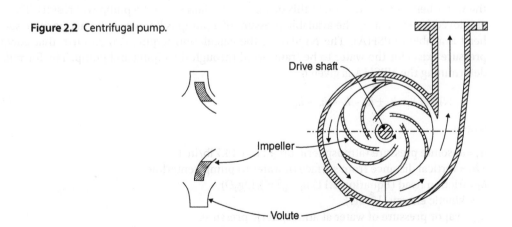

Drive shaft

Impeller

Volute

Centrifugal pumps basically consist of a stationary pump casing and an impeller mounted on a rotating shaft. The pump casing provides a pressure boundary for the pump and contains channels to properly direct the suction and discharge flow. The pump casing has a suction and a discharge nozzles for the main flow path of the pump with small drain and vent fittings to remove gases contained in the casing. Normally, there is a small removable bolt or bolt-like fitting that when loosened or removed serves as a drain for trapped water. This fitting is primarily employed when maintenance for the pump is required. The fitting is also critical if, for some reason, the pump is placed in a low temperature area (<33 °F) for an extended period. Often, when located in a lower temperature environment, a failure to drain water will crack the pump casing because of the freezing and expansion of any water that has not been drained properly.

Figure 2.4 shows the typical locations of the pump suction, impeller, volute, and discharge. The pump casing channels the pharmaceutical water from the suction connection to the center, or eye, of the impeller. The vanes (i.e. blades) of the rotating in power create a radial and rotary motion to the water, forcing the water to the outer boundary of the pump casing where the water is collected in this casing boundary referred to as the volute. The volute is a region that expands in cross-sectional area as it wraps around the pump casing. The purpose of the volute is to collect the water discharged from the periphery of the impeller at high velocity and gradually cause a reduction in the fluid velocity by increasing the flow area. This transforms the velocity head to the static pressure. The fluid is then discharged from the centrifugal pump through the discharge connection.

When referring to the inlet (suction) and discharge piping size, in most cases, the inlet and outlet diameters are designated, for instance, as 3×2 where 3 is the inlet size in inches and 2 is the discharge size in inches. Another detail concerning pump sizing is the impeller sizing. When specifying the impeller, it should be understood that the supplied impeller size is not the maximum size. Typically, the supplied impeller is one or two sizes smaller than the maximum for the pump selected. Should an increased flow rate be needed at some point, it is much easier and less expensive to accomplish this by installing a larger impeller to the existing pump than the option of replacing the original pump.

While there are several modifications to the centrifugal pump described (e.g. a double volute design), the single volute centrifugal pump is the type most commonly employed for pharmaceutical water applications.

Another important element of centrifugal pump sizing, design, and specification is verifying that the pharmaceutical water at the pump entrance possesses enough pressure to overcome the friction losses and yet preclude the pressure in the pump from falling below the pressure of the water being pumped (should this occur, severe damage to the pump will result). The term employed for evaluating the available pressure of a pump is known as the net positive suction head available (NPSHA). The NPSHA is the calculation required to confirm that adequate pressure exists for the water to be transferred through the piping and pump. The formula for determining the NPSHA is given by

$$\text{NPSHA} = h_a \pm h_z - h_f + h_v - h_{vp} \qquad (2.25)$$

where

h_a = absolute pressure (atmospheric pressure + 14.7 lb/in^2)
$\pm h_z$ = vertical distance from surface of water to pump centerline
h_f = friction head (Equation 2.11) $h_f = (fLv^2)/(2g_c D)$
h_v = kinetic energy ($V^2/2g_c$)
h_{vp} = vapor pressure of water at atmospheric pressure.

Although, for project engineering and design purposes, engineering units are most often used regardless of the system employed, the dimensions must be consistent (e.g. pounds, feet minutes, gallons must not be used with kilograms, or meters). While this may appear to be intuitively obvious, in a rush to get a calculated result, errors such as mixing dimensions are not uncommon.

A typical pump performance curve has a curve denoted as NPSHR (net positive head required). The NPSH is the value needed to ensure the pump will have enough lift to transport the water being pumped. In order for this to occur, the calculated value of the NPSHA must always be greater than the NPSHR shown on the performance curve.

Pharmaceutical water installations using centrifugal pumps have several terms that are employed that project and maintenance personnel should familiar with. For new installations, the system vendor should supply the data for these terms. The first term to be noted is related to a pump's output and is referred to as the water horsepower and given by

$$\text{Whp} = \frac{(H)(Q)(\text{SG})}{3960} \tag{2.26}$$

where

Whp = water horsepower, hp
H = pump head, ft
Q = volumetric flow, GPM, or gal/min
SG = specific gravity.

The Whp can be expressed by another useful relationship:

$$\text{Whp} = \frac{Q\Delta p}{1714} \tag{2.27}$$

where

Q = flow rate, gal/min
Δp = pump operating pressure difference.

Another term employed in pump terminology and related to work horsepower is brake horsepower (Bhp). The Bhp is similar to Whp for pumping water. The Bhp is given as

$$\text{Bhp} = \frac{(H)(Q)(\text{s.g.})}{(3960)(\text{eff})} \tag{2.28}$$

where eff = pump efficiency, %.

For water, the difference between Equations (2.26) and (2.28) is the pump efficiency. The pump efficiency is commonly shown as a part of the pump chart. Typically, pump efficiencies for pharmaceutical applications range from 60 to 75%. The value is predicated on pumping or recirculating water in the 80–160 gal/min range. For a general value, 68 or 70% efficiency is reasonable for most purposes when dealing with water at a viscosity of about 1 cP.

When evaluating operating costs, pump performance can exhibit a significant savings when the efficiency is considered. Consequently, it is important for planned pump installations to include the pump efficiency along with the flow and the pump operating head when preparing pump specifications for project or replacement procurement.

Example 2.6 A purified water circulation pump is to be installed as part of a new installation intended to package and deliver USP water to hospitals, laboratories, and manufacturers. The water is transported from the circulation loop to a packaging line a short distance from the loop. The pump is sized to circulate purified water a rate of 80 gal/min with a pump head of 67 ft. The pump pressure is designed to circulate water at a pressure of 29.0 lb/in^2. A change order has been approved to add an additional filter operation to the circulation loop. The filter installation will increase the pressure to 35 lb/in^2 (80.9 ft of head). The current P&ID specifies a 3 hp pump. The pump efficiency is 68%. The new pump has an efficiency of 75%. With the new change order, is the current design satisfactory?

The pump has a 230 V AC motor running on 4.5 A of current. The pump operates continuously for an annual period of 7900 h. At $0.10/kWh electric cost, estimate the annual electric cost of the pump.

Solution
Employing Equation (2.28) and inserting the existing operating variables, the Bhp required is

$$Bhp = \frac{(67)(80)(1.0)}{(3960)(0.68)} = 1.99 \approx 2.0 \, Bhp, \text{Pump currently employed}$$

Similarly, inserting the values for the proposed pumping operation in Equation (2.28), the calculation yields

$$Bhp = \frac{(80.9)(80)(1.0)}{(3960)(0.70)} = 2.33 \, \textbf{Bhp, Pump size required for new operation.} \leftarrow \textbf{Solution}$$

The new configuration requires additional horsepower to accommodate the increased operating pressure. Since pumps are not available in fractional horsepower, the existing pump should be replaced by a new 3 hp pump employing the same motor requirements (230 VAC, 9 A). The 3 hp pump is about $200 more than the smaller pump.

Both pumps maintain the same motor requirements. Thus, the annual operating requirements for the existing pump can be determined. Recalling Equation (3.1) (kW = 0.746 × hp), the annual power consumption and operating cost can be calculated as

$$(0.746)(3hp) = 2.24 kW \times 7900h / yr = 17680 kWh \times \$0.10 / kWh = \$1768. \leftarrow \textbf{Solution}$$

The annual electrical cost to continuously operate the pump is $1768.

A reasonable check for the accuracy of the estimated annual cost can be obtained by employing Equation (3.1) ($P = I \times V$, where P = watts, I = amps, and V = volts) and recalling that 1 kW = 1000 W. By inserting the given values in Equation (3.2), the calculation yields

$$P = (9A) \times (230V) = \frac{2070W}{1} \times \frac{1kW}{100W} = 2.07kW \times 7900h / yr = \textbf{16353kWh} \leftarrow$$

Equation (3.2) indicates the annual power consumption for the pump is 16 353 kWh. Consequently, the annual electric cost is (16 353 kWh) × ($0.10/kWh) = **$1635.30** ← **Solution.**

Using Equation (3.2) as a basis of calculation, the annual electrical cost is $1635.30. This value is 7.5% below the value obtained when Equation (3.1) was the basis of calculation. These values are a close match. In reality, variables such as voltage and amperage are not fixed constant readings. They are actually an average of the currents and amperages and can differ by as much as ±10% from the identified values.

2.2.2 Centrifugal Pump Installation Considerations

While being one of the most common types of equipment found in a pharmaceutical facility in general and a water processing operation in particular, a great deal of maintenance time is consumed by centrifugal pump maintenance and repair. One difficulty particularly when dealing with an as-built design is the failure to specifically address the facility P&ID and associated layout for the installed system. Often a poor layout is the primary cause of centrifugal pump malfunctions and failures. For the project engineer with little or no significant pump installation or maintenance experience, this can often be an unsolvable situation. A paucity of pump piping installation and design information can often contribute to the problem.

Hence, it is not uncommon for many pumping and associated piping configurations to be improperly designed and installed. Here, a poor installation and design can create continuous failures with no apparent resolution. By considering factors such as pump location and flow patterns during and prior to installation, many pump difficulties can be averted. Many problems are not clearly addressed in typical operations and maintenance (O&M) manuals. The following information can preclude most of the mistakes often encountered in pump piping configurations, regardless whether or not the centrifugal pumps are a cGMP classification:

- *Position the pump in an easily accessible location.* In addition to easy access, the location should provide enough working space for maintenance, repair, and inspection and component replacement activities to take place in an ergonomically effective manner. Consideration should also be given to adequate overhead distance, should a crane be needed for activities such as pump removal, impeller replacement, or motor repair. Also, piping installation should not interfere with the maintenance, repair, inspection and routine lubrication, and cleaning of the centrifugal pump.
- *Verify the pipe diameter on both the inlet and outlet besides of the pump are, at a minimum, at least one size larger than the pump nozzle.* Typically, for the horizontal inlet, an eccentric reducer should be installed to assure the inlet diameter will accommodate the larger diameter pipe and smaller inlet. On the discharge side, it would be advisable to consider installation of a throttling valve to maintain a flow of 15 ft/s or less. This velocity should preclude the possibility of inside pipe erosion due to excess velocity.
- *There should be no elbows and installed on the inlet nozzle of the pump.* Because of the uneven flow in an elbow, caused by formation of eddies and turbulence, this uneven flow goes to the eye of the pump impeller. This action in turn creates an air entrainment, which creates pump cavitation, impeller pitting, and vibration and ultimately causes pump failure. A common failure mechanism resulting from this incorrect configuration is an uneven load on the bearings and the shaft leading to failure of these components. Just as one elbow is detrimental, two or more elbows in the suction line are also problematic.
- *The suction line should have a 5–10 diameter straight run prior to entering the pump.* This is a standard procedure when designing and specifying in-line instrumentation. The purpose is to have the flow characteristics (i.e. flow rate, pressure) recording or reading unbiased flow that can occur from elbows, branches, tees, etc. that can affect the flow patterns as a result of being too close to the gage or instrument. The same rationale applies for fluid (water in this case) entering the pump. It is most important to have uniform flow characteristics for the flow entering the suction side.

- *Eliminate potential sources of air entrainment in the suction line.* When the piping configuration allows for air pockets to enter the suction side, the entrained air can cause significant impeller damage. Similarly, pump installations that discharge to a sump or tank should be designed to ensure there will be no entrainment during discharge. By confirming the design will always maintain an adequate NPSH or an automatic shutoff for low level or flow, the potential problem can be obviated.
- *Configure the piping with a design that minimizes or eliminates any strain on the pump casing.* This task can best be accomplished by careful alignment of the pump and motor coupling as well as flanges and gaskets. Another important design consideration is the need to have a stand-alone support installation for piping, valves, and line mounted gages and instruments. Supports, such as pipe hangers, should have dedicated hangers for pipes, and valves, if necessary.
- *If the potential for a low flow to the suction side of the pump exists, installation of a bypass valve on the discharge line should be considered.* At a lower flow, the bypass valve will minimize the need for the pump to exhibit erratic characteristics as potential entrainment since the bypass can direct a portion of the discharge flow back to the suction line and enhance the low flow rate.

2.3 Thermodynamics Interlude

Thermodynamics is predicated on three fundamental concepts identified as the three laws of thermodynamics. The three laws coupled with the ideal or perfect gas law can be applied to understand and compute one of the most important areas in science and engineering. The perfect gas law is germane since a great deal of thermodynamics involves gas-phase physical transformations and processes. For purified water manufacturing, water in the gas phase (steam) is widely used in many manufacturing operations. Also, a great deal of basic thermodynamics knowledge is based on gas-phase operations since there exists a basic equation of state (the perfect gas law) that often can be readily used in performing thermodynamics calculations and evaluations for process and project applications.

The perfect gas law has been identified as a useful complement to the laws of thermodynamics. The perfect gas law is identified by Equation (2.29).

A section on the fundamentals of thermodynamics could also be relevant if it were included in Chapter 3 or Chapter 8 of this text. However, by presenting this material, a more in-depth understanding of the principles and applications of "thermo" early on should be of greater value since many previously unidentified thermodynamic applications may be observed. So, to fully understand the practical aspects specific to pharmaceutical water production, a basic knowledge and understanding of thermodynamics and some relevant thermodynamic relations is desirable. When covering this section, it will be seen that thermodynamics is involved with the initial and final status of thermodynamic systems. Consequently, as will be noted further, thermodynamics is independent of the time required to achieve a final status. Perhaps the most convenient starting point for thermodynamics is the ideal gas law, given by

$$pV = nRT$$

(2.29)

where

p = pressure of a defined gas system
V = volume of the defined gas system

Table 2.9 Common values of the ideal gas law constant (R).

System dimensions	Units	R
1) Metric units	g cal/(g mol)(K)	1.987
2) Engineering units	Btu/(lb mol)(°R)	1.987
3) Metric units	J/(g mol)(K)	8.314
4) Engineering units	(ft-lb$_f$)/(lb mol)(°R)	1545.3
5) Engineering units	lb$_f$/in^2 (ft^3)(lb mol)	10.73
6) Engineering units	(atm)(ft^3)/(lb mol)(°R)	0.730 23

n = number of moles in the defined gas system (for a perfect gas, one mole of a gas occupies 359 ft^3 at standard conditions. In metric units, the value is 22.4 l.)

R = gas constant

T = absolute temperature, °R for engineering units and K for metric systems

The gas constant, R, is often encountered in chemistry, physics, and introductory thermodynamics courses when introducing the gaseous phase. Most often, the units employed are metric units. However, in most engineering applications, engineering units are still employed. For the gas constant value, R, a value of 0.0821 (l)(atm)/g mol is the most common value encountered. However, for engineered systems, Table 2.9 illustrates a list of gas constants identified in engineering and metric units often employed in project and process applications.

With the variable of Equation (2.29) identified, further relevant information from the equation can be obtained. By recalling the definition of a mole, $n = g/M$, where n is the number of moles of gas and M is the molecular weight of the gas, Equation (2.29) thus becomes

$$pV = \left(\frac{g}{M}\right)RT \qquad (2.30)$$

For the gas phase, this relationship, Equation (2.30), is quite helpful for various calculations.

Equation (2.30) is applicable to single gases. If the density of a mixture of gases such as air (taken as 79% nitrogen – 29 lb/lb mol and 21% oxygen – 32 lb/lb mol for engineering calculations), the weighted average molecular weight is used.

In addition to the perfect gas law, a useful relationship also commonly employed in thermodynamic operations as well as other gas-phase systems is a relationship known as the general law. The general law is given as

$$\frac{p_1 V_1}{T_1} = \frac{p_2 V_2}{T_2} \qquad (2.31)$$

Here, p_1 and p_2 are initial and final pressures, respectively, and T_1 and T_2 are initial and final absolute temperatures with units given in °R for engineering units and K in metric units.

Example 2.7 A low pressure boiler is being procured to produce equipment cleaning steam at a temperature of 212 °F. A small manual gate valve will be installed in the steam line. To size the valve, the steam density is required. Determine the density of the steam for this application. Assume atmospheric conditions (i.e. 14.7 lb/in^2) apply. Assume one pound mole of steam as a basis of calculation.

Solution

Steam has a molecular weight of 18 lb/lb mol. The steam temperature is 212 °F, which when converted to the Rankine or absolute temperature becomes

$$\left(212°F + 460\underset{1}{°F}\right) \times \frac{(1°R)}{\left(\underset{1}{1°F}\right)} = 672°R$$

Inserting the known values in Equation (2.30), it is possible to now calculate the steam density. By examining the units of the variables of Equation (2.28), a gas constant value of 10.73 $(lb/in^2)(ft^3)(lb\,mol)(°R)$ is consistent:

$$\rho = \frac{\left(14.7\underset{1}{lb/in^2}\right)\left(18\underset{1}{lb/lb\,mol}\right)}{10.73\left(\underset{1}{lb/in^2}\right)\left(ft^3\right)\left(\underset{1}{lb\,mol}\right)\left(\underset{1}{°R}\right)\left(672\underset{1}{°R}\right)} = 0.0367 lb / ft^3 \leftarrow \text{Solution}$$

The calculated value for the density of steam at 14.7 lb/in² at atmospheric pressure is 0.0367 lb/ft³. Employing Ref. [2], *Cameron Hydraulic Data*, page 5–3, the published value at identical conditions for steam is 0.037 31 lb/ft³. This represents a difference of 1.64% between the two values.

While there are several techniques available to correct for the deviation of Equation (2.30) (compressibility factors and van der Waals equation, for instance), for pharmaceutical water manufacturing purposes, equation is acceptable for gas-phase operations.

Thermodynamics is primarily concerned with the transformation of heat into mechanical work, W, as well as the reverse operation of converting mechanical work into energy, often in the form of heat. Thermodynamics primarily deals with energy in a defined portion of space of space identified as a system. Typical thermodynamic systems include refrigeration units, boilers, and iron-containing alloys exposed to the environment (this thermodynamic activity is identified as corrosion and addressed in more detail in Chapter 8). Thus, it is critical that thermodynamic systems, to operate as designed, have clearly defined system boundaries. Using the refrigeration system, the system boundaries would be the compressor, associated recirculation piping, the condenser, and installed valves. Purified water operations are, for the most part, closed systems. This means that matter remains within the system but not necessarily energy in the form of heat, in this case the refrigeration unit. For this type of unit, the refrigerant is the major component, or material of concern. Probably the most common type of refrigerant recognized is Freon (chlorofluorocarbons, a class of chemicals that is being phased out due to the propensity to cause environmental damage by depleting the ozone layer surrounding the Earth).

While the concern of purified water manufacturing is the closed system, another thermodynamic system is referred to as an open system. As might be suspected, an open system is one that allows matter to flow across boundaries. Allowing a block of ice on the sidewalk to melt would be a simple open system.

The third type of thermodynamic system is the isolated system. An isolated system is one that permits no loss of matter or energy from the system boundary. Ideally, an example of an isolated system would be the block of ice encased in a box that is totally insulated in such a manner that the ice will not melt and would remain frozen for an indefinite period, thus not having water drain out of the ideal insulated box or allowing heat to enter the box. While well-insulated systems, such as a thermos bottle, can maintain a constant temperature for very long periods, ultimately, the ice will melt and the temperature will change.

A thermodynamic system and the surrounding environment are known as the universe. A refrigeration system and the rooms and hallways that provide cooling are known as the universe of the refrigeration system.

In addition to the thermodynamic definitions presented, a few other definitions and conventions are needed before specifically defining the three thermodynamic laws. While referred to in a rather general manner, a specific definition of energy is needed. Energy is the capability to perform work, while work (W) is defined as exerting a force through a given distance or length. A force is defined as a pressure multiplied by an area or $F =$ pressure $\times$ area ($p \times A$), where F is the applied force and A is the area. By slightly rearranging the terms, we can define pressure as an applied force per unit area, or F/A. In this chapter, probably the most commonly used force is probably the atmospheric pressure at varying altitudes given as lb_f/in^2.

The term total energy of a system is also critical to understanding the fundamentals of thermodynamics. Total energy, E or U (employed interchangeably), is understood to be the total of the forms of the types of energy contained in a system. These common forms of energy include not limited to:

U, the internal energy of a system (this type of energy is a result of molecular motion, intermolecular attraction, molecular orientation and related factors)
E_p, the potential energy of a system
E_k, the kinetic energy of a system

In a mathematical form, the total energy of a system can be defined as $E = U + E_p + E_k + \cdots$.

Another important thermodynamic concept is the term adiabatic flow. Simply defined, adiabatic flow is an operation where an increase in the volume of a fluid (liquid or gas) occurs with no heat transfer occurring in or out of the system boundaries.

A very important term employed in both thermodynamics and heat transfer is enthalpy. Enthalpy is mathematically defined as

$$H = U + pV \tag{2.32}$$

where

$H =$ enthalpy
$U =$ internal energy
$p =$ pressure of the system
$V =$ volume of the system.

Heat capacity at a constant pressure is related to enthalpy by the following relationship:

$$\left(\frac{\partial H}{\partial T} \right)_p = C_p \tag{2.33}$$

Equation (2.33) states that the variation of enthalpy with a variation of temperature at a constant pressure is equal to the heat capacity C_p.

For a condition where the pressure varies with a constant volume being maintained, a similar term can be applied. In this case, the heat capacity at constant volume is defined as

$$\left(\frac{\partial H}{\partial T} \right)_v = C_v \tag{2.34}$$

The heat capacity at constant volume, C_p, is primarily employed when dealing with pharmaceutical operations. The alternate heat capacity, C_v, is often employed when dealing with gas-phase operations.

With the relevant terms and background addressed, the first law of thermodynamics can now be presented. The first law of thermodynamics is

$$\Delta E = E_2 - E_1 \tag{2.35}$$

where

ΔE = change of internal energy
E_2 = final state of the system
E_1 = initial state of the system.

If energy in the form of heat (Q) is added to the closed system, Equation (2.35) can be stated by

$$\Delta E = Q - W \tag{2.36}$$

Here, W is the symbol for work, as noted earlier. Further, it is noted that Equations (2.35) and (2.36) are identical.

Often, ΔU is employed as an alternative to ΔE. Both symbols are used to denote the change in the internal energy of the system.

While energy is most often encountered in units of calories or Btu, for engineering calculations, it is not uncommon to deal with other factors when dealing with thermodynamics calculations. Some conversion factors often encountered are shown as follows:

$$1\text{Btu} = 1.055 \times 10^{10}\,\text{erg}\left(1\text{erg} = \left(1\text{g-cm}/\text{s}^2\right)\right)$$
$$= 0.293\text{w-h}$$
$$= 778\text{ft}/\text{lb}_f$$
$$= 0.3676\text{ft}^3\text{-atm}$$
$$= 10.41\text{l-atm}$$
$$= 2.93 \times 10^{-4}\,\text{kw-h}$$
$$= 3.93 \times 10^{-4}\,\text{hp-h}$$

Once again, it is emphasized that thermodynamics is only concerned with an initial state and a final state and independent of the path employed to reach the final state. When using Equations (2.23) and (2.24), it is important to employ the standard sign notation. If the system absorbs heat from the environment, the sign is positive, +. A simple example of this sign convention is heating water in a pot. Heat must be added to get the water to boil. Consequently, the number of Btu's or calories required is a positive number of Btu's or calories. Work done by a system is also positive. A simple example would be the water heating in the pot. If a top cover were on the pot, the top would gradually rise from its original position as the water expands in volume. The force per unit area exerted by the expanding water multiplied by the vertical distance the pot cover travels amounts to the total positive, +, work performed.

One of the most widely used thermodynamic compounds is steam. A convenient summary of the relevant properties of saturated steam (in this case, saturated steam is steam that cannot further absorb water) is presented in Table 2.13.

The water in the pot example can be modified to obtain another important thermodynamic concept, specifically work. Consider a long cylinder, similar to a pot with one end sealed, only longer and in a horizontal plane with the cylinder mouth parallel to the x axis or →, or the horizontal direction, for this example. Now, a cover for the cylinder is inserted such that the cover fits tightly inside the cylinder and moves in the x direction, →, only. Going just a bit further in our mental picture, assume the piston is moved a small distance, Δl, in length. In fact, assume the piston is moved an even smaller length, in this case a differential distance, dl ($\Delta x \approx dl$, an infinitesimally small value). Also, by recalling the definition of work a force multiplied by a distance, presented earlier, in differential form, work can be expressed as

$$dW = Fdl$$

Recalling that pressure is defined as a pressure per unit area, it follows that

$$dW = pAdl$$

where, once again,

dW = differential work
p = applied pressure, lb/in^2 in this case
A = piston area, in ft^2 or in^2
dl = differential distance, in inches or ft.

If the volume of the air or other gas in the cylinder is allowed to expand with the pressure remaining constant, it can be stated that

$$dW = pdV$$

where

p = applied pressure on the gas in the cylinder due to piston movement
dV = differential (very small volume displacement).

Assuming the original volumes and final volumes are known, or can be calculated, it can then be noted that

$$W = \int_{v_1}^{v_2} pdV$$

which, upon integration and evaluation at the boundary conditions, yields the resulting relationship:

$$W = p\Delta V = p(v_2 - v_1) \tag{2.37}$$

Since thermodynamics is concerned with an initial and final state of a system, v_2 is assumed to be the final state and v_1 is taken as the initial state.

It might have been observed that dW was not solved by a standard integration format. Simply stated, it was stated previously that thermodynamics is independent of the path and only the initial and final states of a system were relevant. In this case, there is no equation that can be integrated mathematically to obtain a final value based on the sum of infinitely small areas, as per the definition of an integral. Rather, because of fact that work only appears at the system boundary, and the path is not germane, the final integral value of the dependent variable, work (W), is W and not ΔW.

Equation (2.29) is a fundamental relationship associated with the field of thermodynamics and can be employed to obtain useful information.

Example 2.8 One lb mol of water is vaporized at 212 °F and 14.7 lb/in² pressure (atmospheric). At these conditions, the heat of vaporization (H_{vap}) for water is 970 Btu/lb. With this data, determine values for:

A. Q, the heat given off
B. W, the work performed
C. ΔE (or ΔU), the energy change in the system

Solution
A. One lb mol of water contains 18 lb/lb mol. Thus,

$$\frac{\left(18\text{lb H}_2\text{O}\right)}{\cancel{\text{lb mol H}_2\text{O}}} \times \left(\overset{1}{\cancel{\text{lb mol H}_2\text{O}}}\right) = 18\text{lb H}_2\text{O}$$

$$\left(18\overset{1}{\cancel{\text{lb H}_2\text{O}}}\right) \times \frac{\left(970\text{Btu}\right)}{\underset{1}{\cancel{\text{lb H}_2\text{O}}}} = 17460\text{Btu} \leftarrow \text{Solution.}$$

B. Equation (2.29) can be used for this application. Recall that Equation (2.37) involves Δv, which in this case represents the volume change from water in the liquid state to gas (steam). Since the volume of the steam is much greater than the volume of the same amount of water in a liquid state, the volume of the water in a liquid state can be neglected. As a result, Equation (2.37), for this problem, becomes

$$W = p\Delta v$$

Here, v is the steam volume at 212 °F and atmospheric pressure (14.7 psi). Recall Example 2.6, where a density for stem at the same conditions was calculated. For this example, the steam density was calculated as 0.0367 lb/ft³. The reciprocal for this value is 27.248 ft³/lb. Since the basis of calculation is 2 lb mol or 72 lb, then

$$W = \frac{27.248\text{ft}^3}{\cancel{\text{lb}}}\left(18\cancel{\text{lb}}\right)\left(14.7\text{lb}/\cancel{\text{in}^2}\right)\frac{1\text{ft}^3}{1728\cancel{\text{in}^3}}\left(12\overset{1}{\cancel{\text{in}}}/\text{ft}\right) = 50.07\text{ft}/\text{lb}_f \leftarrow \text{Solution}$$

In the previous solution, it should be noted how the dimensions are employed to obtain the solution. While most technically trained individuals involved in design and process operations are knowledgeable in dimensional analysis, individuals not as familiar with employing units to solve calculations should consider using the dimensional analysis technique in more depth and more frequency.

Part C presents a slight difficulty. The energy of the system was determined to be Btu's and the work was calculated as 50.07 ft/lb_f.

C. lb/ft. The dimensions are not consistent. However, the situation is easily resolved by referring to a book of common conversion factors, such as those listed earlier, also usually available in a variety of hard copy sources at the facility, or by referring to an online conversion factor routine. For this solution, the desired conversion factor is 1 Btu = 778 ft/lb$_f$. With this information and Equation (2.36) ($\Delta E = Q - W$), the solution to C is

$$\frac{50.07\,\text{ft}}{\text{lb}} \times \frac{1\,\text{Btu}}{778\,\text{ft}/\text{lb}} = 0.0643\text{Btu} = \textbf{Work as Btu's}$$

$$\Delta E = 17460\text{Btu} - 0.0643\text{Btu} = \textbf{17459.9Btu} \leftarrow \textbf{Solution}$$

The work performed on the process is a small fraction of the heat added to the system.

The second law of thermodynamics deals with a concept referred to as equilibrium and the approach to equilibrium. A thermodynamic system is in equilibrium when its various forms of energy, such as internal energy (U), potential energy (E_p), and kinetic energy (E_p), are all in a state of balance. This means, in effect, the system undergoes no changes. For this to occur in a thermodynamic system, there is no temperature change. A thermodynamic system that is altered without a temperature change is identified as an isothermal process. To understand the second law, it is important to understand that it applies to isothermal processes that are reversible. Since thermodynamics is predicated on the initial and final states and independent of the path, a reversible process can be identified as an operation that exhibits the same properties at the endpoint of the process as the initial or start of the process.

When dealing with purified water production, three irreversible processes are often employed. RO (Section 2.7) is one irreversible process, while condensing of supersaturated vapors and freezing of a supercooled liquid such as water (Chapter 3) are the other irreversible processes that can be encountered.

For purified water operations, reversible processes that may be addressed include vaporization and fusion (water turning to ice), while the latter is more frequently encountered in lyophilization (freeze-drying) operations for vaccine production and other biotech products.

When dealing with the second law of thermodynamics, it is again important to recall that the third law is applicable only to reversible processes. With the previous information being described, a definition of the second law of thermodynamics can be put forth. The basis of the third law is a term known as entropy. Entropy and the second law can quantitatively be defined as

$$\Delta S = \frac{q_{\text{rev}}}{T} \tag{2.38}$$

where

q_{rev} indicates that the heat contained in the system is reversible
ΔS is the entropy
T is the absolute temperature.

In a qualitative sense, entropy can be viewed as technique to measure unusable energy. Equation (2.31) notes that an increase in heat content of system increases entropy at a constant temperature. However, the unavailability of energy is also an important component of entropy. An example of this is that entropy is increased if heat in a closed system flows from a location of higher temperature to a region of lower temperature. In this case, the primary task of the heat is to seek an equilibrium temperature and is thus not available to perform work. An excellent definition of entropy is given by the authors of *Chemical Process Principles, Part II*,

Thermodynamics: "Entropy is an intrinsic property of matter so defined that an increase in the unavailability of the total energy of a system is quantitatively expressed by a corresponding increase in its entropy" [3].

2.4   Heat Transfer for Pharmaceutical Water Production

For pharmaceutical manufacturing, chemical reactions, product separation via distillation freeze-drying, and other operations involve large operating and capital costs. While the heat transfer costs for pharmaceutical water manufacturing are not as great, knowledge of heat transfer requirements in the manufacture of high quality water requires the same design and operating knowledge of more complex systems. As is the case with much of the pharmaceutical industry, an in-house design capability is rapidly disappearing. The role of pharmaceutical engineers is more often working as a project engineer and not a process design engineer. The project role is more involved in verifying that project costs are in line with project schedule and the installation is being fabricated in accordance with contractual specifications and requirements. However, it is also most important that vendor supplied systems conform to the design and installation required by the system owner. Scheduled vendor presentations are often the only method that company staff can be apprised of project status and system details. P&ID and associated data (e.g. hard copy handouts, PowerPoint presentations, etc.) are, on occasion, displayed with assumption that few project personnel are knowledgeable about the technical aspects of the pharmaceutical water operation. That is, the system vendor often knows that team members such as purchasing personnel and often quality assurance individuals have little or no working knowledge of pressure drop, heat transfer, heat exchangers, or pumps. So, although the owner has a team in place to monitor the project, many aspects of the project remain unaddressed because, in many cases, the engineering group is often the only technically knowledgeable team member. This situation often occurs when a project involves heat transfer equipment such as heat exchangers, jacket vessels, and evaporation systems. For this reason, it would be advantageous for pharmaceutical project personnel to have a technical understanding of the salient aspects of heat transfer. This section is also applicable to process engineers involved in design or design verification activities in pharmaceutical and biotech operations, although the trend to outsourcing of projects appears to be rapidly expanding.

Simply stated, heat transfer is primarily focused on two specific items: temperature and the flow of heat. Temperature is the quantity of thermal energy in a system and heat transfer is the movement of thermal energy from one locale to another.

By viewing heat transfer on a microscopic view, it can be shown that a thermal energy is a function of the kinetic energy of molecules. A higher molecular temperature results in greater agitation (movement) of the molecules contained in the system of interest. Consequently, in order to conform to thermodynamic criteria, in this case the zeroth law[2] of thermodynamics, it is standard for regions containing greater molecular kinetic energy to transfer this energy to regions with lower kinetic energy levels.

---

2 Simply put, the zeroth law (or third law) states, "Systems in thermal contact are not in complete equilibrium with one another until they have the same temperature."

The rate of heat transfer is dependent on specific properties of materials involved in the heat transfer process between two regions of different temperatures. Among the properties determining the rate of heat transfer are:

- Thermal conductivities
- Heat capacity (engineering units) or specific heat (dimensionless) – The specific heat is actually the ratio of the heat capacity of a fluid to the heat capacity of water for pharmaceutical water production. Since the dimensions are the same, the units will cancel out. For heat transfer calculations, the units of heat capacity are $Btu/ft^2 \, °F \, h$.
- Density
- Fluid velocity
- Fluid viscosity
- Surface emissivity – While for pharmaceutical water manufacturing equipment, this is also usually a very small fraction of the overall heat transfer process, it is a small contributor and should be noted (generally, for pharmaceutical applications, the emissivity, or radiative heat transfer, is generally ignored).

In heat transfer calculations, there are three types of mechanisms employed to design heat transfer equipment. One mechanism heat transfer due to radiative effects is rarely more than 2% of the total heat transfer operation; the value is often much lower for properly designed and insulated systems. The three mechanisms for heat transfer are:

1) Conduction – As noted, regions possessing high molecular kinetic energy pass their thermal energy to regions with lower molecular energy through direct molecular interaction or collisions.
2) Convection – When heat is conducted into a stationary, non-agitated fluid, the fluid will expand in volume. As a result, the heated fluid will be less dense and gravitate upward, and the molecular motion of the fluids will cause a mixing of the fluid and heat transport throughout the fluid as a result of the mixing of high and low temperature fluid mixing. As might be expected, a greater degree of mixing of the fluid will result in greater convection and consequently a higher heat transfer rate, a desirable characteristic.
3) Radiation – This heat transfer mechanism is dependent on the temperature of the material, the fluid in this case. As is the case with conduction and convection, radiation always from a higher temperature material to a lower temperature material; in this case the material is water (of varying purity or quality). In this case, the heat is transferred through electromagnetic waves. A material above absolute zero (the lowest temperature at which minimal molecular activity occurs, $-273 \, °C$ or $-460 \, °F$) generates electromagnetic waves (radiation) due to temperature. As the temperature of the material increases, the quantity of energy emitted as radiation increases and the nature of the radiation also changes. As the temperature increases, the energy given off as a result of radiation emission also increases. Because heat transfer in pharmaceutical water operations occurs at relatively lower temperature ranges, above freezing to slightly above the boiling point of water for most cGMP applications, radiation heat transfer is, as noted previously, not a significant variable in pharmaceutical heat transfer and generally not a salient issue.

In Section 2.3, the focus was on thermodynamics. This section focuses primarily on heat transfer operations. By observation, the difference between heat transfer and thermodynamics for project and process tasks is the time factor. Thermodynamics is involved largely in mechanical and heat transformations with no regard to the time frame in which a transformation is occurring.

For pharmaceutical water operations employing heat transfer operations and equipment, one common component is usually employed. That unit of equipment is the heat exchanger. A heat exchanger is basically a static installed device designed to transfer heat from a hot phase to a cold phase with the hot fluid and the cold fluid being separated by a solid boundary, usually a pure metal such as copper or a metal alloy, often a grade of stainless steel.

The study of heat transfer in general and many pharmaceutical operations in particular focuses on three basic equations as starting elements in heat transfer calculations. These two fundamental equations are:

$$Q = -kA\left(\frac{\Delta T}{\Delta x}\right) \tag{2.39}$$

where

Q = rate of heat transfer, Btu/h
k = thermal conductivity, the capability of a material to conduct heat, Btu/h-ft °F
(Thermal conductivity is commonly encountered when evaluating and designing heat exchangers for pharmaceutical water manufacturing operations. Thermal conductivities are encountered as materials of construction, heating, and cooling fluids in both the vapor phase (steam) and liquid phase. Table 2.10 provides values for liquids, gases, and construction materials (solids) often encountered in heat transfer applications when dealing with purified water systems.)
A = area where conduction is occurring, ft^2
ΔT = temperature difference between the higher temperature in the material and lower temperature in the material, °F or °R
Δx = distance between higher temperature location and lower temperature location and lower temperature location, ft ($\Delta T/\Delta x$ is called the gradient)

$$Q = mC_p\Delta T \text{ (required)} \tag{2.40}$$

where

Q = heat load, Btu/h
m = mass flow rate, lb/h
C_p = heat capacity at constant pressure
$\Delta T = T_2 - T_1$ where T_2 is temperature of the entering fluid and T_1 is the temperature of the exiting fluid from a unit (in this case, as will be explained briefly, the unit referred to is a heat exchanger).

$$Q = UA(\text{LMTD})(\text{actual}) \tag{2.41}$$

Table 2.10 Thermal conductivity of purified water system components.

Material	Temperature, °F (°C)	Thermal conductivity, Btu-h-ft^2°F (°F/ft)
Stainless steel	212 (100)	26.0
PVDF	32 (0)	0.09
Pure steam	212 (100)	0.0136
USP water	100 (38)	0.363

where

U = overall heat transfer coefficient, Btu/ft^2°F h
A = heat transfer surface area, ft^2
LMTD = log mean temperature difference, used in heat transfer calculations to determine the temperature driving force for heat transfer systems, particularly for heat exchangers

The LMTD is a natural logarithmic average of the temperature difference between hot and cold fluid streams at each of the ends of the heat exchanger. In mathematical terms, the LMTD or ΔT_{lm} can be expressed as

$$\Delta T_{lm} = \frac{(\Delta T_2 - \Delta T_1)}{\ln(\Delta T_2 - \Delta T_1)} \tag{2.42}$$

where

ΔT_2 = the larger temperature difference between the two fluid streams at either the entrance or the exit of the heat transfer device (heat exchanger)
ΔT_1 = the smaller temperature difference between the two streams at either the entrance or the exit of the two fluid streams of the heat transfer device (heat exchanger)

Perhaps an easier form of working with the LMTD (Equation 2.42) is to apply an alternate format where the LMTD (ΔT_{lm}) can be expressed as

$$\Delta T_{lm} = \frac{(T_{Hin} - T_{Cout}) - (T_{Hout} - T_{Cin})}{\ln(T_{Hin} - T_{Cout}) / (T_{Hout} - T_{Cin})} \tag{2.43}$$

where

T_{Hin} = temperature of the hot fluid (water) entering the heat exchanger
T_{Cout} = temperature of the cooling water exiting the heat exchanger
T_{Hout} = temperature of the hot fluid (water) exiting the heat exchanger
T_{Cin} = temperature of the cooling water entering the heat exchanger

It is important to note that unlike thermodynamic calculations that use absolute temperatures in most calculations, heat transfer operations employ predominately Fahrenheit and centigrade temperature scales.

Recall earlier in Chapter 2 some relevant definitions were presented. One definition was flux. Equation (2.41) can be rearranged to be described by $Q/A = U\Delta T_{lm}$. The relevance of this rearrangement is that the flux can be described as two quantities: one version that describes flux in terms of energy (Q/A) and the other that uses the log mean temperature for part of the calculation ($U\Delta T_{lm}$). Flux is often overlooked as a key parameter in heat transfer. If the flux value is too high (i.e. the amount of heat per unit area time is higher than planned), system upsets such as runaway reactions due to unexpectedly high reaction rate (Section 4.1.3) and pressure vessel failures can occur. Such an event is more probable if a safety analysis such as a HAZOPS study is not performed.

The following example is a typical application where the log mean temperature is employed to help resolve a heat transfer operation.

Example 2.9 A distilled water is being designed to leave an evaporator and enter a storage tank. The water is intended to enter a heat exchanger before going to the storage tank. The water (process water) is entering the heat exchanger at 210 °F. After exiting the heat exchanger, the process water is to be cooled to 68 °F. The cooling water is intended to enter the shell side of the heat exchanger at 68 °F and leave the heat exchanger at 78 °F. In order to properly size a heat exchanger for the cooling operation, it is required to determine the LMTD (ΔT_{lm}) for the operation. For this process, calculate the LMTD required.

Solution

Equation (2.43) is most applicable for this calculation. By identifying the variables, the following values are obtained:

$$T_{Hin} = 210°F$$
$$T_{Cout} = 78°F$$
$$T_{Hout} = 78°F$$
$$T_{Cin} = 68°F$$

Inserting the variables in Equation (2.43) yields

$$\Delta T_{lm} = \frac{(210-78)-(78-68)}{\ln(210-78)/(78-68)}$$
$$\Delta T_{lm} = 54.9°F \leftarrow \textbf{Solution}$$

The resulting LMTD, 54.9 °F, is typical for fluids being cooled or heated with no change of phase. However, dealing with a heat transfer situation where there is a change of phase where water is heated to steam, the calculation requires the addition of the latent heat of vaporization (ΔH_{vap}) to the calculation. For water, the standard value used in engineering calculations for ΔH_{vap} is 970 Btu/lb.

Example 2.9 presents a typical calculation where the phase change must be considered. Example 2.9 also introduces some concepts that are used in heating, ventilating, and air conditioning (HVAC) used in Chapter 3. The following example identifies some thermodynamics concepts.

Example 2.10

A. As part of an evaporation operation, purified water will be heated until it reaches the boiling point. The water is being fed into the process at a rate of 15 gal/min. Once the water is vaporized, the vapor will be compressed to transform the atmospheric steam to a more energetic steam by increasing the temperature and pressure. The steam will be used to heat purified water, which in turn will humidify an aseptic syringe filling and packaging area in the facility. The manufacturing location is located at sea level. The purified water feed is 20 °C (68 °F), and at these conditions, the heat capacity of the water (C_p) is 1 Btu/lb °F. At the given conditions, determine the hourly Btu heating requirements and compressor discharge temperature of the compressed steam and the compression ratio required to achieve the calculated discharge temperature of below 275 °F. Since this is a preliminary calculation, assume a compressor efficiency of 80%. A reciprocating compressor is planned to be used.

B. The operation also has an identical facility in Albuquerque, New Mexico, at an altitude of 5250 ft. For this facility, determine the hourly heating requirements, the compression ratio, if the same compressor is used, and the discharge temperature.

Solution

The basis of calculation for this calculation is Equation (2.40), $Q = mC_p\Delta T$. Since the feed rate is given as 15 gal/min and relationship (2.40) requires the units in lb/h, the first step is to convert from gal/min to lb/h. Recalling that 1 gal of pure water weighs 8.34 lb, the result is

$$\frac{\left(15\frac{1}{\text{gal}}\right)}{\frac{\text{min}}{1}} \times \frac{(8.34\text{lb})}{\frac{\text{gal}}{1}} \times \frac{\left(60\frac{1}{\text{min}}\right)}{\text{h}} = 7506 \text{ lb/h} \quad \text{hourly mass flow rate}$$

C_p for water at 68 °F (20 °C) = 1 Btu/lb°F:

$$\Delta T = 212 - 68°F = 144°F$$

$Q = (7506 \text{ lb/h}) \times (1 \text{ Btu/lb°F}) \times (144°F) = \textbf{1 080 864 Btu/h} - \textbf{energy required to heat the water to 212 °F}$. This value, 1 080 864 Btu/h, is required to achieve 212 °F. However, to vaporize the 7506 lb/h of water, the latent heat of vaporization ($\Delta H_{vap} = 970$ Btu/lb) must be included in the calculation:

$$\frac{\left(7506\frac{1}{\text{lb}}\right)}{\text{h}} \times \frac{(970\text{Btu})}{\frac{\text{lb}}{1}} = 7280820\text{Btu} / \text{h} - \textbf{Energy required to vaporize 7506lb/h}$$

(It can be seen from the result that heating the water to the boiling temperature requires 1 080 864 Btu's versus 7 280 820 Btu's, fewer Btu's than the heat input required to vaporize the water, about 15% of the total heat input. Consequently, unless dealing with process design requirements, the weight of water multiplied by the heat of vaporization is within the range of accuracy (~15%) to serve as a reasonable result for project engineering calculations and/or contractor calculation confirmation.)

The purpose of the compressor is to compress the generated steam. At a higher pressure, the temperature also rises. A higher temperature increases the ΔT of the heating medium, which is the steam. As the temperature of the steam rises, the energy content of the steam also rises. The higher ΔT, referred to as the driving force, permits better heat transfer as well as a reduction of the size of the heat transfer area. This reduction results in a reduction in the size and cost of heat transfer equipment (heat exchangers) required.

At boiling, the pressure of the steam, at atmospheric conditions (sea level), and boiling at 212 °F, the pressure is 14.7 lb/in^2 (actually, the atmospheric pressure is 14.696 lb/in^2). The compression ratio is given by the following relationship:

$$r = \frac{(P_D + P_S)}{P_S} \tag{2.44}$$

where

r = compression ratio
P_D = compressor gage discharge pressure, 30 lb/in^2
P_S = compressor inlet (suction) pressure, 14.7 lb/in^2

$$r = \frac{\left(30\text{lb} / \text{in}^2 + 14.7\text{lb} / \text{in}^2\right)}{14.7\text{lb} / \text{in}^2} = 3.04 \approx 3 = \textbf{Compression ratio}$$

The discharge temperature can be determined by the following thermodynamic relationship:

$$T_D = \frac{T_S\left(R^{(k-1/k)} - 1\right)}{\eta} + T_S \tag{2.45}$$

where

R = ratio of the compressor discharge pressure to the inlet pressure = P_D/P_S = 30 psig/ 14.7 psig = 2.04

k = ratio of specific heats for steam = C_p/C_v = 1.33

η = compressor efficiency, 80% = 0.8

T_D = compressor discharge temperature, °F

T_S = compressor inlet temperature = 212 °F.

Substituting values in previous equation yields

$$T_D = \frac{212°F\left(2.04^{(0.249)} - 1\right)}{0.8} + 212°F = 262.5°F \sim \mathbf{263°F} \leftarrow \mathbf{Solution}$$

The discharge temperature of 263 °F is the temperature of the steam intended to heat and humidify the proposed aseptic location in the facility. Although vapor compression appears to be a convenient heat transfer mechanism, it has several drawbacks. A very important factor is the discharge temperature. If the temperature is too high, compressor component material damage is a serious consideration. At a higher temperature, the coefficient of expansion becomes a serious consideration. Also, a higher temperature can overly stress the materials of construction, particularly if a sudden temperature increase occurs. For these reasons, the compressor discharge temperature is commonly limited to 275 °F. With a limiting discharge temperature of 275 °F, there is also a limiting compression ratio when considering a vapor recompression operation. The general guideline is to operate at compression ratios between 2 and 6.

B. Referring to Table 2.3, the boiling point of water at 5250 ft elevation is 202.0 °F. Also by referring to standard steam tables, the absolute pressure is 12.010 lb/in² for steam at the given altitude. For preliminary calculations, it is assumed the Btu hourly heating requirements are identical to the value determined in Part A (e.g. 7 280 820 Btu/h).

The procedure for determining the compression ratio is, again, predicated on Equation (2.44) with the major difference being the inlet or suction pressure, which, in this case, is 12.01 lb/in². Solving for the compression ratio,

$$r = \frac{\left(30\text{psig} + 12.01\text{psig}\right)}{12.01\text{psig}} = 3.49 \approx \mathbf{3.5\ Compression\ ratio}$$

The difference in altitude has very little effect on the compression ratio.

The discharge temperature for Part B is determined by employing Equation (2.45) using the following input variables:

R = ratio of the compressor discharge pressure to the inlet pressure = P_D/P_S = 30 psig/ 12.01 psig = 2.497 psig ≈ 2.50

k = ratio of specific heats for steam = C_p/C_v = 1.33

η = compressor efficiency, 80% = 0.8

T_D = compressor discharge temperature, °F

T_S = compressor inlet temperature (from the steam tables) = 202.0 °F

Substituting values in Equation (2.45) yields

$$T_D = \frac{\left(202.0°F\right)\left(2.50^{0.249} - 1\right)}{0.8} + 202.0 = 266.7°F \approx \mathbf{267°F} \leftarrow \mathbf{Solution}$$

The temperature difference at higher altitude is about 4 °F greater than employing the same operation at atmospheric conditions. This is, in effect providing at higher altitude, a greater ΔT, or driving force.

A more applicable vapor compression application will be detailed when addressing evaporators, specifically mechanical vapor recompression systems (see Figure 2.6), a common operation employed in the manufacture of WFI.

A term used often in this section is "heat exchanger." In fact, the heat exchanger is perhaps the most employed equipment in purified water heat transfer operations. Some basic material was already presented on the most common type of heat exchanger, the shell and tube design. A more detailed explanation and analysis of heat exchanger design principles would be of value at this point.

While there are many types of heat exchangers that can be used for purified water and clean steam applications, the most widely used and probably the most cost-effective design for these applications is the shell and tube heat exchanger, which has been alluded to earlier in this section. The shell and tube design is preferable for several reasons:

- In general, water and steam are the major components of pharmaceutical water components. This raw material composition does not create significant buildup or fouling of the shell and tube exchanger components. Impurities such as high salt concentration or viscous feed are either removed or not present in the process train by the time the process feed reaches the heat exchanger.
- For virtually all standard purified water manufacturing operations, shell and tube exchangers can be procured as built equipment requiring no special fabrication criteria. While for some applications, sanitary connections may be required and the tubing fabricated of stainless steel (most often type 316 or 316L), many vendors have sufficient inventories to minimize lead time (the time required from actual ordering to site delivery).
- With a large number of suppliers available, pricing should be competitive enough to obtain several quotes. If the system involves several heat exchangers and is part of a built installation, the system vendor should be able to provide quotes from the heat exchanger vendor selected.
- The calculations to verify the adequacy of a shell and tube heat exchanger are straightforward, and sufficient data is available to perform a design evaluation with a minimum of calculations.
- Shell and tube designs are very similar regardless of whether or not the application is for heating or cooling. The only significant difference may be the heat transfer surface area specified.

Figure 2.3 is a diagram of a basic single-pass shell and tube heat exchanger. Single pass indicates the tube side fluid enters the heat exchanger and is either heated or cooled by the shell side fluid, depending on service, and exit the heat exchanger. To obtain better heat transfer using the same shell diameter, it is common to have the straight tubes, as shown in Figure 2.3, formed into U-shaped tubes with the tube side fluid circulating through the tubes for several passes before exiting the heat exchanger. This multipass tube circulation or U tube bundle allows for fluid to be heated or cooled, depending on application, to be exposed to the shell side temperature for a longer effective heat transfer length and, consequently, a greater heat transfer time. Figure 2.4 illustrates a U tube bundle that is inserted in the shell of a multipass (two passes in this case) heat exchanger (note the baffles in the bundle).

When performing heat exchanger calculations, often the surface area of the heat exchanger tubing will be required. If the tube bundle is of a U tube configuration, the number of tubes for

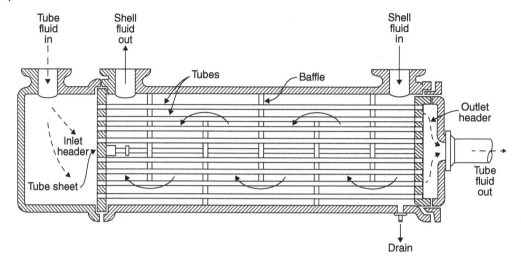

Figure 2.3 Single-pass heat exchanger. *Source:* Faust et al. [4]. Reproduced with permission of John Wiley & Sons, Inc.

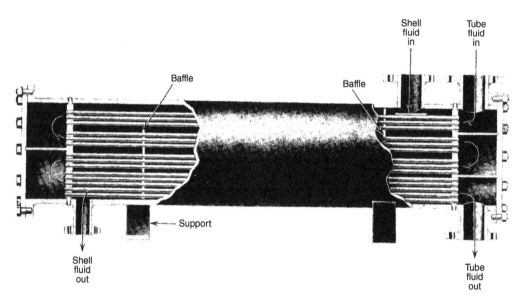

Figure 2.4 Multipass U tube heat exchanger.

calculations is double the actual number of tubes used for fabrication. For instance, if 10 pieces of 6 ft tubing are employed to fabricate a U tube bundle, calculating the heat transfer surface area will be predicated on a basis of 20 tubes, not 10. The baffles are also shown for this particular bundle.

While U tube designs offer an advantage in heat transfer, it is standard practice to make corrections for U tube exchanger designs. Typically, for pharmaceutical water processing, a correction is made to account for flow patterns in the exchanger. A rigorous calculation technique is used to add correction factors to most heat exchangers. However, for project applications, it is reasonable to use a fraction to correct for flow geometries in shell and tube exchangers. Consequently, a factor of 0.9 is reasonable. By using the 0.9 correction factor, the basic heat transfer equation (Equation 2.41) can now be expressed as

$$Q = 0.9UA(\text{LMTD}) \tag{2.46}$$

For water and steam service, the typical configuration is to introduce the steam on the shell side of the heart exchanger and have the water enter the tubes of the exchanger. The diagram exhibited in Figure 2.4 displays a series of semicircular plates or metal baffles mounted on the tube sheet of the exchanger. The baffles have hole in them through which the tubes pass. The purpose of the baffles is to provide an enhanced heat transfer coefficient (Equation 2.41). The baffle also helps to provide structural support the tubes mounted in the shell.

For a given heat exchanger design, the distance between baffles can be determined by the following relationship:

$$B = \frac{a_s\, p}{D_s C} = \frac{a_s}{D_s} \frac{[1 - D_o]}{P} \tag{2.47}$$

where

B = baffle spacing, ft
a_s = open area of shell (the length of the tubes multiplied by the cross-sectional area of each tube multiplied by the total number of tubes), ft^2
p = tube pitch (the distance between the center of one tube to the center of the nearest adjacent tube), ft
D_s = outer shell diameter, ft
C = tube clearance (the distance from the outside tube wall to the nearest tube wall), ft
D_o = outside diameter

Equation (2.36) is an empirical relationship. It does not take into account shell side component factors such as velocity and shell side component condensation on the baffles, tubes, and shell. Also, this is a cumbersome calculation particularly if the intent is to verify the adequacy of the heat exchanger components, in this case the baffles, and not for a process design. Another very important element concerning baffle design is the time required to rigorously verify the baffle design. As a result, several accepted values for baffle configurations are generally employed and can easily be confirmed by evaluating the vendor drawings. For baffle designs for water and steam service, the following is recommended:

- Baffle spacing should be about 0.4 times the inside diameter (in the range between 0.3 and 0.5 times the diameter, inside pipe diameter [ID] or outside diameter [OD]).
- The minimum baffle spacing is 0.20 times the shell diameter or 2 in., whichever is the greater of the values.
- The maximum distance between baffles should be 30 in. or the shell diameter, whichever is smaller.

In addition to the baffles, the other major components of shell and tube heat exchangers are the long tubes surrounded by the exchanger shell. These tubes are specifically referred to as tubes rather than pipes. Figure 2.4 exhibits a basic heat exchanger designed to allow the tube side fluid to undergo several passes. The tubes discharge to a header that separates the tube fluid from the steam or shell side of the exchanger. The major difference between pipes and tubes (a bundle of tubes, such as the configuration shown in Figure 2.4, is referred to as tube sheet and is often employed to obtain a more efficient flow than the header configuration) is that the measurement basis for tubing is the OD, whereas piping identification is predicated on the ID. Also, because relatively pure water and steam are primarily used, U tube bundles are preferable because there is very little accumulation in both the tubes and the shell of an exchanger. This being the case, the need to clean heat exchangers requires minimal maintenance other than routine physical inspection and monitoring of the temperature. Because of the minimal potential for accumulation of buildup in the heat exchanger, it is, as described earlier, common to use a U tube bundle design in purified water operations. Another reason for using bundles is that most heat exchangers involved in pharmaceutical water processes are usually not large-scale operations. Typical flows are less than 10 000 gal/h and require relatively small heat exchangers, particularly when compared with large-scale heat exchanger units for petroleum refining operations where flows through heat exchangers are order of magnitudes greater.

When specifying, designing, or installing shell and tube heat exchangers, the following preferential flow through the tubes is recommended:

- A corrosive fluid or a fluid likely to deposit sediment or sludge
- Any fouling fluid
- The less viscous of two fluids
- The fluid under higher pressure
- The hotter fluid
- Lower flow quantity

For preliminary heat exchanger layout analysis or design, the following fluid velocities are also recommended (assuming minimum vapor phase flow):

- Normal tube side velocities for pharmaceutical fluids, including water, 3–6 ft/s.
- Tube side velocities for cooling water, 5–8 ft/s.
- Maximum tube side velocity (to minimize fouling), 10–12 ft/s. The maximum velocity is a very important value. While, for fluid flow and heat transfer applications, a high fluid velocity in pipes is desirable, high velocity flow can degrade the piping. A phenomenon known as erosion corrosion can occur inside the piping if the liquid in the tube contains solids, even in small concentrations. This topic is expanded in Chapter 8, specifically Section 8.3.7.9.
- Shell side liquid velocities, 1–3 ft/s.

Heat exchanger tubing is classified by an identification format known as BWG (Birmingham Wire Gage). For heat exchanger applications, the important element that makes tubing desirable for heat exchanger applications is the wall thickness or BWG size.

While 10S walled steel pipe has a relatively thin wall, when compared with 1 in. BWG 16 tubing, the difference is quite pronounced. For 1 in. Schedule 10S piping, the wall thickness is 0.109 in. compared with 0.065 in. for BWG 16 tubing, a wall thickness difference of 1.68. This wall thickness ratio is similar for most pipe and tube wall thicknesses of these sizes of piping and tubing. As will be seen, a thinner wall thickness is very desirable for heat transfer applications, particularly when shell and tube heat exchangers are of interest.

For purified water shell and tube heat exchanger design, tube diameters generally range from 3/4 in. OD to 1¼ in. OD with the gage ranging from BWG 18 to BWG 14. The tube length depends on the calculated surface area that is required.

Table 2.11 identifies the dimensions of the most commonly employed heat exchanger tubing.

Tube wall thickness (Δx) and the thermal conductivity (k) are important factors in determining the value of an overall heat transfer coefficient (U), which, as defined in Equation (2.31), is a key variable in determining the rate of heat transfer in a heat exchanger. As will be shown, the overall heat transfer is determined by obtaining and summing several factors that calculate the

Table 2.11 Dimensions of common heat exchanger tubes for pharmaceutical water.

Outside diameter (in.)	Wall thickness		Inside diameter (in.)	Inside cross-sectional area (ft²)	Circumference (ft) or surface (ft²/ft of length)	
	BWG	in.			Outside	Inside
3/4	14	0.083	0.584	0.001 86	0.1963	0.1529
	16	0.065	0.620	0.002 10	0.1963	0.1623
	18	0.049	0.652	0.002 32	0.1963	0.1707
1	14	0.083	0.834	0.003 79	0.2618	0.2183
	16	0.065	0.870	0.004 13	0.2618	0.2277
	18	0.049	0.902	0.004 44	0.2618	0.2361
1¼	14	0.083	1.084	0.006 41	0.3271	0.2839
	16	0.065	1.120	0.006 84	0.3271	0.2932
	18	0.049	1.152	0.007 24	0.3271	0.3015

convective heat transfer and the conductive heat transfer contributions to the overall heat transfer coefficient (U). In fact, for purified water manufacturing, when water and steam are the primary process raw materials, it will be possible to simplify the calculations, as will be shown.

The overall heat transfer coefficient is given by

$$\frac{1}{U} = \frac{1}{h_i} + \frac{\Delta di}{kD_m} + \frac{D_i}{h_o D_o} + R_d \tag{2.48}$$

where

U = overall heat transfer coefficient, Btu/ft²°Fh
h_i = heat transfer coefficient for the inside tube wall, Btu/ft²°Fh
h_o = heat transfer for the outside tube wall, most often the wall that interfaces with the steam or shell side of the heat exchanger
Δx = tube wall thickness, ft²
k = thermal conductivity of the tube, Btu/hft² (°F/ft)
D_o = outside heat exchanger tube diameter, ft
D_i = inside heat exchanger tube diameter, ft
D_m = mean heat exchanger tube diameter, ft $\left[D_m = \left(D_o + D_i\right)/2\right]$
R_d = $1/h_i$, fouling factors, 0.000 066 h °Fft²/Btu for steam and 0.001 h °Fft²/Btu for water

Since the outside of the tube is exposed to the steam on the shell side of the exchanger, it is reasonable to expect very little effect from condensing steam and water (R_d) on the outside of the tube.

While this is generally the case, condensation of steam is a factor of the steam temperature and pressure. Since most pharmaceutical process steam is kept below 225 °F, it is reasonable to expect a small amount of the process steam to condense to water, particularly forming on the outside surface of the tubing. If not removed, the shell side will eventually fill with condensed steam that transforms to water by giving off the latent heat. As the condensate in the shell side accumulates, the shell side heat transfer coefficient, h_o, gets lower in value and reduces the performance of the heat exchanger. To preclude this "flooding" from occurring, it is common to install a device to discharge the water from the shell side (steam side) of the exchanger. Usually located at the bottom or discharge of a vertical heat exchanger and at the discharge end of a horizontally mounted exchanger, such a device is identified as a steam trap. There are several types of steam traps with the three most common types being the inverted bucket trap, the ball float type, and the thermostatic trap. While each has a particular operating mode, heat exchangers intended for purified water applications are generally unaffected by the type employed. The major point for heat exchanger steam traps is to remove the condensate buildup on the steam side (usually the shell side) of the exchanger. The condensate produced by heat exchangers for purified water applications is not significant, rendering the selection basically a cost consideration. The same holds true for condensate disposal. Economics should determine whether the condensate should be reintroduced into the steam feed or discharged to a drain.

For as-built or contractor-designed systems employing heat exchangers, steam trap specifications should be part of the installation package and available for project review and verification. While detailed design calculations may not be supplied as project deliverables, the approximate amount of condensate can be quickly estimated. For heat exchangers employed to heat water with shell side steam, the applicable relationship is

$$\text{Lb}/\text{h Condensate} = \frac{(G)(\text{s.g.})(C_p)(\Delta T)(8.34)}{\lambda} \tag{2.49}$$

where

G = flow rate, lb/h or gal/min, determined by dimensions required
SG = specific gravity of water = 1 for most cases
C_p = heat capacity of water = 1 Btu/lb°F for most cases
ΔT = temperature difference between water flow entering and exiting heat exchanger
λ = latent heat of vaporization, Btu/lb.

The difference is that Equation (2.49) involves the latent heat of vaporization to account for condensate formation.

Often, an estimate of the condensate formation from a heat exchanger is adequate. A typical case would be a project engineer responsible for confirming the supplier installed heat exchange system will perform as intended. For such cases, a reasonable approximation relationship exists. This approximate relationship is given by

$$\text{Lb}/\text{h Condensate} = \frac{G}{2} \times \Delta T \tag{2.50}$$

where

G = flow of water (gal/min)
ΔT = temperature rise (°F).

Example 2.11 A tablet manufacturing operation maintains an area to clean stainless steel bins used to dry mix vitamin and mineral powder blends for tableting operations. Water is used to clean the bins as the final wash after completion of cleaning operations. The water originates from a storage tank that maintains the water at 68 °F. The rinsing operation SOP requires the final rinse water to be 149 °F (65 °C). To achieve this temperature, water is fed through a shell and tube heat exchanger with shell side steam at a temperature of 219 °F and the USP feedwater at a temperature of 68 °F (20 °C). The feedwater flow rate is 33 gal/min. The heat of vaporization for water is 970 Btu/lb°F and the density is 8.33 lb/gal. The heat capacity for the feedwater is 1 Btu/lb (i.e. a specific gravity of 1).

Using the information provided, determine the hourly condensate formation in lb/h. employing the methods described (Equations 2.49 and 2.50).

Solution

The first method employed is Equation (2.49). The flow rate is given as 33 gal/min. The temperature rise is 81 °F ($\Delta T = 81$ °F). Inserting the values in Equation (2.49) yields

$$\text{Lb / h Condensate} = \frac{\left(33 \, \overset{1}{\cancel{\text{gal}}} / \overset{1}{\cancel{\text{min}}}\right)(1)\left(\overset{1}{\cancel{1\text{Btu}}} / \text{lb}°\text{F}\right)\left(81 \overset{1}{\cancel{°\text{F}}}\right)\left(8.33 \, \text{lb} / \overset{1}{\cancel{\text{gal}}}\right)\left(60 \text{min} / \text{h}\right)}{970 \, \cancel{\text{Btu} / \text{lb}}}$$

$$= \mathbf{1377 lb / h}$$

The more rigorous result, 1377 lb/h, can now be compared with the approximation given by Equation (2.50). Inserting the values given, the result is

$$\text{Lb / h Condensate} = \frac{33}{2} \times 81 = 1336.5 \approx \mathbf{1337 lb / h} \leftarrow \mathbf{Solution}$$

The difference between the formal relationship (Equation 2.49) and the approximation (Equation 2.50) is 2.91%

Either result is not a significant amount (about 2.5 gal/min) when estimating if a condensate recycle is economically viable. Since the final wash is the last step of the cleaning operation, the final wash does not occur on a continuous basis. Typically, large operations employ many mixing bins, and it is not standard to employ only one or two mixing bins for a key operation. Also, installing a condensate recycle would require installing a piping system to the condensate recycle area and insulating the piping (to maintain the condensate temperature to the recycle operation), shutoff and/or bypass valves, and a pump, if the distance is significant as well as the installation labor costs. Consequently, an a priori estimate indicates that it is probably more viable to discharge the condensate to the nearest usable floor drain.

The previous example has a final rinse temperature of 149 °F (65 °C). This temperature is actually a bit higher than would normally be used for rinsing purposes. Unless personal protective equipment such as thermal gloves and safety glasses is part of the worker standard, this elevated temperature could present a safety issue because of prolonged use of this higher temperature rinse water. Additionally, if the condensate water is discharged to a drain, it is advisable to have the drain pipe insulated for safety purposes.

Steam trap selection is an item that should be addressed for project applications. Since heat exchangers for purified water processing generally do not operate at extreme temperatures and pressures, the type of steam trap selected is not as critical as the materials employed. Steam used for heating pharmaceutical-grade waters is not covered by cGMP criteria, this steam is facility utility steam and non-critical since it is not contacting any product or intermediate compound. Consequently, there is no need to use steam traps fabricated of stainless steel. A steam trap fabricated of iron or a non-stainless alloy or iron is satisfactory.

While steam trap installation is relatively straightforward, when evaluating a steam trap installation, it is important to note that the trap should be installed as far below the heat exchanger as possible while providing sufficient headroom and space for maintenance activities to be performed.

While steam traps are generally quite reliable, a new installation or modified installation may encounter a problem if the unit fails to drain. Some of the common causes for the trap not draining are as follows:

- The steam trap is air bound, preventing the condensate to discharge.
- The differential pressure from the inlet to the outlet is negative, often a result of failing to mount the steam trap at the lowest point possible.
- The trap is undersized for the differential pressure.
- A vacuum exists in the steam space (typically, the heat exchanger shell side).

Determining the heat transfer coefficient for fluid inside the tube, h_i, is obtained by employing dimensionless numbers. Previously, another dimensionless grouping was applied in fluid hydraulics calculations; this term was the Reynolds number. In addition to the Reynolds number, two other dimensionless groups are employed in heat transfer involving, in this case shell and tube exchangers. The first dimensionless number of interest is the Nusselt number. The Nusselt number (Nu) is given as

$$\text{Nu} = \frac{h_i D}{k} \qquad (2.51)$$

Here, h_i is the heat transfer coefficient for the inside wall of the heat exchanger tube (Btu/h ft^2 °F) and k is the thermal conductivity (Btu/ft°F h) of the fluid in the tube, water or steam in this case.

D_i is the inside tube wall diameter, ft. It is important to note that the inside diameter values for Table 2.11 are given in inches. This requires that values given must be converted to feet for certain heat transfer calculations. Unless otherwise identified, for purified water operations involving the Nusselt number, when a diameter is encountered, the diameter, D, is defined as the inside tube diameter and the subscript is not required.

The Nusselt number can be viewed as a ratio of two heat transfer coefficients. One term in the ratio is the contribution of convection to the fluid in the tube, while the other variable is the contribution of heat conduction in the fluid. Accordingly, the Nusselt number can also be expressed generically as

$$\text{Nu} = \frac{hD}{k} = \frac{\text{Convective heat transfer coefficient}}{\text{Conductive heat transfer coefficient}} \qquad (2.52)$$

For most liquid heat transfer applications, including purified water, convection is a much more rapid process than conduction. Consequently, the value of the Nusselt number is most often 1 or greater.

As will be shown, the Nusselt number is a critical value when determining U, the overall heat transfer coefficient.

While not generally encountered in fluid heat transfer calculations, there exists a dimensionless number very similar to the Nusselt number in mathematical notation identified as the Biot (Bi) number. The difference between the Nusselt (Nu) number and the Biot number is that the Nusselt number is applicable for fluids, whereas the Biot number is used for solids, such as solid metal rods. The Biot number is employed for heat transfer purposes when a system is being evaluated for conductive heat transfer evaluations very often for unsteady-state processes. The Biot number is a measurement of heat transfer at the surface of a solid compared with internal heat conduction of the same solid. It is important to note that the Nusselt number uses thermal conductivity for fluids in the tube and the thermal conductivity of the fluid.

Once again, the Biot number is employed for solids. This is relevant since heat transfer calculations employ conductivity for both fluids and alloys and metals and, on occasion, the two conductivities are accidentally interchanged when performing a heat transfer calculation.

The dimensionless Biot number (Bi) and is given as

$$\text{Bi} = \frac{\text{Heat transfer on the surface}}{\text{Internal heat conduction}} = \frac{hL}{k} \tag{2.53}$$

where

h = heat transfer coefficient of the solid, Btu/ft^2°F h
L = the volume of the solid divided by the surface area of the given solid, ft
k = thermal conductivity of the solid, Btu/ft°F h.

Because the Biot number may be applicable in heat transfer operations where conduction is of concern, it rarely will be encountered in purified water processes involving heat transfer.

The other dimensionless number used in heat transfer is the Prandtl number (Pr). The dimensionless Prandtl number is given as

$$\text{Pr} = \frac{C_p \mu}{k} = \frac{\text{Momentum diffusivity}}{\text{Thermal diffusivity}} \tag{2.54}$$

where

C_p = heat capacity, Btu/lb °F
μ = viscosity, lb/ft h
k = thermal conductivity, Btu/ft °F h.

Momentum diffusivity is a term defined by two known quantities, viscosity and density. The two terms when combined are given by the ratio μ/ρ, recalling that ρ is density (lb/ft^3). The ratio μ/ρ is identified as the kinematic viscosity (also referred to, when dealing with mass and heat transport calculations, as diffusivity) and for purified water operations is rarely encountered. For thermal diffusivity, the symbol α is employed. The thermal diffusivity is defined by

$$\alpha = \frac{k}{\rho C_p} \tag{2.55}$$

where

k = thermal conductivity, Btu/ft°F h
ρ = density, lb/ft^3
C_p = heat capacity Btu/lb°F.

Equation (2.37) identifies parameters that are given quantities (e.g. diameter, tube wall thickness, thermal conductivity) when considering the calculation of an overall heat transfer coefficient. Heat transfer coefficients, h, as opposed to the overall heat transfer coefficient, U, can be determined by considering h as a function of the Prandtl number (Equation 2.54) and more relevantly the Nusselt number (Equation 2.51). The Prandtl number and Nusselt number determinations are actually the relevant factors in determining the value of the overall heat transfer coefficient, U, for heat exchanger service, as will be explained.

A general relationship has been developed to relate the Nusselt number, the Prandtl number, the Nusselt number, and the Reynolds number into a model that allows for determining the overall heat transfer coefficient, U, with reasonable accuracy. The relationship was developed by using a concept known as the Buckingham's Pi (Π) theorem (just as the term Σ is a term indicating a series of terms intended to be added, the Π denotes a series of terms to be multiplied). The theorem is based on organizing a group of dimensions and arranging them into dimensionless numbers or dimensionless groups, a large number of which have already been identified and defined (e.g. Reynolds number, Prandtl number, Nusselt number). Many texts prepared for chemical engineering operations and texts specific to scale-up of processes have detailed sections describing the theorem and how to use it. References [5, 6], as well as online searches, exhibit detailed procedures for obtaining dimensionless groups for heat and mass transfer applications.

A particularly important correlation, derived in part by the Buckingham Π theorem, is a relationship known as the Dittus–Boelter equation, given by

$$Nu = 0.023Re^{0.8}Pr^{n} \tag{2.56}$$

- For this relationship, ~~2.42,~~ the Reynolds number (Re) is >10 000.
- The Prandtl number (Pr) for this formula ranges from $0.7 \leq Pr \leq 160$.
- For operations requiring the fluid to be heated, $n = 0.4$.
- For operations requiring the fluid within the tubes to be cooled by the heat exchanger, $n = 0.3$. Often, even the most experienced engineers neglect to differentiate between a cooling application ($n = 0.3$) and a heating scenario ($n = 0.4$). The result is a significant numerical error that goes unnoticed until the error is discovered during start-up, or shortly thereafter.

Although Equation (2.56) is a good method for a more rigorous evaluation of heat transfer in circular tubing, it has an important limitation. It should not be employed where there is a large temperature difference between the tube side fluid and the heating media. If the temperature difference is ignored, the calculation should use a constant viscosity.

One important design variable that is relevant to Equation (2.56) is the ratio of tube length to the inside diameter of the selected tubing (L/D). For most liquids, including water, the ratio should be $L/D \geq 10$, where L is the tube length and D is the tube diameter. Additionally, the dimensions of both the length and diameter must be the same so that the L/D ratio is dimensionless. The L/D ratio is, for the most part, irrelevant when dealing with non-turbulent or laminar flow (i.e. flow in a regime where the Reynolds number is <2100, but not >4000); but it is of importance in the laminar flow range.

For heat exchangers involving purified water production, Equation (2.56) can be modified to more accurately represent pharmaceutical water/steam process.

With the terms and concepts being defined, it is now possible to calculate overall heat transfer coefficients and related required quantities for a crucial component of purified water operations, the heat exchanger.

While it is evident that heat exchanger tubing is the major component of heat exchangers, an important differentiation must be noted prior to designing or evaluating heat exchangers. Previously, when alluding to heat exchanger area, the reference was for cross-sectional areas. That is, the area referenced was the flow area, or the area using the inside diameter as a basis. Example 2.1 was identified ($A = \Pi D^{2}/4$). For heat exchangers, the surface area of the tubing must also be evaluated. While the cross-sectional area of the tubes is critical for heating and cooling of the tubes, the tube surface area is also important to evaluate the heating or cooling of the fluid flowing through the fluid inside the tubing. The relationship for calculating the total surface area of the tubing is given by Equation (2.57):

$$A_s = (\pi)(D_{av})(L)\left(\text{no. of tubes in the heat exchanger}\right) \tag{2.57}$$

where

A_s = total surface area of the heat exchanger tubing, ft^2

D_{av} = average diameter of the tubing, ft. (This value is obtained by adding the inside diameter of the tubing and the outside diameter of the tubing and dividing by two (i.e. $(A_o + A_i)/2$). While it would appear that Equation (2.39) $[Q = -kA(\Delta T/\Delta x)]$ would be the relationship to use, because the tube wall for heat exchanger tubing is so thin, Equation (2.43) is usually employed)

L = length of the tubes, ft.

Example 2.12 A one-pass heat exchanger uses steam to heat a recirculation line from a temperature of 25 °C (77 °F) to a circulating temperature of 60 °C (140 °F). The steam for heating the water is supplied at 20 lb/in^2 (227.96 °F). The water is being circulated through the purified water loop at a flow of 60 gal/min. The steam condensate is discharged to a drain at 130 °F. The heat exchanger tubing is 1 in. BWG 16 and fabricated of 316L stainless steel. The exchanger is a one-pass tube design containing a total of 10 tubes, each tube having a length of 3 ft (36 in.) with the tubes entering and discharging through headers (the heat exchanger does not contain a U tube bundle). The heat exchanger was installed as an as-built component. The generally accepted value for the outside coefficient (h_o) is between 2000 and 2050 Btu/ft^2 °F h, with the average being 2025 Btu/ft^2 °F h.

A. Determine the overall heat transfer coefficient, U, of the heat exchanger using simplified calculations.

B. Compare the results using a more rigorous design calculation.

Solution

The simplified calculation commences with performing a calculation for the quantity of heat or energy required. This is achieved by employing Equation (2.40) ($Q = mC_p\Delta T$). Additionally, a modified version of Equation (2.41) ($Q = UA\Delta T_{lm}$) will be used as the basis of the calculation. In lieu of the log mean temperature (ΔT_{lm}), ΔT or 63 °F will be used. The linear temperature difference is used to simplify the calculation. Additionally, the modified temperature difference is more commonly employed than the more rigorous log mean temperature difference.

A. Since the flow is volumetric and the feed is in lb/min, the volumetric flow is converted to lb/h:

$$\left(30\text{gal}/\text{min}\right)\left(1\text{ft}^3/7.48\text{gal}\right)\left(62.3\text{lb}/\text{ft}^3\right)\left(60\text{min}/\text{h}\right) = \mathbf{14992\,lb/h}$$

Using a value of 1 Btu/°F for C_p, a temperature difference of $\Delta T = 63$ °F, the calculation is now

$$Q = \left(14992\text{lb}/\text{h}\right)\left(1\text{Btu}/\text{lb°F}\right)\left(63°F\right) = \mathbf{944496\,Btu/h}$$

The heat transfer area for the heat exchanger is required. Referring to Table 2.11, the required data for 1 in. BWG 16 tubing is

$$OD = 1\text{in} = \left(1\text{ft}/12\text{in}\right)\left(1\text{in}\right) = 0.0833\text{ft}$$

where ID = 0.870 in = (1 ft/12 in) (0.870 in) = 0.0725 ft. With this information, the average tube diameter can be determined:

$$D_{av} = \frac{\left(D_{od} + D_{id}\right)}{2} = \frac{\left(0.0833\text{ft} + 0.0725\text{ft}\right)}{2} = \mathbf{0.779\,ft}$$

With the information provided, the total heat transfer surface area can now be ascertained through the use of Equation (2.57):

$$A = (3.142)(0.779\text{ft})\frac{(3\text{ft})}{\text{tube}}\frac{(10\text{ tubes})}{1} = 73.4\text{ft}^2$$

It is now possible to determine the overall heat transfer coefficient by the simplification of Equation (2.31) to $U = Q/A\Delta T$. Inserting the calculated values in the altered form yields

$$U = \frac{944496\text{Btu}/\text{h}}{(73.4\text{ft}^2)(63°\text{F})} = 204.25 \text{ Btu}/\text{ft}^2\text{°Fh} \approx \textbf{204Btu}/\textbf{ft}^2\textbf{°Fh} \leftarrow \textbf{Solution}$$

The overall heat transfer coefficient using the simplified calculation is 204 Btu/ft^2°Fh.

B. The more rigorous calculation of the overall heat transfer coefficient can now be determined. The basis of calculation is the Dittus–Boelter equation (Equation 2.56). In order to employ the Dittus–Boelter equation, it can be seen that three dimensionless groups must be determined to perform the calculation, the Nusselt number (Nu), the Reynolds number (Re), and the Prandtl number (Pr).

The second solution commences with a determination of the Reynolds number defined in Equation (2.14) as $N_{Re} = Dv\rho/\mu$. For this configuration, it is noted that the flow is through 10 tubes. Thus, the solution involves defining the diameter as the sum of the 10 tube diameters in the heat exchanger:

$$D_{av} = 0.779\text{in}$$
$$\mu = 1.14\text{lb ft°Fh}$$
$$\rho = 62.3\text{lb}/\text{ft}^3$$

where flow = 30 gal/min. This calculation involves determining the cross-sectional area of the tubes (BWG 16):

$$D_{av} = 0.779\text{in} = (0.779\text{in})/(1\text{ft}/12\text{in}) = 0.0649\text{ft}$$
$$D = (0.0649\text{ft}/\text{tube})(20 \text{ tubes}) = 1.298\text{ft}$$
$$A = \frac{\pi D^2}{4} = \frac{3.142(1.088\text{ft})^2}{4} = 1.323\text{ft}^2$$

The hourly linear velocity is

$$(30\text{gal}/\text{min})(60\text{min}/\text{h})(1\text{ft}^3/7.48\text{gal})(\text{ft}^3) = 727\text{ft}/\text{h}$$

The Reynolds number can be calculated as

$$Re = \frac{Dv\rho}{\mu} = \frac{(0.725\text{ft})(727\text{ft}/\text{h})(62.3\text{lb}/\text{ft}^3)}{1.14\text{lb}/\text{ft°Fh}} = 28804$$

While not germane to the problem solution, the calculated velocity, 241 ft/h, is about 0.2 ft/s and is well below the recommended maximum rate of 15 ft/s.

The Dittus–Boelter equation defines a Reynolds number to the 0.8 power, and the result is $(28\,804)^{0.8} = \textbf{3695}$.

The Prandtl number calculation employing Equation (2.54) and the following values is determined by

$$C_p = 0.999\text{Btu} / \text{lb°F}$$
$$\mu = 1.14\text{lbft} / \text{h}$$
$$k = 0.363\text{Btu} / \text{ft°Fh}$$

With the input variable being identified, the Prandtl number is

$$\text{Pr} = \frac{C_p \mu}{k} = \frac{\left(0.999\text{Btu} / \text{lb°F}\right)\left(1.14\text{lbft} / \text{h}\right)}{0.363\text{Btu} / \text{ft°Fh}} = \textbf{3.14} \leftarrow N_{\text{Pr}}$$

The Dittus–Boelter equation defines an exponent to the fourth power for heating operations:

$$\left(3.14\right)^{0.4} = \textbf{1.58}.$$

Recalling that $(h_i D_i / k)$ is the definition of the of the Nusselt number for the inside diameter of the exchanger tubing, enough data is available to determine h_i, the inside heat transfer coefficient. By inserting required values in Equation (2.42) and rearranging, the calculation is

$$h_i = \frac{0.023(3695)(1.58)\left(0.379\text{Btu} / \text{ft°Fh}\right)}{0.779\text{ft}} = \textbf{65.3Btu} / \textbf{ft}^2 \textbf{°Fh} \leftarrow \textbf{Solution}$$

While the calculated value of $65.3\,\text{Btu/ft}^2\text{°Fh}$ may appear to be low, a long tube single-pass heat exchanger often has a U of about 110–$120\,\text{Btu/ft}^2\text{°Fh}$ when newly installed. For multipass heat exchangers performing similar service, the value for U, newly installed, can be in the 175–$200\,\text{Btu/ft}^2\text{°Fh}$ range. It should be recalled that $\Delta x / \Delta k$ was assumed to be negligible due to the low wall thickness of BWG tubing and thus not included in the calculation. The Boelter equation determines the Nusselt Number and not U, the overall heat transfer coefficient, primarily due to the fact that assumptions were made to allow for more speedy result. The purpose of this exercise is to confirm what the system contractor has specified and to ensure it conforms to system specifications and design criteria. The intent is not to design a heat exchanger, but to confirm that contractor unit is a reasonable design for the application.

2.5 Evaporation

Evaporation is one of the oldest manufacturing operations and also one of the most energy-intensive processes.

In the chemical process and waste processing industries, evaporation is a common unit operation used to obtain a concentrated bottoms product from a liquid–solids mixture. The concentrated solids recovered from the evaporation operation are most often considered as the product. Unlike waste processing and process applications, pharmaceutical water production considers the water produced in the evaporation process as the product. The product water is used as the purest water required for pharmaceutical operations, WFI. Evaporation is the most frequently unit operation employed for WFI manufacture. In fact, certain pharmacopeias other than the USP place a constraint on the permitted purification techniques as part of the permitted purification techniques specified for WFI production. The technique allows only distillation (evaporation) as the final purification step. Consequently, in some countries that manufacture pharmaceuticals, operations such as ultrafiltration and RO cannot be employed as a final manufacturing step. Japan, for instance, is one country conforming to this standard. Indeed, most pharmaceutical operations employ evaporation technology for manufacture of ultrapure water (i.e. WFI) applications.

While manufacture of WFI is a major application of evaporators for pharmaceutical applications, evaporators are also used to concentrate aqueous waste streams for off-site disposal. The evaporators produce a smaller volume of waste for disposal with the water available for utilities applications or disposal through the facility water treatment system.

Evaporators are also widely used in the food industry where typical applications include cheese whey evaporators and soy whey evaporators. These products are widely employed as high protein feed additives for cattle and hogs. Often, the evaporator bottoms are further concentrated by spray drying to produce a dry powder additive.

Two evaporator types are most commonly used in purified water manufacturing:

1) A forced circulation design
2) Multiple-effect evaporators
3) A mechanical vapor recompression design

A diagram of a forced circulation evaporator is shown in Figure 2.5. A mechanical vapor recompression evaporator is shown in Figure 2.6.

Evaporators intended for pharmaceutical water manufacturing are composed of three principal design components:

1) Heat transfer equipment
2) Vapor–liquid separation
3) Energy economy

As Figure 2.5 illustrates, there are three basic components to an evaporator, and in this case, a forced circulation design is shown. The component providing energy to the evaporator in the form of heating is the heat exchanger. The heat exchanger is referred to as the calandria, a term rarely used now. In this case, the heat exchanger is vertically or horizontally mounted. Steam is the heating source for evaporators.

The heat exchanger position is determined by two factors: available overhead space and the degree of boiling in the tube. If the heat exchanger is horizontally mounted, the pressure of the heated water entering the vapor body will depress the boiling of water in the heat exchanger. The result is a somewhat more efficient heat exchange in the tube since there is more water in contact with the heat exchanger tubing and there is minimum boiling on the inner tube surface.

Once the solution (in this case the solution is water) is heated to boiling temperature (212 °F or 100 °C), the boiling feedwater is discharged into the large vessel referred to as the vapor body. The term vapor body is somewhat of a misnomer.

One or more vapor bodies boiling at the same pressure are often referred to as an effect. A multiple-effect evaporator is an evaporator system in which the overhead vapor from one effect serves as the heating medium for a second effect. Often, the second effect is under vacuum. This permits boiling to occur at a lower temperature difference (ΔT). As the heated water

Figure 2.5 Forced circulation evaporator.

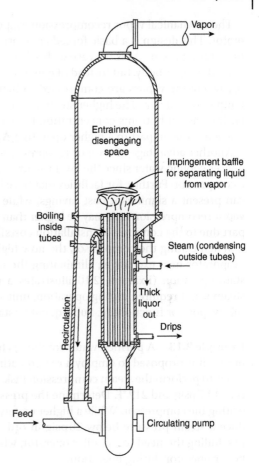

Vapor

Entrainment disengaging space

Impingement baffle for separating liquid from vapor

Boiling inside tubes

Steam (condensing outside tubes)

Thick liquor out

Drips

Recirculation

Feed

Circulating pump

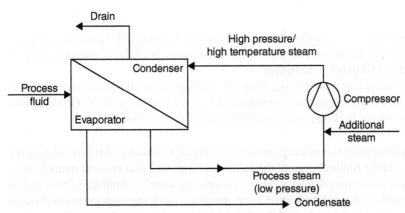

Drain

High pressure/ high temperature steam

Condenser

Process fluid

Compressor

Evaporator

Additional steam

Process steam (low pressure)

Condensate

Figure 2.6 Mechanical vapor recompression evaporator.

enters the vapor body, flash evaporation occurs. Flash evaporation is the process where the heated water undergoes an adiabatic expansion once entering the vapor body. This expansion causes the water to vaporize with the vapor going to a condenser, in this case, or to another process operation such as a compressor or another vapor body (effect). The bottoms products are, as noted, waste products for pharmaceutical water purity operations.

The mechanical vapor recompression evaporator is, perhaps, the most detailed type of evaporator. This design is a basic forced circulation evaporator with a significant equipment addition, a compressor. As is the case with a forced circulation evaporator, the steam produced rises from the vapor body, but instead of going to a cooling condenser, the steam is fed to a compressor where the vapors are compressed to higher pressure and, consequently produces higher temperature steam. The higher steam temperature allows for a greater evaporation and a similar increase in bottoms concentration higher purity overhead product and higher evaporator efficiency due in part to a higher operating ΔT (the driving force) for the process.

Another advantage of a vapor recompression system is the elimination of the need for condenser cooling water since the low pressure steam generated in the vapor body is introduced to a compressor. Further, for facilities that have high water and steam costs, vapor recompression can present a significant cost savings, while the installation cost for installing a mechanical vapor recompression unit may be higher than a standard forced circulation evaporator (in large part due to the compressor installation costs).

When exiting the compressor, the now higher energy steam is fed into the shell side of the evaporator heat exchanger, eliminating the need for steam once the evaporator is in steady-state operation (Example 2.10 illustrates a typical compressor application). While process steam is not required during operation, unit start-up requires the introduction of steam to get the evaporator functioning. Subsequent to start-up, the initial steam injection can be stopped.

Example 2.13 A purified water solution is to act as the feed for a vapor recompression evaporator. It is proposed to employ a reciprocating compressor with a compression ratio of 1.3 in order to perform the steam compression task. The feed from the vapor body to the compressor is at 14.7 psig and 212 °F. Determine the pressure and temperature of the steam subsequent to exiting the compressor. While a higher energy steam is desired (higher ΔT), the unit must produce a steam pressure below 20 psig in order to be classified as an atmospheric boiler, thus precluding the need for a boiler operator, when operating. Determine the need for a licensed boiler operator during operation.

Solution
For this example, Table 2.12 serves as the basis of calculation. At 0 psig (14.7 psia) the temperature is 212 °F. For a compression ratio of 1.2, the pressure coming of the compressor is (1.2) (14.7 psia) = **17.64 psia (2.94 psig)**. ← **Solution**

The resulting pressure is less than 20 psig, thus eliminating the need for licensed operator. The steam temperature off the compressor is about 222 °F, or a ΔT of about 10 °F. This temperature is about the limit for the boiling point rise (BPR) for evaporator design.

When dealing with design aspects of evaporators, it is critical to have key data correlating the solids concentration and the boiling point of the solution at a given solids content using laboratory equipment. As the concentration of solids increases, the solution boiling point will also increase. Data is recorded at predetermined time intervals until the concentration/boiling point plot starts to level off. At this point, while a large quantity of heat is added to the solution, there is a minimal concentration change. This is because the heat input is essentially a process of heating the solids with little water evaporation occurring.

While not a definite correlation, a temperature rise of 5–7 °F often corresponds to a 50–55% solids (500 000–550 000 ppm) concentration for organic feeds such as cheese whey. At this range, the economics of evaporation usually indicates this is the endpoint of the evaporation operation. When the material of interest is the bottoms product, such as concentrated cheese whey, it can then be spray-dried or used as a highly viscous feed additive or for similar applications.

Table 2.12 Feedwater concentrations for fouling and scaling minimization of RO membranes.

Parameter	Guideline value	Comments
Suspended solids	<1 NTU	NTU = national turbidity units. A measurement that determines the level of light transmitted through a water column containing varying levels of suspended organic and inorganic matter
Colloids, primarily silicates	SDI < 5	SDI = silt density index. The rate at which a 4 µm RO filter will plug when under a pressure of 30 psig. For spiral wound filters, the value is 5, and for hollow wound fiber, the value is 3, each under defined conditions
Microbes	<1000 cfu	For USP water, the cfu level is 500 cfu/100 ml, and for WFI, the level is 10 cfu/100 ml (Table 2.1)
Organic compounds	TOC < 3 ppm	The TOC value for USP water (Table 2.1) is <500 ppb, and for WFI, the value is also <500 ppb
Color	<3 APHA color units	APHA = American Public Health Association. A measurement technique based on ASTM STD D 1209 where the standard is distilled water with a value of 0
Metals (Fe, Mn, Al)	<0.05 ppm	
CaCO$_3$ (calcium carbonate)	LSI < 0	LSI = Langelier saturation index. A measure of the solubility of calcium carbonate in water and scale potential on RO membranes with the following values:
		LSI > 0, the CaCO$_3$ is over saturated LSI = 0, the CaCO$_3$ is in equilibrium with the water LSI < 0, the CaCO$_3$ in water is under saturated and has limited scaling potential on a RO membrane
Silica	200 ppm	USP recommendation <0.1 ppm silica
Trace metals (Ba, Sr)	<0.05 ppm	

The standard technique employed in obtaining the optimum boiling point elevation, the temperature where the feed concentration has little change even though the sample feed continues to be heated, is to perform bench-scale tests using a jacketed 3.0 l vessel surrounded by an electric heating mantle. The feed concentration is then plotted with the corresponding temperatures as abscissa and ordinate until the concentration basically levels off. Once the bench-scale data is reduced, it is possible to calculate the operating temperature for a full-scale evaporator system since parameters such as ΔT, LMTD, and viscosity can be included in heat transfer coefficient and pump calculations.

In addition to the need for laboratory studies to determine the optimum boiling point range and final bottoms concentration, it is also critical to determine the viscosity of the feed solution. Viscosity determinations are taken at intervals corresponding to when concentration readings are recorded. One part of the sample is used to determine the feed concentration at a specified time interval. The other sample is used to record the viscosity. Further, it is important that the viscosity measurement is recorded at the same temperature of the bottoms solution. This should be part of the SOP relating to bench-scale boiling point rise (BPR) determination.

While there are several techniques employed to determine the feed concentration at the sample points, most involve weighing a dried sample to constant weight. While small drying ovens are commonly employed, other less time-consuming techniques may be employed.

When high purity water, such as WFI, is the desired end product, the bottoms concentration is not germane to the finished product (i.e. the overhead product water, WFI, in this case).

The acceptability criterion, in this case, is the ability of the product to conform to the criteria set forth in the facility and corporate standard operating procedures (SOPs), often predicated on the data shown in Table 2.1.

Calculations required for performing evaporator design and performance are based on the standard format for a material balance. The basic relationship for an evaporator material balance is given by

$$E = \left(\frac{1 - x_p}{x_f} \right) F \tag{2.58}$$

where

F = feed rate of water to be evaporated, lb/min, gal/min, or ppm
E = quantity of water recovered as WFI, lb/min or ppm
x_f = fraction of the feed composed of solids
x_p = fraction of the product overheads composed of solids.

When dealing with product and feed fractions, it is important to note that the quantities are dimensionless, and as a result, units should be consistent.

Example 2.14   It is required to obtain a WFI product for a syringe filling operation employing USP grade water as the feedstock. To obtain the requisite purity (1.3 μs/cm), laboratory quality control studies have determined that a purity level of 15 ppm is required. The water produced at the facility is USP grade and, on average, contains 400 ppm impurities, primarily metallic ions and cfu material. The USP water is fed to a forced circulation evaporator at the rate of 30 gal/min at ambient conditions. Calculate the hourly WFI volume produced via the evaporation operation.

Solution
Equation (2.45) and a feed rate of 30 gal/min will serve as the basis of calculation for this example. With a feed fraction of 400 ppm and a product of 15 ppm, the ratio of x_f/x_p is 25 ppm/400 ppm = 0.0375. Since the dimensions cancel, the value 0.0375 is dimensionless. With a feed of 30 gal/min, insertion of variables in Equation (2.45) yields $E = (1 - 0.0375)30$ gal/min = **28.875 gal/min** ≈ 29 gal of WFI is produced via evaporation. For an hourly amount,

$$\left(29 \text{gal} / \text{min} \right) \left(60 \text{min} / \text{h} \right) = 1740 \text{gal} / \text{h} \leftarrow \textbf{Solution}$$

While the previous example is not very difficult to solve, it points out an operation for a typical process design scenario. Even the most complex process design systems are actually composed of individual problem-solving procedures that, when addressed in an orderly format, will result in a system when once installed can operate within the specified user requirements. This is particularly relevant in pharmaceutical water system design and installations. The era of large pharmaceutical in-house process design groups is rapidly disappearing, as pointed out earlier. Consequently, as-built systems (often modularized, depending on size) are becoming the standard. Successful installation and operation is very much dependent on the facility personnel being capable of evaluating the fundamental elements of the installation (e.g. evaporators, ion exchange units, reverse osmosis units) in a judicious and expeditious manner. Often, the project engineer is the only facility representative capable of reliably performing an accurate, timely evaluation. In fact, if a new installation or extensive project improvement is planned, the staff project engineer or project manager may be the only individual with the experience and broad knowledge base to understand the total project venue.

2.6   Ion Exchange Systems

Ion exchange resins are usually spherical, amber-colored synthetic polymers that, though insoluble, mimic or react much as acids, bases, or salts. The resins differ from acids, bases, or salts in that only cations or anions are free to take part in chemical reactions. Those ion exchangers in which the anionic portions are able to react are anion exchangers (negatively charged ions such as SO_4^{++}, Cl^-, and NO_3^-); those in which the cationic (positively charged ions/metal ions such as Na^+ and Mg^{+2}) portions are able to react are cationic exchangers. In aqueous media, such as pharmaceutical water, ion exchange resins are able to exchange their cations (or anions) for other cations (or anions). For cation exchange resins, the chemically active constituent of the resin is often a sulfonic acid derivative. And for anion exchangers, the active constituent is often a quaternary ammonium compound.

The nature of the exchangeable ions associated with a given ion exchange is the responsibility of the user. However, with a large number of ion exchange systems being supplied by vendors as part of "as-built" installations, the user often has no input as to the efficacy of the ion exchange installation supplied. Regardless of the type of ion exchange system used (i.e. in-house design or vendor supplied), ion exchange installations have certain standard design criteria. For instance, an ion exchange resin is required to exchange sodium ions with other cation, such as magnesium, in solution. To accomplish this, the ion exchange solution (ion exchange resins in aqueous solution) is initially injected with a solution containing a concentrated dose of sodium ions (a concentrated solution of sodium chloride – brine solution – is most common for this application). This operation is referred to as regeneration. Similarly, if an anion exchange operation is desired that will replace ions such as carbonates (CO_3^{+2}), the resin is first treated with a basic solution such as caustic soda (NaOH (sodium hydroxide)). By repeating the operation at predetermined intervals, it is possible to use the ion exchange continuously by employing the ability of ion exchange resins to regenerate the exhausted species, whether it is a cation or an anion.

With the advent of smaller pharmaceutical water production units, regeneration is becoming a less viable option, and, in many cases, the operator employs a maintenance contractor, often the installation vendor, to be responsible for ion exchange system replacement and maintenance using a contract arrangement. Subsequent to ion exchange unit replacement, the contractor will, typically, regenerate the resin unit at the contractor, or contractor designated facility, and places the unit back in operation.

An understanding of ion exchange design principles requires the introduction of some basic conversion factors and terms that will be briefly reviewed. A fundamental relationship for ion exchange studies is the association between parts per million and percent. A solution containing 10 000 parts per million (ppm) is the equivalent of a 1% solution. Similarly, a solution containing 20 000 ppm is a 2% solution and a solution of 50 000 ppm is a 5% solution. Another important relationship is the relationship between the grain (gr) and the pound. One pound of material contains 7000 gr of the same material. Mathematically, the conversion factor would be

$$\frac{7000\,gr}{lb} \tag{2.59}$$

Other concepts required for ion exchange calculations are equivalent weight, equivalents, molarity, and normality.

Perhaps the most effective technique applicable to understanding the ion exchange operation is to use a common operation, in this case water softening by ion exchange.

There are two basic types of water hardness: bicarbonate hardness, or temporary hardness, and permanent hardness. Temporary hardness, the most common type, is identified by the presence of either magnesium bicarbonate or sodium bicarbonate in water intended, in this case for pharmaceutical applications. Without treatment, magnesium and sodium bicarbonates tend to build up insoluble scale in pipes, valves, and other critical components. Also, without treatment, calcium bicarbonate or magnesium bicarbonate can become impurities in pharmaceutical preparations using pharmaceutical water as a solvent or cleaning agent. The scale formation occurs when untreated water is heated. The chemical reaction for this phenomenon is

$$Ca(HCO_3)_{2+} \text{ Heat } \rightarrow CaCO_3 + H_2O + CO_2$$

The process involved in softening water is simple and straightforward. Hard water flows through an ion exchange column consisting of sodium containing cation exchange resin beads. The sodium replaces the magnesium and calcium ions in the hard water. If the ion exchange resin is identified by the term IXR, then the following relationship applies:

$$CaSO_4 + 2IXR(Na) \rightarrow IXR_2(Ca) + Na_2SO_4$$

For removal of magnesium, the applicable chemical operation would be

$$MgSO_4 + 2IXR(Na) \rightarrow IXR_2(Mg) + Na_2SO_4$$

When the exchange capacity of the ion exchange resin is exhausted (i.e. the ability to exchange calcium or magnesium is gone), the ion exchange column is then regenerated using the sodium chloride (brine) solution. When used, regeneration of the resin allows for continuous reuse of the ion exchange rein without significantly affecting ion exchange resin performance for an extended period. For this ion exchange operation, it is not uncommon for ion exchange resins to have an exchange capability of over 35 kgr of hardness (measured as calcium carbonate-CaCO3) per cubic foot of resin; greater efficiency is available when the ion exchange is achieved if the resin is regenerated.

2.7 Reverse Osmosis

RO is a common unit operation often employed in the manufacturing of pharmaceutical waters. While it is often used to produce grades of pure water, its limitations appear to be limited to water below the WFI grade. In fact, water produced through RO is prohibited from use as WFI in several countries outside the United States. However, because of the prevalence of installed RO operating units, it is important to understand the principles and operation of RO systems. Also, it would be of value to understand RO system basics when new systems are to be installed and/or designed. A capability to perform basic calculations associated with RO design and operation is also germane to understanding RO as a unit operation in pharmaceutical water systems.

Essentially, RO is an operation intended to remove dissolved ions from a solution. As is the case with most RO operations, RO pharmaceutical operations also involve removal of ions from aqueous solution. RO systems are not intended to remove solids or suspended solids from an aqueous solution. In fact, exposing the major component of RO systems, the module membrane, to solids in the solution severely degrades the performance of the RO system. In fact, before introducing feedwater into the RO system, it is common practice to have the pharmaceutical feedwater pass through several filtration operations that produces a feedwater of a solids content of, often, less than 0.3 μm.

RO membranes are fabricated as cylindrical modules, usually spiral wound (sheets of membrane sandwiched with mesh spacers, connected with and wound around a centered permeate tube). The membrane can be made of cellulose acetate, polysulfonates, or aromatic polyamides. While cellulosic membranes are the most inexpensive and the oldest type currently in use, some of the newer RO membranes offer several advantages depending on the application. For instance, cellulose-based membranes are the lowest in cost but not as effective in removal of certain organics. Figure 2.7 exhibits a typical spiral wound RO membrane.

The membrane selection is also a factor of the feed pH. In general, the pH for RO membranes is in the 4–11 range. This presents no difficulty for pharmaceutical waters since the feed and product pH is much narrower limits, usually 6–8. As for temperature, a cellulose acetate membrane and aromatic polyamides should not exceed 35 °C (95 °F). Some newer membrane types, such as thin film composites, can operate at higher temperatures, but not exceeding 45 °C (113 °F). While cellulose acetate membranes have a high tolerance for chlorides in the feedwater, newer membrane materials such as aromatic polyamides and thin film composites do not exhibit such chloride and other oxidant chemical tolerances. However, as will be covered, an important performance factor, biodegradability, is considerably higher for cellulose acetate membranes. This means that for pharmaceutical operations that exhibit higher cfu values, the use of cellulose acetate membranes should be carefully evaluated prior to acceptance as an RO membrane. Other membrane materials such as polyamides and thin film composites have lower tendencies to biodegrade. While this does not appear to be significant for pharmaceutical applications, it should be taken into consideration if the installation is new. While membranes are a very small part of a total RO installation, typically on the order of $200 000 or more for large-scale systems (in the 200–300 gal/day range), the membrane selection can be an indication of the quality of the subcontractor selected to install the system. Similarly, since maintenance of the RO system is usually performed by contractors, the quality of membrane selection can be an important factor in maintenance organization performing the assigned task.

In addition to membrane biodegradation as an important failure mechanism, the critical issue of membrane fouling is also most important. In fact, because of the tendency for membranes to foul during processing, it is standard practice to place RO systems at or near the end of the pharmaceutical water process. At this point in the operation, the water entering the process is already quite pure, when compared with typical minimum required RO feedwater.

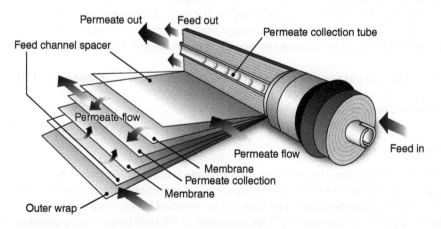

Figure 2.7 Typical reverse osmosis membrane. *Source:* Puretec Water.

Figure 2.8 Flow diagram of a reverse osmosis system. *Source:* Culligan.

RO systems are very basic installations. A typical configuration consists of several cylindrical membrane modules mounted in a parallel or combination of series and parallel modules. The membrane diameter can vary from 2 to 4 in. in diameter. For pharmaceutical water manufacturing, a 4 in. diameter is common. While the length of the membrane cylinder varies, membranes with a length of about 40 in. are typical. Basically, the feedwater is pumped into the modules. The feedwater is then pumped into the membrane, and the feedwater is then forced radially through the membranes with the water flowing through the membranes and the impurities failing to go through the membranes. The impurities are discharged as waste from the membrane modules. The now purified flow is discharged through the center cylinder for further processing or transported to a storage tank. Generally, the purified water is exposed to ultraviolet (UV) light to destroy any surviving microbes prior to further processing or storage.

For operations and design purposes, a typical RO operation designed to manufacture USP purified water, 800 gal/day, can be used as a basic design criterion.

Figure 2.8 illustrates a typical RO system flow diagram.

2.7.1 Principles of Reverse Osmosis

The term reverse osmosis implies the existence of a process known as osmosis. Osmosis is defined as "The passage of a solvent through a semipermeable membrane from a less concentrated solution to a higher concentration." The flow can be reversed by applying external pressure on the volume of higher concentration. This operation is referred to as reverse osmosis. For pharmaceutical water applications, RO, when applied to dilute water solutions, is of importance since RO is a widely employed unit operation. An important phenomenon associated with osmosis and RO is identified as osmotic pressure. Perhaps the best definition of osmotic pressure is the one formulated by Enrico Fermi and can be stated as

> The osmotic pressure of a dilute solution is equal to the pressure exerted by an ideal gas at the same temperature and occupying the same volume as the solution and containing number of moles equal to the number of moles of the solutes dissolved in the solution.

RO, as noted, is a relevant operation in producing purified water. For design and project purposes, a quantitative relationship is needed to perform these tasks. A reasonably accurate paradigm to identify the relationship between osmotic pressure and solute concentration could serve as an effective basis of calculation. Such a relationship does exist. This formula, developed by van't Hoff, was developed in 1901, for which he was awarded the Nobel Prize. The van't Hoff relation is given as

$$\pi = cRT \tag{2.60}$$

where

π = osmotic pressure
c = concentration
R = ideal gas constant, 0.080 203 (liter bar)/g mol K
T = temperature, K.

It can be observed that the osmotic pressure relationship defined in Equation (2.60) is very similar to the ideal gas relationship, $PV = nRT$, where n is the number of moles under consideration. If the concentration is defined as c/V or moles per unit volume, the ideal gas law would be equivalent to the osmotic pressure. Additionally, it should be understood that the calculated osmotic pressure, π, is the applied pressure required to have the RO theoretically proceed. In reality, the designer would specify a larger pressure to overcome details such as membrane efficiency, pumping efficiencies, and membrane fouling and scaling.

Membrane fouling is arguably the major cause of RO system failure. Fouling and scaling are generally caused by a partizcular level of total dissolved solids (TDS) in the feed solution. While there are no fixed levels of particulates that are defined in the RO feed, guidelines have been recommended to minimize fouling of the membrane [2]. Table 2.12 identifies guidelines that are intended to minimize membrane fouling and scaling in RO membranes.

Aside from the basic design equation, given as Equation (2.60), there are several terms associated with RO. The most common terms are:

- Feed – Water entering the RO system
- Permeate – Portion of the feed that passes through the semi permeable membrane and is collected as the product for water treatment
- Concentrate – Also identified as reject, the portion of the feed that does not pass through the membrane and is processed as a separate stream containing impurities and generally discharged through the facility nonhazardous drains
- Recovery – Ratio of permeate to the feed rate

Perhaps the best format to understand the basic equation (2.60) while being an uncomplicated linear relationship is to perform a straightforward non-pharmaceutical situation. As will be shown shortly, using RO for purified water production has some special feedwater details, which should be evaluated prior to implementing an RO system.

Example 2.15 A study is being performed to use a reverse osmosis (RO) system to desalinate seawater. Samples of the seawater indicate that the major constituent is sodium chloride (NaCl). On average, the NaCl content is 3.1%. The other constituents of the seawater are not economically important for the operation. The preliminary evaluation of the proposed system requires that the approximate osmotic pressure be determined in order to approximate the pump requirements. Determine the pressure required to operate the RO system. Assume the operating pressure is 68 °F (20 °C) and the density of seawater as 1.028 g/ml.

Solution

To determine the minimum osmotic pressure, the basis of calculation will be Equation (2.60), $\pi = cRT$. With two variables identified, the concentration must be determined. For osmotic pressure, the concentration is given as molality, moles of solute/kg solvent. As identified previously, the molecular weight of sodium chloride (NaCl) is 58.5 or 58.5 g NaCl/g mol NaCl. Also, 3.1% sodium chloride content translates to 31 000 ppm sodium chloride or 31 000 mg NaCl/kg or 31 g/1000 l of seawater.

31 g NaCl/58 g/g mol = 0.534–0.53 = molality of 3.1% seawater solution

Molality is a much less commonly used parameter than a related measurement, molarity. Consequently, it would be of value to determine the difference between molarity and molality. The molality is 0.53 for the seawater solution. The molarity defined as moles/liter of solution is the difference between a kilogram of solvent (1000 g) and a liter of seawater solution 1000 ml. Thus, the molarity of a liter of seawater is basically the difference in densities since the molarity is temperature dependent as opposed to molality. Thus, the molarity of the seawater solution is merely the molality multiplied by the density of the seawater or

$$0.53 \times 1.028 = 0.545 \sim \textbf{0.55 M}. \text{ The percentage difference is } \frac{0.55 - 0.53}{0.55}(100\%) = 3.64\%$$

This is a small difference when dealing with large quantities of water. For pharmaceutical applications, the impurities' level for feedwater entering RO systems is significantly less than the seawater example, as will be shown. Thus, when dealing with RO operations for pharmaceutical applications, it is more convenient to use molarity (0.55 M in this case) rather than molality since the deviation is very small. Actually, in the case of pharmaceutical water, the density of the RO feedwater will be closer to the density of pure water, making the deviation even smaller. Thus, for design and operating purposes, molality and molarity can be used interchangeably when dealing with dilute RO feedwater.

The required variables for determining the osmotic pressure are now in place:

$R = 0.0823$ liter bar/g mol K
$T = 68\,°F = 293.15–293.2\,K$
$c = 0.55$ mol/l
$\pi = (0.55\,\text{g mol/l})\ (0.0823\,\text{liter bar/g mol K})\ (293.2\,K) = 13.3\,\text{bar} = 13.3\,\text{atm} \times (14.7\,\text{lb/in}^2\,\text{atm}^{-1})$
$= 195.5 \approx \textbf{196 lb/in}^2 \leftarrow \textbf{Solution}$

(While a bar is considered to be the equivalent of 1 atmosphere or 14.7 lb/in², the actual conversion is 1 bar is the equivalent of 14.504 lb/in². This would result in a value of 193.4 lb/in² which there is a difference of less than 1.4%, a negligible value when dealing with relatively high pressures.)

The previous example is given as an example of the computational method employed to determine the osmotic pressure of a solution undergoing an RO operation. Actually, seawater undergoes significant pretreatment prior to the actual RO operation. As will be seen, RO feed must be relatively free of solutes (i.e. the feedwater is relatively free of significant impurities) to operate effectively.

Since pump curves express pressure in feet of head, to size the pump required, the pressure must be converted from lb/in^2 to ft of head. By rearranging Equation (2.8) the result is

$$\text{Fluid head, ft} = \frac{\left(lb/in^2\right) \times (2.31)}{S.G.}$$

Since the pumping fluid is purified water, the specific gravity (SG) is 1.00. Substituting values results in

$$\text{Fluid head} = 196 lb/in^2 \times 2.31 ft\ head/lbin^{-2} = 452.76 \approx \textbf{453 ft of Head} \leftarrow \textbf{Solution}$$

The calculated result identifies a pump with a rather large head requirement.

The value of 196 psi (453 ft) is the theoretical value required to achieve operation of an RO system. This pump pressure is required to specify the size of a centrifugal pump needed to operate the system.

For purified water applications, the pump requirements for an effective RO system are significantly different than that of a seawater system. The difference is due to the solids level (molality and molarity of the solutes in purified water is significantly lower than seawater feed). For purified water manufacturing, the values of molarity and molality are considered to be identical, when dealing with dilute feed systems, generally less than 1000 ppm dissolved and suspended solids

2.7.2 Reverse Osmosis Installation and Operational Costs

For virtually all industrial processes, a basic understanding of installation and O&M costs is an imperative reporting requirement. Without such recorded data, it would be virtually impossible to ascertain if the process is economically profitable. Most often, this type of information is provided on an annualized basis or a fraction of this variable, e.g. $/year, $/week, etc.

Another format in which operating costs can be presented is as an annual fraction or percent of total installed cost of the process or the total capital expense (Capex) is also a convenient presentation format.

A practical example of O&M costs for an installed RO operating system is to focus on a 367 gal/min industrial RO unit with an estimated 10-year operating life. The unit is designed to produce a 75% product recovery during the unit's operating lifetime. Over a 10-year operating period, the total critical expenses can be estimated as [7]

Capex (installed cost) = $521 195
Energy = $480 727
O&M = $1 148 252
Reject (waste) disposal cost = $6 436 578
Total RO system cost = $8 586 752

The total O&M cost over the unit lifetime can be given by O&M/total cost = $1 148 252/$85 86 752 = 0.13 × 100% = 13%. This value, 13%, represents the approximate percent of the total cost required to maintain the RO system in a production mode over a 10-year period, or about $114 148 yearly, or about 1.3% annually. A large fraction of these costs are items such as membrane replacement, typically performed at scheduled intervals, membrane replacement labor costs, pump maintenance including scheduled preventive maintenance inspections, pump part replacement, and routine maintenance and instrument inspection and calibration. Routine maintenance commonly involves checking the piping connections since the RO unit operates at a significantly higher pressure.

Since many of these costs are categorized as overhead or non-chargeable (labor maintenance cost is a common example), the true costs are often not known in many actual O&M reporting results.

This O&M cost is significant since, often when total project cost estimates are prepared, provisioning for spares and maintenance support may not be addressed in the original cost estimates and consequently final funding and operating requirements.

Chapter 10 identifies the need to provide system spare parts lists for an installed system, such as an RO operation. However, the spares list should also provide preliminary costs for the parts and components identified. Delivery schedules for long lead time items would also be an important asset.

2.7.3 Reverse Osmosis Design Hint

One of the positive aspects of RO system specifications and design is the fact that system scale-up is essentially a linear paradigm. That is, typically sophisticated hydraulic and similitude calculations are not required to scale up a smaller operation, such as a pilot plant to an actual large production operation. The primary rationale for this conclusion is predicated on the basic design equation for an RO operation. This relationship is given by Equation (2.60) ($\pi = cRT$). By observation, it can be observed that the relationship is linear. As such, estimating variables such as flow rates, production output, and even maintenance costs can be estimated, with reasonable accuracy, by simple "ratio and proportion" or linearity (actually, virtually any continuous function can be approximated by a straight line, linearity, if the difference between two values is small enough).

Because of the basic linearity of RO scale-up, one can estimate, with reasonable accuracy, the approximate annual O&M costs for such a system. A reasonable estimate is to use about 5% of the total system installation as an annual O&M expense. Since many installations use a 10-year depreciation, this value, 5%, is a reasonable value for budgeting purposes.

2.8 cGMP Design and Facility Maintenance Considerations for Pharmaceutical Water Systems

In addition to verifying that a pharmaceutical manufacturing installation has been designed with cGMP guidelines as a priority, it is also important to ensure maintenance operations adhere to the same criteria.

Once the cGMP installation has been accepted by the owner representative, effective O&M is no longer the responsibility of the vendors involved in the original unit, unless post-installation maintenance support arrangements have been put in place. Regardless, the owner is responsible, subsequent to installation turnover. Subsequent to turnover, the facility is responsible for maintaining the performance and reliability of the system (further information on this topic is found in Chapter 5). By owner acceptance of the completed installation, the system has been confirmed, by the owner, as conforming to cGMP design criteria specified. It could be useful if a set of factors were available to consider cGMP issues relating to design and maintenance. Consequently, itemized below is a listing of cGMP considerations that should be considered in the design and maintenance of USP water systems.

Table 2.13 Properties of saturated steam.

Gauge pressure (psig)		Temperature (°F)	Heat in Btu/lb			Specific volume (ft³/lb)
			Sensible	Latent	Total	
INS. VAC.	25	134	102	1017	1119	142.0
	20	162	129	1001	1130	73.9
	15	179	147	990	1137	51.3
	10	192	160	982	1142	39.4
	5	203	171	976	1147	31.8
	0	212	180	970	1150	26.8
	1	215	183	968	1151	25.2
	2	219	187	966	1153	23.5
	3	222	190	964	1154	22.3
	4	224	192	962	1154	21.4
	5	227	195	960	1155	20.1
	6	230	198	959	1157	19.4
	7	232	200	957	1157	18.7
	8	233	201	956	1157	18.4
	9	237	205	954	1159	17.1
	10	239	207	953	1160	16.5
	12	244	212	949	1161	15.3
	14	248	216	947	1163	14.3
	16	252	220	944	1164	13.4
	18	256	224	941	1165	12.6
	20	259	227	939	1166	11.9
	22	262	230	937	1167	11.3
	24	265	233	934	1167	10.8
	25	267	234	933	1169	10.3
	26	268	236	93.3	1169	10.3
	28	271	239	930	1169	9.85

Source: Data from Spirax Sarco, Inc.

These considerations are necessary to ensure the cGMP water installation conforms to the provisions of 21 CFR 211.113, which states:

a) Appropriate written procedures, designed to prevent objectionable microorganisms in drug products not required to be sterile, shall be established and followed.

b) Appropriate written procedures designed to prevent microbiological contamination of drug products purporting to be sterile, shall be established and followed. Such procedures shall include validation of any sterilization process.

In recent years, due to several reasons, the pharmaceutical industry has experienced the evolution of shrinking facility size, the growth of biologics, and an emerging market demand requiring a modification of standard design and maintenance practices. While design criteria such as facility scale-up paradigms may vary, the elements identified later are applicable to all pharmaceutical facilities covered by cGMP criteria. In fact, with facility downsizing becoming more common, the importance of addressing these facility and maintenance considerations may be more germane since these considerations may be more relevant since the smaller operation may magnify a defect that could go unnoticed in a larger operation:

1) Evaluate the piping installation for the presence of any dead legs that may be present. Any pipe length in excess of six pipe diameters with no branches or tees should be evaluated. Dead legs are undesirable, because they present the possibility of water stagnation, and potential biological builds up. Also, when performing periodic sanitization and flushing procedures, avoid disconnecting piping or system components. Exposing a closed system to the operating environment introduces the possibility of contamination due to bacteria and particulates in the environment (while generally recognized as poor practice, it must be emphasized that dead legs should not be employed to introduce cleaning and sanitization chemicals into the pharmaceutical water system). Also, it is worth noting that the FDA has issued warning letters concerning piping systems that have dead legs as part of the installation, and this is a common FDA inspection item.

2) If the system contains a carbon bed, a storage tank, and associated distribution and transmission lines, procedures and equipment should be available to permit scheduled sanitization and finishing of the system. If required, back-washing operations should be provided.

3) A sufficient number of sample points should be considered for the system. The sample points should be provided for the incoming city water (or facility provided water), and each individual operation employed in the water manufacturing operation. The sample points should also be located after each filter and use point.

4) Overflow lines from storage tanks, carbon beds, ion exchange units, and similar equipment should be connected to the floor drains and equipped with a backflow preventer, if necessary. Similarly, items such as cGMP laboratory sinks should also be equipped with backflow prevention.

5) Routine maintenance inspections should be performed to check for leaking connections, another entry point for microbial buildup.

6) If a carbon bed is in the system for chlorine removal, the unit should be equipped with a backwash capability designed to perform the backwash function at designated intervals.

7) The pharmaceutical water line segments should be sloped with a drain capability at the lowest points.

8) Potable water hoses connected to cGMP components should be capped when not in use.

9) Piping should be fabricated of approved materials. Approved materials include stainless steel, PVDF, and polypropylene. Polyvinyl chloride (PVC) is not employed in pharmaceutical applications because of the leaching potential and fume hazards posed by fire.

10) In-line instruments should have sanitary connections to minimize the chance of contamination. This is particularly germane if the finished product is higher purity such a WFI. In this case, threaded connections should be avoided. It is advisable to avoid possible gaps in the instrument connection.

11) Consider installing diaphragm valves in the process lines and sanitary valves for finished water lines.

12) All lines should be identified as well as indicate the direction of flow.

13) Fine filters, such as $0.2\,\mu m$ filters, should not be employed in recirculation lines of finished product. Filters smaller than $2\,\mu m$ should be used for manufacturing pharmaceutical water specifically and should not be part of the recirculation system.

14) Installations such as ion exchange columns and carbon beds should have bypass capabilities. A bypass configuration is desired so that water, during processing, will not cause reduced flow that can cause stagnation and potential microbial growth. A bypass can also allow maintenance or inspection of the unit without significant reduction in the operation.

15) If the water system is a once through operation (i.e. a system that is not equipped with a storage tank or a recirculation mode), the system design should preclude locations where water stagnation can occur. This is particularly relevant at locations upstream of the chlorine removal operation (i.e. activated carbon units).

16) The system should be configured to prevent backflow of water at process locations where different water qualities intermingle.

17) Recirculation pumps for water to and from storage tanks should be equipped with sanitary design features to minimize microbial buildup.

18) Atmospheric vents (i.e. breather vents) on the water storage tank should be equipped with filters to minimize microbe entry via the ambient environment. The one component that generates considerable concern is the vent filter. A testing program should be established to confirm the integrity of the filter and verify the filter construction is intact. For ease of testing and inspection, the vent filter should be located in a position on the holding tank that allows for convenient accessibility. Evaluating a vent filter containing a desiccant capability to absorb moisture should be considered.

19) The pharmaceutical water production location should be of the same sanitary level as the water produced. Smaller water manufacturing units should not be located in isolated rooms/locations where microbe accumulation is not monitored. This detail is more relevant if the water unit undergoes periodic sanitization and line flushing. During the sanitization and flushing of the lines, the piping is often opened at line joints. This could expose the piping to microbial infestation from the surrounding environment.

20) Feedwater should be tested daily to confirm the quality conforms to acceptance criteria. Any change in water quality or impurity content and concentration should be recorded and reported. On occasion, municipal water may come from a different source (e.g. changing from a reservoir source to an underground well or receiving water from a different reservoir in the system, not uncommon during a drought); this change can alter the performance of the ion exchange units. A change in feedwater can alter the performance of the ion exchange units since they are usually designed to process water with a standard ion concentration. Cations and anions of a higher than anticipated concentration can cause early exhaustion of the columns and can cause significant quality problems. While this element may be an accepted approach for larger manufacturing operations, smaller suppliers, such as manufacturers of active pharmaceutical ingredients (API) and excipients, should be aware of these subtle factors and requirements.

21) In addition to an effective incoming water sampling, documenting, and sampling plan, the SOP should contain procedures for cleaning, maintenance, testing, and calibration, as required in 21 CFR 211.67.

22) The water quality plan should include alert levels for microbial content in the water as well as a microbe testing program. Particular attention should be given to the quality of the manufactured water at use points and sampling ports. Also, sampling ports should be at locations that are representative of the use points (i.e. sampling ports should not be in an isolated location).

23) An approved cleaning/sanitization or flushing procedure should be in place with results documented.

24) For processed water storage tanks, a temperature monitoring system should be installed. This is particularly relevant if the storage tank is located outside a building. The temperature

monitoring system should be equipped with an alarm system intended to actuate when the water temperature not within the acceptable range.

25) Ensure that the equipment used in the manufacture, processing, packing, or holding of drug products is of appropriate design and adequate size and suitably located to facilitate operations for its intended use and for its cleaning and maintenance as required by 21 CFR 211.63.

26) The pharmaceutical water system P&ID should be current and readily available with all changes documented. Any FDA warning letters relating to water systems installed in company facilities should be introduced to operations personnel and apprised of the defects to make necessary changes to facility operations, if applicable. This design element is relevant since facility operations can employ corporate engineering to provide or assist in pharmaceutical water projects at facilities. Should this be the case, there exists the possibility that corporate engineering failed to properly address the 483 issues uncovered.

27) Finished water piping lines should not have sagging piping. Pipe supports should be located at points consistent with approved spacing locations as specified in Chapter 9.

28) Scheduled monitoring and maintenance on the pharmaceutical water system should include replacement of prefilters, mixed bed filters, carbon filters (primarily for removal of chlorine and organics), ion exchange resin units (including units performing resin regeneration service), and filters. Replacement of ion exchange units should be predicated not only on conductivity of the effluent but also on microbial loading and scheduled testing and analysis. Scheduled or unscheduled filter replacement should also include microbial testing and analysis.

29) Installed UV lamps should have a maintenance schedule identifying documented daily, weekly, monthly, and annual maintenance procedures to be performed. Scheduled activities should include:
   - Cleaning of the glass lamp and protective cover.
   - Scheduled calibration check, usually every 6 months.
   - UV lamp replacement, annually, unless a problem emerges.
   - Measurement of lamp intensity.
   - Documentation of approved maintenance and preventive maintenance activities. If the facility is equipped with a computerized maintenance management system (CMMS), UV maintenance (including preventive maintenance) should be a standard scheduled activity. Documenting of maintenance procedures is specifically identified in 29 CFR 211.58 ("Maintenance") and 21 CFR 211.100 ("Written procedures; deviations")

30) In addition to maintaining documentation for equipment and procedures for UV components, additional documentation (validated CMMS inputs and formal logs) should be maintained for cGMP components, equipment, and procedures. Also, approved documentation must be maintained for sanitization, equipment replacement (i.e. change control and training), and maintenance activities. This documentation must be easily accessible to the FDA or FDA representatives upon request.

31) If chemical sanitization is employed (typically, a solution of 6% hydrogen peroxide, formalin, or an approved equivalent, an antimicrobial agent, or a steam cleaning operation), an analysis should be performed for chemical residue, subsequent to flushing.

32) If a significant procedural or system modification is performed, the pharmaceutical water system must be revalidated following the implemented change. Additionally, as is the case with any cGMP change, training updates must be provided for personnel involved with the change.

33) Ball valves should not be incorporated in pharmaceutical water or any cGMP installation. They present the possibility of water staying in the crevices of the ball and the wall when

the valve is closed. The situation is exacerbated if the valve remains closed for a period of time, permitting the buildup of microbes and a biofilm.

34) The minimum velocity of water through the piping and equipment configuration should be about 6 ft/s, and to prevent against erosion, the maximum velocity should not exceed 15 ft/s.

35) If the feedwater is from a municipal water system, reports from the municipality testing should be considered in lieu of in-house testing.

36) If washer hoses are employed, filter attachments should be affixed to the end of the hoses after use. If a filter is not employed, the inside of the hose is exposed to ambient air and, depending on area classification, may be exposed to microbial buildup for the inactive period. Another viable solution is to have a secondary valve in the hosing line and approved operational procedures that provide for opening the secondary valve before the primary and flush prior to use.

37) Incoming municipal water should be sampled, periodically, to determine the silt density index (SDI) of the samples during seasonal changes. Sample temperature should also be recorded, if suspended solids content is not provided by the municipality (see item 35).

38) When installing a water system, it is most important that the product water conforms to the actual need. If a water system produces a WFI grade for an operation for which, at most, a deionized water quality is needed, this activity is, again, pouring money down the drain. Such a situation can occur when an older unit, originally intended for a high purity application, is no longer used for the original intended application. For such a situation, the quality specification is often not altered for a variety of reasons. One reason is the often draconian task of preparing a change control document for a reduction of the water quality specification for the operation.

39) As-built installations are not custom designed to conform to the user's needs. While, in most cases, the installed system meets the operating requirements, it is most important to maintain involvement in the installation implementation from the design, construction, and installation phases of the project and culminating in commissioning and validation. Extensive reliance on the equipment supplier should be carefully evaluated.

40) If an RO installation is part of the operating system, including cGMP installations, the operating flux should be producing product within the specified flux parameters (gallons of permeate per square foot of membrane surface area per day). It is important to note that flux varies by the type and temperature of the water being treated. Also, a higher operating flux increases the probability of membrane fouling, a major cause of RO system failure. With a higher membrane fouling frequency, the costs for system maintenance and membrane replacement increase.

41) Proper sizing of water system components is a critical design consideration. For instance, water exhibiting high turbidity or temperature excursions during a cold spell can degrade water treatment system components that are installed prior to the ion exchange and/or RO units. To minimize the possibility of these units creating a bottleneck, it would be advisable provide a design capable of easily adding extra capacity. Typical expanded capability could include installation of extra ion exchange resin columns if there is no ion exchange regeneration. For resin systems with a regeneration capability, the flux through the column should not exceed 12 gal/ft^2 min.

42) Filter selection is an important consideration in cGMP water production operations. In addition to removing suspended solids, the filtration equipment should be capable of removing matter to about 0.3 μm. Usually, a water filtration train employs a 5 μm filter subsequent to the mixed bed filter, if installed, and the activated carbon filter, primarily for chlorine removal. A 2 or 3 μm cartridge filter should be installed at the discharge of the ion exchange units. The primary purpose of these filters is to capture any ion exchange resin

beads or fines that may be released into the water treatment steam. The filter units usually terminate with a 0.3 µm filter prior to entering the RO installation.

43) Confirm the RO pump is adequately sized for the unit. Just as a pump that significantly exceeds actual operating requirements, a pump that is undersized can cause operating damage. An oversized pump often uses excess electric power, and if significantly oversized, pump cavitation can occur if the treatment train flow, over time, is reduced due to diminished water demand. If reduced production does occur, pump or parts (smaller impeller installation) change-out is rarely addressed.

44) Confirm new installations are compatible with existing units. If a new RO installation is planned or currently installed, it is most important that the new system operates efficiently with the existing unit or units. If a new RO system has a significantly larger input flow, verify that additional flux needed for the ion exchange units (installed downstream of the RO unit) are within the specified ion exchange column operating specifications. Failure to address this item can result in premature resin exhaustion and more frequent resin replacement or regeneration, depending on operating mode. If a new unit is equipped with an automated control system, it is very important that new automated system operates compatibly with the older control system.

45) Regardless of the claimed degree of system automation, all pharmaceutical water systems require a degree of personnel involvement. At a minimum, trained operating personnel and maintenance personnel will always be required. This should be a standard procedure regardless if cGMP or non-cGMP systems are at issue. Maintenance-related downtime (i.e. equipment repair and/or parts replacement) is very often the major impediment to operations and causing significant downtime.

46) When considering a water purification system, consideration should be given to the extent of point of integration of the system with existing operations. For instance, if a laboratory water system is to be installed, the pretreatment operation could be very important. While the current system may have a pretreatment deionization capability as a standard operation, reliance on others to maintain the pretreatment deionization and related operations may be problematic. Consequently, integration of a cGMP water system should be evaluated regarding the point of system integration. For a new installation, such as laboratory water, the pros and cons of a total stand-alone system are advisable. Additionally, if a system retrofit is implemented, it is most important that details such as pipe insulation and hot weather water condensation on chillers and/or condensers be addressed. Failure to incorporate these considerations can result in significant added maintenance expense, particularly if bare piping is in an overhead location in a cGMP location.

47) Avoid stagnation of water in storage tanks for purified water and WFI. Purified Water should be in a continuous circulating mode. In this scenario, the possibility of microbe buildup in a static storage situation is minimized. The possibility of leaching from a container wall, depending on the tank material of construction, is also minimized. In a constant circulation mode, the residence time that allows possible leaching also is minimized. While a storage tank is a standard component in many purified water systems, the need for continuous recycle is a design consideration.

References

1 Smith, R.N. and Pierce, C. (1955). *Solving General Chemistry Problems*, 1e. San Francisco: Freeman and Company.
2 Westaway, C.R. and Loomis, A.W. (1979). *Cameron Hydraulic Data*. Woodcliff Lake, NJ: Ingersoll-Rand.

3 Hougen, O.A., Watson, K.M., and Ragatz, R.A. (1959). *Chemical Process Principles, Part II, Thermodynamics*. New York: John Wiley & Sons, Inc.

4 Faust, A.S., Wenzel, L.A., Clump, L.A., et al. (1967). *Principles of Unit Operations*. New York: John Wiley & Sons, Inc.

5 The Barnstead Company (1971). *The Barnstead Basic Book on Water*. Boston: The Barnstead Company.

6 Zamansky, M.W. (1957). *Heat and Thermodynamics*. New York: McGraw-Hill.

7 Stover, R.L. (2014). A primer on reverse osmosis technology. *Chemical Engineering* 121 (7): 38–43.

Further Reading

Adams, J.N. (1997). Quickly estimate pipe sizing with Jack's cube. *Chemical Engineering Progress* 93 (12): 55–58.

American Institute of Chemical Engineers (AIChE) (1962). *Student Member's Handbook*. New York: American Institute of Chemical Engineers.

American Institute of Chemical Engineers (AIChE) (1978). *Practical Aspects of Heat Transfer*. New York: American Institute of Chemical Engineers.

Anderson, H.V. (1955). *Chemical Calculations*. New York: McGraw-Hill.

Avallone, H.L. (1986). High purity water. *Pharmaceutical Engineering* 6 (1).

Bennett, R.C. (1964). Recompression evaporation. *Chemical Engineering Progress* 71 (7): 67–70.

Brush, H. and Zoccolante, G. (2009). Methods of producing water for injection. *Pharmaceutical Engineering* 29 (4): 20–28.

Castellan, G.W. (1983). *Physical Chemistry*, 3e. Reading, MA: Addison-Wesley.

Challener, C.A. (2014). Pharma investments reflect key industry trends. *Pharmaceutical Technology* 38 (8): 18–24.

Daniels, D.G. (2009). Avoid these 10 mistakes when selecting your new water treatment system. *Power* 153 (9): 18–20.

Fermi, E. (1956). *Thermodynamics*. New York: Dover Publications.

Glaser, V. (2006). Minimizing the cost and use of purified water. *Genetic Engineering News* 26 (12): 38–40.

Granger, R.A. (1995). *Fluid Mechanics*. New York: Dover Publications.

Hitchcock, T. (2010). Impact of single-use biomanufacturing systems. *Genetic Engineering & Biotechnology News* 20 (10): 44–47.

Johnstone, R.E. and Thring, M.W. (1957). *Pilot Plant Models and Scale-up Methods in Chemical Engineering*. New York: McGraw-Hill.

Kittsley, S.L. (1955). *Physical Chemistry*. New York: Barnes & Noble.

Kucera, J. (2008). Understanding RO membrane performance. *Chemical Engineering Progress* 104 (5): 30–34.

Lee, D.A. (1976). Handy formulas for balancing materials. *Chemical Engineering* 83 (22).

Mackay, R. (2004). Preclude the pitfalls of pump piping. *Chemical Processing* 67 (3): 29–32.

Perry, R.H. and Green, D. (1987). *Perry's Chemical Engineers' Handbook*, 6e. New York: McGraw-Hill.

Riche, E.L. and Mabic, S. (2006). Basic considerations for laboratory water. *Genetic Engineering News* 26 (12): 42–43.

Robinson, R.N. (1987). *Chemical Engineering Reference Manual*. Belmont, CA: Professional Publications.

Rohm and Haas (1974). *If You Use Water*. Philadelphia, PA: Rohm and Haas.

Romanyshyn, G. (2010). Optimizing pumping systems. *Flow Control* 26 (7): 20–25.

Sandler, H.J. and Lukjewicz, E.T. (1987). *Practical Process Engineering*. New York: McGraw Hill Book Company.

Simon, A.L. (1976). *Practical Hydraulics*. New York: John Wiley & Sons, Inc.

Tankha, A. and Williford, H. (1987). GMP considerations in the design and validation of a deionized water system. *Pharmaceutical Engineering* 7 (4): 17–20.

The Crane Company (1988). Flow of Fluids Through Valves, Fittings and Pipe, Technical Paper 410, King of Prussia, PA.

U.S. Food and Drug Administration (FDA) (1993). *Guide to Inspections of High Purity Water Systems*. Silver Spring: U.S. Food and Drug Administration, Office of Regulatory Affairs (June).

3

Heating, Ventilating, and Air Conditioning

The basic HVAC system, including pharmaceutical operations, is composed of:

- Units and components that have provisions for heating specific facilities and/or areas within the facilities.
- Specific locations of a facility that require air conditioning.
- Locations requiring air filtration for particulates, vapors, gases, or a combination of these factors.

The establishment of basic engineering HVAC design criteria is fundamental to successful planning, implementation, design, and post-installation tasks (i.e. commissioning, validation, and operation and maintenance). While design criteria are critical to virtually all capital programs, it is particularly relevant for pharmaceutical HVAC and other pharmaceutical projects because of the cGMP regulations that must be addressed. Because validation activities, required for cGMP programs, are often a very detailed procedure, an extra cost factor must be considered. As such, guidance is often required to provide design consultants, vendors, project engineers, facility personnel, validation personnel, and, to a lesser extent, quality assurance operations with requirements and design conditions for an HVAC construction program.

The earlier the user requirements can be incorporated in the design and specification aspects of an HVAC cGMP project (or any cGMP project), the better it is from a cost savings standpoint. It is advantageous to implement and evaluate performance and design enhancements in the early project phase. Early recognition of potential improvements can produce significant savings. If design modifications must be performed during the latter phases of the program, costly change orders may be required to remediate unforeseen flaws. If such errors are noted in the early program phase, cost savings and reputation can be enhanced with a low risk. Thus, proposals to effect cost or operational savings at the early project phase, whether accepted or rejected, can only be viewed in a positive light since the design alteration proposal is being evaluated at basically no cost to the project since the time expended is only a few overhead hours.

Practical Pharmaceutical Engineering, First Edition. Gary Prager.
© 2019 John Wiley & Sons, Inc. Published 2019 by John Wiley & Sons, Inc.

Successful pharmaceutical HVAC programs rely on an understanding of several electrical, thermodynamic, and mechanical elements. While an in-depth knowledge of these specialties is most desirable, a working familiarity of the major operating principles and related equipment involved in pharmaceutical HVAC programs can be of immense value, particularly when dealing with vendor-supplied equipment and "as-built" designs and installations.

It has been estimated that HVAC equipment demand will increase at an annual rate of 4.5%. Cooling equipment is expected to increase at an annual rate of 4.6%, while heating equipment will increase at a rate of 3.9%.

A good deal of this increase is due to the impetus for more energy-efficient operating levels including criteria such as the Leadership in Energy and Environmental Design (LEED) program. The LEED initiative is a major international program involving third-party confirmation implement design factors such as enhanced energy efficiency, carbon dioxide emission reduction, and efficient use of required assets and resources.

Another significant reason for increased cooling requirements is the phaseout of halogenated refrigerants because of their ozone-depleting properties. Replacement refrigerants are less efficient than halogenated compounds and will require enlargement of major refrigeration components such as compressors. A significant portion of the increased HVAC demand is also the result of a significant increase in the installation of more efficient heat pump systems.

As-built HVAC installations will take on more relevance as a number of generic and American Petroleum Institute (API) operations proliferate, all of which require cGMP adherence to design and operating criteria; a key to generic profitability is maintaining low capital, manufacturing, and maintenance costs. One approach to achieving this goal is to employ available HVAC systems requiring a minimum of design input. While as-built designs offer a potentially lower construction cost, they may require cGMP considerations normally implemented as standard elements in traditional pharmaceutical HVAC units.

A key requirement of an HVAC installation, which encompasses cGMP components, is the need to prepare a documented program that confirms that the installed system, equipment, and/or instrument were designed, installed, and operated in accordance with the specifications and requirements set forth. This documentation is referred to as the qualification process. The qualification process primarily is composed of a series of documented procedures known as validation. There are four major documents that comprise the validation program:

1) Design qualification (DQ) – This document defines the specifications of the system, equipment, and/or instruments that will conform to the user needs.
2) Installation qualification (IQ) – The purpose of the IQ is to verify that the components are installed in accordance with manufacturer's requirements, specifications, and applicable drawings. The IQ also documents and verifies that instruments and alarms are calibrated and documented.
3) Operational qualification (OQ) – The OQ documents and verifies that the installed equipment is capable of operating within the approved limits and required tolerances.
4) Performance qualification (PQ) – The PQ confirms and documents that the installed equipment consistently produces, under normal operating conditions, an acceptable product.

A more detailed coverage of the validation process is found in Chapter 10.

3.1 Fundamentals of HVAC Electrical Systems

The mechanical components of an HVAC system such as fans, compressors, blowers, variable speed drive units, and motors all require electrical power to function. In addition to an understanding of how these components operate, it is also desirable to possess an understanding of how to verify the myriad of components in modern HVAC, operate, or perform as per contract and design criteria. This aspect is normally addressed in the HVAC system commissioning, validation, and maintenance activities normally required for pharmaceutical operations. Additionally, a basic knowledge of HVAC systems and components can be used as a technique to enhance operations and save significantly on capital and maintenance costs.

The material contained in this section will also be of value for elements of Chapters 2, 4–7, and 10 since they also contain significant electrical and electric motor operation and specification criteria.

Pharmaceutical HVAC projects are, subsequent to installation, evaluated by independent commissioning agent. This agent or company is often referred to as an "air balancing contractor" but more commonly referred to as a "test, adjust, balance (TAB)" contractor. The task of this contractor is to verify contract design items such as airflows, filter performance, and compressor operation, and fan and/or blower specifications, system pressure drop, and motor performance initially conform to design and contract specifications.

3.1.1 Electric Motors

The fundamental component of HVAC systems is the electric motor. An electric motor is, basically, a rotating device that converts electrical power into mechanical power. There are two types of electric motors that are employed most often in pharmaceutical operations, direct current (dc) and alternating current (ac) motors. In the pharmaceutical industry, dc motors are most often employed in locations where ample manufacturing space is available. A typical application of dc motors is in large pharmaceutical prescription packaging and nonprescription packaging operations. However, in most manufacturing and HVAC pharmaceutical operations, ac motors are the common power source.

The specific type of ac motor employed is referred to as an induction motor. In particular, squirrel-cage, three-phase induction motors are the most widely used motors in most pharmaceutical applications and HVAC operations in particular. For HVAC applications, an induction motor can be defined as a motor that produces voltage by the motion of a magnetic field across a conductor, typically a bundle of copper wiring.

While it is beneficial to possess knowledge of the operation and components that constitute electric motors, an understanding of the key terms and basic mathematical relationships employed in motor HVAC installations and other pharmaceutical applications employing electric motors is more relevant when involved in the project engineering programs that focus primarily on successful installation and operation.

An ac motor is composed of two major assemblies, a stator and a rotor (shaft). The interactions between the electrical currents and the stator rotating magnetic field generate a turning effort or twist (torque), which the rotor exhibits when transmitting power. Because the electromagnetic field is a physical field produced by electron flow and involves no mechanical action, such as shaft rotation, the rotor speed will lag the magnetic field speed. This lag allows the rotor to cut the magnetic field, or lines of force, and produce useful torque, or work.

An important conversion factor that is useful in motor evaluations is the relationship between power, as a mechanical function, and electrical power, expressed in watts or kilowatts. A common relationship between the two forms of power is given as

$$\text{Kilowatts}(kW) = (0.746) \times (\text{horsepower}, hp) \tag{3.1}$$

When dealing with Equation (3.1), another important relationship is that of power (watts), current (amps), and electromotive force (volts). It is important to note that 1 kW is equivalent to 1000 W, an important detail when performing motor calculations. These factors are related by Equation (3.2) as

$$P = IV \tag{3.2}$$

where

P = power, watts
I = circuit current, amps
V = voltage, volts

Since power in large electricity-consuming operations is used in large quantities, kilowatts are employed as the unit of power.

HVAC motors are designed primarily for air moving and other light to medium duty applications such as fans and, in some cases, centrifugal pumps. The most commonly used motor for pharmaceutical and related HVAC operations is the squirrel-cage motor. The name is derived from the motor rotor, which, supposedly, resembles a squirrel, or other rodents, cage. It is observed that the rotor windings are composed of copper bars, which are connected by metallic rings at the ends. For squirrel-cage motors, the voltage level an ac motor operates on is an important factor. The voltage an ac motor uses is called a phase. For pharmaceutical HVAC applications, ac motors are specified to operate in one of three phases, 120, 240, or 480 VAC (volts of ac electricity). As stated above, for pharmaceutical HVAC applications, the most commonly used motor type is a squirrel-cage induction motor, specifically a 240 VAC type.

Another important motor term is Hertz (Hz). A Hertz can be defined as the frequency or time required for an ac current pulse to complete one cycle. The standard frequency used in the United States and Canada is 60 Hz (cycles per second).

For motor operating criteria, the specified frequency in Hz must exactly match the motor plate information, which for the US or Canadian operation is, as noted, 60 Hz.

3.1.2 Motor Plate and Associated Data

Because this type of data is critical for the program, it is most important to have access to much of this information in a format that is readily available. Thus, a significant quantity of the required motor installation and operation data is found on the motor data plate, which is affixed to every electric motor. This metal data plate or nameplate identifies, among other items, the manufacturer, motor type, horsepower, and electrical requirements for the motor. It is also germane to understand specific electric motor concepts that, while not identified on the motor plate, are also important when evaluating the effectiveness and suitability of proposed or installed motors.

It is not uncommon for motors employed in air-conditioning and refrigeration applications (compressors) not to indicate the motor horsepower. Since the compression operation is only one variable of the refrigeration cycle (there are also evaporation, condensation, and metering operations), the motor is but a small part of the total system and is usually the most reliable component. However, for certain refrigeration applications, it is possible to reasonably estimate the motor horsepower via a "rule of thumb" estimate. This rule of thumb estimate is that "for air conditioning compressor motors, there is one ton of refrigeration per ton." A ton of refrigeration is defined as heat removal at the rate of 12 000 Btu/h (200 Btu/min). This value equates to the quantity of heat required to melt a ton (2000 lb) of ice in 24 h. Thus, a 25 ton refrigeration unit would be about 25 horsepower (hp) and would melt approximately 25 tons of ice during a 24 h period.

The above value is most applicable for air-conditioning installations intended for personnel occupancy and not for certain pharmaceutical manufacturing applications where a constant monitored environment is required. For instance, this rule of thumb estimate would not be employed in cGMP environments such as manufacturing areas or packaging areas, but rather it should serve as an order of magnitude estimate that can be applied when troubleshooting a malfunctioning refrigeration system.

A term also identified on the motor data plate, or the motor specification, is the torque. Simply defined, the torque is the twist or turning ability as applied to a motor shaft. Torque, for electric motors, is most commonly given in units of foot-pounds (ft-lb). For situations where the motor horsepower is not readily available while other variables such as horsepower and shaft speed are known, the shaft torque can be determined via the relationship

$$\text{Torque in ft-lb} = \frac{(\text{hp})(5250)}{\text{rpm}} \tag{3.3}$$

where rpm = revolutions per minute.

The voltage listed on the motor plate is the voltage the motor was designed to use when installed. Since the supplied voltage to the motor is rarely exact, the motor normally operates within ±10% of the voltage identified on the motor plate.

The National Electrical Manufacturers Association (NEMA) has classified electric motors in several categories based on shape and service application. There are four basic classes of NEMA motors used for pharmaceutical applications:

1) NEMA class A
2) NEMA class B
3) NEMA class F
4) NEMA class H

Each class is specified for a particular application such as powering a vacuum pump, compressor use, liquid mixers, etc. For pharmaceutical applications, including HVAC installations, the most commonly used motor is a NEMA class B. A NEMA class B motor is classified by the insulation used in the motor windings. Class B motors are insulated to operate at a specific maximum temperature. Maximum operating temperatures for motors are:

- Class A – 105 °C (221 °F)
- Class B – 130 °C (266 °F)
- Class F – 155 °C (311 °F)
- Class H – 180 °C (356 °F)

The operating temperature of motors is related to a term found on the motor data plate often identified by the term "thermally protected" or similar nomenclature. Motors that start automatically (e.g. thermostat controlled) after tripping and are located out of operator's sight must be protected against overheating due to failure to start or overloading.

Electric motors are encased in protective enclosures. For most pharmaceutical applications, there are six types of NEMA enclosures:

1) Totally enclosed – Intended to protect atmospheric exposure (this enclosure is not airtight).
2) Totally enclosed no ventilated (TENV) – No provisions for cooling the motor are provided.
3) Totally enclosed fan cooled (TEFC) – The motor is equipped with a fan that blows air over the enclosure. Most motors used for pharmaceutical applications are TEFC designs.
4) Explosion proof – The enclosure is designed to survive an internal explosion caused by the ignition of flammable vapors or gases that are present in the vicinity of the motor.

5) Dust explosion proof – Motor enclosures designed for operation in locations where explosions or ignition of dust can present a hazard.
6) Waterproof – Designed to preclude water, in liquid state, to enter the motor enclosure.

For motors that are not thermally protected, the insulation criteria are most effective when this motor type is incorporated into a well-documented preventive maintenance program employing principles such as those found in Chapter 5.

Another relevant motor plate term is noted by either the words service factor or the initials SF. The service factor of a motor is a number that indicates how much more work a given motor can do beyond the rated horsepower. This is a safety factor and is not to be considered as part of the motors normal useful horsepower. A motor may have no service factor and thus has no safety factor in the event the motor becomes overloaded. A common service factor on motors is an SF of 1.15. This number multiplied by the rated horsepower yields the actual horsepower the motor could operate in an emergency. For example, 25 hp motor with an SF of 1.15 could actually operate at 28.75 hp for a brief time period. A motor with a high service factor is used in applications where the load may vary and may occasionally be confronted with an unexpected overload in horsepower. Air-conditioning systems often use motors with the SF rating of 1.15.

The SF should serve as a guideline and is dependent on actual facility operating environments when specifying time and operating limits. The values are, however, excellent baselines for establishment of effective motor operating standards and maintenance inspection procedures.

Motor temperature verification is a significant evaluation tool to verify proper motor installation and operation. It is useful to determine if, upon installation and operation, the motor is of the correct horsepower for the proposed task to be performed. A high initial operating temperature can often indicate an undersized motor or an incorrect wiring installation (of course, improper insulation can also be the cause and should be analyzed subsequent to design, maintenance, and operational malfunctioning as failure modes) is being used.

The motor plate also identifies the various types of motor enclosures associated with a specific motor and identified with a NEMA enclosure type. A motor enclosure often encountered in HVAC service is the open air over (OPAO) design. This enclosure type is intended for fan and blower service. When dealing with OPAO installation, it is critical that the motor be located near the fan blade air stream in order to provide motor cooling. The totally enclosed air over (TEAO) is also intended primarily for HVAC applications. In this case, cooling airflow is provided by an external device. Also employed for HVAC and other pharmaceutical applications are two other commonly used enclosures: the TEFC (as identified above) motor enclosure and the open drip proof (ODP) enclosure. As the name implies, the TEFC enclosure employs a fan affixed to the enclosure and functions by blowing cooling air throughout the enclosure. There are also internal fans affixed to the shaft of the motor. These internal fans are essentially attached to the front and rear of the motor shaft. The discs have slotted fins that are distributed radially from the rotor shaft. Cooling is required since motor windings generate considerable quantities of heat due to resistance of the electrical flow through the conducting motor windings. The ODP enclosure is constructed such that drops of liquid or solid particles falling on the enclosure, at an angle of less than 15 degrees from the vertical, cannot enter the motor directly.

While not commonly employed for HVAC use, the totally enclosed nonventilated (TENV) and washdown enclosure types may be encountered. The TENV model is not equipped with an external cooling fan and is neither airtight nor waterproof; it relies on natural convection for capability. The washdown enclosure type is designed for use in wet areas or applications that require frequent cleaning such as batch mixers.

Another key motor operating element, noted on the motor plate, is a term identified as full load amperage (FLA) as it is most commonly noted on the motor plate. FLA is the amount of current or amperage (amps) the motor draws when operating at the horsepower identified on the motor plate. As one might suspect, the amperage used is lower if the motor is operating at less than the rated horsepower. Similarly, the motor will require higher amperage if operating above the rated horsepower. New motor installations should operate at or near the FLA indicated on the motor plate or in some cases the installation manual. If the current draw, as measured by a calibrated instrument such as a multimeter, is higher than the identified FLA, this is an indication that the motor is overloaded.

Operating motor temperature should be a consideration in HVAC facility-scheduled preventive maintenance activities.

The way motors are integrated with the electrical power source is called the wiring configuration. All three-phase ac systems can be interconnected between the individual three-phase power lines installed in most pharmaceutical manufacturing facilities. These connections are identified in the United States as a delta or wye configurations. The reason for these configurations is economic and relevant because wiring configuration must be identified in motor specifications for all HVAC installations as well as other motor applications.

Another item found on the motor plate is the efficiency. While major motor manufacturers present the efficiency on the motor plate, occasionally there will be a motor plate lacking this data. Typically, this may be an indication that the motor does not conform to efficiency standards normally expected.

The motor efficiency identified on the motor plate can be given in terms of percent or as a fraction. If a motor efficiency is given as 0.885, this identifies the efficiency as 88.5%. Motors with higher efficiency ratings usually contain more copper or other metals (e.g. aluminum or cast iron) than lower efficiency motors. A higher copper content usually causes the motor to operate at a lower temperature, thus increasing the operating efficiency.

Copper and other metals dissipate heat more quickly than nonmetal motor components. If the motor continuously operates at a higher than specified temperature, components can start to deform and can alter the motor performance and, thus, will not operate in accordance to performance specifications since components will be out of tolerance.

There are NEMA standards that give minimum operating efficiencies for various motor sizes as a function of horsepower. Table 3.1 presents these values. The values presented are for the most common type and class of motors used in pharmaceutical applications.

Table 3.1 Efficiency of various induction motors.[a]

Power (hp)	Efficiency (%)
1–4	78.8
5–9	84
10–19	85.5
20–49	88.5
50–99	90.2
100–124	91.7
>125	92.4

[a] NEMA Class B, single speed, >1 hp, >500 operating h/year, ODP, TEFC.

When performing HVAC project evaluations employing motors, the term "number of poles" arises. Poles are in the number of sets of three-way electromagnetic windings that a motor has. Most HVAC-related motors have two sets of three-phase windings and are called four-pole motors.

Motors are used as a power source for HVAC components such as fans and blowers. A key operating variable for motors employed in this type of configuration is the rotating speed of the motor shaft in revolutions per minute (rpm). The speed of an installed motor is given by the relationship

$$\text{rpm} = \frac{120 \times \text{Frequency}}{\text{No. of poles}} \tag{3.4}$$

Most electric motors specified for HVAC use are of the four-pole design.

The value given on the motor plate or determined by employing Equation (5.1) is a theoretical number. It fails to account for factors such as the loss encountered when converting electrical energy to mechanical energy. Consequently, a small deviation from the motor plate data can be expected. This deviation is referred to as slip. For induction motors, slip is a necessary action for proper operation.

The actual slip speed of a motor can be described by a mathematical paradigm of the form

$$\text{Actual speed} = \frac{100\% - \% \text{ slip}}{100\%} \tag{3.5}$$

Table 3.2 presents values for motor slip as a function of motor size. HVAC applications, the values given in the table, are sufficient for evaluation purposes. It should be noted that while not often used in large HVAC applications, a 7.5 hp motor is commonly employed in pharmaceutical water systems where the flow rate is in the 30–70 gal/min range. Pharmaceutical water systems are addressed in further detail in Chapter 2.

For motors used in HVAC and other pharmaceutical applications, the slippage should not exceed 5%.

A typical use of the information available from motor plate data and Tables 3.1 and 3.2 is shown in the following example.

Example 3.1 A firm has been selected to install a 25 hp motor as a component of an exhaust system. The selected motor is a 240 VAC four pole, 60 Hz, NEMA class B induction motor. Determine the theoretical rpm output of the motor as well as the expected speed subsequent to installation.

Table 3.2 Motor slip and motor size relation.[a]

Power (hp)	Slip (%)
0.5	4.6
7.5	3.3
10	4.4
15	2.2
25	3.0
50	1.7
≥250	0.7

[a] Operating at FLA rpm.

Solution

Referring to Equation (3.4), where the following data is given:

Number of poles = 4
Frequency = 60 Hz

$$\text{rpm} = \frac{120 \times \text{Frequency}}{\text{No. of poles}}$$

$$\text{rpm} = \frac{(120)(60)}{4} = 1800 \, \text{rpm} \leftarrow \textbf{Solution}$$

Referring to Table 5.2, for a 25 hp motor at full load amperage (1800 rpm), the slippage is given as 3%. Thus

$$100 - 3\% = 97\% = 0.97 \left(\text{in decimal format}\right)$$

$$\left(1800 \, \text{rpm}\right)(0.97) = 1746 \, \textbf{rpm} \leftarrow \textbf{Solution}$$

If slippage values are not available, 4% slippage serves as a reasonable value.

Another important motor plate term is noted by the word "frame." An important contribution by NEMA has been the standardization of motor frames. As a result of this effort, electric motors are now standardized. This means that horsepower, speed, and enclosure have the same frame size even though they are produced by different manufacturers. This frame standardization makes it possible for a motor from one manufacturer to be replaced with a similar motor from another company, assuming they are both in standard frame sizes. Current motor frame nomenclature was standardized in 1964. Subsequent to that date, frame sizes are designated with the letter "T" in their identification. When the letter T is identified on the motor plate, this assures a common installation for all NEMA motors. Smaller motors (≥ 1 hp) often do not have a T designation. But, since most HVAC motors are rarely less than 5 hp, this letter designation is encountered for virtually all HVAC applications.

While not normally encountered in HVAC units employing belt drives, the letter designation "TS" may occur on some frame notation. The S denotes a short shaft, relative to the standard shaft size. Should the TS identification be encountered on the motor plate, the design specifications and piping and instrumentation diagrams (P&IDs) should be checked to confirm the design. Table 3.3 identifies typical HVAC motors and frame sizes. These frame sizes are applicable for TEFC, ODP, and TENV motors.

Table 3.3 Horsepower and motor horsepower and associated frame size.

Motor horsepower (hp)[a]	NEMA frame designation
5	184T
7.5	213T
10	215T
25	284T
50	326T
100	445T

[a] Values are based on a four-pole AC motor.

While the relevance of this data may be questioned, failure to address this item during the design and early installation phase can result in a schedule delay as well as a costly retrofit/replacement.

3.2 Design Considerations

The information in the following pages includes typical HVAC design considerations employed in pharmaceutical operations. The material encompasses aspects of design, installation, operation, and maintenance that are common for pharmaceutical facilities. When reviewing the material, relevant 21 CFR 210, 211, and 600 provisions must also be evaluated in the design.

It is most important to note that, as is emphasized elsewhere, the term pharmaceutical engineering is differentiated from other engineering specialties, primarily due to the fact that drug manufacturing operations are regulated by the provisions of 21 CFR with 21 CFR parts 210 and 211 being of specific importance.

As a general regulatory classification, many design considerations are referred to as aseptic processes. An aseptic process is a manufacturing operation where the product (in this case, the item is a drug, but aseptic processes can also involve food and beverage manufacturing) is exposed to a temperature ranging from 91 °C (196 °F) to 146 °C (295 °F) and thereby producing a sterile product. The manufacturing operation is usually performed in a cleanroom environment, the design elements of which are detailed in Section 3.3.1.

Aseptic processing requirements are identified in 21 CFR 211 with the primary subparts being contained in subparts D and E. The following requirements are specific to aseptic operations:

- 21 CFR 211.67(a) – Equipment cleaning and maintenance. In addition to requiring scheduled maintenance, cleaning, and inspection, the need to ensure sanitization and/or sterilization activities must be performed at specified intervals.
- 21 CFR 211.84(d) – Testing and approval or rejection of components, drug product containers, and closures. Included in the testing provisions is the requirement to test each lot of a component, drug product container, or closure, possessing the possibility for microbiological contamination, to be subjected to microbiological testing prior to use.
- 21 CFR 211.94(c) – Drug product containers and closures. The need to remove pyrogenic (a compound capable of producing a fever) materials from containers and closures containing drug products.
- 21 CFR 211.110(a) – Sampling and testing of in-process materials and drug products. Item 6 of these criteria specifically identifies the need to perform testing to assure that the microbe presence on critical surfaces is at or below the specification limit. Verifying this specification is referred to as bioburden testing. In addition to having established and approved procedures for bioburden testing, the need to validate such testing is also identified, along with other in-process testing.
- 21 CFR 211.113(b) – Control of microbiological contamination. Paragraph (b) specifies the need to have "appropriate written procedures designed to prevent microbiological contamination of drug products purporting to be sterile...."

Other design elements involving cGMP criteria but not necessarily requiring an operating cleanroom are elements such as:

- Space temperature and humidity (21 CFR 211.46).
- Ventilation (21 CFR 211.46).

- Filtration (21 CFR 211.72, 211.46, and 211.42).
- Internal load (21 CFR 211.42).
- Air distribution (21 CFR 211.46 and 211.42).
- Room pressurization (21 CFR 211.46).
- Design versus operational ranges (21 CFR 211.58).
- Cleanrooms (21 CFR 211.28, 211.42, and 211.67).

Air handling units (AHU) generally do not exceed 45000 cubic feet per minute (cfm) ($21.5\,m^3$/s) for supplying outside air to the facility. Separate AHU may be required for heating/ cooling applications required for specific processes and/or locations such as cleanrooms or aseptic operations.

Laboratory exhaust systems (fume hoods) must be provided in lab areas (additional information concerning laboratory fume hoods can be found in Section 10.2). Normally, auxiliary air will not be required for this unit; the fume hood operates using standard laboratory air as part of the AHU supplied air. If, subsequent to a functional requirements assessment, it is determined that a variable volumetric flow is required, this option should be evaluated as a design feature. Also, a two-position (occupied/unoccupied) operating mode should be a fume hood design feature. The fume hoods must meet the testing requirements of ASHRAE standard 110-95 as well as applicable regulations of OSHA, as prescribed in 29 CFR 1910.94. For operations that contain a large number of fume hoods, variable air volume (i.e. two-position hoods), and occupancy control, in addition to exhaust and supply air units, should be evaluated.

If dust collection installations are required, these systems should be operable independent of the HVAC system. To avoid costly change orders subsequent to system installation, a careful analysis should be performed to verify that the collection system is designed and specified to collect the size and quantity of particulates/dust produced. An analysis of dust and/or particulate collection should also include a hazards analysis of the materials to be collected and the collection system to be employed. The analysis should include evaluation criteria such as the relevant material safety data sheets (MSDS), current National Electrical Code (NEC) classifications, National Fire Prevention Association (NFPA), National Institutes for Occupational Safety and Health (NIOSH) standards, and explosive venting requirements as put forth by several recognized organizations such as the API and the American Society of Mechanical Engineers (ASME).

While much of this information will be furnished subsequent to project initiation, it should be available for review and evaluation from the project/contractor specifications.

The project team selected for the HVAC program should consider retaining a health and safety representation as part of the project team. As will be shown, there are several aspects of HVAC planning and operation that transcend the traditional mechanical, electrical, and plumbing (MEP) specialties.

Outside air intakes should be positioned in the lowest reasonable locations. Air exhaust units should be located at the highest reasonable locations.

Installed ductwork shall conform to Sheet Metal and Air Conditioning Contractors' National Association (SMACNA) standards, including:

- SMACNA – *HVAC Duct Construction Standards: Metal and Flexible*, 2005.
- SMACNA – *HVAC Air Duct Leakage Test Manual*.
- SMACNA – *Architectural Sheet Metal Manual*, 6th Edition.

These documents should, as a minimum, be referenced in construction specifications and contractor work plans if duct installation is included in the planned HVAC operation.

As part of the facility sustainability program, energy recovery opportunities should be evaluated in any proposed HVAC capital project.

When possible, chilled water, or brine, intended for air-conditioning applications will originate from water- or air-cooled chillers.

Where possible, hot water (<200 °F) should be used as the heating media in lieu of low pressure (<20 psig) steam. Plant steam can contain chemicals that help reduce corrosion and foaming in the steam system. If these chemicals are not employed, the corrosion rate of the system can increase, causing:

- The life of the steam system to shorten.
- Corrosion products iron oxide (rust) to increase in concentration in the humidified air.

Additional information on corrosion can be found in Chapter 8. Methods of preventing and minimizing corrosive activity in metals are also discussed.

A possible alternative to chemical free steam systems is fabricating the system using stainless steel. This option is very expensive and is usually not considered as viable unless an economic justification can be determined.

However, low pressure steam is the preferred heating medium if 100% outside air is FED – through preheat coils. The temperature differential is the driving force for effective heat transfer. Consequently, the highest ΔT that can be supplied is desirable.

The establishment of the basic engineering design criteria is fundamental to the technical planning and schematic design phase activities for the development of the basis of design (BOD) for any project.

This section is intended to provide further guidance to facility engineers and design consultants relating to the HVAC design conditions and requirements for a project. Therefore, this section is specifically relevant to new capital construction and significant cGMP and aseptic renovation projects.

Virtually all pharmaceutical manufacturing operations have engineering standards and specifications (Chapter 10) defining the operating and design criteria for new and existing facilities. These standards are often contained within the organization engineering department. As a result, an individual is assigned to an HVAC team and not in the engineering department. Often, the assigned individual has no knowledge of HVAC.

In this case, while being a team member, his contributions to the program may be minimal. Often, the individual's involvement consists of merely attending project meetings with no significant project contribution expected. However, if serious problems arise in the program, he too will be held responsible. Consequently, it is desirable to possess a fundamental understanding of design parameters used to specify and design pharmaceutical HVAC projects.

While many items are involved in preparing design criteria or BOD for pharmaceutical HVAC programs, the minimum items requiring a design basis or design criteria are:

- Ambient weather data.
- Space temperature and humidity.
- Ventilation.
- Filtration.
- Internal loads.
- Air distribution.
- Room pressurization.
- Sound criteria.
- Control system tolerances.
- Building or room operating schedule.
- Energy conservation.
- Basic systems.

- Space criteria requirements.
- Design versus operational range parameters.
- Electrical equipment.
- Reliability.

A brief explanation of the minimum HVAC parameters described above and used in these types of projects is in order.

There are some basic considerations that should be addressed at the beginning of an HVAC installation project. Items that should be detailed by the construction manager and the owner should include:

- Duct sizing
- Equipment
- Control system
- Grill and register flow
- Noise level
- Energy requirements
- Ventilation
- Air balancing requirements
- Commissioning plan (Chapter 10)

3.2.1 Weather Data

Weather conditions employed for design purposes, usually based on the peak or maximum loading requirements for HVAC installations, are predicated on published data for the nearest city with an available weather data collection capability. A common source of this data is the *ASHRAE Fundamentals Handbook*. Although the latest revision would be most helpful, for this type of data, earlier editions should suffice since this type of data is usually constant on an edition to edition basis.

Standard cooling design temperatures (dry-bulb and corresponding wet-bulb temperature) are based on values that equal or exceed an average of 35 h period during the months of June through September in the Northern Hemisphere and December through March in the Southern Hemisphere for cGMP applications. For non-cGMP locales (e.g. offices and warehouses), the design standard need not be as critical.

Heating design temperatures represent values that equal or exceed an average of 35 h during the months of November through March (Northern Hemisphere) and May through September (Southern Hemisphere). As is the case for cooling applications, for non-cGMP applications, less critical designs are normally employed.

Typically, heating and cooling degree days and "design days" are temperature profiles and should be based on currently available hourly plotted data with the nearest recorded city included in the selected software package commonly used in HVAC design packages such as Carrier's E2O-II Systems Software Network, Trane Trace Programs, DOE-2.1E, Elite software, or other similar programs. ASHRAE/IESNA Standard 90.1-1999 can be used for heating and cooling degree days.

3.2.2 Temperature and Humidity

In pharmaceutical operations humidification processes are used to control the humidity of a space (e.g. a warehouse area, a tableting operation). Also, dehumidification is commonly practiced in air-conditioning systems.

When involved with humidity, some basic definitions of humidity terms are in order. Key humidity terms are:

- Humidity – Vapor content of a gas (common units are lb vapor/lb noncondensable gas).
- Absolute humidity – Amount of water vapor in a given quantity of air (common units are grains/pound), where 7000 grains = 1 pound.
- Molal humidity – Vapor content of a gas (common units are moles vapor/mole noncondensable gas).
- Relative humidity – Ratio of partial pressure of vapor to partial pressure of vapor at saturation (often expressed as a percent).
- Molal humid volume – Volume of 1 lb mole of dry gas plus its associated vapor (units are ft^3/lb mole of dry gas).
- Molal humid heat – Heat required to raise the temperature of 1 lb mole of dry gas plus its associated vapor 1 °F (units are Btu/lb mole of dry gas, °F).
- Adiabatic–saturation temperature – Temperature that would be attained if the gas were saturated in an adiabatic process (units are given as °F or °R).
- Dry-bulb temperature – The actual temperature of air as measured by an ordinary thermometer.
- Wet-bulb temperature – Steady-state temperature attained by a wet-bulb thermometer under standard conditions (units are °F or °R).
- Dew-point temperature at which vapor begins to condense when the gas phase is cooled at constant pressure (units are °F or °R).
- Psychrometer – Device used to measure the relative humidity of air.

The role of humidity in pharmaceutical operations has been briefly identified, as well as the humidity terms defined above. To effectively use this information for design and operating parameters relating to the properties of air and water vapor (humidity), a chart has been developed to assist in engineering scenarios requiring solutions dealing with aspects of air involving absolute humidity, temperature (dry-bulb and wet-bulb), dew point, vapor pressure, total heat (enthalpy), and relative humidity. This chart, identified as a psychometric chart, is shown in Figure 3.1.

Figure 3.1 can be used to resolve a myriad of design, operations, and maintenance issues specific to air and humidity. The salient feature of Table 3.1 is that by knowing two of the variables about an air volume found on the chart, it is possible to obtain any other variable from the chart.

By viewing the psychometric chart, significant information involving humidity and temperature design requirements can be obtained. Basic information and understanding of the material contained in Figure 3.1 commences with an explanation of the psychometric chart.

The dotted lines indicate humidity in percent. The dry-bulb temperature (db) is shown on the bottom axis (the abscissa). The wet-bulb temperature (wb) is on the uppermost curve. This curve is the saturation curve, the temperature at which the air is saturated with moisture. The scale on the right side provides moisture content of the air in grains/pound (gr/lb) of dry air. The scale to the left of the wet-bulb temperature curve (saturation curve) is the enthalpy or energy content, given in Btu/lb.

Example 3.2 A drug warehouse maintains a temperature of 75 °F dry-bulb temperature. The humidity is maintained at 40%. Determine *A*, the dew point of the operation; *B*, the wet-bulb temperature; and *C*, the enthalpy

Solution
For Example 3.2, Figure 3.1 is the basis of calculation.

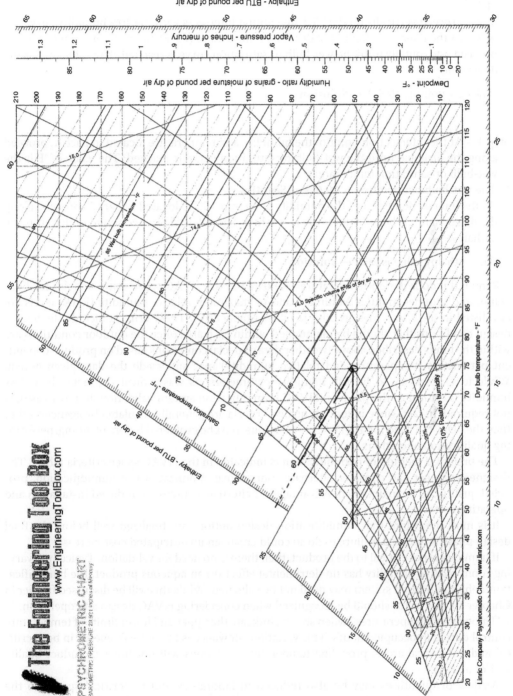

Figure 3.1 Standard psychometric chart. *Source:* Courtesy of the Engineering Toolbox.

A.

Step 1 – On the bottom axis (db, or dry-bulb temperature), locate the dry-bulb temperature of 75 °F.

Step 2 – Proceed vertically up the 75 °F line until it intersects with the 40% humidity curve.

Step 3 – At the intersection point of dry-bulb temperature (75 °F) and percent humidity curve, proceed horizontally until the wet-bulb or saturation curve is reached. The value of °F is the dew point.

A. The dew point is **49.5 °F** ← **Solution**

B.

Step 1 – At the intersection point of the humidity and temperature located in part A, proceed along the diagonal until the wet-bulb or saturation curve is reached. The value of °F is the wet-bulb or saturation temperature.

B. The wet-bulb temperature is **59.6 °F** ← **Solution**

C. At the intersection point of the humidity and temperature located in part A, proceed along the diagonal until the enthalpy scale is reached. The value of Btu per pound of dry air is the enthalpy.

C. The enthalpy value is **26.2** ← **Solution**

Required indoor design temperatures are listed in the HVAC space criteria tables for both winter and summer months. Minimum and/or maximum relative humidity (RH%) levels are also listed in the HVAC design criteria tables (Table 3.10). These criteria include the indoor design conditions that should be maintained within each area once the outdoor conditions are within the limits described above (weather data). Special conditions based on product-dependent parameters (temperature and relative humidity) may supersede the data given herein. While it is necessary to furnish humidity parameters, as was indicated previously, unless humidity is critical to the product, HVAC humidification system validation, in most cases, is not required. In fact, since indoor air quality criteria are generally regulated by agencies other than the FDA, the need for all HVAC installations must be evaluated before planning/performing qualification operations.

The preferred method of humidification is indicated in the HVAC space criteria tables. The designer/engineer can consider alternate sources (e.g. atomization) of humidification, provided quality can be maintained. Advantages of alternate sources are reduced installation and operating costs.

It is most important that humidification design options are finalized well before the final design phase is initiated. Failure to do so could create an unanticipated cost increase.

If humidity is not critical to the product, then there is no need for validation. Typically, if varying room relative humidity has no detrimental effect on an aqueous product, the humidification/dehumidification system may need not be validated. While this will be discussed further in Chapter 10, this factor should be recognized when considering HVAC design and operation.

"Unoccupied temperatures," when shown, indicate the upper and lower limits of temperature control during unoccupied hours. This variation allowance is to conserve energy in both critical and noncritical areas, providing temperature variations will not impact product quality and/or a safety.

Airflow or quantities may be also reduced in laboratory and noncritical spaces to the minimum required to maintain space pressurization and to a zero airflow in other spaces. This variation allowance is to conserve energy in both critical and noncritical areas, providing space condition and variations will not impact quality and/or safety.

It is important to emphasize the fact that temperature and humidity requirements are defined by the user and are detailed in the user requirements specification (URS) with the room/space requirements predicated on product requirements. For instance, pharmaceutical powders normally have an upper humidity requirement to prevent "clumping" or agglomeration. Similarly, lower humidity requirements are based on safety when static buildup coupled with electrostatic discharge is a consideration. Temperature requirements are also based on product as well as operator comfort.

3.2.3 Ventilation

The recommended minimum outside air quantities for selected areas is in accordance with AHS/ASHRAE 62-1999. The actual outside air quantities for all areas should be based on occupancy, pressurization criteria, minimum dilution ventilation requirements, and exhaust requirements. The quantities of outside air required for individual operations should be determined by corporate or facility environmental health and safety operations. It is most important that these standards be transmitted to the company or organization responsible for installation, commissioning, validation, startup, maintenance, and operation of the HVAC system.

Supply air quantities should be based on cooling loads, minimum dilution ventilation requirements, and/or required makeup air for exhaust or dust collection systems, whichever is greater.

Ventilation design features for pharmaceutical HVAC systems most often follow AHS/ASHRAE 62-1999 guidance. Typical design options include:

- 6–14 air changes per hour (ACPH). It is important to note according to 21 CFR 211.42(10) (iii) that there are no air change requirements for pharmaceutical facilities.
- Cooling loads.
- Minimum dilutions.
- 100% outside air is normally used for vaccines and pharmaceuticals (again, not a cGMP requirement).

While, as noted, there are no specific air change rates, there exists a relationship that is useful in determining air change rates for defined areas such as conventional cleanrooms, details of which are addressed further in this chapter. This rate relationship is given as

$$\text{ACPH} = \frac{\left(\text{Average airflow velocity}\right) \times \left(\text{Room area}\right) \times 60\,\text{min/h}}{\text{Room volume}} \tag{3.6}$$

As a project engineer, or team member assigned to such a task, air change out requirements should be detailed by the HVAC contractor. Similarly, the project team should confirm that the specification and contract requirements conform to the design basis and installation design.

Example 3.3 It is desired to construct a small visual vial inspection room as part of a cGMP cleanroom clinical manufacturing operation. The length of the proposed room is 40 ft, the width is to be 10 ft with a height of 12 ft. The average velocity into the room is designed to be 90 ft/min. To conform to the design basis, the facility must have at least 400 ACPH. Does the proposed design meet the hourly air change out rate?

Solution
Formula 3.5 is employed for this example.
The calculated area is

$$\left(40\text{ft}\right) \times \left(10\text{ft}\right) = \mathbf{400\,ft^2}$$

The calculated volume of the room is

$$\left(400\,\text{ft}^2\right)\times\left(12\,\text{ft}\right)=\textbf{4800}\,\textbf{ft}^3$$

Inserting the given values in Equation (3.6) yields

$$\text{ACPH}=\frac{\left(90\,\text{ft}/\min\right)\left(400\,\text{ft}^2\right)\left(60\,\min/\text{h}\right)}{4800\,\text{ft}^3}=\textbf{450\,ACPH}\leftarrow\textbf{Solution}$$

The proposed design conforms to the design criteria.

Summer outside air quantities for ventilation for nonmechanically cooled areas should be based on a maximum temperature rise of 5 °C (10 °F) above design ambient conditions in nontropical locations. These conditions may require cooling for mechanical and electrical spaces to meet manufacturer's equipment ratings. An important consideration is the need to evaluate the dehumidification requirements and associated potential for corrosion of HVAC components.

Two major devices used for air movement, including change outs, are blowers and fans. The blower, commonly referred to as a crossflow or tangential flow fan, is essentially a series of large curved blades that rotate axially, via a motor, commonly in a cylindrical-shaped enclosure housing the rotating blades. Air enters the opening and is discharged through an exit orifice at a much higher flow rate to process ducts for further processing, such as air heating for facility temperature control. A fan is also intended to process air but is more commonly used for air circulation in a working volume, such as an office or a manufacturing area.

For fans and blowers, a major design requirement is the volumetric flow rate. When the volumetric flow rates are known, the blower or fan can be sized to determine the required fan/blower motor horsepower, in addition to the volumetric flow, the static head, usually given in inches of water, and the fan efficiency. The design equation can be expressed as

$$\text{Hp}=\frac{\text{cfm}\times\text{piw}}{6343\times\text{Fan efficiency}} \tag{3.7}$$

where

cfm = flow rate, ft/min
piw = static pressure head, inches of water
fan or blower efficiency is dimensionless

Example 3.4   A tangential flow blower is sized for a total airflow of 9100 ft^3/min of return air and 4100 ft^3/min of outside air for a total flow of 13 200 ft^3/ min. The static pressure is designed to have a total static head of 6.74 in. of water. The blower efficiency is specified as 0.7. Determine the design horsepower required for the blower.

Solution
A design volumetric flow of 13 200 ft^3/min is the basis of calculation for this example. In addition to Equation (3.7), and the design flow (13 200 ft^3/min), the variables required are

Static pressure head = 3.30 in. H$_2$O
Blower efficiency = 0.7

Inserting the given values in the equation yields

$$Hp = \frac{(13200)(6.74)}{(6343)(0.70)} = \textbf{20.0 hp} \leftarrow \textbf{Solution}$$

The calculated value of 20.0 hp is a calculated result. As a margin of error for a large flow rate such as this, it would be advisable to use a larger motor. On occasion, once installed, additional lengths of ductwork may be required to provide flow to locations not previously considered in the original design. This would require a higher static pressure head. The cost of using a 25 hp motor, when amortized over a period of time, becomes negligible. Consequently, good engineering practice would probably specify a 25 hp motor in lieu of a 20.0 hp motor.

3.2.4 Air Filtration

Most pharmaceutical and biotech facility operations involve manufacturing and packaging operations. These requirements are defined in a document (ASHRAE 52.1-1992) prepared by the American Society of Heating, Refrigeration and Air-Conditioning Engineers (ASHRAE). This manual identifies design criteria specific to air filtration requirements, including locations involved in manufacturing, packaging, and research operations.

All systems should be designed to perform at the minimum filtration requirements identified in Table 3.9. In areas where HEPA (High Efficiency Particulate Absorber) filtration is required for the specific application, it is most important that installation testing be performed prior to system operation. The HEPA filter has an efficiency of 99.97% for removal of particles as small as 0.3 μm. Because of these stringent purity requirements, required HEPA integrity testing must be documented and available for commissioning and qualification procedures. Additionally, placement of HEPA filters should also be evaluated for cost considerations. For instance, HEPA filters are best positioned with the filters horizontally as opposed to a vertical placement. The reason for this is that in the vertical position, the filter material tends to separate from the filter frame. The force of gravity tends to pull the filter material downward, whereas the rigid frame is not deformed to the downward pull of gravity. It is not uncommon for pharmaceutical operations to spend large sums of maintenance funds because of routine noncritical filter inspections. Often, these inspections are performed several times a year. In reality, there is only one recognized HEPA inspection procedure that occurs prior to the HEPA filter leaving the manufacturer's facility. To perform an evaluation/inspection in accordance with the manufacturer's quality control and quality assurance procedures would require use of a test facility too large to viably employ for this type of maintenance procedure. As a result, filter inspection vendors employ methods that are not necessarily representative of an actual HEPA filter test procedure. Further, it is not uncommon to spend a large sum for inspection procedures and result in relatively few actual annual HEPA filter replacements and record an inspection cost that is significantly higher than replacement costs.

The efficacy of extensive HEPA filter inspections is additionally questionable when considering the HEPA filter inspection programs at most nuclear utilities. Here, HEPA filter inspection usually occurs during routine shutdowns for nuclear facility maintenance, which usually occurs at intervals between 18 and 24 months. Because of the high reliability of HEPA filters, a formal program, such as that employed in the pharmaceutical and biopharmaceutical industry, may be questioned as a degree of "overkill" when economics are investigated. Further, HEPA filter maintenance vendors are not prone to provide extensive explanations as to the nature of the repair or replacement performed on defective units. For instance, a common filter failure mode is the replacement of the filter if a tear in the filter material is greater than 5%. The quantitative value (5%) is normally accepted by the owner without question. Generally, the degree of filter damage is often a judgment made by the technician performing the inspection.

In lieu of frequent HEPA filter inspections, consideration should be given to employing effective prefilters, also referred to as a roughing filter, installed in line with the HEPA filter. The prefilter is employed to remove all common larger dusts in order to remove the load on the relatively expensive HEPA filter. The prefilter unit should be Underwriters Laboratory, Class I, and have a 90% discoloration rating in accordance with the NIST dust spot procedure on atmospheric dust. This configuration, coupled with horizontal HEPA filter placement, as opposed to vertical HEPA filter placement, should significantly affect air filtration costs.

HEPA filters are most often installed at the HVAC unit, terminal diffusers, and exhaust. Of course, in addition to effective installation, the number of filters and efficiency are also based on product safety and product requirements.

When evaluating air filters, ASHRAE 52.1-1992 specifies the minimum filtration requirement as 30% efficient, based on the ASHRAE test described.

A reliable method to evaluate the actual need for frequent HEPA filter inspection is to implement a Six Sigma program or, at a minimum, significant elements of the program. While it is difficult to employ a related change control program, often due to ignorance of the subject or middle management fear to delve into the unknown, an organized Six Sigma team, or a group of personnel knowledgeable in the Six Sigma program elements, could implement an improvement program in an efficient and timely manner. Chapter 11 presents a brief introduction of the Six Sigma basics as well as concepts potentially germane to the issue.

3.2.5   Internal Loads

Internal heat gains from lighting, electric power, or other heat-generating equipment and from people or animals occupying the space used for design (peak load calculations, HVAC equipment, and component sizing) should be based on actual process or activity requirements for each identified area. Appropriate load factors should be established at the beginning of each project. Again, it is most important that these factors be clearly transmitted to the company or organization responsible for the HVAC project. While in many cases, personnel assigned as project team participants may lack the design background for in-depth analysis of the HVAC project, verifying elements such as internal loads will assure that the responsible company or organization has addressed the items identified (e.g. ventilation, filtration, and internal loads).

While values may vary, the following information (Tables 3.3–3.5) should serve as guidelines when determining, evaluating, and/or specifying occupancy requirements, lighting, and equipment loads when designing, specifying, and/or evaluating installations employing HVAC systems or components.

Table 5.4 should serve as a method to verify that vendor-supplied data and specifications are reasonably conformant to the user requirements when presented to the system owner.

As is the case with Table 5.4, Table 5.5 should be used to verify that the vendor design is conformant to the owner/user requirements.

3.2.6   Air Distribution

As a general guideline, supply and exhaust air systems must be designed to minimize product contamination and protect personnel. Rooms/spaces must be designed to achieve defined pressurization levels (usually contained in the URS). Locations processing pharmaceutical powders and clean areas are typical examples of locales requiring special consideration.

Recommended minimum room air changes in occupied air-conditioned areas are located in Section 3.3.1 and identified as "ACPH." This information should serve as a design guide and subject to a more rigorous design modification as the design phase of the project proceeds.

Table 3.4 Heat generated from lighting sources.

Building/facility type	Lighting heat gain[a]
Manufacturing/packaging	1.8 W[b]/ft^2 (19.4 W/m^2)
Laboratories	2.0 W/ft^2 (21.5 W/m^2)
Offices	1.5 W/ft^2 (16.1 W/m^2)
Receiving/shops	1.0 W/ft^2 (1.8 W/m^2)
Toilet/change rooms	1.0 W/ft^2 (10.8 W/m^2)
Corridors	0.8 W/ft^2 (8.6 W/m^2)
Warehouse	0.25 W/ft^2 (2.7 W/m^2)

[a] Heat gain values are based on fluorescent lighting, warehouses areas based on halide lighting.
[b] W indicates that the power is given in watts.

Table 3.5 Equipment heat load.[a]

Equipment	Heat load
Offices	2.5 W/ft^2 (27 W/m^2)
Chemical laboratories	4 W/ft^2 (43 W/m^2)
Warehouse	1 W/ft^2 (10.8 W/m^2)
Manufacturing/packaging/shops	5.5 W/ft^2 (59.2 W/m^2)

[a] The above values are guidelines and can vary with individual equipment and layouts.

It is important to note that there is no minimum cGMP requirement for ACPH in the United States. Airflow into and out of a space should be based on providing the required cooling, heating, relative humidity, pressurization, particulate control, dilution ventilation, recovery time from an upset condition, and non-cGMP codes.

In all cases, air supply and exhaust air circulation systems should be designed to minimize product contamination and maximize employee protection. Additionally, rooms must be designed to achieve and maintain specified pressurization levels. This is particularly relevant where pharmaceutical powders and "clean" areas are employed.

3.2.7 Room Pressurization

Systems should be designed to maintain specified pressures within the ascribed locations as per pharmaceutical regulations. Regardless of the room/area pressurization levels employed, the pressure levels must be maintained when personnel enter or depart from a room, where there are an equipment movement through an entrance or exit and a significant personnel movement in the room/area. If the room/area contains laminar flow hoods or sensitive instrumentation, the design should accommodate for these equipments and/or instruments. Manufacturing areas should be under constant monitoring and pressurization levels. For laboratories and other non-sterile areas, supply and exhaust air units can operate at reduced levels during unoccupied periods. As a general rule, pressurized rooms/areas should maintain a positive pressure relative to hallways, change areas, and noncritical locales, with the idea being to have airborne contaminants discharging from critical areas.

Airflows and required differential pressures should be defined as early as possible in the design phase and clearly identified in the design specifications and system P&IDs. Ideally, the pressurization design should be in working order prior to commissioning activities, details of which are presented in Chapter 10.

3.2.8 Sound and Acoustic Criteria

While seemingly unimportant, excess noise generated from the workplace environment is a serious concern for office and manufacturing activities. A January 2009 study by Health Consultants, Inc., estimates that the annual cost due to loss of productivity, special education, and medical care because of hearing loss is $56 billion. The average disability for noise-induced hearing loss costs $17 000 per ear. Accordingly, addressing this element of HVAC is warranted.

Simply explained, sound is a rapidly repeated fluctuation in air pressure, kinetic energy in the form of pressure. If the frequency of change is consistent, it is possible to detect the sound of a particular pitch or frequency. When pressure changes occur at random frequencies, this sound is often referred to as "broadband" or more commonly identified to as "noise" [1].

Noise-induced hearing loss is a disorder associated with repeated trauma. The degree to which hearing is affected depends on the intensity, frequency, duration of exposure, and individual susceptibility. Noise-induced hearing loss occurs gradually and can cause irreversible damage to the inner ear. Research has shown that exposure to excessive noise also causes stress to other parts of the body, resulting in increased muscle tension, a quickened pulse rate, and increased blood pressure. Workers exposed to noise may experience nervousness and extreme fatigue. Thus, when an HVAC installation, modification, or design is encountered in a pharmaceutical operation, it is critical that noise levels be considered, evaluated, and design modifications implemented, if required, in an HVAC system.

HVAC systems ducts distribute more than air; they also distribute noise produced by fans and blowers, which are encountered by introducing outside freshly heated or cooled air and recycled air. Often these fans and/or blowers are antiquated and undersized (frequently a result of duct expansion without a corresponding duct flow increase) and as a result can be more of a noise concern in older facilities. Duct liners have been specifically developed to absorb, and thereby reduce, the noise before it emanates from ductwork. Designers apply different thicknesses and lengths of duct liner to reduce noise of various levels. Understanding the acoustical benefits of duct liners and available choices in liner materials helps in specifying a system that is efficient and effective, both in terms of noise performance and associated HVAC performance.

Fan noise can create acoustic conditions that are undesirable and unacceptable for many types of occupancy, including pharmaceutical operations. It might interfere with human speech, for example. However, a low level of background noise is often considered beneficial for ordinary activities, such as those in an office, because it improves speech privacy by interfering with the audibility of more distracting noises such as a talkative coworker [1].

Duct liners are an effective technique to significantly reduce the kinetic energy, manifested as noise in this case, that is often associated with the movement of air by absorbing the excessive sound produced. Duct liners are generally composed of synthetic materials or fiber glass. The synthetic materials are in the form of polyimide foams and elastomeric foams. Fiber glass acoustic duct liners are composed of bulk-packed glass fibers, which allow air to easily pass through the loosely packed fibers. The motion of the air through the packing causes the sound to be absorbed within the fiber glass duct lining, thus attenuating the noise accompanied by the airflow.

Polyimide linings are fabricated in the form of flexible, buoyant spongelike foam that allows for sound absorption and a resulting sound attenuation. Polyimide foam is available as flexible sheets or strips with varying thickness and can be easily cut to specified dimensions of length and width. For duct acoustic liners, a width of 1 in. thickness is most common.

The third type of acoustic duct liner is fabricated of elastomeric materials also in the form of foam with the exception that these elastomers are denser and resemble lumber stock in appearance and machining properties. Elastomeric duct liners are fabricated of synthetic rubber material that is cross-linked to maintain stability. As a rule of thumb, when considering duct liner material, elastomeric foam is the least efficient material when compared with fiber glass and polyimide foam.

As a general guideline, sound absorption is directly proportional to the exposed duct liner surface area and duct liner thickness. For more detailed calculations, such as the length of ducting required, a standard design reference is the ASHRAE publication, ASHRAE *Handbook – HVAC Applications* [2].

For retrofit applications, it is important to consider the fact that reduction of the duct cross-sectional area, due to addition of an acoustic duct liner, can significantly increase the volumetric flow rate when compared with the original design prior to the duct liner installation.

Example 3.5   28 × 8 in. exposed air duct is installed below the ceiling of a former storage room in a pharmaceutical manufacturing facility. The room is currently used as a non-cGMP administrative office. By employing the appropriate references, the facility engineering and facility maintenance operations were assigned the task of attenuating the noise transmitted by the overhead duct through installation of a given length (10 ft) of polyimide foam of 1 in. thickness and starting at the duct exhaust grill. An analysis of the noise generated indicated that a 15–20% noise reduction was required (Table 3.6). The duct currently transfers air at the design rate of 1050 cfm.

Determine the difference in airflows between the existing duct installation and the ducting subsequent to installing the inch-thick polyimide foam.

Solution
Existing airflow in duct calculation (cfm):

Table 3.6 People density.[a]

Building/area type	People density
Offices	150 ft^2/person (14 m^2/person)
Laboratories	200 ft^2/person (18.5 m^2/person)
Warehouses	750 ft^2/person (70 m^2/person)
Manufacturing/packaging	75 ft^2/person (7 m^2/person)
Dining	75 ft^2/person (7 m^2/person)
Kitchen	250 ft^2/person (23 m^2/person)
Serving/vending	50 ft^2/person (5 m^2/person)

[a] Values are for cooling calculations.

$$\text{Cross-sectional area of the duct} = (28\text{in.}) \times (8\text{in.}) = \frac{224\,\cancel{\text{in}}^{2}}{1} \times \frac{1\text{ft}^{2}}{144\,\cancel{\text{in}}^{2}} = 1.56\text{ft}^{2}$$

$$\frac{1050\text{ft}^{3}}{\text{min}} \times \frac{1}{1.56\text{ft}^{2}} = 673.08\text{ft}/\text{min} \text{ current duct velocity}$$

Calculated duct velocity, subsequent to addition of 1 in. duct lining:

Cross-sectional area of duct, including installed polyimide liner = (27 in.) × (7 in.) = 189 in^{2},

$$\frac{189\,\cancel{\text{in}}^{2}}{1} \times \frac{1\text{ft}^{2}}{144\,\cancel{\text{in}}^{2}} = 1.31\text{ft}^{2}$$

$$\frac{1050\text{ft}^{3}}{\text{min}} \times \frac{1}{1.31\text{ft}^{2}} = 802\text{ft}/\text{m proposed design velocity}$$

$$\frac{(802\text{ft}/\text{min} - 673.08\text{ft}/\text{min})}{802\text{ft}/\text{min}} \times 100\% = 16\% \text{ duct flow velocity increase with duct liner} \leftarrow \textbf{Solution}$$

The velocity with the acoustic liner installed is in the range of the recommended maximum allowable velocity for ducts with airflows of less than 1200 cfm. The recommended airflow is 785 ft/min.[1] However, since the total duct length is relatively short, requiring only 10 ft of acoustic insulation, this minor excess can be accepted and the planned modification should significantly mitigate the noise problem.

When designing for noise attenuation in existing operations where ducts are responsible for significantly contributing to excessive noise, it is important that duct support brackets also be evaluated. Changes made by ductwork attached to walls or ceiling hangers that support the duct may cause vibrations due to poor attachment, incorrectly sized hanger supports, or supports that have been in place for a longer time. As a result, the once secure hangers can cause vibrations over time as the hanger fixture loosens. These vibrations cause resonance and an associated amplification of the vibrations creating an undesirable noise in the ductwork configuration. While acoustic insulation may mitigate the problem, replacing or firming the support bracket can permanently remediate the noise, often at a lower cost. Consequently, prior to implementing duct lining activities or duct support replacement, a thorough study should be performed by qualified personnel.

In addition to designing for a target noise level, proper duct insulation, and hanger integrity, other HVAC components that are part of the HVAC installation should also be evaluated. Typically, these associated components would include reviews of fans, blowers, air handlers, and refrigeration units that operate in the vicinity of the duct or ducts being reviewed. This extra effort may prove worthwhile in verifying the suspected cause of the HVAC noise. Often, a third-party contractor might be considered for this type of study.

It may be interesting to note that Example 3.3 uses a relationship that is essentially identical to Equation (2.3). This equation is commonly employed in duct design, sizing, and testing. It is also widely used in fluid flow applications such as pipe sizing, as is demonstrated in Chapter 2.

---

1 Source: www.engineeringtoolbox.com/duct-velocity-d_928.html (accessed 13 December 2017).

Example: Noise criteria (NC) or levels are a common method of specifying the background noise in rooms. The volume (loudness) is determined by the sound pressure, p, and expressed as sound pressure level L_p in dB (decibel, a logarithmic unit of sound intensity). The sound intensity is normally given in watts/m^2; the intensity reference level, I, is $I = 10^{-12}$ W/m^2. For sound criteria determinations, the reference sound pressure is $20\,\mu$Pa (Pa = pascal = 1 N/m^2). The relationship between sound intensity and pressure is given by the approximate relationship

$$I \sim p^2 \sim \frac{1}{r^2} \qquad (3.8)$$

where r is the distance from the sound source. From relationship (3.8), it follows that

$$p \sim \frac{1}{r} \qquad (3.9)$$

The above relationship states that the sound pressure is linearly inversely proportional to the distance from the sound source.

The often used term "intensity of sound pressure" is not correct. In lieu of this term, strength, magnitude, or level should be employed. "Sound intensity" is sound power per unit area, while "pressure" is a measure of force per unit area. Intensity is not equal to pressure.

As noted, intensity is defined as power (watts) per unit area (meters). For determining sound intensity, the reference level is 1.0×10^{-12} W/m^2 (9.29×10^{-14} W/ft^2).

For comparison applications, the following relationship can be employed:

$$1\text{Pa} = 1\text{N}/\text{m}^2 = 94\text{dB} \qquad (3.10)$$

$$\text{where } 1\text{bar} = 10^5\,\text{Pa} = 14.50\text{lb}/\text{in}^2$$

While it is relevant to have an understanding of noise design standards and how they were formulated, practical application is, in this case, more germane to HVAC evaluation. For this purpose, several design and evaluation techniques are available. The most widely used method is the NC, which is used widely in architectural and mechanical designs common in HVAC installations and most common for ventilation systems. A valuable tool would be a correlation for NC values and decibel levels in common pharmaceutical operating locales. Table 3.7 presents such values. Table 3.6 also serves as a guide for expected values in common pharmaceutical facility areas. Actual values can vary from those given in the table but should be reasonably accurate for project engineering, commissioning, and most validation applications.

The NC consist of a family of curves that define the maximum allowable octave-band sound level corresponding to a chosen NC design goal. Table 3.8 displays the NC curves in a tabulated format. Table 3.8 exhibits the relationship of NC curves to sound pressure levels (employing 20 dB or $20\,\mu$Pa as the reference) at selected frequencies.

As noted in Table 3.8, the NC significantly diminish as the frequency increases.

Because of the costs involved in industrial-related hearing loss, maintaining NC is critical in design, installation, and operating phases of HVAC systems. In addition to design and operation, employee monitoring and education is most important to preclude hearing-related losses from occurring. Nevertheless, failure to adequately address NC during the design phase can result in significant costs subsequent to the initial operation.

Table 3.7 Noise evaluation variables.[a]

Space	NC rating	Noise rating (dBA)[b]	Common classification
Laboratories/manufacturing and packaging locales	NC-50	55	Noisy
Warehouse/workshop	NC-55	50	Noisy
Meeting rooms/dining	NC-40	45	Quiet
Offices	NC-40	45	Quiet
Reception areas	NC-45	55	Moderately noisy
Restrooms, corridors, and stair towers	NC-45	55	Noisy
Mechanical rooms	NC-85	80	Noisy
Cleanrooms	NC-55-70	80	Noisy
Animal holding areas	NC-45	45	Quiet
Open offices	NC-40	40	Quiet
Ancillary areas	NC-45	55	Moderately noisy

[a] Specific locations may be somewhat depending on the type of equipment installed.
[b] dBA is a weighted method compensating for the human hearing of sound pressure at different frequencies.

Table 3.8 Noise criterions (NC) at selected frequencies.

Criterion curves	63 Hz	125 Hz	250 Hz	500 Hz	1000 Hz	2000 Hz	4000 Hz	8000 Hz
NC-15	47	36	29	29	22	17	14	11
NC-20	51	40	33	33	26	22	19	16
NC-25	54	44	37	37	31	27	24	21
NC-30	57	48	41	41	35	31	29	27
NC-35	60	52	45	45	40	36	34	32
NC-40	64	56	50	50	45	41	39	37
NC-45	67	60	54	54	49	46	44	42
NC-50	71	64	58	58	54	51	49	47
NC-55	74	67	62	62	58	56	54	52
NC-60	77	71	67	67	63	61	59	57
NC-65	80	75	71	68	68	66	64	62

Table 3.9 displays the allowable exposure time for typical decibel levels as defined in OSHA 1910.95(b)(1).

HVAC installations should be carefully evaluated, as indicated previously, during all phases of the project. Particular attention should be given to HVAC programs involving new installations in existing facilities. This is especially relevant in older facilities where significant HVAC infrastructure installation is planned. One very important design consideration, when dealing with new installations in an older facility, is to verify that all drawings, procedures, and specifications and revisions are current.

Table 3.9 Permissible noise exposures.

Duration per day (h)	Sound level (dBA)
8	90
6	92
4	95
3	95
2	100
11/2	102
1/2	110
1/4 or less	115

In an era where major corporate dynamics such as mergers and takeovers are increasingly common, key information is often lost because of personnel downsizings, transfers to other operations, and failure to effectively implement document control procedures. Neglecting to maintain updated design and/or operating information can be the victim of such organizational alterations. The expense involved in correcting outdated or undocumented system information can be quite costly, particularly if the modifications have to be made by vendors who must implement required and unplanned design alterations by means of costly engineering change orders.

The most effective technique to minimize these unanticipated costs is to perform an actual "walk down" of the existing installation/area where the proposed HVAC modification will occur. This activity should be performed prior to releasing bid proposals or bid quotes. The key element of this analysis would involve comparing existing drawings and related documentation with what actually exists. It would also be helpful to compare details such as design temperatures and pressures with actual results. Equipment and/or locations that perform lower than the design standards (e.g. airflow lower than design or area temperatures higher than originally specified) may indicate that changes, such as undocumented duct installations, were performed.

One common point of resistance to performing the walk down and associated analysis is the manpower requirement. Tying up maintenance and/or engineering personnel is often perceived as a waste of assets and time. It should be noted, however, that contractors anticipating involvement in the project will most likely perform similar walk downs prior to the initiation of the project. Generally not included in the contractor price will be required engineering change proposals to remediate inconsistencies noted in the contractor walk down, existing drawing, and specifications reviews. These costs are commonly not included since many bidders will not address the required changes so that their bid will remain low, with the needed changes incorporated as change orders subsequent to winning the task.

Again, in lieu of staff personnel performing the walk down, it might be advantageous to consider having a consultant perform the task. The cost of this action may prove to be worthwhile if only a few drawings/specifications are found to be in error.

One general consideration for implementing design changes during the early project phase is to incorporate potential design, specification, and drawing changes prior to the formal approval of the URS. The URS is a document that defines the constructability of an installation and provides specific system performance and design parameters that include installation criteria such as:

- Functionality details
- Critical parameters and operating ranges
- Cleaning and maintenance requirements

If relevant design modifications can be suggested and approved prior to the final approval of the URS, it is possible to create significant savings in both time and cost. Consequently, presenting potential improvements, regardless of how unfeasible they may appear, can be a valuable project benefit. And since the URS is an approved document and not a hardware item, proposals put forth during this phase will generally not significantly affect cost or time if deemed to be not a viable proposal.

The URS is prepared with input from the system architect and engineering firm as well as the construction management firm. For large installations, such as an extensive HVAC operation, detailed documentations such as the URS and the FRS are most critical to efficient design, installation, and operation.

It is most important to distinguish between the FRS and the URS. The FRS is generally a product of the user. Once the need for a new system or upgrade is identified, the user groups (typically composed of the end user, engineering, maintenance, and quality assurance) agree on the basic needs required to satisfy the new or enhanced installation. The FRS is often prepared by the user staff personnel, but often a consultant is utilized to assist in FRS preparation.

Table 3.10 briefly summarizes the salient features involved in pharmaceutical design criteria. It serves as a guideline or basis and is not intended to be design standards.

3.2.9 Building Control Systems

Building control systems, also referred to as building management systems (BMS), are installed to maintain the HVAC systems within the specified operating parameters. Additionally, the BMS is designed to:

- Alert personnel of critical failures.
- Track and trend critical parameters such as temperature and humidity.
- Support data for product investigations.
- Check performance and troubleshoot equipment.

3.3 Cleanrooms

Defining, designing, and installation of cleanrooms are an integral requirement of effective drug manufacturing. An understanding of the regulations and accepted design standards is critical to safe manufacturing operations of API and finished products.

3.3.1 Cleanroom Design Fundamentals

Cleanrooms could arguably be classified as an element that should properly be addressed as an aseptic operation since the purpose of a cleanroom is to ensure an area for performing sanitary or sterile operations is properly employed and designed. But, since a major element of

Table 3.10 HVAC design criteria.

Standard no.	General HVAC design criteria		Page
Effective date	Revision no.	Supercedes document no.	Supercedes date

<table>
<tr><td colspan="3" align="center">Facilities providing products for international distribution HVAC
space criteria table • GSC-INT-STD-1</td></tr>
<tr><td>Area class</td><td>ISO: EU:</td><td>5 A</td></tr>
<tr><td rowspan="4">Typical application</td><td rowspan="4">Space description:</td><td>Point of fill or other aseptic manipulations areas with open product</td></tr>
<tr><td>Sterile and laminar flow applications</td></tr>
<tr><td>Sterile filtration, sterilization/depryogenation of components – unloading, filling and stoppering, lyophilization transfer, assembly of sterilized equipment parts, sterile powder dispensing/weighing</td></tr>
<tr><td rowspan="6">Air quality</td><td>Temperature (°F):</td><td>66 ± 2 (19 ± 1 °C)</td></tr>
<tr><td>RH, summer (%):</td><td>40 + 5</td></tr>
<tr><td>RH, winter (%):</td><td>40 − 5</td></tr>
<tr><td>UNOCC temperature:</td><td>N/A</td></tr>
<tr><td>Air changes per hour:</td><td>Laminar flow, site sop work level requirements, typically 90 FPM ± 20% (0.457 + 20% m/s) at work surface</td></tr>
<tr><td>Pressure gradient:</td><td>0.05″ wg (12.4 Pa) (minimum required) differential between classifications. 0.03″ wg (7.5 Pa) setpoint, 0.02″ wg (5.0 Pa) minimum differential between like classification if pressure gradient required

Provide pressure gradient as required by assigned RBEC level if greater; refer to RBEC criteria – table II of RBEC guideline, April 2001, AHP EH&S Department</td></tr>
<tr><td rowspan="4">Recirculation criteria</td><td>Supply 100% outside air:</td><td>No</td></tr>
<tr><td>Exhaust all air:</td><td>No</td></tr>
<tr><td>Recirc allowed:</td><td>Yes</td></tr>
<tr><td>Min OA air (% of total):</td><td>• Provide OA volume to comply with ASHRAE Standard 62-1999 or as required to maintain pressurization

• Provide OA based upon assigned RBEC level; refer to RBEC criteria – table II of RBEC guideline, April 2001, AHP EH&S Department

• 100% If solvents are present</td></tr>
<tr><td rowspan="5">Filtration required</td><td>Supply:</td><td>• 30%, 65%, and 95% in AHU with terminal 99.995%

• Series 99.995% HEPA filters may be required; refer to GES Standard STD-1-002 section – series filters Grade A/Class 100 areas</td></tr>
<tr><td>Exhaust:</td><td>• Normally none

• Provide filtration as required by assigned RBEC level; refer to RBEC criteria – table II of RBEC guideline, April 2001, AHP EH&S Department</td></tr>
<tr><td>Return:</td><td>• 99.995% terminal filters if cross contamination potential

• Provide filtration as required by assigned RBEC level; refer to RBEC criteria – table II of RBEC guideline, April 2001, AHP EH&S Department</td></tr>
<tr><td>Dust collection:</td><td>• Provide conventional dust collection system

• Provide exhaust filtration as required by assigned RBEC level; refer to RBEC criteria – table II of RBEC guideline, April 2001, AHP EH&S Department</td></tr>
</table>

(Continued)

Table 3.10 (Continued)

Miscellaneous	**Monitoring:**	Individual measurement and local light indication is required for each room for temperature, relative humidity, and differential pressure. Operating, alert, and alarm status must be indicated. Coordinate with design, normal operating, and operating range values as explained in global engineering standard STD-4-001 design versus operating range parameters
	Humidification: **Dehumidification:**	Pure steam Dependent upon site location and available chilled water temperature. Dessicant dehumidification may be required
	Remarks:	Provide additional 15% coil, fan, filter, and OA intake reserve system capacity

Standard no.		**General HVAC design criteria**	Page
Effective date	Revision no.	Supercedes document no.	Supercedes date

		**Facilities providing products for international distribution HVAC space criteria table ● GSC-INT-STD-6**
Area class	**ISO:** **EU:** **US:**	N/A N/A **Pharmaceutical grade**
Typical application	**Space description:**	*Manufacturing:* Oral solid dosage Blending, drying, granulation, milling, compression, primary packaging rooms and spaces with similar high dust producing activities
Air quality	**Temperature (°F):**	70 ± 2 $(21 \pm 1\,°C)$
	RH, maximum (%):	$50 + 5$
	RH, minimum (%):	$35 - 5$
	UNOCC temperature:	N/A
	Air changes per hour:	15 ACPH or higher if required to compensate for calculated heat load
	Pressure gradient:	0.03″ wg (7.5 Pa) setpoint, 0.02″ wg (5.0 Pa) minimum differential between corridors and process rooms, cGMP and non-cGMP spaces, and between like spaces if pressure gradient required Provide pressure gradient as required by assigned RBEC level if greater; refer to RBEC criteria – table II of RBEC guideline, April 2001, AHP EH&S Department
Recirculation criteria	**Supply 100% outside air:**	No
	Exhaust all air:	No
	Recirc allowed:	Yes
	Min OA air (% of total):	● Provide OA volume to comply with ASHRAE Standard 62-1999 or as required to maintain pressurization ● Provide OA based upon assigned RBEC level; refer to RBEC criteria – table II of RBEC Guideline, April 2001, AHP EH&S Department ● 100% if solvents are present

Table 3.10 (Continued)

Filtration required	**Supply:**	• 30%, 95%, 99.995% IN AHU
	Exhaust:	• Normally none • Provide filtration as required by assigned RBEC level; refer to RBEC criteria – table II of RBEC guideline, April 2001, AHP EH&S Department
	Return:	• None if dedicated to a single product with no RBEC level assigned • 99.995% terminal filters if cross contamination potential • Provide filtration as required by assigned RBEC level; refer to RBEC criteria – table II of RBEC guideline, April 2001, AHP EH&S Department
	Dust collection:	• Provide conventional dust collection system • Provide exhaust filtration as required by assigned RBEC level; refer to RBEC criteria – table II of RBEC guideline, April 2001, AHP EH&S Department
Miscellaneous	**Monitoring:**	Individual measurement and local light indication is required for each room for temperature, relative humidity, and differential pressure. Operating, alert, and alarm status must be indicated. Coordinate with design, normal operating, and operating range values as explained in global engineering standard STD-4-001 design versus operating range parameters
	Humidification: **Dehumidification:**	Clean steam. If in direct contact with product, pure steam is required. Dependent upon site location and available chilled water temperature. Dessicant dehumidification may be required
	Remarks:	Provide additional 15% coil, fan, filter, and OA intake reserve system capacity

cleanroom operation is dependent on effective HVAC design and operation, including cleanrooms in HVAC considerations is an acceptable approach.

The US standard for cleanrooms is defined by ISO 14644-1. Prior to 1 November 2001, the accepted reference was FED-STD 209E. While ISO 14644-1 is the current regulation, FED-STD 209E is still often used.

ISO 14644 defines a cleanroom as:

> A room in which the concentration airborne particles is controlled and which is constructed and used in a manner to minimize the introduction, generation and retention of the particles inside the room and in which other relevant parameters, e.g. temperature, humidity and pressure are controlled as necessary. Figure 3.2 illustrates a typical small modular cleanroom configuration. It is important to note the unit is a stand-alone facility.

As per ISO 14644-1, Section 3.2, cleanroom designation is predicated on:

- Classification number expressed as ISO class.
- Occupancy state.
- Particle size and their related concentration.
- Example – ISO Class 5 is approximately equivalent to Class 100, as per the former FED-Standard 209E, which translates to less than 3530 particles/m^3 or equal to greater than 0.5 μm at rest.

Determining the number of sampling locations is important in cleanroom design and specification. Accordingly, ISO 14644-1 provides a guideline for determining the approximate

Figure 3.2 Modular cleanroom. *Source:* Courtesy Clean Air Products.

number of sampling locations required for cleanrooms of a known area. Section B 4.1.1 notes that the number of cleanroom sampling locations is calculated by determining the square root of the cleanroom area with the area dimensions in m^2. It is, again, most important to note that the square root of the area is given in metric units (m^2). As an example, if a cleanroom has an area of $104 \, m^2$, the minimum number of sampling locations would be $(104)^{1/2} = 10.2 \approx 11$. Here, the value is rounded up to the next highest whole number if the value is fractional. The number of sampling locations calculated is a conservative value. A more representative value can be obtained by increasing the value by a factor of 1.2 or 1.3. For instance, a more realistic number of sampling locations would be $11 \times 1.3 = 14.3 \approx 15$. Since it is relatively inexpensive to install particle detectors, using sampling locations at somewhat higher than calculated values should not significantly affect the project cost.

Another general guideline would be to locate a minimum number of five sampling locations for the cleanroom area, even if the square root calculation indicates that fewer sampling locations are required. Required sample volumes per location (in liters) are given in B 4.2.1 of ISO 14644-1. Further, exceptions to the sampling volume per location are given in B 4.2.2 of ISO 14644-1. While the number of sampling locations required may vary by selected vendor/supplier, the values employed should reasonably approximate the number of sampling locations given by these guidelines.

When the number of sampling locations have been determined, it is now possible to ascertain the cleanroom classification and particle size requirements, the key criteria for cleanroom design.

Table 3.11 presents a summary of the ISO 14644-1 particulate cleanroom standards for most common particulate pharmaceutical cleanroom scenarios.

ISO 14644-1 cleanroom criteria for certain particulates defined Classes 5, 6, 7, 8, and 9 as the levels required for cleanrooms or aseptic areas. These are the design criteria levels that are specified in the design phase of the project.

Table 3.11 ISO 14644-1 selected cleanroom standards for particles/m^3.

ISO class	FED-STD 109E	Particle size (μm)					
		≥0.1	≥0.2	≥0.3	≥0.5	≥1	≥5
1	—	10	2	—	—	—	—
2	—	100	24	10	4	—	—
3	1	1000	237	102	35	8	—
4	10	10000	2370	1020	352	83	—
5	100	100000	23700	10200	3520	832	29
6	1000	1.0×10^6	29000	102000	35200	8320	293
7	10000	—	—	—	352000	83200	2930
8	100000	—	—	3.27×10^6	3.52×10^9	832000	29300
9	—	—	—	—	3.52×10^8	8.32×10^6	293000

While Table 3.9 is a basis for cleanroom standards, generally, there are no particulate classifications for non-aseptic facilities in the United States.

While there are no specific linear airflow velocities defined for ISO classes 8, 7, and 6, Class 5 has a range that varies from 40 to 90 ft/s (0.2–0.5 m/s, respectively), and Class 4 is in the range of 60–90 ft/s (0.3–0.5 m/s, respectively). As design guidelines, the following hourly air changes can be used as a design basis for the common ISO classes:

ISO Class 8-10-20 ACPH
ISO Class 7-30-70 ACPH
ISO Class 6-70-160 ACPH

ISO Class 5 to ISO Class 1 cleanrooms are predicated on linear airflow through the room measured in meters/second. For ISO Class 5, it is recommended that a range of 300–400 ACPH be used.

For more specific design applications, the following relationships are useful [3]:

$$\text{Cleanroom recirculation rate, ft}^3 / \min = \left(\text{Room width, ft}\right) \times \left(\text{Length, ft}\right)$$
$$\times \left(\text{Height, ft}\right) \times \left(\text{Air change rate, min}^{-1}\right) \tag{3.11}$$

$$\text{Filter count}\left(\text{number of filters}\right) = \frac{\text{Cleanroom area}\left(\text{ft}^2\right) \times \left(\%\,\text{coverage}\right)}{8} \tag{3.12}$$

For ISO Class 4 (FED-STD 109E, Class 10), use an ultralow penetration air (ULPA) filter, 99.9997% ≥0.12 μm particle size
For Class ISO Class 5 (FED-STD 109E, Class 100), use a HEPA (High Efficiency Particulate Absolute) filter, 99.99% ≥0.3 μm particle size.
For ISO Class 6 (FED-STD 109E, Class 1000), use a 99.99% HEPA filter, ≥0.3 μm particle size.
For ISO Class 7 (FED-STD 109E, Class 10000), use a 99.97% HEPA filter, ≥0.3 μm particle size.

Note that for this calculation, the percent coverage is given as the fractional equivalent (i.e. 50% = 0.50).

When using Equation (3.8), the following values can be used for % coverage:

ISO Class 4 (FED-STD 109E, Class 10) Coverage = 100%
ISO Class 5 (FED-STD 109E, Class 100) Coverage = 50–70%
ISO Class 6 (FED-STD 109E, Class 1,000) Coverage = 20–30%
ISO Class 7 (FED-STD 109E, Class 10,000) Coverage = 7–15%

$$\text{Percent}\,(\%)\,\text{filter coverage} = \frac{\text{Filter count}}{\text{Cleanroom area}\,(\text{ft}^2)} \times 8 \times 100\% \tag{3.13}$$

$$\text{Air changes per minute} = \frac{\left(\text{Filter count}\right) \times \left(720\text{ft}^3\,/\,\text{min}\right)}{\text{Cleanroom volume}\,(\text{ft}^3)} \tag{3.14}$$

The information provided above, while valid for most design applications, is intended to be used by specialists experienced in cleanroom design. However, this information should be sufficient for the customer to use this material to confirm that vendor designs are reasonable and conform to user specifications and design intent.

Example 3.6 A generic manufacturer has determined that there is a need for a small instrument calibration and instrument standard storage cleanroom for a vial fill and finish line. The cleanroom dimensions are 40 ft × 25 ft × 8 ft in height. It was concluded that an ISO Class 6 cleanroom would be acceptable as a design criteria. Prior to going out for bid, the user desires approximate requirements for the project. These requirements include:

1) Number of filters required
2) Air changes per minute

Solution
Prior to performing the required calculations, it would be advantageous to determine the cleanroom area and volume:

A. Cleanroom area $= \left(40\text{ft}\right) \times \left(25\text{ft}\right) = \mathbf{1000\,ft^2}$

B. Cleanroom volume $= \left(40\text{ft}\right) \times \left(25\text{ft}\right) \times \left(8\text{ft}\right) = \mathbf{8000\,ft^3}$

Number of filters required
Equation (3.12) is the applicable relationship. For this situation, 25% (0.25) coverage is to be used:

$$\text{No. of filters required} = \frac{\left(1000\text{ft}^2\right)(0.25)}{8} = 31.25 \approx \mathbf{32\ filters\ required} \leftarrow \mathbf{Solution}$$

Hourly air changes
For this determination, a modified form of Equation (3.11) will be used:

$$\text{Hourly air changes} = \frac{\left(32\,\text{filters}\right)\left(720\text{ft}^3\,/\,\text{min}^{-1}\right)\left(60\text{min/h}\right)}{8000\text{ft}^3} = 172.8 \approx \mathbf{173\ changes\,/\,h} \leftarrow \mathbf{Solution}$$

The result conforms to the recommended hourly air change rate of 70–180 ACPH for a Class 6 cleanroom as given above.

It is important to note that the majority of HVAC calculations, as is the case with most pharmaceutical engineering problems, involve basic relationships that can be solved by most technical individuals. For more difficult scenarios, consultants with a more in-depth background are often employed. Unlike graduate school problems, in-depth solutions are often not required since it is not cost effective to rely on basic technical personnel to solve esoteric problems, which rarely occur in standard manufacturing, operating, and design functions.

Depending on the type of air quality that must be produced, a relationship exists, which allows for a calculation of the particle concentration in a cleanroom volume. The empirical relationship is given as:

$$C_n = 10^N \left(\frac{0.1}{D} \right)^{2.08} \tag{3.15}$$

$$C_n = \frac{\text{Concentration of particles}}{\text{m}^3 (\text{air})}$$

where

N = ISO classification less than or equal to 9 (Table 3.9)
D = particle size, mm (for most bacteria, the range is 0.2 mm–2.0 μm (0.002 mm); for most fungi, 3.8 μm (0.0038 mm); and for most viruses, 0.07 μm (0.00007 mm)–0.11 μm (0.00011 mm))

Cleanrooms are designed to maintain a slight positive pressure, relative to the surrounding area. One exception to these criteria is the case where a negative pressure room is installed in locations where infectious diseases, certain pathogens, biocontaminants, identified hazardous chemicals, flammables, identified explosive compounds, and powders are used as part of the manufacturing operations. In this scenario, the concern is containment but rather keeping these "bad actors" from rapidly discharging.

A fundamental design installation and operating standard is to remember that a major factor in obtaining a desirable cleanroom class is airflow. Introducing clean air through proper use of HEPA filters and proper airflow (introduced from the ceiling and vented through the lower wall or floor) and designing for the highest air change rate are standard techniques that can achieve the goal of a properly designed cleanroom.

While there are no regulations regarding the level of positive pressurization required, pressurization levels are important to prevent cross contamination of products in a multiproduct facility of concern. In such a case, room relative pressure may be critical to the product or constituents if

- Product or components are in a dry or powder form; it is possible that the material may be transported, as particulates, using room air to nearby or adjacent product areas. Also, transport of some vapors (e.g. vapors discharged from residual solvents) can also be carried into the air stream to adjacent or nearby product manufacturing operations such as concurrent tablet presses or pan coating operations. While this is a low probability for newer facilities, older operations with planned HVAC installations must use careful planning to preclude this scenario. It is also very important to closely coordinate all stages of cleanroom construction with the vendors or in-house operations performing the task.
- Presence of airborne particulates or volatiles that may present a personnel exposure risk. While such a scenario seems implausible, repetitive operations over an extended timeframe enhance the possibility.
- Failure to properly monitor the room prior to air balancing.
- Failure to properly monitor airflows at room entry areas.

When designing and planning a cleanroom, many difficulties can be avoided by considering and/or implementing the following construction and cleanroom installation issues [4].

- Application – A thorough analysis should be performed when committing to a cleanroom space. As a general rule, there are only two reasons for a cleanroom installation, better quality and better yield.
- Again, it is common practice to maintain a positive pressure difference in cleanrooms. While newer installations are designed to introduce airflow to enter at the ceiling level and discharge, subsequent to circulating throughout the cleanroom volume, it is good design practice to locate the discharge vents at the lower portions of the room so that contaminants will discharge with the inlet airflow.
- Cleanroom component placement and modularity – If the possibility exists that a cleanroom installation may be expanded at some time, employing a modular design might be economically and operationally worthwhile. Modular design elements could include factors such as mobile equipment with lock-in-place casters allowing for easier movement, enclosing a larger space than currently anticipated to allow for future expansion, yet keeping the expanded space separate from the existing space via an easily disassembled barrier or possible tax advantages by installing a modular cleanroom (7-year amortization vs. a longer period for an "as-built" permanent operation). For a new cleanroom operation, modular installations could prove to be quite effective, particularly if the manufacturing operation is not well defined when dealing with production quantities in the planned cleanroom. Modularization can also be worth considering for specific operations that are not planned to be permanent. For instance, if a small generic manufacturer or contract research operation (CRO) is tasked to do early phase clinical testing on a vaccine product, a manual vial inspection line meeting ISO Class 5 standards may be required. Since the line may only be required for a couple of years, it might be advisable to consider a small modular cleanroom for this particular operation. While cGMP criteria such as a biologics license application (BLA) and qualification would be required, the cost savings presented by the modularization and associated tax benefits may make the activity cost effective.
- Approved operating procedures – As is the case with all cGMP activities, personnel should be trained and certified for cleanroom operations. Entry locations, changing rooms, and cleanroom clothing locations must be clearly identified. This is most important since workers represent the largest source of contamination for cleanroom spaces. Regarding operating procedures, comprehensive cleaning procedures should be given highest priority before, during, and subsequent operating periods. Also, in addition to having properly installed utilities such as lighting and other fixtures, approved flooring and ceiling materials should also be incorporated into construction in specifications. Particular attention should be given to ensure proper materials are planned for specified cleanroom classes. For instance, Class 10 ceiling panels should be fabricated of vinyl or epoxy-painted components, whereas Class 100 tiles should be a material identified as cleanroom tile. This tile, cleanroom tile, is also the type to be used in Class 1000 and Class 10 000 cleanrooms. Another installation that should be installed according to ISO class is floor coverings. Here, attention should be given to details such as using sheet vinyl or poured epoxy for Classes 10, 100, 1000, and 10 000 installations.
- Operation and maintenance – Maintenance personnel should be an integral part of the user's project team during the planning and installation phase. Their input is necessary to assist in confirming the equipment layout. Previously noted ergonomic factors, such as clearances and maintainability, should be reviewed for operational efficacy. Maintenance personnel should also be involved in the specifications and details such as location of installation

components such as windows, doors, and cleanroom utilities use points. While this is generally the responsibility of the construction manager and the architect, the user/owner is ultimately responsible for approval and signoff of the installation.

- Carefully review user requirements and specifications – It is not uncommon for installed cleanroom components to be at odds with the user requirements. Failure to adequately address these issues in the design and planning phase can ultimately result in an installation that will not conform to performance criteria. As a result, additional installation costs requiring engineering change orders can often occur.
- Off-site construction – For some installations, employing off-site or prefabricated components (modular) should be evaluated. Of course, a decision such as this must be considered early in the design and user requirements phase of the project. Such an approach offers several advantages. One major positive is the reduction of the need for component and warehousing of certain equipment. Through advanced planning and logistics, system components can be provided at the time of module assembly, thus eliminating or significantly reducing the aforementioned storage requirements.
- Floor plans – In addition to being an important accoutrement to office walls, the floor plan serves several key functions. One important activity is to verify that walls, ceilings doorways, and similar architectural components do not interfere with equipment installation. The layout drawings should be employed to confirm that electric vehicles carrying components to construction/installation locales have enough clearance to allow the forklift to maneuver while carrying equipment on the lift. Here, cardboard or plastic cutouts made to scale can be useful when evaluating equipment delivery, movement, and placement. Of course, 3D CAD drawings in real-time manipulation capability would be a more effective technique, if costs allow.
- Continuous review of cleanroom utilities components – Installing a 25 hp motor may cost less than a 30 hp motor, but it may ultimately create an operating problem by delivering a lower operating pressure head than required. Changing equipment specifications subsequent to specification issuance can be very costly unless carefully evaluated. If components such as variable frequency drives (VFD) are integral to the 30 hp specifications, the motor change may require expensive reprogramming of the VFD control software. More importantly, this motor modification can create a significant project delay. Consequently, utilities and equipment alterations should be carefully evaluated before implementing significant changes during the construction phase. For a cleanroom application, any change could be a significant unplanned cost increase.
- Verify materials of construction specifications – For example, if the specifications require that flooring materials must conform to the Americans with Disabilities Act (ADA) guidelines for the coefficient of friction, approved documentation confirming this requirement should be readily available from the OEM or system construction management. A statement provided by the subcontractor and not an approved source, other than the manufacturer, should not be accepted as an equivalent. Again, for a cleanroom, a specification modification, though minor, can result in a significant unplanned for cost increase.
- Carefully evaluate interfaces between cleanroom classifications – If going from one class to another, verify that the design includes change out facilities, if required, that do not compromise the classification. This is particularly important if a barrier isolation installation is involved, and a maintenance operation must be performed within the location of concern.

Another important source describing recommended practices for the design, testing, and product reliability, as well as contamination minimization practices for cleanrooms, is a non-profit organization known as the Institute of Environmental Sciences and Technology (IEST). As part of their function, IEST publishes and disseminates a number of documents that provide guidance to organizations involved in the various aspects of cleanroom operations described above.

The guidance given above serves as a basic checklist for cleanroom construction design and construction considerations. However, while cleanroom construction design and criteria are extremely important, continued maintenance and satisfactory operation, subsequent to initial startup, is critical to successful facility operation and profitability. Consequently, a brief incorporation of relevant standard guidance and practices would be helpful in the operation and maintenance of cleanroom installations. Several of the IEST documents deal with contamination control and parallel the standards set forth in ISO 14644-1 and ISO 14644-2. These publications include:

- *IEST-RP-CC001.5 – HEPA and ULPA Filters* (HEPA, high efficiency particulate air; ULPA, ultra low penetration air).
- *IEST-RP-CC002.3 – Unidirectional-Flow Clean-Air Devices.*
- *IEST-RP-CC006.3 – Testing Cleanrooms.*
- *IEST-RP-CC007.3 – Testing ULPA Filter.*
- *IEST-RP-CC012.2 – Considerations in Cleanroom Design.*
- *IEST-RP-CC0018.4 – Cleanroom Housekeeping Operating and Monitoring Procedures.*
- *IEST-RP-CC0019.1 – Qualifications for Organizations Engaged in the Testing and Certification of Cleanrooms and Clean-Air Devices.*
- *IEST-RP-CC0021.3 – Testing HEPA and ULPA Filter Media.*
- *IEST-RP-CC022.2 – Electrostatic Discharge in Cleanrooms and Other Controlled Environments.*

IEST also has documentation designed to assist in the implementation of maintenance and operation of functioning cleanrooms. Technical guides and documents intended to enhance cleanroom testing and operations are:

1) *IEST-RP-CC026.2 – Cleanroom Operations.*
2) *IEST-G-CC1002 – Determination of the Concentration of Airborne Ultrafine Particles.*
3) *IEST-G-CC-1004 – Sequential-Sampling Plan for Use in Classification of the Particulate Cleanliness of Air in Cleanrooms and Clean Zones.*
4) *IEST-G-CC-1003 – Measurement of Airborne Macroparticles.*
5) *IEST-G-CC-1001 – Counting Airborne Particles for Classification and Monitoring of Cleanrooms and Clean Zones.*
6) *IEST-RP-CC031.2 – Method for Characterizing Outgassed Organic Compounds from Cleanroom Materials and Components.*

Item 6, IEST-RP-CC031.2, has taken on added significance of late since several organizations question the health effects of compounds that may be emanating from specific construction and installation materials containing quantities and concentrations of known hazardous materials such as residual formaldehyde.

While it is unlikely that the site engineering, operations, and maintenance organizations in a facility will be significantly involved in the guidelines and documentation described above, contractors, subcontractors, and the construction management firm should have an in-depth knowledge of the germane aspects of the documentation required for the selected crafts performing the operation or installation task. Additionally, The IEST and associated materials described above should be clearly referenced in bid specification documents prior to contract award. Potential contractors and subcontractors should be capable of providing valid qualifications relating to their ability to perform the work scope required in the reference of IEST documents [5].

3.3.2 Cleanroom Monitoring, Maintenance, and Design Considerations for USP <797> and USP <800> Facilities

This section addresses two USP monographs that will affect handling, storage, and packaging of drugs that are compounded in specific healthcare facilities rather than pharmaceuticals manufactured in FDA-regulated operations.

3.3.2.1 USP <797>

While there is a significant body of data and design information relating to cleanroom particulates, little information is provided concerning microbial monitoring and analysis, specifically the growth of colony-forming units (CFU) for facilities not covered by cGMP and related FDA regulations. Operations involved in compounding of pharmaceuticals, as opposed to manufactured pharmaceuticals, are prepared for direct use by patients and are compounded by a state-registered pharmacist in facilities designed for that purpose. Nevertheless, addressing compounding considerations for direct use by patients should be a topic for consideration in pharmaceutical manufacturing, albeit in limited quantities of product.

No formal cleanroom CFU regulatory information/criteria are in place for compounding preparation cleanrooms; there are recommended guideline action levels for CFU presence specific to ISO Classes 5, 7, and 8 (or greater) for airborne sampling and surface sampling [6]. Table 3.12 presents recommended action levels for microbial contamination based on air sampling using collection plates. The CFU count is based on a cubic meter as a standard volume.

Recommended procedures are also in place for monitoring microbial activity for cleanroom surface samples. Table 3.13 provides guidelines for microbial surface sampling. The recommended action levels are put forth in USP <797> (specifics of <USP> 797 are provided in the paragraph below): semiannual microbial and particulate sampling for cleanroom venues. However, the action levels are recommendations and actual specification ranges and limits that are the responsibility of the facility owners, and monthly microbial testing should be considered.

While basic cleanroom design fundamentals have been addressed primarily from the engineering aspect in Section 3.3.1, two USP monographs exist to verify compliance and certification of cleanroom design and operation. The two monographs are identified as:

USP <797>, *Pharmaceutical Compounding-Sterile Preparations*, May 2016, and USP <800>, *Hazardous Drugs – Handling in Healthcare Settings*, April 2016. Essentially, the intent of these documents is to confirm the design adequacy for cleanroom measurements involving airflow, leakage testing, and environmental measurements (differential pressure, temperature, and relative humidity) for operations specific to drug compounding facilities

Table 3.12 Recommended action levels for microbial contamination.

Classification	CPU count
ISO Class 5	>1
ISO Class 7	>10
ISO Class 8 or worse	>100

Source: Courtesy of Champion Aire.

Table 3.13 Surface sampling.

Classification	Fingertip sample	Surface sample
ISO Class 5	>3	>3
ISO Class 7	N/A	>5
ISO Class 8 or worse	N/A	>100

Source: Courtesy of Champion Aire.

prepared by state-licensed pharmacists. As was briefly alluded, USP <797> and USP <800> are not FDA regulations, implying recognition, monitoring activities the responsibility of local, state, and federal regulatory agencies to effectively implement these monographs (USP <797> and USP <800>).

In addition to addressing preparation of compounded sterile pharmaceuticals (CSPs), USP<797> includes commercial and hospital pharmacies, doctor's offices and other clean area designs, storage specifications, and quality assurance plans, which include environmental monitoring to include the safe handling of these preparations. USP <797> specifically addresses:

1) Microbial contamination and USP sterility test.
2) Endotoxins, USP bacterial test.
3) Physical or chemical contamination.
4) Preparation of incorrect potency confirmation.
5) Incorrect ingredient.

As can be seen, bacterial contamination is a critical variable for safe handling operations. Effective evaluation of potential contamination and contamination routes should involve, as a preliminary procedure, an effective compounding cleanroom sampling plan, using Tables 3.12 and 3.13 as references.

Subsequent to sample collection, the sample then undergoes incubation for a specific time period and then analyzed as to identity (bacillus, streptococcus, etc.) with the identification performed by a qualified professional.

Sampling procedures are common techniques employing collection plates such as RODAC plates or equivalent petri dish microbe collection units. For surface monitoring, swabs are often the preferred sampling tool.

3.3.2.2   USP <800>

- USP <800> (with an implementation date of 1 July 2018) is intended to describe quality standards for handling hazardous drugs (HDs) in healthcare settings and help promote patient safety, worker safety, and environmental protection. USP <800> identifies HDs by using the NIOSH drugs that are potential threats to human and animals. The HDs are based on:
- Carcinogenicity.
- Teratogenicity (developmental toxicity).
- Reproductive toxicity.
- Organ toxicity at low doses.
- Genotoxicity (chemical agents that cause cell mutations).
- Structure and toxicity profiles of new drugs that mimic existing HDs.

The responsible entity must maintain a list of HDs, antineoplastic (preventing or inhibiting the growth and spread of tumors and malignant cells), and other HDs used in healthcare. The entity must maintain a list of HDs, which may include items on the current NIOSH list in addition to other agents not on the NIOSH list. The list must be reviewed at least annually and whenever a new agent or dosage form is used.

The responsible entity that handles HDs must incorporate the standards of USP <800> in their occupational safety plan. Additionally, the entity's health and safety plan should, at a minimum, include:

- Engineering controls.
- Competent personnel.
- Safe work practices.
- Proper use of personal protective equipment (PPE).
- Safe work practices.
- Policies for waste segregation and disposal (this activity should be performed in accordance with approved SOPs).

UPS <800> recommends but does not require sampling and analysis for HD presence. If sampling is performed, it is suggested that wipe sampling kits be used.

One important detail noted in <USP> 800 is the importance of compounding HD procedures. While UPS <797> identifies compounding in healthcare environments, the emphasis is on microbial analysis. For UPS <800> compounding criteria for HDs is most germane. When dealing with <USP> 800, compounding and other HD manipulations include:

- Crushing or splitting of tablets.
- Opening capsules.
- Pouring oral or topical fluids from one container to another.
- Weighing or mixing components.
- Constituting or reconstituting powder or lyophilized HDs.
- Withdrawing or diluting injectable HDs from parenteral containers.
- Expelling air or HDs from syringes.
- Contacting HD residue present on PPE or other garments.
- Deactivating, decontaminating, cleaning, and/or disinfecting contaminated or suspected contaminated HD areas.
- Maintenance activities for potentially contaminated equipment and devices.

For HDs that require refrigeration, the HDs should be stored in a refrigerator dedicated to HD storage and located in a room with a negative pressure differential.

External ventilation units should be installed in affected areas.

Existing cleanrooms complying with UPS <797> or UPS <800> were generally not designed with centralized cleanroom monitoring.

> Small cleanrooms also did not invest in smart fan filter units (FFU) centralized air controls (a typical FFU is less than 1 hp). Air exchanges (ISO 7 or ISO 8 levels) or exhaust (for negative pressures as specified for USP 800) hazardous material compounding are manually adjusted by the installer/certifier to conform to compliance standards [7].

For venues lacking airflow monitoring devices, continuous differential pressure monitoring is an effective monitoring tool. Frequently scheduled maintenance and inspection of the FFU(s) for blockage or an unexpected large differential pressure drop is often a problem. The maintenance SOP should indicate the differential pressure range where filter replacement should be performed. Maintenance activities should include documenting differential pressure drops during the monitoring activity. Installation of a PC-based control, monitoring, and alarm system should be evaluated if maintenance frequency, repair frequency, or parts replacement becomes a significant cost consideration. However, for smaller cleanroom monitoring requirements that use only a few FFUs, manual monitoring with frequent differential pressure checks and documentation may be sufficient.

As is the case with UPS <797>, biannual certification of existing facilities is recommended.

References

1 Ayers, S. and Fullerton, J.L. (2012). Reducing HVAC noise with duct liner. *HPAC Engineering* 84 (1): 58–63 (January).
2 ASHRAE (2011). *ASHRAE Handbook – HVAC Applications.* Atlanta: American Society of Heating Refrigerating and Air Conditioning Engineers.
3 Simplex Isolation Systems (2012). What to know when considering a cleanroom. *Controlled Environments* 84 (2): 15–19 (February). SimplexIsolationSystems.com (accessed 12 December 2017).
4 Parsons, H. (2011). New ISO cleanroom standards: What will they mean for pharma? *Controlled Environments* 14 (3): 12–19 (March). www.cemag.us (accessed 12 December 2017).
5 Eudy, J. (2011). An overview of contamination control guidance documents. *Controlled Environments* 14 (5): 8–11.
6 Morris, S. (2016). Microbial monitoring in cleanroom settings. *Controlled Environments* 19 (5): 14 (November/December).
7 Abramowitz, H. (2016). Monitoring considerations for pharmaceutical cleanrooms. *Controlled Environments* 19 (2): 14–15 (March/April).

Further Reading

Abramowitz, H. (2016). Outfitting a cleanroom with monitoring systems. *Controlled Environments* 19 (3): 27 (May/June).
ASHRAE (1996). *ASHRAE Guideline 1-1996, HVAC Commissioning Process.* Atlanta: ASHRAE.
Brumbaugh, J.E. (2004). *HVAC Fundamentals*, vol. 2, 4e. Indianapolis: Wiley Publishing, Inc.
Burroughs, H.E. and Hansen, S.J. (2008). *Managing Indoor Air Quality.* Lilburn, GA: Fairmont Press, Inc.
Curtiss, P.S. and Breth, N. (2004). *HVAC Instant Answers.* New York: McGraw-Hill.
Department of the Navy (1969). *Basic Electricity*, NAVPER 10086-B. Washington, DC: Department of the Navy.
Flowstar Inc. (2011). *Cleanroom Design Guidelines.* Gilroy, CA: Flowstar Inc. www.flowstar.com/cleanroom_design_criteria.html (accessed 12 December 2017).
HVAC equipment to see 4.5% gains thru 2013 (2009). *Flow Control* 15 (11): 10 (November).
Koller, N. (2002). Presentation for the North Jersey Section, American Institute of Chemical Engineers, CE & IC, Burlington, NJ (April 2002).
Kravitz, R. (2016) A new way to clean cleanrooms. *Controlled Environments* 19 (4): 8–9 (July/August).

Lish, K.C. (1972). *Nuclear Power Plant Systems and Equipment.* New York: Industrial Press.

Mansoor, A. (2002). Understand AC induction motor vibration. *Hydrocarbon Processing* 81 (3): 71–78 (March).

Marks, L. and Baumeister, T. (1979). *Standard Handbook for Mechanical Engineers*, 8e. New York: McGraw-Hill.

Office of the FED-Register, US Government Printing Office (1984). *Code of FED-Regulations, 29, 1900–1910.* Washington, DC: US Government Printing Office.

Spradlin, M. and Rose, L. (2011). Cleanroom construction. *Pharmaceutical Manufacturing* 10 (3): 31–34 (March).

Strong, M. (2016). Design and construction standards when building a cleanroom. *Controlled Environments* 19 (5): 6–9 (September/October).

Tannenbaum, D. (2000). *The Air Conditioning/Refrigeration Toolbox Manual.* Lawrenceville, NJ: ARCO.

What you need to know about humidity (2011). Vaisala Corporation, Humidity 201 Seminar, Fairfield, NJ (22 June 2011). www.vaisala.com (accessed 12 December 2017).

4

Pressure Vessels, Reactors, and Fermentors

A significant quantity of pharmaceutical and biotech equipment is intended for holding, processing, or transferring purposes, often in the liquid or gas/vapor phase. This chapter is intended to present fundamental design and operating information to understand and apply the significant elements involved in these widely used components for general and pharmaceutical processing applications.

4.1 Introduction

While pressure vessels, reactors, and fermentors may appear to be little more than tanks or tanks with agitators in the body, there are detailed design and operating requirements employed to verify successful installation and performance of these vessels. In addition to desiring economic operation and design of these components, there is the overriding element of safety, a most important element of pressure vessels, reactors, and fermentors.

4.1.1 Pressure Vessels

Pressure vessels are employed in both cGMP pharmaceutical applications and non-cGMP pharmaceutical manufacturing operations. Non-cGMP installations located at pharmaceutical manufacturing facilities in which pressure vessels are most likely to be encountered include HVAC systems (Chapter 3) and facility utilities such as nitrogen tanks and inert gases. Here, the HVAC systems can be classified as both cGMP and non-cGMP. Boilers that provide central heating facilities such as to general office space and warehouse operations are not regulated by cGMP regulations. Similarly, production reactors and laboratory reactors, under pressure, are regulated by cGMP criteria. Nitrogen and inert gases are used as "Blanket gases to prevent oxygen from coming in contact with constituents of pharmaceutical manufacturing."

Practical Pharmaceutical Engineering, First Edition. Gary Prager.
© 2019 John Wiley & Sons, Inc. Published 2019 by John Wiley & Sons, Inc.

While not always referred to as pressure vessels, many pharmaceutical systems are designed and operated in accordance with regulations and criteria set forth in a national set of regulations identified as the American Society of Mechanical Engineers Pressure Vessel Code, or more commonly referred to as the ASME Code. This code, or series of regulations, was promulgated in 1914. Currently, this is revised every 3 years and addenda are published annually.

Unlike most standards and specifications produced by many professional organizations, many sections of the ASME Code have the effect of law in of the United States and Canada, depending on the state. As a result, organizations, such as insurance companies, are usually relied upon to enforce pressure vessel design, specifications, installation, and inspections. Inspection of pressure vessels is the responsibility of the National Board of Boiler and Pressure Vessel Inspectors (NBBI). Each state has a responsibility for vessel inspections in their jurisdictions. Pressure vessels that have been approved by the NBBI have a unique number associated with the vessel and is identified on the vessel and all relevant vessel drawings.

The Pressure Vessel Code is divided into 11 sections, each encompassing specific elements of pressure vessel design, fabrication, installation, and inspection. The 11 sections are:

I – Power Boilers
II – Material Specifications
    Part A – Ferrous
    Part B – Nonferrous
    Part C – Welding Rods, Electrodes, and Filler Metals
III – Nuclear Power Plant Components
    Subsection NA – General Requirements (Including Appendices)
    Subsection NB – Class 1 Components
    Subsection NC – Class 2 Components
    Subsection ND – Class 3 Components
    Subsection NE – Class MC Components
    Subsection NF – Component Supports
    Subsection NG – Core Support Structures
IV – Heating Boilers
V – Nondestructive Examination
VI – Recommended Rules for Care and Operation of Heating Boilers
VII – Recommended Rules for Care of Power Boiler
VIII – Pressure Vessels, Division 1
    – Pressure Vessels, Division 2, Alternative Rules
IX – Welding and Brazing Qualifications
X – Fiberglass Reinforced Plastic Pressure Vessels
XI – Rules for Inspection of Nuclear Power Plant Components

In addition to pressure vessels, the code also provides design criteria and regulatory provisions for components commonly associated with pressure vessel installations. These additional calculations and criteria are:

- ASME B31.1 – Power Piping
- ASME B31.3 – Process Piping
- ASME B31.4 – Hydrocarbon Piping
- ASME B31.4 – Refrigeration Piping
- ASME B31.8 – Gas Transmission Piping
- ASME B31.11 – Slurry Piping
- ASME B16.5 – Flange and Flange Fittings

- ASME B16.47 – Large Steel Flanges
- ASME B16.1 – Grey Iron Flange & Flange Fittings
- ASME B16.3 – Malleable Iron Flange & Flange Fittings
- ASME B16.11 – Forged Fittings
- MSS-SP-67 – Butterfly Valves

Pressure vessels and associated components employed in pharmaceutical cGMP applications are regulated by the criteria set forth in ASME Section VIII, Division 1.

For pressure vessels, including heat exchangers operating between 15 and 3000 psig and a temperature range that varies between –50 and 1000 °F, there are code sections that are frequently employed in pharmaceutical vessels. For the most part, pressure vessels designed for pharmaceutical service operate at pressures up to 300 psig and at maximum temperatures of about 300 °F (150 °C). For vessels that are designed to operate at below 15 psig, commonly referred to as storage tanks, there is another guideline prepared by the American Petroleum Institute and tanks operating under this guideline and are referred to as API-620 tanks, named after the American Petroleum Institute guideline, API-620. However, the concerns of this section are pressure vessels that are designed and operated in accordance with provisions of Section VIII. These Section VIII subsections include:

- Subsection A – This part of the code identifies general requirements, including relief devices.
- Subsection B – This subsection primarily involves fabrication methods such as welding (Part UW), riveted fabrication (Part UR), and brazing (Part UB).
- Subsection C – Subsection C deals with pressure vessel materials of fabrication. Applicable areas include UCS (carbon and low alloy steels), UNF (nonferrous metals), UHA (high alloy), UCI (cast iron), UCD (cast ductile (metals that can be deformed)), UCL (clad), and UHT (heat-treated iron-containing (ferrite) materials).

When dealing with pharmaceutical applications, the most commonly encountered pressure vessels are reactors and certain types of storage vessels that are regulated by the ASME Code. Specifically, vessels that operate under a pressure of greater than $15 \, lb/in^2$ are regulated by the ASME Code.

An important point concerning relief devices is to note that relief devices are not required for vessels less than 6 in. in diameter and a volume of less than $10 \, ft^3$. For vessels, tanks, and other similar components operating under an external pressure (i.e. a vacuum), the ASME Code is also applicable.

As is the case with most specialties, pressure vessels have a specific terminology. Some of the relevant terms for pressure vessels are given below. A more comprehensive compilation of terms can be obtained in ASME Section VIII, Division 1 of the code. The terms below will be further explained as they are used in the chapter. Relevant pressure vessel terminology includes the following:

- Required thickness – The minimum vessel thickness is determined by application of the formulas contained in the code. The calculated thickness is exclusive of any corrosion allowance.
- Design thickness – The minimum required thickness plus the corrosion allowance.
- Design pressure – The intended operating pressure of the vessel; the design pressure is based on the minimum allowable thickness. The design pressure is also the basis for determining at what pressure relief devices actuate.
- Maximum allowable working pressure (MAWP) – The maximum allowable pressure at which a code vessel can operate. It must be emphasized that, as is the case with most code pressures, the MAWP is noted as gage pressure, not absolute pressure. For absolute pressure an additional, $14.7 \, lb/in^2$ (15 psi) must be added to the MAWP value.

- Nominal thickness – A commercially available plate pipe or sheet that identifies the vessel fabrication material. The code specifies the nominal thickness selected to conform to the pressure requirements of the vessel.
- Operating pressure – The normal pressure at which the vessel operates. The operating pressure is measured at the top of the vessel. Normally, the operating pressure is below the set pressure of the protective devices such as relief devices and/or rupture discs associated with the vessel. This information (the set pressure and/or the operating pressure as well as the relief device setting) should be identified in P&ID's supplied by the system vendor.

For pharmaceutical applications, there are often used parts of Section VIII, Division 1 that are most commonly employed for pressure vessels and other pressurized components such as heat exchangers (Section 2.2.2). These code regulations include:

- Subsection A – Part UG that describes general provisions of vessel requirements such as materials, external loading considerations, and relief devices.
- Subsection B – This part of the code provides details covering fabrication techniques, welds (UR), and riveting techniques. Also covered in this subsection are forged criteria (UB) and brazing procedures (UB).

4.1.2 Basics of Pressure Vessel Design and Specifications

Design of pressure vessels is most often performed by firms specializing in the fabrication of pressure vessels. Subsequent to installation and operation, the operator/owner is responsible for maintaining the pressure vessel. Similarly, scheduled maintenance, inspection, and repairs are operator/owner responsibilities. While design is most often contracted to vessel fabricators, it is germane that pharmaceutical manufacturing (typically, the facility engineering and maintenance operations) staff have a basic understanding of the design elements involved in the pressure vessel.

Perhaps the most important design requirement for pressure vessels is the vessel wall thickness, since the function of the vessel is to safely contain fluids at an elevated pressure (>15 psig). The ASME Pressure Vessel Code provides calculations to achieve these tasks in a uniform and consistent manner that, if correctly designed and implemented, are legally operable in the planned environment.

A primary design criterion of pressure vessels is determining the vessel wall thickness. This result is basically a function of two variables, the material properties of the vessel material and the stress or force per unit area exerted by the fluid in the vessel. Figure 4.1 is a typical pressure vessel often seen in industrial operations such as pharmaceutical and biotech operations. On occasion, pressure vessels can be located outdoors. Typically, the outdoor pressure vessels contain inert gases such as nitrogen at cryogenic temperatures or argon and helium at non-cryogenic temperatures. These gases often serve as "blanket gases" that are intended to cover, or blanket, specific compounds that, when exposed to the environment, can oxidize or otherwise alter the constituents.

The critical material value required to determine the allowable stress is the yield strength of the design material. For cGMP pressure vessels, the material employed for fabrication is generally a stainless steel. While there are several types of stainless steel, the two most commonly employed stainless alloys are 304L and 316L. The term employed for the strength of alloys and metals such as 304L and 316L is the yield strength or ultimate yield strength of the alloy for metals intended for pressure vessel service; the term stress is employed in lieu of yield strength. The units of stress are, in engineering units, generally recorded in values of psi (lb/in^2), but often the term ksi or thousand lb/in^2 is used. Thus, a stainless steel such as 301 stainless steel plate can have a recorded yield strength of 40 000 psig or 40 ksi.

Figure 4.1 Typical pressure vessel.

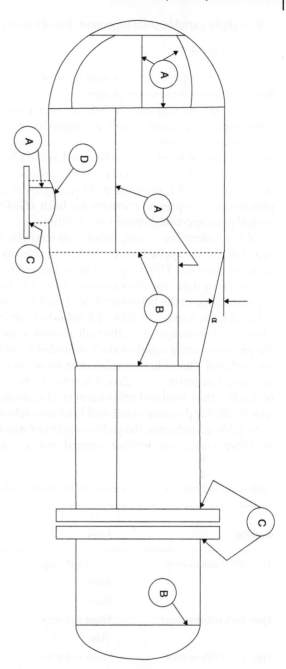

For pressure vessels, the stress is defined by the familiar expression:

$$\sigma = \frac{F}{A} \tag{4.1}$$

σ = applied stress, psi (lb/in^2) – The term, σ, is also the applied pressure
F = applied force, lb$_f$
A = cross-sectional area of the applied force, in^2

By a slight variable rearrangement, Equation (4.1) can be defined as

$$F = p \times A \tag{4.2}$$

where p is the applied pressure with units of lb/in^2 or psig. For pressure vessels, the term applied force is more commonly employed.

Safe design of many components requires a safety margin be incorporated in certain critical components; pressure vessels are considered critical components when referring to safe operations. As would be expected, materials for pressure vessel design have a required safety margin. This safety margin is identified as the allowable stress.

Since most pharmaceutical pressure vessels operate well within the temperature limits of steels, in Table 4.1, presents yield strengths of common metals used in the fabrication of pharmaceutical pressure vessels for both cGMP applications and non-cGMP uses under typical pharmaceutical operating conditions.

Table 4.1 identifies a term, offset yield strength. When determining the actual yield strength of a material, the applied stress vs. strain applied to the material usually is accurate for a limited range of the data. This range is linear. However, when dealing with the upper ranges of the stress–strain data, the data deviates from the linear range. The offset yield strength is an approximation of the behavior of the data if it were truly linear at the higher values.

The data presented in Table 4.1 identifies the maximum yield strength of the metal alloys identified. For design safety, the code identifies the allowable stress for several alloys, including the pharmaceutical stainless steels intended for cGMP applications in Table 4.1. Data for yield strength and allowable stress also exist for many other materials such as plastics, rubbers, and reinforced materials. The allowable stress is obtained by dividing the yield strength by a factor of 3.5. It is the calculated value for allowable stress that is used in pressure vessel calculations, specifically for pressure vessel wall thickness calculations.

As Table 4.1 indicates, the yield strengths of stainless steels are significantly lower than that of the other steels. More detailed material on steel and steel properties can be found in Chapter 8.

Table 4.1 Yield strengths of selected steels for pharmaceutical pressure vessels.

Material	Form	Yield strength, (0.2% offset) (psi)	Allowable stress (psi)
Type 304 stainless steel	Sheet and strip	35 000	10 000
	Plate	30 000	8 572
	Bars	30 000	8 572
Type 304L stainless steel	Sheet and strip	28 000	8 000
	Plate	28 000	8 000
Type 316 stainless steel	Sheet and strip	40 000	11 428
	Plate	35 000	10 000
	Bars	30 000	8 572
Type 316L stainless steel	Sheet and strip	32 000	9 143
	Plate	32 000	9 143
Type A653 steel	Sheet	53 700	15 343
Type A366/1008 steel	Sheet	72 000	20 572

Source: Adapted from International Nickel & onlinemetals.com.

Example 4.1   Type 309 stainless steel plate has a yield strength of 40 000 psi. Determine the allowable stress for this alloy.

Solution
Referring to Table 4.1, the yield strength is 40 000 psi, and the allowable stress for ASME Code use is:

$$\frac{40000\,\text{psi}}{3.5} = 11429\,\text{psi} \leftarrow \textbf{Solution}$$

Pressure vessel fabrication consists of several components welded to a basic cylinder or sphere designed to perform a specific function. Welded components are often the weakest points or locations of pressure vessel integrity. Consequently, to address this element of pressure vessel fabrication, the code provides guidelines for pressure vessel welding procedures. Included in these guidelines are weld efficiency values; these values are based on radiographic inspection of the welded joints of a pressure vessel. In general, depending on the type of welded joint, weld efficiency varies from a value of 0.6 to 1.0. Weld efficiencies are factors of the type of weld and the degree of inspection, typically the percentage of radiography the welded joints undergo. In order to determine weld efficiency, which is a major variable required for calculating vessel wall thickness, described earlier in the chapter, a familiarity with weld identification is very germane for project and process engineering purposes.

The information given below is intended to provide an understanding of the calculations and procedures required for pressure vessel wall thickness determination.

Before delving further into pressure vessel wall thickness calculations, the salient factor in vessel design and fabrication, a comment on pressure vessel failure modes, is worthwhile. A simple comparison can be made with a garden hose with flowing water. If for various reasons, the hose ruptures due to aging, undue bending, or other means, the rupture in the hose is longitudinal and not circumferential. Pressure vessels also fail, primarily, in that mode. The reason for this failure mode can be seen by using the following relationships derived in most texts dealing with mechanics, specifically strength of materials and thin shelled cylinders and not considering the end closures. The two important relationships are:

$$\sigma_c = S_c = \frac{pr}{t} \tag{4.3}$$

$$\sigma_l = S_l = \frac{pr}{2t} \tag{4.4}$$

σ_c, S_c = internal stress in the circumferential direction, psi
σ_l, S_l = internal stress in the longitudinal direction, psi
pr = internal pressure, directed outwards, psi
t = wall thickness, in.

As can be seen from Equations (4.3) and (4.4), the circumferential stress is twice the longitudinal stress. Referring to the garden hose rupture scenario, it can be seen that because the allowable circumferential stress is significantly greater than the allowable longitudinal stress, the failure mode is longitudinal due to the lower allowable stress in the longitudinal mode. Based on the preceding analysis, the circumferential stress is the basis of the design for performing pressure vessel thickness calculations.

The pressure vessel code provides criteria for weld identification. ASME VIII, Division 1 denotes the types of welds for pressure vessel by category and type. Weld categories are defined in UW-3 of the code. Weld joint categories specify the locations where specific welds are applicable. A most convenient technique for remembering the purpose of weld categories is to use the basic relationship: weld category = weld location. That is, the location of the weld specifies the category. There are four categories of welds identified are:

- Category A – This is a longitudinal weld that not only joins the main sections of the vessel body, most often metal plate, but also includes hemispherical heads welded to the cylindrical body of the vessel. Nozzles welded to the vessel body are also classified as Category A joints. Longitudinal welds are important since overpressure failures are most apt to occur at this orientation. An example of this failure mode is recalling garden hose ruptures are most often axial (linear) and not radial as the failure mode.
- Category B – This category involves circumferential welds that are part of the vessel shell. Category B welds also includes welds that connect formed heads, specifically, hemispherical and ellipsoidal designs. These heads are normally butt-welded to the shell body.
- Category C – Welds that connect flanges to nozzles or welding a flange to the main shell of the vessel are classified as Category C weld joints.
- Category D – Welded joints connecting nozzles to the pressure vessel body shell.

Pharmaceutical and biotech pressure vessel applications are primarily of Categories A, B, C, and D.

Figure 4.2 illustrates the various categories of welds as applicable to pressure vessels.

In addition to weld joint categories, the ASME Code also identifies six types of welded joints. These weld joints are described in table UW-12 of ASME VIII, Division 1 of the code. The six types identified are:

1) Double sided butt joint welded on one side with no support or backing strip (a backing strip or welding strip is basically a metal strip that increases the weld penetration and weld strength). The butt weld is commonly employed in pharmaceutical pressure vessels.
2) Single-sided butt joint with backing strip in place (basically, a butt weld is welding two pieces of metal together with no overlap of metal).
3) A single butt weld with no examination (primarily radiographic imaging inspection).

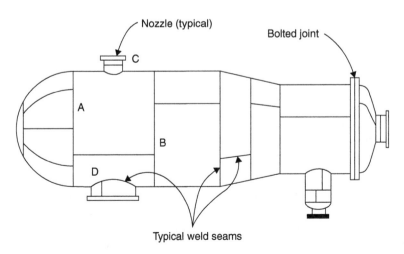

Figure 4.2 Weld joint categories.

4) Double fillet lap joint weld (this type of welded joint is basically two pieces of metal that overlap one another with a fillet weld on each end of the overlapping metal). Not typically employed in pharmaceutical pressure vessels.
5) Single full fillet lap joint with plug weld (generally not used for pharmaceutical pressure vessels).
6) Single full fillet lap joint. This weld, not typically used in pharmaceutical pressure vessels, is essentially a small hole (about a ¼ in. diameter) that is over the top metal piece, and the hole is filled with welding material and welded to the bottom metal of the lap joint. The technique is often used when there is limited space for spot welding machines (not normally employed for pharmaceutical applications).

For virtually all pressure vessels intended for pharmaceutical and biotech use, Types 1, 2, and 3 are used.

Just as there is a simple relationship for the relationship between weld category and geometry, there is also a convenient relationship to correlate weld joints and weld geometry. This relationship is given by weld type = joint geometry. By relating the relationships between weld category and location and weld type and joint geometry, it is significantly more convenient to understand pressure vessel calculations involving weld category and weld type.

There are several formulas that are used to determine the thickness of a pressure vessel. The ASME Code identifies the following formulas that are used to calculate the wall thickness. The formulas are specific to various vessel designs and are given by the following relationships:

$$1.\ \text{Cylindrical shell } t = \frac{PR}{SE - 0.6P} \tag{4.5}$$

$$2.\ \text{Hemispherical head or spherical shell } t = \frac{PR}{2SE - 0.2P} \tag{4.6}$$

$$3.\ \text{2:1 ellipsoidal head } t = \frac{PD}{2SE - 0.2P} \tag{4.7}$$

$$4.\ \text{Flanged and dished head } t = \frac{1.77\,PL}{2SE - 0.2P} \tag{4.8}$$

$$5.\ \text{Flat head } t = D\sqrt{\frac{CP}{SE}} \tag{4.9}$$

$$6.\ \text{Spherical } t = \frac{PR}{0.2SE - 0.2P} \tag{4.10}$$

where

t = Minimum required thickness, in.
P = Design pressure, psi
R = Inside radius, in.
S = Allowable stress, psi (refer to Table 4.1 for stainless steel values)
D = Inside diameter, in.
L = Inside spherical crown radius, in.
E = Weld joint efficiency factor, determined by weld joint location and degree Radiography examination of the weld
C = A factor depending upon method of head to shell attachment, for most applications, a value of 0.33 is reasonable.

If it is necessary to determine the surface area of the pressure vessel heads, a table presenting the formulas for surface areas of ASME vessel heads is given in Table 4.6. It is common to neglect the volume and area of the vessel heads since the actual level rarely goes beyond the straight cylindrical sections of the pressure vessel and the vessel volume does not contain a significant volume when compared with the straight section of the vessel.

The ellipsoidal head is often identified as an elliptical head (Equation 4.7). The 2:1 semi-elliptical design is the most common type of head used for pressure vessels. The thickness of the pressure vessel shell is normally equal to the thickness of the wall of the head. If the head thickness is thicker than the shell thickness, the head must be tapered so that the head, when joined to the shell, will be of identical wall thickness. In addition to being an ASME code requirement, for pharmaceutical applications, this detail is very important since any bore or thickness difference presents the possibility of accumulation of material in the crevice; this presents potential pressure vessel cleaning problems.

Using flat heads for pressure vessels is normally not an option often considered. This is due to the fact that hemispherical or elliptical heads withstand greater stress than equivalent flat head designs. The difference between head thicknesses is shown in Example 4.2.

Spherical vessels (Equation 4.10) are used for liquids and gases that are under high pressures. In addition to being expensive to fabricate, pharmaceutical processes rarely require containments at such high pressures. Equation is included for primarily information purposes.

Nozzles that are welded to the shell wall present a special situation. Welded nozzles are formed by removing shell material (the hole) from the basic shell configuration. The stresses in the immediate area of the welded nozzle are a high stress area, relative to the larger shell. While the stress decreases radially with distance from weld location, a higher stress remains at the weld site due to the hole in the shell. As a result, there is a risk of vessel failure at the weld orifice at the wall. To ameliorate this potential problem, three options are commonly evaluated:

1) Increase the nozzle wall thickness
2) Increase the shell wall thickness
3) Employ a combination of options 1 and 2.

Since option 2 could be quite expensive (increasing the shell wall thickness) [1] and option 3 could also prove costly, it is evident that option 1 (reinforcing the nozzle area) should be evaluated first. For nozzle area calculations, the relationship $A = Dt$, where D is the nozzle inside diameter in inches and t is the nozzle wall thickness, also in inches, can be used. For this evaluation, the shell thickness should be less than the nozzle wall thickness. More detailed information concerning determination of the nozzle thickness can be found in ASME VIII, Division 1, UG 45. Nozzle configurations often have unique requirements depending on the design configuration.

One most important requirement when dealing with pressure vessel thickness calculations is to be aware that the minimum wall thickness for ASME Code vessels is 1/8″ or 0.125 in.

All the relationships for determining the thickness requires a value for weld efficiency. Table 4.2 displays efficiencies for various welds identified above. Also, as would be expected, the efficiency is also a function of the degree of radiographic inspection performed on a weld. There are three categories of weld inspection using radiographic inspection:

1) Full radiographic weld – The weld of concern, longitudinal or circumferential, are fully radiographed allowing the entire weld to be viewed and inspected.
2) Spot weld – Random radiographs of small areas of the overall weld.
3) None – Weld that have not been inspected by radiography.

Table 4.2 Weld joint efficiency values.

Weld type	Joint category	Extent of radiography	Efficiency
Type 1 – butt weld, double sided with no backing strip left in place	A,B,C,D	Full	1.00
Type 1 – butt weld, double sided with no backing strip left in place	A,B,C,D	Spot	0.85
Type 1 – butt weld, double sided with no backing strip left in place	A,B,C,D	None	0.70
Type 2 – single-sided butt joint, backing strip left in place	A,B,C,D	Full	0.90
Type 2 – single-sided butt joint, backing strip left in place	A,B,C,D	Spot	0.85
Type 2 – single-sided butt joint, backing strip left in place	A,B,C,D	None	0.70
Type 3 – single-sided butt joint with no radiography examination	A,B,C	N/A	0.65
Type 3 – single-sided butt joint with no radiography examination	A,B,C	N/A	0.60
Type 3 – single-sided butt joint with no radiography examination	A,B,C	N/A	0.60

While Equations (4.5)–(4.9) identify E, the weld efficiency, it is important to note that a weld efficiency value applies primarily to weld types 1, 2, and 3 directly (butt welds). For other butt weld types that do not undergo full radiography or no radiography at all, a more detailed design analysis is required. But since the majority of pharmaceutical pressure vessels intended for cGMP or non-cGMP service is butt welded, the design techniques employed herein are quite adequate for project engineering calculations. Further, observing Figure 4.1 in more detail, the angle α is identified as less than or equal to 30°. This value identifies the maximum angle one plate can have when it is to be butt welded as part of the shell body.

Table 4.2 itemizes some typical weld joint efficiencies. A cursory analysis of Table 4.2 indicates some relevant details. For instance, the efficiency, E, is defined as a decimal or fraction. This, in turn, indicates that the efficiency is a ratio of two variables. In this case, the variable is the strength of the metal to be welded (parent metal) and the strength of the weld. In this case, the efficiency definition is

$$E = \frac{\text{Strength of weld}}{\text{Strength of the parent weld}} \tag{4.11}$$

The units employed are not germane to this relationship since the dimensions will cancel. However, the dimensions employed in both the numerator and denominator must be consistent. Based on Equation (4.11), it can be noted if the weld is weaker than the parent metal, the efficiency, E, will be less than 1.00 ($E < 1.00$). But a more practical concept of weld efficiency is to understand that the efficiency is determined by the weld type and degree of radiography performed.

Example 4.2 A specification is to be prepared for a 100 gal pressure vessel intended for transfer of a heated fluid to a reactor. The design temperature is 450 °F with a design pressure of 55 psig. The pressure vessel head is hemispherical with a 15° radius, measured from the centerline of the reactor. The vessel is fabricated of 316L stainless steel. All longitudinal welds are Category A, Type 1 and circumferential shell welds are Category B, Type 1. The hemispherical heads (top and bottom) are butt-welded to the shell and will be fabricated as Category B, Type 1 configurations. All welds are RT-1 (full radiography, $E = 1.00$). The inside diameter of the shell is 48 in. There will be three 3 in. diameter nozzles Category D, Type 1 also fabricated of 316L stainless steel butt-welded to various locations on the cylindrical shell. Calculate the minimum thickness for:

A. The pressure vessel shell.
B. The hemispherical heads.
C. The nozzles.
D. The nozzles are welded with no radiographs for the nozzles.
E. Fabricating the pressure vessel with flat heads in lieu of hemispherical heads.
F. Would Equation (4.3) serve as a quick and reasonably accurate estimate for determining the shell wall thickness for this pressure vessel?

Solution
A. Shell thickness calculation:

Equation (4.2) is the applicable relationship for this calculation. Employing Tables 4.1 and 4.2 and the given data:

Design pressure = 55 psig
Shell radius = 24 in.
Shell allowable stress = 8000 psi
Weld efficiency, $E = 1.00$

Recalling Equation (4.5) and inserting the required values yields

$$t = \frac{PR}{SE - 0.6P}$$

$$t = \frac{(55)(24)}{(8000)(1.00) - 0.6(55)} = 0.166 \text{ in.} - \text{Use } 0.375 \text{ in. } (3/8) \text{ stock} \leftarrow \text{Solution}$$

B. Head thickness calculation:

For these sections of the pressure vessel, Equation (4.6) is the applicable relationship. Once again, inserting the given variables yields

$$t = \frac{PR}{2(SE) - 0.2P} = \frac{(55)(15)}{2(8000)(1.00) - 0.2(55)} = 0.052 \text{ in.} - \text{Use } 1/8 \text{ in. stock} \leftarrow \text{Solution}$$

C. Nozzle thickness calculations:

The applicable equation for the butt welded nozzles is Equation (4.5):

$$t = \frac{PR}{SE - 0.6P} = \frac{(55)(1.5)}{(8000)(1.00) - (0.6)(55)} = 0.0104 \text{ in.} - \text{Use } 1/8 \text{ in. stock} \leftarrow \text{Solution}$$

D. Nozzles with no radiographic examination:

While this configuration is still a Category A, Type 1 weld, the weld efficiency E becomes 0.70. Inserting the given values in Equation (4.5), the thickness calculation is:

$$t = \frac{PR}{SE - 0.6P} = \frac{(55)(1.5)}{(8000)(0.70) - 0.6(55)} = 0.0149 \text{ in.} - \text{Use } 1/8 \text{ in.} \leftarrow \text{Solution}$$

E. Head thickness calculation (flat heads):

Equation (4.9) is selected for this calculation.

$$t = D\sqrt{\frac{CP}{SE}} = 48\sqrt{\frac{0.33(55)}{1.00(8000)}} = 2.28\text{in.} - \textbf{select } 2\tfrac{3}{8}\textbf{in. plate} \leftarrow \textbf{Solution}$$

F. Estimated wall thickness using Equation (4.3):

Recalling Equation (4.3) is given as $S_c = (pr/t)$. Rearranging, Equation (4.3) becomes $t = pr/S_c$. Inserting the given value in the modified version of the equation yields:

$$t = \frac{(55)(24)}{8000} = 0.165\text{in.} - \textbf{specify } 0.380\left(\tfrac{3}{8}\right)\textbf{in. plate} \leftarrow \textbf{Solution}$$

$$= 0.165\text{in.} - \textbf{use } 0.380\left(\tfrac{3}{8}\right)\textbf{in.} \leftarrow \textbf{Solution}$$

The results for Parts C and D of Example 4.2 indicate the nozzle will have thin wall thicknesses. Further, since the example doesn't indicate whether a flange will be added to the nozzle end or if a valve is to be butt-welded to the nozzle (unlikely), it is common practice to add reinforcement to the nozzle. This is commonly done by using a larger gage steel for the nozzles. Also, the wall thickness calculation results are for the minimum required wall thickness for the pressure vessel constituents. By employing standard stainless stock (commonly supplied in 1/8 in. intervals) delivery times and availability factors can be avoided. Depending on schedule and cost, this could significantly affect project cost.

The result for Part E of Example 4.2 verifies why flat heads are not often employed in pressure vessels, the additional thickness, particularly when the vessel is fabricated of stainless steel, adds a significant cost to the pressure vessel.

It would appear the result for using Equation (4.3) for Part F is a reasonable technique to estimate thin wall cylinder thickness since the calculated estimate varies by about 1%. However, it is important to note Part A of the example problem assumed an efficiency of 1.00. This is rarely the case when dealing with actual pressure vessel design and specification procedures.

While stainless steel is used in cGMP pressure vessels, non-stainless grades are also employed in non-cGMP operations in non-critical applications such as water treatment systems used for boiler steam water supply. For non-stainless steel grades, such as those identified in Table 4.1, a corrosion allowance is added to the pressure vessel wall thickness. Consequently, an additional 0.125(⅛") is commonly added to the wall thickness, for, as noted, a corrosion allowance.

While the above example is intended to provide a familiarity with the design calculations used to determine pressure vessel wall thickness, another important variable must also be addressed. This variable is weld efficiency. While several weld efficiencies have been used in the above example problem (Example 4.2), in actual practice, fabrication cost is a critical concern in pressure vessel design and specification. One significant cost factor is weld radiography. Performing a full radiograph is expensive and time consuming, although recent technology changes have shortened the task time. For this reason, with few exceptions (e.g. pressure vessels that are used in the processing of hazardous liquids, gases, or solids in solution, ASME Section VIII nuclear pressure vessels – or other hazards identified in the facility Process Safety Management Plan – Chapter 1), the standard approach to efficiency evaluation is to use a spot weld and associated efficiency of 0.85. This value ($E = 0.85$) is generally the most cost efficient value in maintaining pressure vessel safety considerations.

When preparing or evaluating a pressure vessel design or specification, it is of utmost importance to include the facility insurers since their input determines the adequacy of the specification and/or pressure vessel design.

4.1.3 Pharmaceutical Reactors

While the task of designing reactors is often done by firms specializing in chemical kinetics and pharmaceutical and bioreactors, it is relevant for personnel involved in engineering, maintenance, and manufacturing operations to possess familiarity, with certain aspects of reaction chemistry and basic design fundamentals. This section is intended to equip involved personnel with a basic understanding of pharmaceutical reactors, bioreactors, and fermentors.

4.1.4 Kinetics and Reactor Fundamentals

The emergence of biotechnology has significantly altered the stereotype of reactors for pharmaceutical applications. Within the last few years, reactor yields were often in hundreds of pounds or greater, or hundreds of kilograms, or greater with some reactor volumes of 200001 (~65 000 gal). With the emergence of biotechnology, extremely profitable yields are often determined in grams or liters of product. While the traditional chemical reactor is still employed for many pharmaceutical formulations, change is rapidly occurring. As a result of the new manufacturing paradigm being created, the emergence of biotechnology, a new generation of smaller, more efficient bioreactors has been and is being rapidly implemented.

The exponential growth of biotechnology that can be appreciated by noting the estimated market for biotech products was $105 billion in 2010 and by 2015; sales are expected to exceed $149 billion. This represents a nearly 30% market growth [2].

While there are continuously evolving designs for biotech reactors, the principles of reactor kinetics are basic and can serve as a basis for understanding standard pharmaceutical reactors as well as evolving and existing biotech reactor concepts and design.

The distinguishing feature of pharmaceutical reactors is that they are primarily employed to produce single batches of a product or an intermediate product. Unlike many chemical manufacturing operations that continuously produce a product such as many petroleum refining operations, pharmaceutical reactors, including biotech reactors, create a fixed amount of product in a given time period. As a result, both pharmaceutical reactor and biotech reactors are similar in design and function.

Since there are many similarities between pharmaceutical reactors, biopharmaceutical reactors, and fermentors, an overview of traditional pharmaceutical reactors and reaction kinetics should be beneficial.

Figure 4.3 illustrates a schematic diagram of a typical pharmaceutical reactor. In actual practice, it is not uncommon to have several reactors of this type arranged in series. This is due to the fact that manufacture of a drug product may require several reaction steps during the manufacturing phase. Also, it is not uncommon to employ operations such as filtration and solvent extraction before the intermediate product is transferred to another reactor before further reactions are undertaken.

Figure 4.3 is intended to represent a fraction of the entire manufacturing operation. Consequently, some explanation of the terms, symbols, and abbreviations would be helpful to some. The sump is the drain line. RD is the abbreviation for a rupture disc, a device that protects the reactor from an overpressure rupture in the reactor. The term PRV (PRV-1) identifies a pressure relief valve; in this case, the steam jacket is protected from a rupture by the relief valve

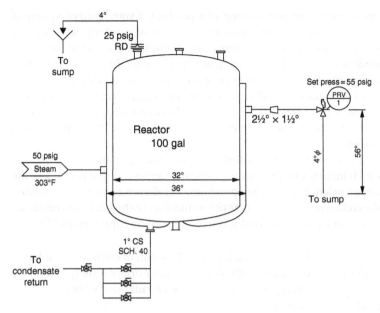

Figure 4.3 Schematic diagram of a pharmaceutical production reactor.

(Section 4.2). The abbreviation CS identifies that the drain piping is composed of carbon steel. The carbon steel is acceptable since the steam condensate is not part of a cGMP system since the steam jacket is not in contact with a critical component of the system (e.g. the reactants and products in the reactor).

In addition to being fabricated of stainless steel, the reactor can also be glass lined. A glass-lined reactor is desirable due to the fact that the reactants would be less likely for the reactant components and products to react with the stainless steel reactor wall. The glass lining can also make the cleaning operation easier. On the other hand, glass-lined reactors are sensitive to sudden shocks such as a forceful impact on the reactor shell or a sudden temperature increase or decrease. Further, while not shown in Figure 4.3, the reactor would typically be installed with an electrically powered mechanical agitator affixed to a vertical shaft, typically installed in the middle nozzle of the reactor head. The agitator is either rotating at a fixed rate (rpm) or equipped to vary the number of rotations per minute.

The size of a reactor is, ideally, determined by the quantity of product or yield required. The required product is, in turn, usually a function of product market demands. However, the practical reactor implementation is often a decision made based on the ability to locate a reactor that conforms to the approximate requirements. Often, as long as operating criteria such as temperature and pressure requirements are acceptable, an oversized reactor is often selected, with the assumption or rationalization that product demand may increase.

Pharmaceuticals, as opposed to biotech operations, are manufactured via chemical processes involving chemical reactions as part of the operations. This element of the manufacturing process involves an understanding of the mechanisms governing chemical production, specifically the time required to produce a desired product at a required concentration. These required elements, reaction time and final concentration, are obtained by employing a technology commonly referred to as chemical kinetics. Chemical kinetics is concerned not only with the velocity of the reactions (the rate at which a given mass of reactants are transformed, via chemical reaction, to a mass of product) but also with the intermediate steps through which the reactants are converted to the desired product.

In order to produce a consistency in manufacturing of a product, a standardized system is required to ensure an ability to manufacture consistent batches of product, whether the product is an intermediate chemical or the final pharmaceutical product. As part of the standardization procedures, chemical reactions have been generally classified by a term known as the reaction order. Basically, there are three different orders of chemical reactions, first order, second order, and third order reactions (generally not covered in manufacturing operations).

Since a pharmaceutical product often involves several manufacturing steps, it's not uncommon to have several reactions of different orders and, consequently, several reactors involved in a drug manufacturing process.

The reaction rate, a key parameter in reaction kinetics, is determined by many factors. The rate of organic reactions, which includes many pharmaceutical processes, is determined by the energy of the reactants, proper orientation of the constituents (referred to as the probability factor), rate at which specific collisions occur, the quantity or number of the effective collisions, and the energy of the reactants. In a more quantitative form, this can be expressed as [3]:

Number of effective collisions/cm^3/s (rate)	= Total number of collisions/cm^3 (collision frequency)	× Fraction of collisions that have sufficient energy (energy factor)	× Fraction of collisions that have proper orientation (probability or orientation factor)

Of the factors identified above, the most important variable is the fraction of collisions possessing sufficient energy to initiate a reaction (energy factor); this is often referred to as energy of activation and denoted as E_{act} (also identified in thermodynamic terms as ΔH_{act}; refer to Section 2.3) with common units of Btu or kcal, depending, in large part, on whether the reaction is in a laboratory or manufacturing environment. The energy of activation E_{act} can, in many cases, be lowered to create a more rapid reaction rate. This phenomenon is accomplished by incorporating a material known as a catalyst in the reaction scheme. Catalysts are materials that, while lowering the activation energy, are not affected by the reaction it is involved in altering. That is, the quantity and physical state of the catalyst remains constant throughout the kinetics and reaction steps of the chemical process. Catalysts are often chemicals such as nickel in a modified form, platinum, palladium, and other materials. The configuration can also enhance catalyst performance. For instance, catalysts are often fabricated in the form of a mesh with reactants passing through the small openings in the mesh to increase surface exposure of the reactants to the catalyst. Small spherical beadlike shapes are often used.

Catalytic reactions are very important in pharmaceutical manufacture. One of the most common catalytic reactions in pharmaceutical manufacture is hydrogenation. Hydrogenation is a reaction where double bonded carbon atoms are reduced to single bond carbon atoms. Hydrogenation is accomplished by reacting hydrogen gas with alkenes under high pressure (often in excess of 60 atm of pressure) and in the presence of a catalyst such as those described above. Subsequent to the hydrogenation, the double bonded alkenes are reacted to create single bond alkanes (i.e. $C=C \rightarrow C-C$). Hydrogenation can also be performed with aromatic compounds (benzene base) to produce a less toxic product.

The collision frequency is, primarily, a function of proximity of the particles to one another (concentration), size of the particles, and the velocity of the particles in solution (assuming most pharmaceutical reactions are in solution). Particle velocity in solution is very dependent on temperature of the reactants. In fact, for many organic reactions, such as pharmaceutical constituents, the reaction rate is doubled for every 10 °C (18 °F).

The probability factor, the least relevant constituent, is a term that encompasses the geometry of the particles composing the constituents and the type of reaction occurring.

When dealing with chemical kinetics and reaction orders, it is important to note that the order of a reaction cannot be evaluated from the chemical equation that identifies the stoichiometric balance as well as reactants and products. The reaction kinetics is independent of the reaction.

A first order reaction is the type of reaction where the reaction rate is proportional to the concentration of the reactant. This is given by the generic reaction

$$A \rightarrow R$$

where

A = the initial concentration of reactant (the concentration units are commonly given in terms of g/cm^3, lb/ft^3, or mol/cm^3). However, concentration is most commonly given with the units of mol/l (N/V)

R = concentration of the product

Reactions can be given in the form of a rate equation. For a first order reaction, the rate equation can be in the form

$$-r_A = \frac{-dN}{V\,dt} = \frac{\text{Moles reacting}}{(\text{Volume of reactor})(\text{Time})} \tag{4.12}$$

Another first order reaction can take the form

$$A \rightarrow R + S$$

where R and S represent the concentration of the products. A typical first order reaction is the decomposition of hydrogen peroxide, a liquid, decomposing to water, also a liquid and gaseous oxygen. The reaction is:

$$H_2O_2 \rightarrow H_2O + \tfrac{1}{2}O_2 \uparrow$$

In a batch reactor, the reactants and product are both constant volumes, as is the case with many first order and higher reaction orders. As such, the rate equation can be given as:

$$\frac{-dA}{dt} = kA \tag{4.13}$$

where

A = concentration of the reactant
t = time
k = the proportionality constant that, in this case, is identified as the rate constant
$-dA/dt$ = rate of concentration decrease with time (note the minus sign).

By slightly modifying Equation (4.13) with the addition of specific value terms and rearranging Equation (4.13) such that

A_0 = initial concentration of reactant A
A_f = final concentration of reactant A subsequent to reaction at time t

Equation (4.13) can be put in integral form such that

$$-\int_{A_0}^{A} \frac{dA}{A} = k\int_{t_0}^{t} dt$$

Subsequent to integration within the range identified, Equation (4.11) can be given as

$$kt = \ln\left(\frac{A_0}{A_f}\right) \qquad (4.14)$$

An alternative form of Equation (4.14) can be obtained by replacing the natural logarithm, ln, with the base 10 logarithm to obtain a more recognizable form:

$$kt = 2.303\log\left(\frac{A_0}{A_f}\right) \qquad (4.15)$$

Equation (4.12) is the integrated form of the rate equation for a first order reaction such as the types shown above. Equation (4.15) can also be expressed in exponential form as:

$$A = A_0 e^{-kt} \qquad (4.16)$$

Equation (4.12) is also used to verify a reaction is first order and to confirm the reaction rate constant by plotting time (usually on the x axis or the abscissa) vs. the log (or the natural logarithm, ln) of the concentrations at specific times (usually plotted on the y axis or the ordinate). In fact, all reaction orders must be determined by actual experimental data. Fortunately, with the availability of automated data analysis, determination of reaction order is much more efficient than manual analysis.

It should also be noted that the slope of the first order equation is negative. This is to be expected since the initial concentration of the reactant is decreasing with time.

The following example is intended to understand the basic technique of determining a first order reaction based on experimental data [4].

Determining the applicability of a first order reaction, based on experimental data, can be ascertained when

- The values for the reaction constant, k, are essentially identical when experimental data are used in Equation (4.12)
- A straight line is observed when the log or the natural logarithm (ln) of A is plotted against time (t).

Example 4.3 An intermediate reaction is required to produce an active pharmaceutical ingredient for dietary consumer product. Laboratory reactions have produced the following reaction concentrations at the time sequences indicated:

Time (s)	0.0	900.0	1800.0
Concentration, reactant A (g)	50.8	19.7	7.62

Confirm the results that are those of a first order reaction.

Solution

Assuming the reaction is unimolecular and one of the following forms

$$A \rightarrow R + S$$
$$A \rightarrow R$$

Also, the value for k, the slope of the line when time is plotted against the concentration, A, should be linear.

By substituting concentration and time values in Equation (4.1), the equation can be expressed as:

$$k\left(900s\right) = \ln\left(\frac{50.8}{19.7}\right) = 0.00105s^{-1} \leftarrow \textbf{Solution}$$

Employing Equation (4.1) and using a reaction time of 1800 s, the solution is

$$k(1800) = \ln\left(\frac{50.8}{1800}\right) = 0.00105s^{-1} \leftarrow \textbf{Solution}$$

Since the reaction rate constants are identical, the reaction is, indeed, a first order reaction. It is worth noting that the reaction rate constant for a first order reaction is an inverse time unit, s^{-1}, in this case. The above answer is applicable only for first order reactions. The units for the rate constant k will vary with the order of the reaction.

While batch reactors are more independent of reaction time than continuous reactor designs such as plug flow reactors or continuous stirred tank reactors (CSTR), it would be useful to know the time required for a particular concentration to be obtained. Such a result can be particularly desirable when dealing with bench-scale or scale-up procedures. Specifically, in these types of studies, it would be useful to know when a given reactant fraction is consumed at a given time. Knowing the time required to achieve a known reactant concentration at a specific elapsed time could be very helpful for bench studies and scale-up procedures. One convenient calculation would be to know the time required to consume half the initial concentration. This term is referred to as the reaction half-life. The reaction half-life is calculated by the relationship:

$$t_{1/2} = \frac{0.693}{k} \tag{4.17}$$

t = the reaction half-life
k = reaction rate constant

To appreciate the concept of the reaction half life, consider Example 4.3. For this example, it would be desirable to know the time required to obtain half the original concentration of reactant A. Substituting the calculated reaction rate constant into Equation (4.17) yields:

$$t_{1/2} = \frac{0.693}{0.00105s} = 660s \leftarrow \textbf{Solution}$$

As the result indicates, half of reactant A is consumed in 660 s, or 11 min. It is interesting to note that, according to the experimental results obtained in Example 4.3, after 1800 s, 7.62 g of the reactant remain unreacted. As the concentration of reactant A becomes more dilute, it will take a longer time for fewer molecules to react, thus lengthening the reaction time.

Another type of reaction that can be identified by the reaction rate constant is a second order reaction. Second order reactions can be found in two forms:

$2A \rightarrow R$ or the form

$A + bB \rightarrow R$

A typical second order reaction is the reaction of butyl chloride (C_4H_9Cl) with water. The reaction is:

$$C_4H_9Cl + H_2O \rightarrow C_4H_9OH + HCl$$

For this second order reaction, the products are butyl alcohol and hydrochloric acid; this reaction is of the form $A + B \rightarrow AB$.

For a second order reaction of the form $2A \rightarrow R$, the rate equation is given as:

$$-r = k[A]^2$$

When integrated with respect to time, the rate law for a second order reaction becomes:

$$\frac{1}{[A]} = \frac{1}{[A_0]} + kt$$

where

$[A]$ = concentration of reactant A
$[A_0]$ = initial concentration of A at time $t = 0$
k = reaction rate constant, mol/l
t = time, s or min.

For second order reactions, the rate constant is determined by plotting time on the abscissa (x axis) and the reciprocal of the concentration ($1/[A]$) on the ordinate (y axis), with the slope being the reaction rate constant k. One major difference between the slopes of the first order reaction and the second order reactions is the slope for the first order reaction is negative, whereas the slope of the second order reaction is positive. This is because the plot for the second order reaction is the reciprocal of the concentration when plotted against time.

Just as a first order reaction has a half-life (Equation 4.17), the half-life of a second order reaction is

$$t_{1/2} = \frac{1}{k[A_0]} \tag{4.18}$$

Yet another type of reaction is the zero order reaction. The basic zero order reaction is of the form $A \rightarrow R$. An example of a zero order reaction is the reaction of iodine (I_2), in the presence of an acid (H^+) at a pH below 4, with acetone (CH_3OOCH):

$$CH_3COOH + I_2 \xrightarrow{H^+} ICH_2COCH + HI$$

Basically, a zero order reaction occurs when the reaction rate is not dependent on the reactant concentration. In this case, the rate law for a zero order reaction is given by:

$$-D\frac{[A]}{DT} = k \tag{4.19}$$

In its integrated form, the result is:

$$[A_0] - [A] = kt \tag{4.20}$$

Here, the units of k are moles/time (e.g. g mol/min).

By observing Equation (4.20) in a bit more detail, it is seen that the relationship is linearly proportional to time, and the coordinates $x = 0$ and $y = 0$ are valid, unlike the previous kinetics mechanisms thus far encountered. It is important to note that while a first order reaction is also proportional, the proportionality lies with the reactant concentration and not with reaction time.

It is most important to emphasize that when attempting to determine the reaction rate constant k via plotting concentration vs. time, the format must be such that the plot is linear when employing standard Cartesian coordinate graph paper, regardless if the task is performed manually or through the use of electronic or software programs. This means that regardless of the reaction order, the plot should be linear. This often requires plotting the $\log_{10}$ or the natural log (ln) of the concentration vs. time, as can be observed from a closer examination of the reaction rate constants previously addressed. While it may appear to be repetitive, obtaining the reaction rate constant is the key to effective understanding and manipulation of reaction kinetics.

As is the case with first and second order reactions, zero order reactions also have a half-life. In this case, the half-life ($t_{1/2}$) is given by

$$t_{1/2} = \frac{[A_0]}{2k} \tag{4.21}$$

$[A_0]$ = initial concentration, mol/l
k = mol/min, typically

Zero order reactions are often encountered in catalytic reactions. Also, reactions exhibiting zero order behavior (i.e. linear proportionality with respect to time and concentration) indicate that a more complex reaction mechanism is occurring involving a number of steps. This situation, several reactions occurring in succession, can create a significant bottleneck in the kinetics. The scenario can be viewed as an equilibrium reaction, where reactants and products shift back and forth between the product and components. As a result of this behavior, the reaction slows significantly and exhibits zero order behavior even though the actual kinetics is significantly more complicated.

While not specifically noted, many zero order reactions may be involved in equilibrium type reactions that delay completion of the reaction. As the concentration of reactants is diminished, it is conceivable, as speculated in several proposed zero order reactions, the actual reaction rate will be lower than the equilibrium mechanism(s). Should this occur, it is possible the reaction order may change from the zero order reaction. However, for most pharmaceutical manufacturing operations, product yield, and quality, it is more relevant than actual reaction mechanisms. This is one reason why pharmaceutical manufacturing can be time consuming and more capital intensive than often assumed.

Operating procedures involving reaction kinetics can appear to conform to accepted formats and yet result in product with significant deviations from manufacturing batch requirements [5]. Also, to maximize product yield, it is not uncommon to have a batch run last longer than the kinetics indicate, provided the longer reaction time does not harm the product. Similarly, by adding excess reactants in quantities exceeding the theoretical stoichiometric values, an enhanced yield can occur. Of course, the reactants added in excess must not compromise the economics of the process. If required, separation of unused reactants can be performed by employing unit operations such as crystallization, extraction, distillation, etc. Since many of the reactants are organic and not ionic, ion exchange, in this case, is not viable.

Reaction rates can also be affected by the occurrence of side reactions encountered in pharmaceutical reactions. These side reactions often result in the formation of isomers. Isomers are compounds that have identical molecular weights, but differ in molecular structure. Perhaps the most commonly recognized isomer is the cyclohexane isomer that can exist in either a "chair" or "boat" configuration, depending on the carbon to carbon bond angle. Other common isomers are normal and iso forms of alkanes as well as cis and trans forms of organic compounds [6]. Several isomers are used as bases for pharmaceuticals and the structure is a variable of the drug's effectiveness.

For reaction kinetics, each isomer has a different reaction rate constant. For manufacturing operations, these variables, reactant masses, and reaction time should be accounted for in the SOPs identified for the reaction process. However, these variables should be recognized as important when dealing with batch rejections involving manufacturing reactor operations.

The fist, second, and zero order reactions are the orders most commonly used generally in organic reactions and specifically pharmaceutical production. While higher order reactions do exist, most mechanisms employ combinations of the three kinetic paradigms described thus far. Additionally, formulating an effective rate equation must be determined through experimental methods, as previously described. Fortunately, these efforts are usually performed by research operations both within the manufacturing organization and, often, by contract research organizations (CROs) where facility engineering organizations have no role in the task, other than, perhaps evaluating the resulting SOPs and other regulatory proposals.

While the reaction rate is determined, in large part, by the reaction rate constant, there is a relationship that can, in many cases, change the reaction constant. By changing the reaction rate constant, it is possible to shorten or increase the reaction rate.

It was noted previously that a 10°C (18°F) increase in temperature can, theoretically, double the reaction rate. This observation was confirmed with the development of a reasonably straightforward relationship based on the Arrhenius equation and modified to the following form:

$$\text{Log}\frac{k_2}{k_1} = \frac{E_{act}(T_2 - T_1)}{2.303RT_2T_1} \tag{4.22}$$

k_1 = reaction rate constant at lower temperature, K
k_2 = reaction rate constant at higher temperature, K
R = ideal gas constant, 1.99 cal/g mol K (1.99 Btu/lb mol °R).

While Equation (4.22) is a reasonable estimate, it is predicated on the assumption that $E_{act} \gg RT$.

Example 4.4   For the reaction

$A \rightarrow B + C$

$k = 1.0 \times 10^{-4} \, s^{-1}$ at 25°C (298K)

$k = 2.0 \times 10^{-4} \, s^{-1}$ at 35°C (308K)

A. Determine E_{act} in the temperature range given
B. Calculate the reaction rate constant, k, at 40°C

Since the relationship employed to solve this problem is well defined and the input variables are all given, a dimensional analysis is not required since the input variables and the solution units are known.

Solution
A. Substituting the variables in Equation (4.22) yields

$$\log\frac{2.0 \times 10^{-4}}{1.0 \times 10^{-4}} = \frac{E_{act}(10)}{(2.303)(1.99)(308)(298)} = 12600 \, \text{cal/g mol} \leftarrow \textbf{Solution A}$$

B. $$\log\frac{k}{2.0 \times 10^{-4}} = \frac{(12600)(5)}{(2.303)(2.99)(313)(308)} = 2.78 \times 10^{-4} \, s^{-1} \leftarrow \textbf{Solution B}$$

Table 4.3 Summary of pharmaceutical reaction kinetics information.

Reaction order	Reaction	General rate equation	Type of plot to determine rate constant, k	Rate constant, k, units	Reaction half-life
Zero order	$A \rightarrow R$	$-r_A = k$	$[A]$ vs. t	mol/t	$[A_0]/2k$
First order	$A \rightarrow R$	$-r_A = \ln([A]/[A_0])$	$\ln[A]$ vs. t	t^{-1}	$0.693/k$
	$A \rightarrow R + S$				
Second order	$A + B \rightarrow AB$	$-r_A = k[A_0][B]$	$(1/[A])$vs.t	1/(moles t)	$1/k[A]$
	$2A \rightarrow R$	$-r_A = k[A_0]$			

As stated, Equation (4.22) is assumed to be valid if $E_{act} \gg RT$. In this case, at the highest temperature, $RT = (1.99 \text{ cal/g mol K}) (308 \text{ K}) = 612.92 \approx 613 \text{ cal/g mol}$.

The product 613 cal/g mol represents 4.87% of the energy of activation, 12 600 cal/g mol. This is a negligible value and the results of Example 4.4 are valid.

While reaction kinetics for pharmaceutical manufacturing may appear detailed and complex, the engineering aspects are quite straightforward. Operations are defined via approved procedures and records. The challenge is found in the development and initial production operations for a product (i.e. operations subsequent to Phase III and FDA approval) where a background in chemical kinetics can be quite helpful. Knowledge of chemical kinetics can also be useful in production problems where a root cause analysis and/or a corrective and preventive action (CAPA) analysis is needed. To assist in tasks such as these and to provide a general summary of the topics germane to basic chemical kinetics, Table 4.3 is presented.

4.1.5 Bioreactor Principles

Within the past few years, a significant change in the design concept of a bioreactor has evolved and is continuing to undergo significant state-of-the-art use and design. What was formerly looked upon as merely a fermentation process (to be addressed in the next section) has, with the evolution of genetic engineering, become a new field of biological science, often referred to as biotech or biotechnology.

It is important to note that as a significant design difference between bioreactors and fermentors. The purpose of a bioreactor is to significantly increase the mass of mammalian cells or insect cells. Fermentors are devices designed to increase the mass of bacteria or fungi in a controlled environment. Fermentors have a greater height than bioreactors and are intended to optimize the rate of oxygen mass transfer within the reactor. Bioreactors are lower in height than a fermentor and designed to enhance various cell cultures (i.e. mammalian cells or insect cells) and bioreactor mixing operations.

Essentially, a bioreactor is very similar to a typical pharmaceutical reactor, as shown in Figure 4.4. While there are several types of bioreactors (e.g. bubble column bioreactors, fluidized bed bioreactors, packed bed bioreactors), Figure 4.4 is perhaps the most widely used at this time for manufacturing operations.

A glimpse of both Figures 4.3 and 4.4 exhibits some differences in the equipment. Figure 4.3 has a jacket where steam is introduced. On the other hand, Figure 4.4 has internal cooling coils intended for cooling water or another coolant. This difference indicates that the reactor (Figure 4.3) requires heat input to perform effectively. Similarly, one can conclude the internal cooling coils are used for process cooling, a correct conclusion. As will be seen, bioreactor temperature control, in the form of cooling, is a critical parameter in many bioreactors.

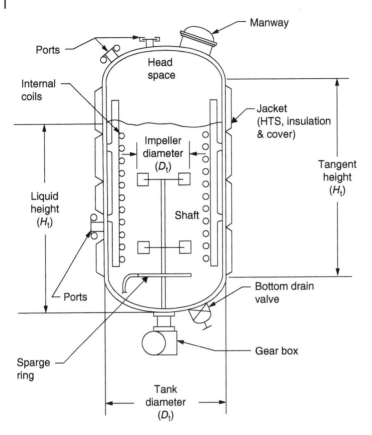

Figure 4.4 Diagram of a typical bioreactor.

Two significant differences between a pharmaceutical chemical reactor and a bioreactor (including fermentors) are the pressure and temperature operating conditions.

In addition to monitoring the bioreactor temperature, monitoring such as pH, dissolved oxygen sensors, partial pressure of the oxygen and carbon dioxide (CO_2) concentration are also common to bioreactors. While pharmaceutical reactors generally perform at lower temperatures and pressures to preserve the properties of both the reactants and the products, bioreactors and fermentors are generally designed to function at higher temperatures (150–1800 °C) and pressures (54.68–59.76 lb/in^2) than typical pharmaceutical reactors.

Another component of the bioreactor identified in Figure 4.4 is the sparge ring at the bottom of the vessel. The purpose of this item is to bubble air into the reactor contents (in biotechnology terms, this mixture is generally referred to as the medium). Regarding the sparger, it is good design practice to have smaller (for larger bioreactors and fermentors that are greater than a 10 000 gal nominal capacity, a sparger hole diameter of 0.25 in. is reasonable) more numerous holes in the sparger since this would allow for more contact with the oxygen through the media. It is also standard design practice to place the sparger ring hole on the top of the ring.

The medium can be composed of mammalian cells or tissues with specific desired protein content, whereas the fermentor is employed primarily for production of bacterial fungi and yeasts that manufacture components such as hormones and other materials intended for pharmaceutical intermediates or end products.

While the sparge ring introduces air to the bioreactor, it is the oxygen content of the air (about 21% oxygen) that is relevant to support bioreactor operations. Referring again to the sparger ring, it is also standard design practice to have smaller, more numerous holes in the sparger since this would allow for more contact with oxygen bubbling through the media, thereby enhancing the production in the bioreactor due to a greater liquid/gas interface. There are also bioreactions that proceed in the absence of oxygen, known as anaerobic reactions; pharmaceutical bioreactors, most often, are not intended for this purpose. In fact, product growth in bioreactors is strongly dependent on the rate of oxygen transfer in the medium.

Another subtle difference between the bioreactor (Figure 4.4) and the pharmaceutical reactor (Figure 4.3) is the location of the impeller. Pharmaceutical reactors, while important for reactant distribution, is not as critical to the organic reaction rate, coupled with the fact that most pharmaceutical reactions are organic. The rate of reaction is not as critical to ultimate yield as a bioreaction is. In general, the reaction rate for a typical bioreaction is orders of magnitudes slower than a typical organic pharmaceutical organic reaction where the energy of activation, E_{act}, is the major reaction rate determinant.

For bioreactors, the scenario is somewhat different. Agitation is an important factor, and, often, placing the agitator head at the bottom of the vessel is often advantageous. One reason for placing the agitator shaft at the bottom is there is often a greater transfer at the sparger than at the top of the bioreactor. Also, the air bubbled through the contents of the bioreactor accumulate at the top of the vessel and not undergo the mixing that can occur in the vicinity of the sparger. In addition to effective agitation, the solubility of oxygen is also critical. Oxygen under ambient conditions has solubility in water of about 8 ppm and decreases as the temperature increases. This low solubility, coupled with the fact that oxygen composes about 21% of air, limits the cell formation rate (i.e. growth rate) of the process. The bioreactor production rate is further complicated because the initial bioreactor mixture is often significantly different than water. For instance, the initial feed composition can be composed of a blend of mammalian cells, other tissue matter, nutrients such as sugars and other materials required to support cell growth. This more viscous mixture requires higher agitation speeds to have a homogeneous composition throughout the bioreactor composition. As a result, it is not uncommon to have a high power requirement for production bioreactors. As a general rule, the horsepower requirements for most bioreactors are in the range between 0.004 hp/1000 gal and 0.006 hp/1000 gal (0.005 is a common value). Additionally, power consumption in bioreactors is proportional to the cube of the agitator speed in rpm [7].

Agitator selection is further complicated by the fact that many organisms involved in bioreactor processing can be damaged if subjected to excessive agitator speed. While agitator speeds, impeller selection and impeller diameters are determined, in large part, by laboratory and pilot plant studies; some other general bioreactor agitator and related component design guidelines include:

- Maintain an impeller diameter/tank diameter (*D/T*) of less than 0.8 (impeller diameter/tank diameter of 0.33 is reasonable; the range of the *D/T* is between 0.17 and 0.80).
- Maintain agitator speeds between 60 and 240 rpm.
- Impeller blades should be located above the bottom sparger ring.
- Locate bottom impeller blades about one impeller diameter from the tank bottom.
- Impellers should be one diameter away from the next vertical impeller on the shaft, if multiple impellers on the same shaft are used in an axial configuration.
- If baffles are to be installed in the bioreactor, carefully assess the cleaning requirements relative to residue buildup on the baffles. It is important to note bioreactor maintenance is

somewhat increased with baffle installation in the bioreactor (standard design employs four baffles in the bioreactor). This is due to the fact that cleaning requirements often lengthen the time required to place the bioreactor back in a production mode.

- If possible, consider using tapered baffle installation in lieu of baffles that are installed at 90° angles to the bioreactor wall. While the agitation may not be quite as efficient with the smooth taper baffles, cleaning procedures are more reliable and less time consuming when using the typical stainless steel bioreactor design. For bioreactor mixes with higher viscosities (>50 000 cP), the need for baffles should be examined. At higher viscosities, the turbulence created by the baffles may have a much smaller effect on mixing than if the bioreactor blend had a viscosity closer to water. In addition to greater cost, the baffles represent a greater downtime due to greater cleaning and sampling time between batches.
- Avoid anchor design impellers, more efficient mixing can be obtained by selecting designs such as helical ribbon or paddle configurations. The anchor design has the disadvantage of inefficient axial (top to bottom) mixing.
- Be consistent with agitator/impeller selection. Because of the performance variation of impeller designs, agitator selection and performance should be evaluated throughout the development phase, starting with bench-scale and pilot phases.

Bioreactors are, as is the case with pharmaceutical reactors, primarily fabricated of either 304 (or 304L) or 316 (or 316L) stainless steel. Bioreactor components such as spargers and agitators are also fabricated of the same grade of stainless steel as the reactor. While the preferred material of construction is a stainless steel for bioreactor fabrication, a rapidly growing trend for using disposable bioreactors and disposable bioreactor components is occurring and will significantly alter the area of bioreactor designs both currently and in the future. Disposable and single use bioreactors will be addressed further in Chapter 9.

As noted, bioreactors are similar to chemical kinetic reactors since they both produce a product or products derived from basic reactants. The difference is that bioreactor components are cellular in nature. Just as chemical reactors have rate equations that determine the order of the reaction, bioreactions have a single basic rate equation defined as:

$$[A] = [A]_0 e^{\mu t} \tag{4.23}$$

where

$[A]_0$ = initial cell mass (dry mass), g/l
$[A]$ = final cell mass at time t (dry mass), g/l
μ = specific growth rate (rate constant), h^{-1}
t = time, h

As can be seen, Equation (4.23) is similar to the first order exponential chemical reaction defined in Equation (4.16). While both relationships are exponential, several important distinctions must be made:

- Because the bioreactor is intended for production, the slope of the line for plotting concentration vs. time (Equation 4.23) will be positive.
- Equation (4.16) measures the consumption of reactants, while Equation (4.23) is intended to measure growth of a product. Consequently, Equation (4.16) is exponentially defined by a minus sign and Equation (4.23) is positive.

For bioreactors and fermentors, the growth rate constant is measured in hours and not seconds or minutes, as is often the case with pharmaceutical chemical kinetics.

Example 4.5 A bioreactor produces a specific protein that is used as the API for a biotech drug. The process data and scale-up data indicating a daily production of 50 g/l of dry mass are required for daily production (24 h of continuous operation). Laboratory studies have determined a specific growth rate (μ) of 0.330 h^{-1}. The laboratory studies did not initialize the starting dry cell mass, since it was decided that a mathematical result would serve as a basic starting cell mass. Determine the initial dry cell mass required to produce 45 g/l of dry cell mass after 24 h of operation. Further, the laboratory and scale-up studies indicate that the process has an efficiency of 65%.

Solution
The basis of calculation for this problem is Equation (4.23). However, it is more convenient to rearrange Equation (4.23) in a more manageable form by a bit of rearrangement.

By recalling $\ln[A] - \ln[A_0] = \ln([A]/[A_0])$, Equation (4.23) becomes $\ln([A]/[A_0]) = \mu t$.
Inserting the given parameters for Example 4.5 yields

$$\ln\frac{50}{[A_0]} = (0.330)(24) = 7.92 = \ln\frac{[50]}{7.92} = 1.204 \text{g}/1\left(\text{dry mass}\right)$$

Recalling that the efficiency is 60%, the result becomes $(0.60)(1.204) =$ **0.7224 g/l of dry product required as starting dry mass ← Solution**

Example 4.5 is useful for several reasons:

1) The example identifies the need for comprehensive data collection when process development using bioreactor design is required.
2) While many software programs are of great value in bioreactor studies, a basic knowledge of bioreactor principles is nevertheless most important.
3) For biotech manufacturing, it is most important that personnel involved in the operations are thoroughly trained and familiar with the process. Since biotech manufacturing is exponential, small deviations can seriously affect the yield and integrity. Consequently, specification limits and tolerances are usually much tighter than typical pharmaceutical manufacturing operations.
4) While an extensive bioreactor knowledge base exists and is rapidly expanding, a great deal of design and operating data is empirically determined and applicable only to specific configurations, starting materials and unique configurations. Much of the analytical information is specific to only one or two designs and often cannot be transferred to other operations, while ostensibly appearing similar. To successfully transfer operations from a smaller operation (i.e. bench scale or pilot plant scale) to a manufacturing scale, several factors must be quantitatively evaluated. Similarity in bioreactors, as well as many engineering operations involving fluid mechanics, identifies four key parameters that must be evaluated. These parameters are:
   • Geometrical similarity – One body is similar to a second body when each point in one body corresponds to a point in the second body that is linearly scaled to the first body by use of a scale factor – a typical example is two holding tanks with a scale of 1 in. = 1 ft (12 in.). Then all corresponding measurements on the tank should have the ratio of 1 in. = 1 ft. Geometric similarity is particularly relevant for proper bioreactor design. Components such as proper scale-up impeller spacing, baffle sizing and location, and correct vessel sizing and scale-up are crucial in successful production scale bioreactor operation.
   • Mechanical similarity – In addition to components that are linearly scaled, mechanical similarity also encompasses dynamic operations, which are typical to mixing operations.

Scale-up for this similitude component, mechanical similarity, is often accomplished through the application of dimensionless groupings such as the Reynolds number. Mechanical similarity is achieved by having dimensionless groupings numerically equal when evaluating a smaller unit and a larger unit. For example, if a smaller laboratory-agitated vessel has an agitated Reynolds number of 1000, the larger production vessel should also have a Reynolds number of 1000. While not specifically addressed, the agitated Reynolds number and its relevance to bioreactors, fermentors, and many mixing systems in general will be addressed shortly in the section covering bioreactor design.

- Thermal similarity – Thermal similarity also uses dimensionless groups to determine scale-up parameters. In this case, the more important dimensionless groups are the Prandtl number ($C_p \mu / k$) and the Nusselt number (hD/k), both of which have been previously defined in Chapter 2. While these dimensionless groups must be identical for thermal scale-up, another important element must also be equal for thermal scale-up. That term is the thermal flux, Q/A (Btu/ft^2°F h). For thermal similitude to occur, the flux value must be identical when sizing for system scale-up. In fact, when dealing with mass transfer systems, the mass fluxes must also be identical for similitude to occur. When dealing with temperature similitude, it is important for the heat transfer characteristics to be either identical or scalable by a fixed ratio of the experimental unit to the full scale design. When dealing with mass transfer in scale-up studies, the mass flux should also be identical (i.e. the flux value for the pilot or laboratory system, lb/ft^2 h, should be identical to the proposed full-scale system).

- Chemical similarity – From the previous section covering chemical kinetics and this section covering bioreactions and bioreactors, it is reasonable to conclude that chemical similarity occurs when the concentrations of reactants and products maintain a constant ratio to one another while undergoing reaction phase. Basically, regardless of the mass of reactants involved, the reaction rate or growth rate proceeds at a constant rate whether it is the reaction rate constant for chemical kinetics or the growth rate for bioreactions.

While identified as an important design factor earlier in the section in a more generic sense, perhaps the most critical element of bioreactor and fermentor scale-up is the oxygen transfer rate, a variable that has been often identified in this section. Another key operating parameter for successful bioreactor and fermentor operation is effective mixing, also addressed in this section. Since mixing generally requires a high energy supply, energy supply and dissipation is a very important operating variable as well.

The importance of the oxygen concentration in biotech reactions was noted when spargers were identified. In fact, the results of Equation (4.23) are very reliant on the oxygen concentration (both the dissolved oxygen in the initial reactants and the mass of oxygen in the bioreactor, as a gaseous constituent of air) introduced into the system. In fact, effective oxygen transfer to the bioreactor constituents is driven by the difference between the oxygen concentrations often expressed in terms of the partial pressure of the oxygen in the reactor. The partial pressure also serves as a concentration difference (i.e. the difference between the higher oxygen concentration and the lower concentration in the bioreactor). Since the driving force for bioreactions is the oxygen pressure difference, it can be correctly concluded that the solute concentration in the bioreactor is negligible compared to the oxygen concentration, the driver of bioreactions. Similarly, since oxygen has a relatively low solubility in water as well as other solutions, it can be expected that bioreactions are time consuming and highly dependent on the dissolved oxygen concentration of the solute, in this case the mammalian cell tissue concentration, usually containing proteins required for biotech manufacturing. This is quite different than pharmaceutical reactions where the reactants and reaction rates are the drivers of the process.

Just as oxygen supply is critical to successful biotech operations, the carbon dioxide (CO_2) by-product must also be removed from the process media to affect successful performance. Consequently, effective CO_2 monitoring and removal are critical.

For bioreactions, the relationship for the oxygen transfer rate in a bioreactor is given by

$$N_g = k_g \left(p_{go} - p_{gl} \right) \tag{4.24}$$

$$N_g = k_g \Delta p_g = k_L \Delta c_g \tag{4.25}$$

N_g = Rate of diffusion of oxygen (mass transfer) in the media mixture, $g\,mol/(s)(cm^2)$ or $lb\,mol/(h)(ft^2)$

k_g = Mass transfer coefficient for gas diffusing into liquid media mixture, $g\,mol/(s)(cm^2)$ or $lb\,mol/(h)(ft^2)$

p_{go} = Partial pressure of the oxygen in the bioreactor, atm.

p_{gl} = Partial pressure of the dissolved oxygen in the liquid cell tissue media

k_l = Mass transfer coefficient of the liquid phase, $g\,mol/(s)(cm^2)$

Δp_g = Difference between partial pressure of oxygen in gas phase and liquid phases in bioreactor, atm

Δc_g = Difference between concentration of oxygen in gas phase and liquid phases in bioreactor, $g\,mol/cm^3$ or $lb\,mol/ft^3$

As can be seen from the definitions of Equations (4.24) and (4.25), rigorous analysis of transfer rates in bioreactors can be quite involved. The situation is further complicated by the fact the transfer rates for the various phases are intended to be measured at the interfaces of the phase boundaries and not in the bulk of the liquid media and the dissolved oxygen media, respectively.

To obtain valid bioreactor data, it is common to employ empirical data as a basis for scale-up data. It is common practice to evaluate the oxygen mass transfer coefficient k_g by using empirical equations via various software methodologies. Empirical curve fit paradigms, coupled with the availability of computerized statistical packages such as Minitab®, analysis of variance (ANOVA), and Excel, can produce an accurate representation of the mass transfer operation in the bioreactor.

While a great deal of emphasis is focused on determining empirical relationships for bioreactors, it has several advantages over developing more detailed paradigms to define bioreactor performance. A most important element is cost. To ultimately produce a cost effective biotech product, the pricing, once approved, must exhibit a profitable yet effective pharmaceutical. Additionally, time to approval is also important; other biotech operations may be developing similar or parallel products. As a result, time, cost, and safety are among the variables that must be considered throughout all phases of a project. Thus, empirical correlations are a very effective technique to obtain reasonable results in a judicious timeframe.

When determining the empirical relationships, it is critical to base the mathematical model on key parameters that can be measured. Since the diffusion rate N_g is a direct function of the mass transfer coefficient, the empirical relationship should include k_g in the relationship. While k_g is the limiting variable in oxygen transfer when dealing with bioreactions and bioreactors, other related variables must also be considered. One such factor is the gas velocity (oxygen) discharging from the sparger, in this case, referred to as the linear gas velocity. The linear gas velocity is defined as:

$$V_s = \frac{Q}{A} \tag{4.26}$$

V_s = linear gas velocity, ft/s

Q = volumetric flow rate, gal/min or l/min (it should be noted that dimensions must be consistent)

A = cross-sectional area of the total sparger holes = number holes × the area of each sparger hole (no. of sparger holes × $\pi D^2/4$, D = total cross-sectional area for all sparger holes). Further, as a design parameter, ZAIN Technologies (Bloomingdale, IL) [8] recommends that the total sparger hole area should be greater than the inside diameter (ID) of the sparger ring.

Equation (4.26) is, incidentally, identical to Equation (2.5). It is also important to note that while the volumetric flow rates are given, most often, in gal/min or l/min, the dimensions must be converted to units in ft/s, ft/min, m/s, or m/min to be dimensionally correct.

The units given in Equation (4.25) are engineering units. However, since a great deal of the information obtained is based on laboratory results where the analytical equipment is measured in metric units, it is common to employ metric units for this data.

Other factors that must be considered in the development of an empirical equation for k_g are horsepower/reactor volume (0.004–0.006) impeller diameter and geometry, flow regime (i.e. the flow in the reactor turbulent or laminar, as described in Section 2.1.1), operating temperature, operating pressure, and other, not so easily defined, variables. Based on the variables identified, it is possible to present a general empirical relationship to describe the mass transfer behavior in an operating bioreactor. The relationship is given as:

$$N_g = (k_g)(\Delta p) \qquad (4.27)$$

N_g = Rate of diffusion

k_g = Mass transfer coefficient (while the dimensions for N_g and k_g are identical, k_g is the rate of diffusion of a gas interfacing with a liquid phase, whereas N_g represents the

Δp = Pressure difference between the pressure in the reactor and the partial pressure of the dissolved oxygen in the bioreactor matrix, the driving force for Equation (4.25). The driving force can also be expressed as Δc, which represents the oxygen concentration difference between the bioreactor and the medium.

Further, k_g can also be defined by [7]:

$$k_g = (c) \left(\frac{\text{Hp}}{V}\right)^g (V_s)^\alpha \qquad (4.28)$$

c = A proportionality constant

Hp/V = Horsepower to volume ratio, previously given as 0.004–0.006 hp/gal

g = An exponent obtained empirically

V_s = The linear gas velocity identified in Equation (4.26)

α = An exponent, also obtained empirically

The term Hp/V is given as 0.004–0.006. The horsepower Hp actually represents a term known as gassed horsepower. That is, during the mixing/agitation operation, bubbles generated from a sparger disperse in the bioreactor liquid. This dispersion action is a form of thermodynamic work. This work is done isothermally. That is, there is no temperature change within the system. The gassed horsepower term is also identified as isothermal expansion horsepower or IEHP. While there is no temperature change, the gas is performing work. The work done by the gas is given by the thermodynamic relationship [9]:

$$\frac{\text{Hp}}{V} = \frac{\left(4.051 \times 10^{-3}\right)\left(P_1 + 62.4 Z \rho\right)}{Z} \ln \frac{P_2 + 62.4 Z \rho}{(P_2)} F_1 \qquad (4.29)$$

Hp/V = Isothermal expansion horsepower (IEHP) or gassed horsepower
P_1 = Pressure at tank bottom, lb/ft^2
P_2 = Pressure at tank top, lb/ft^2
Z = Batch height, ft
Batch volume, thousands of gallons
ρ = Fluid specific gravity, dimensionless
F_1 = Superficial gas velocity, ft/min

As can be seen, Equation (4.29) can be somewhat cumbersome not only in format but also in accumulating the required variables needed to determine a reasonable answer.

As an alternative to Equation (4.29), another effective relationship is also very useful, particularly when a design needs a quick mixer power estimate and not a full process design. The relationship is given as [10]:

$$\frac{\text{IEHP}}{1000\,\text{gal}} = (0.25)\,(V_\text{f}) \tag{4.30}$$

IEHP = Isothermal expansion horsepower = gassed horsepower = work done by the expanding gas = P/V_g
V_f = Superficial gas velocity as defined in Equation (4.25)

For configurations where an approximation or design adequacy is desired, Equation (4.30) is a very reasonable approximation.

Example 4.6   A contract research organization (CRO) has been hired to perform bench-scale studies for a biotech reaction that will have a reactor volume of 1200 gal for manufacturing operations. The laboratory studies coupled with scale-up studies and associated calculations have indicated that the superficial gas velocity should be 6 ft/min (The gas velocity is actually air bubbling through the sparger. But because of the low solubility of oxygen in the bioreactor solution, the oxygen content of air will be in excess and not be a hindering value). Additionally, to achieve the gas flow of 6 ft/min, the total area of the sparger holes will be 1.00 ft^2 for the 1200 gal bioreactor. The laboratory studies also determined that an agitator speed of 100 rpm should provide satisfactory agitation. For preliminary information required for contractor installation cost, it is desired to know what size agitator motor, in terms of horsepower, would be required. The motor gearbox would be sized to provide the needed rpms for the full size unit, once installed. Determine a motor size, in horsepower, to successfully operate the bioreactor agitation unit.

Solution
The quickest way to obtain a reasonable result is to recall that as a general rule, the agitator horsepower for most bioreactors is in the range of 0.004–0.006 hp/1000 gal. Using 0.005 hp/1000 as a constant, the required horsepower is:

$$\frac{0.005\,\text{hp}}{1000\,\text{gal}} \times 1200\,\text{gal} = 6\,\text{hp} \leftarrow \textbf{Solution}$$

The result, 6 hp, appears to be reasonable. However, motor sizes are generally not available as general components. As is noted in Chapter 3, available motors for 6 hp are either sized for 5, 7.5, or 10 hp. Since 5 hp would likely operate either marginally or in a degraded mode, the options are either a 7.5 hp motor or a 10 hp motor. Depending on NEMA requirements, for estimating purposes, a 10 hp motor as part of an agitator system should be considered cost.

In actual practice, a bioreactor with an associated agitator unit would probably be bid out as a single system, and the individual component cost would not be available. It would, in most cases not, be viable to obtain itemized component costs for a rather detailed installation such as a bioreactor unit. Actually, the total project cost is what most vendors bid when competing for a contract, and the agitator motor would be part of the "total as-built" package.

Example 4.6 is useful for verifying results obtained by empirical relationships and laboratory studies. It would also be advantageous if a rigorous technical evaluation that could correlate agitation parameters with power requirements. That is, if an agitator design is selected, it would be possible to determine the power requirements for a reactor system. Fortunately, such a method does exist. The technique involves using a modified form of the Reynolds number and a dimensionless group known as the power number.

In Chapter 2, and briefly in this chapter, the Reynolds number was discussed in some detail. The Reynolds number identified ($Dv\rho/\mu$) is commonly referred to as the linear Reynolds number. The traditional Reynolds number is defined as the ratio of inertial force to viscous force. For mixing systems such as bioreactors, fermentors, and batch chemical reactors, the Reynolds number is expressed as:

$$N_{Re} = \frac{ND^2\rho}{\mu} \tag{4.31}$$

N_{Re} = The agitated Reynolds number (it is most important to differentiate between the standard or linear Reynolds number and the agitated Reynolds number when dealing with this dimensionless number)

N = Shaft revolutions per unit time, revolutions/min for most bioreactors and fermentors and chemical batch reactors.

D = Diameter of the agitator impeller; if the impeller is composed of two blades, the diameter is the total width of both blades in ft.

ρ = Density of the liquid being mixed, lb/ft^3

μ = Viscosity of liquid being mixed, lb/(ft)(s) (as has been noted several times, when dealing with calculations involving dimensionless groups, it is critical that consistent units employing correct conversion factors are used in the calculations)

The linear Reynolds number has been verified by many experimental techniques that confirm the range defining laminar, transitional, and turbulent flow regimes. Such is not the case with the agitated Reynolds number. Because of impeller design, number of blades on the shaft, and presence or nonpresence of tank baffles and baffle design, there are many differing values for the ranges of flow types when calculating the agitated Reynolds number. In fact, in certain systems, turbulent flow ranges can be started with an agitated Reynolds at 1000 or below, compared with a linear Reynolds number, which becomes turbulent in the 4100 range. Similarly, fully turbulent flow, regardless of impeller shape, diameter pitch, baffle, or no baffle, is considered to be fully turbulent at an agitated Reynolds number of 1000 with laminar flow having a value of 10. Consequently, it is of marginal value to formulate quantitative relationships predicated on traditional values such as laminar or nonlaminar ranges when dealing with the agitated Reynolds number. Rather, it is quite important to determine applicable uses for this dimensionless term (i.e. the agitated Reynolds number), regardless of whether the flow is laminar, transitional, or turbulent.

When performing agitated Reynolds number studies, there is one pragmatic technique to ascertain when turbulent flow regime is achieved. This can be determined by increasing the

RPMs and consequently the agitated Reynolds number using the system of interest. When an increase in RPMs resulted in little or no increase of the power number for a baffled tank, it can be generally concluded that the turbulent regime is achieved. It should also be observed that agitation without baffles reduces the power number as the Reynolds number increases. Using the Reynolds number in both baffled and non-baffled configurations is a salient feature when performing scale-up studies in agitated vessels.

Currently, the agitated Reynolds number is widely employed in the area of mixing, particularly in chemical batch reactors, bioreactors, and fermentors, although the number is also employed, on occasion for mixing and blending operations.

A major application of the agitated Reynolds number involves determining the power requirements of an agitator used for mixing/agitating biotech compositions. With many bioreactor and fermentor operations requiring in excess of 24h to complete an operation, power consumption can be a significant part of the overall manufacturing cost.

The second dimensionless number, the power number, is a common application of correlating the agitated Reynolds number with another dimensionless number group to obtain valid values for a reactor system, in this case the power required to operate an impeller agitator. The power number is given as:

$$N_p = \frac{Pg}{N^3 D^5 \rho} \tag{4.32}$$

P = Required agitator power, ft-lb/s – it should be noted that 1 hp = 550 ft-lb/s
g = Gravitational constant, 32.2 ft/s^2
N = Agitator speed, rpm
D = Agitator blade diameter, ft – While there are several types of blade designs, a flat blade mounted in the center of the shaft is often used for preliminary design purposes.
ρ = Fluid density, lb/ft^3

To determine a power value for an agitator, a graphical relationship between the agitated Reynolds number and the power number has been developed. Such a relationship is given in Figure 4.5.

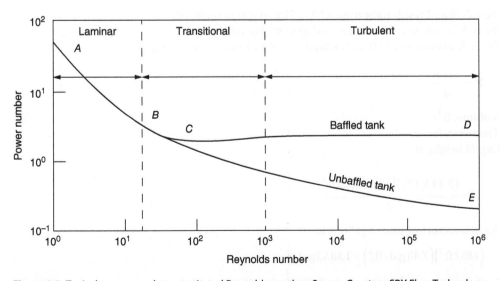

Figure 4.5 Typical power number vs. agitated Reynolds number. *Source:* Courtesy SPX Flow Technology.

Figure 4.5 is a generalized yet accurate form of the Reynolds number power number correlation. For less generalized relationships, rather than using baffle or no baffles, most graphs of this type would show various impeller geometries in lieu of the baffle data.

Example 4.7 A bioreactor with a diameter of 11 ft is to be installed as a manufacturing unit. The height of the tank is 22 ft with a planned liquid height of 18 ft. The tank has a center mounted flat blade impeller with a diameter of 4 ft (the blade diameter is the length of the impeller blade from one end to the other end) with an agitation speed of 100 rpm. The density of the solution is 64 lb/ft^3. At the operating temperature, the viscosity of the solution is 50 000 cP.

Calculate the horsepower of the agitation motor for the unit and compare the result with the estimate of 0.005 hp/1000 gal.

Solution
The first calculation required is the agitated Reynolds number. From the data given,

$N = 100$ rpm $= 1.67$ rps (the units must be dimensionally consistent)
$D = 11.0$ ft
$\rho = 64.0$ lb/ft^3
$\mu = 50\,000$ cP $= 33.6$ lb/ft s (from the conversion factor 1 cP $= 6.72 \times 10^5$ lb/ft h)

Inserting these values in Equation (4.31) to obtain the Reynolds number yields:

$$N_{Re} = \frac{(1.67)(4.0)^2(64.0) = 50.8 = N_{Re}}{33.6}$$

Referring to Figure 4.5, the power number is about 3. Inserting the values given and solving for the power by rearranging Equation (4.32):

$$P = \frac{N_p N^3 D^5 \rho}{g} = \frac{(3)(4.7)(1024)(64)}{32.2} = (28697.5 \text{ ft lb/s}) \, (1 \text{hp/550 ft lb/s})$$
$$= 52.2 \text{hp} \leftarrow \text{Solution}$$

The calculated result indicating a 52–53 hp motor is required.
The task is now to use the relationship 0.005 hp/1000 gal.
The tank diameter is 11 ft with a liquid height of 18 ft. The volume is

$$V = \frac{\pi D^2 h}{4}$$

$V =$ Volume, ft^3
$D =$ Diameter, ft
$h =$ Liquid height, ft

$$V = \frac{(3.14)(121)(18) = 1852 \text{ft}^3}{4}$$

The bioreactor volume in gallons is:

$$\left(1852 \text{ ft}^3\right)\left(7.48 \text{ gal}/\text{ft}^3\right) = 13853 \text{gal}$$

The horsepower is (0.005 hp/gal) (13 853 gal) = **69.3 hp** ← **Solution**

The difference between the calculated values (52.2 hp) and the estimated value (69.3 hp) is 25%, a reasonable quick estimate. Standard AC motors in this power range are 50, 60, and 75 hp. For this application, the 60 hp motor should be sufficient.

In actual design practice, the motor would be of a greater size than typically calculated. Intangible variables such as processing time, motor slip, density, and viscosity variations due to cell culture changes can require additional power requirements than the one specified. While it is difficult to design for unexpected consequences, it is sound practice to include a larger safety margin. Thus, in actual practice, it would not be unusual to specify a motor in excess of 100 hp. Assuming a standard amortization of the motor, the extra cost for a larger motor would be worthwhile.

It is important to consider that while oxygen uptake and oxygen transfer are critical to successful operations, successful scale-up of mixing operations is also a fundamental element for bioreactor and fermentor operation. Oxygen transfer is primarily a function of the intrinsic properties of the broth components (e.g. mass transfer coefficients of the constituents at given temperatures and pressures), while effective mixing is, in large part, determined by the skill employed in mixing and agitation scale-up design criteria. Unfortunately, unlike the large amount of data available for mass transfer operations, there is a paucity of information regarding accurate mixing and agitation operations.

4.1.6 Fermentor Principles

As was alluded to in Section 4.1.5, many of the fermentor design variables are identical to bioreactor design criteria. In fact, bioreactor designs can overlap fermentors when considering larger units (9000 gall). For comparison and informational purposes, Figure 4.6 illustrates a larger fermentor (~18 500 gal) intended to manufacture a fungal cell product API.

One differentiating feature between a bioreactor and a fermentor, noted previously, is that a fermentor is generally taller than a bioreactor, for greater oxygen transfer, whereas a bioreactor often has a larger diameter, for more effective cell culture mixing. While not a "cast in concrete" rule, this generalization is reasonably accurate. To appreciate the scope of this guideline for fermentors, listed below are some typical design specifications for a large production fermentor commonly used for bacterial and fungal manufacturing [7]:

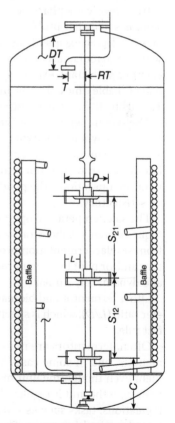

Tank diameter	132 in. (11 ft)
Total height	311 in. (25.9 ft)
Liquid height	214 in. (17.8 ft)
Nominal size	18 500 gal
Total height/diameter	2.35
Liquid height/diameter	1.63
Air velocity	5.91 ft/min
IEHP	300
Heat load	10.9×10^6 Btu/h
Heat transfer area	728 ft^2
Heat flux for the fermentor	14 973 Btu/ft^2°F h

Figure 4.6 Typical industrial pharmaceutical fermentor.

In addition to the typical large fermentor data given above, some other general design guidelines should be noted. When dealing with large fermentors and bioreactors, a small surface area to unit volume ratio usually indicates a cooling heat exchange system is required to keep the temperature within the planned operating range. For the fermentor described above, the flux ($14\,973\,\text{Btu/ft}^2\text{h}$) is within the range that literature standard guidelines specify, which are less than $20\,000\,\text{Btu/ft}^2\text{h}$ for organic solutions, less than $28\,000$ for mixtures containing a high hydrogen content, and less than $35\,000\,\text{Btu/ft}^2\text{h}$. But since the criteria are not a hard and fast rule, a higher flux is generally acceptable as long as the cell cultures are not damaged by the elevated operating conditions. Nevertheless, depending on broth composition, typical flux values for organic solutions should be limited to $28\,000\,\text{Btu/ft}^2\text{h}^4$.

A significant reason for maintaining the heat flux in a lower range, if possible, is to preclude foaming of the fermentor cell broth. Foaming is often caused by the high surface tension fermentor compositions, and bioreactor cultures, possess. The initial compositions can have high surface tension at the beginning of the fermentation operation, or the surface tension can increase as the fermentation proceeds. The high surface tension of the composition permits foaming to occur as the surface tension of the culture blend can resist the pressure exerted on the bubbles forming as boiling proceeds. As the higher boiling occurs, bubbles form at a higher rate than the bubbles can burst since the higher surface tension of the solution slows the bubble bursting. As the fermentation proceeds, massive amounts of bubbles form and create foam. The standard technique to reduce foaming is through the addition of chemical surfactants to the composition, a common technique in the chemical process industries. However, since this activity involves a pharmaceutical product, addition of surfactants, which reduce surface tension, is somewhat problematic. Obviously, the surfactant employed must be safe to use and approved as an excipient and identified in the USP/NF. Also, the surfactant must be compatible with the manufactured product and exhibit no harmful effects. To ensure the safety and efficacy of a surfactant, extensive testing and evaluation could be required at an unanticipated cost and probable time delay. By implementing an operating heat flux in the referenced range, about $10\,000\,\text{Btu/ft}^2\text{h}$, the foaming scenario could be avoided.

The recommended heat flux value is actually the generally employed number, $10\,000\,\text{Btu/ft}^2\text{h}$, used when specifying or designing heat exchangers that process organic materials. While this number is often employed, more accurate data could be obtained by performing laboratory studies at operating temperatures of the fermentation process. For higher temperature operations, the pendant drop method used in conjunction with a goniometer could be satisfactory.

The safe temperature operating range is determined by laboratory studies such as the technique referenced above or by using similar data obtained from previous operating experience from similar fermentation processes. Not to be overlooked is the need to confirm the laboratory and/or pilot level equipment (e.g. reactors, agitators, inlets, and outlets) are geometrically and dynamically similar.

Other fermentor and bioreactor design data such as the design ratio for the length-to-diameter ratio (L/D), which is specified as $2:1$, should serve as a design guideline and not a hard and fast rule.

Often, fermentors are not custom designed for specific products. User/owners select fermentors, and in some cases bioreactors and reactors, which fulfill a manufacturing requirement even though the unit may not perform to optimal design standards. This is particularly relevant if the unit is regulated by cGMP criteria. By using a fermentor that conforms to cGMP regulations, tasks such as validation protocols and cleaning procedures can be greatly simplified, thus resulting in significant cost and time savings. Another fermentation manufacturing option is to contract the fermentation operation to an outside firm. Firms that specialize in manufacturing one specific pharmaceutical component or API are identified as toll manufacturing organizations. Toll manufacturers have specialized equipment, such as fermentors, which can produce a desired product at a cost that is advantageous to the owner of the operation. In a situation such as this, the fermentor may not conform to rigid design standards, such as the *L/D* ratio or impeller to tank diameter ratio, but the lower product yield is compensated by the cost savings by not having to install a new system and the associated regulations required.

Typically, for fermentation operations, once the fermentation operation has completed manufacturing (API), the material is transported to other organizations for fill and finishing operations (e.g. packaging and labeling). This type of operation is commonly performed by companies manufacturing generic products (i.e. pharmaceutical products whose patents have expired and are commonly sold at lower prices than products under patent protection, but required to maintain product quality and integrity as required by FDA regulations).

Another common manufacturing technique employed by pharmaceutical companies is employing a contract manufacturing organization (CMO). The primary difference between a toll manufacturer and a CMO is that a CMO is equipped to perform a much more comprehensive program relating to the proposed product. It is also contracted to perform the full scope of product development, ranging from initial laboratory or bench-scale studies through pilot scale to full manufacturing quantities. The advantage to contracting a CMO for a task such as this is that owner costs and personnel requirements are minimized since the task is not performed by owner staff personnel and the CMO absorbs most of the development costs for a contracted cost. Again, the CMO may have access to fermentors that, while not adhering to rigid design standards, have the advantages of relatively rapid setup time and significant qualification preapproval.

One potential disadvantage of using a toll manufacturer or a CMO is the user has little supervisory authority over the contractor (i.e. the toll manufacturer or the CMO) other than monitoring cost and schedule.

4.1.7 Heat Transfer Aspects of Fermentors

Heat generation and control in a fermentation operation are perhaps the most important aspect of fermentation operations. As has been seen in the fermentor data shown in Section 4.1.6, significant heat generation is common. It should be recalled that fermentation is essentially an organic operation involving cell growth and rapid cell multiplication. This activity releases significant amounts of energy, predominantly given off as heat. Additionally, since many cGMP compliant pharmaceutical fermentors and reactors are fabricated of stainless steel, it is relevant to note stainless is generally a poor conductor of heat when compared to other fermentor and bioreactor materials of construction. This material of construction detail implies significant quantities of heat remain in the fermentor or bioreactor during operations. Additionally, since these processing vessels are also insulated for both economic and safety concerns, the heat loss is minimal. As a result, it can reasonably be expected large quantities of coolant (water) would be required. As the fermentor specification presented above indicates, this is most often the case.

Also a factor in the heat buildup in the vessel is the shell side fouling. Unlike typical shell and tube heat exchangers, fermentors and bioreactors are not surrounded by a shell that limits buildup on the shell side. Instead, the entire fermentor serves as the shell that allows virtually limitless shell side accumulation on the tubes during processing. This also increases the heat buildup in the fermentor because the buildup precludes effective heat transfer from the tube side fluid. For fermentors and bioreactors of more recent design, it is economically practical to equip these units with a clean-in-place (CIP) vessel cleaning operation.

As with many design parameters for fermentor and bioreactor design, many values represent an operating and design range rather than specific values. The heat transfer rate for fermentors and bioreactors is also given in a range. Actual operations often require adjusting the coolant, or in some case the steam, rate until a satisfactory flow condition is achieved. However, it is possible to ascertain a reasonably valid cooling or heating approximation for preliminary design and operation purposes.

The basic design equation for determining the heat transfer rate is given by:

$$Q = UA\Delta T_{lm} \tag{4.33}$$

Q = Heat transfer rate, Btu/h
A = Heat transfer area, ft^2
U = Overall heat transfer coefficient, Btu/h ft^2 °F h
ΔT_{lm} = Log mean temperature difference (defined in Chapter 2)

Recalling Equation (4.33) is identical to Equation (2.41); the basic heat transfer equation can be slightly altered to a more accurately describe heat transfer in fermentors and bioreactors. The heat transfer rate Q can be presented as

$$Q = Q_{fer} + Q_\lambda + Q_{ag} \tag{4.34}$$

Q_{fer} = Heat generated by the fermentation process
Q_λ = Heat of vaporization (evaluated if the fermentor broth undergoes a phase change)
Q_{ag} = Heat generated by the agitation of the fermentor mixture

Because the solids concentration generally increases in the fermentor broth, the heat generated by the fermentation process is absorbed by the solids generated in the fermentation operation and are controlled by large quantities of cooling water. Consequently, this quantity of heat generation is considered as part of the overall fermentor operation and not independently evaluated.

Heat required due to vaporization of the liquid component(s) of the fermentor broth are also not evaluated since as the fermentation proceeds, the solids content also increases and, as noted, the heat is absorbed by the solids in the fermentation broth and the result is a higher operating temperature for the fermentation operation, which is manifested by the higher cooling water flow encountered in most fermentation operations.

Q_{ag}, the heat generated by impellers or agitators is actually a small fraction of the total heat generated in the fermentation process. This energy is also generated as part of the total cooling requirement and is usually not a significant contributor to the overall heat transfer process.

When evaluating an overall heat transfer coefficient for fermentors and bioreactors, it is important to recall the difficulty of applying specific design values for a specific process. The need for an operating range should again be emphasized. The fact is if two identical fermentation or bioreactor is performed, the process parameters will rarely give identical results. Variables such as yields and cooling water temperatures are not identical. For instance, depending on the season and cooling water pretreatment, there could be significant temperature variation with the seasonal cycle; typically, winter cooling water needs will be less than summer cooling water requirements. This cyclic variation can alter the final product yield.

It is because of variations such as these that an operating range is more effective than a fixed heat transfer coefficient. Therefore, it would be desirable to have a reliable overall heat transfer coefficient range for fermentors and bioreactors. Fortunately, there is a value range that is reasonably accurate, if laboratory or pilot plant data are not available [7]. This range assumes heat transfer in a fermentor or bioreactor is akin to a single pass tubular heat exchanger, details of which are provided in Section 2.4, Heat Transfer for Purified Water Manufacturing; details of which are described in the following paragraph.

A specific value for fermentation operations is particularly difficult since the overall heat transfer coefficient can vary significantly from the beginning of the process through the end of operations. A great deal of the uncertainty is due to product buildup or fouling on the outer surface of the heating coils that are, as Figure 4.6 illustrates, located in the body of the fermentor. For one pass heating coils using steam as the heating medium, a reasonable design overall heat transfer coefficient (U), for a clean fermentor, would be 110 Btu/ft^2°F h. For a bioreactor, a reasonable U for a unit with no fouling is also 110 Btu/ft^2°F h. As the fermentation operation nears completion, a heat transfer coefficient for a typical fermentation process would be in the range of 75–90 Btu/ft^2°F h. Of course, variables such as broth density changes, heat builds up due to increasing agitation resistance, viscosity changes as fermentation proceeds, density increases with process, and time can affect the U value.

Both bioreactors and fermentor designs are equipped with heating/cooling coils or heating/cooling jackets, depending on the service. Coils are most often not used for continuous heating or cooling because the liquid flow in the coils not operating at steady state (i.e. [flow in] – [flow out] = [accumulation]). The accumulation is typically a buildup of fluid that impedes the flow and consequently negatively alters the design temperature.

Table 4.4 exhibits the contrast between the overall heat transfer coefficients for jacketed reactors and reactors installed with a coil design for both heating and cooling scenarios.

The values presented in Table 4.4 are for coils and jackets only. The values in the table do not include the effects of agitation on the fluid within the reactor. For reactors with installed coils or jackets and equipped with an agitator for U, dimensions in Btu/ft^2°F h are in the range of 75–175 Btu/ft^2°F h.

The information contained in Table 4.4 indicates the heat transfer coefficient is higher when heating of the system is occurring. The rationale for this result can be explained from basic heat transfer calculations from Chapter 2. Equation (2.44) identifies a basic heat transfer equation, the Dittus-Boelter equation as $Nu = 0.023\,Re^{0.8}\,Pr^n$. Further, Equation (2.38) states the Nusselt number (hd/k) is proportional to the overall heat transfer coefficient, U. In defining the variables of the Dittus-Boelter equation, the exponential value of the Prandtl number was 0.4 for heating and 0.33 for cooling. Consequently, since coils are essentially shell and tube heat exchangers, as described in Chapter 2, the exponents 0.33 and 0.4 are employed. The same analogy is applicable for jacketed reactors.

The U values given above are not intended to be used in lieu of actual process values, but can serve as a reasonable basis when limited data is available, as alluded to previously. For this reason, it is important to perform bench-scale and pilot-scale studies as an augment to the design equations presented. When performing bench-scale or pilot-scale studies, it is most

Table 4.4 Comparison of reactor heat transfer coefficients equipped with jackets and coils.

| | Overall heat transfer coefficient, U (Btu/ft^2°F h) | | | |
| | Heating | | Cooling | |
	Lower value	Higher value	Lower value	Higher value
Jacketed reactor	35	159	18	106
Coils	105	265	35	141

important to, again, emphasize the need to maintain a constant flux (Q/A) value for all cases. That is, if the flux for the laboratory study is $10\,000\,\text{Btu/ft}^2\text{h}$, the same value, $10\,000\,\text{Btu/ft}^2\text{h}$, should be maintained for larger-scale studies.

When considering fermentor heat transfer, it must be emphasized that heat generation is most often the limiting parameter for effective fermentation. The cells are living organisms and subject to survival temperature ranges. This, in effect, indicates that, in addition to maintaining a control over the operating temperature, an effective cooling water flow must be maintained. Consequently, the coils found in both fermentors and bioreactors must be designed to accommodate both a steam supply and a chilled water capability, commonly in the range of 35–44 °F.

4.1.8 Bioreactor and Fermentor Design, Maintenance, Operating, and cGMP Considerations

Ideally, this section should be identified as bioreactor and fermentor design, maintenance, operating, and cGMP considerations. However, as has been noted in several instances, a great deal of fermentor and bioreactor specifications has very large design and operating ranges. It is for this reason that stainless steel fermentors and bioreactors are intentionally overdesigned to assure successful system operation. While this approach may not be a particularly cost-effective approach, it is a proven technique to assure satisfactory production levels.

When dealing with bioreactor and fermentor scale-up, design, installation, and operation, the following design concerns and issues may be worthy of consideration and should be documented as issues to be addressed in procedures, qualification documentation, design criteria, and commissioning documentation, depending on the project status:

- The textbook approach to scale up seldom works [7]. A first reading of this statement would appear to question the veracity of Sections 4.1.5–4.1.7 of this text. It should be noted, however, these sections have emphasized the general scale-up nature of the information presented. While the information presented is relevant, designing fermentors and reactors using the scale-up data presented also requires a degree of engineering judgment (and luck) predicated on related design, installation, and operating experience. A failure to rely on basic engineering principles as a design basis can lead to an improperly designed full-scale unit, regardless of the rigorous laboratory and or pilot plant work performed. It is also important to recall that, unlike chemical kinetics, the reactants involved in bioreactors and fermentors are not dependent on outer electron changes (e.g. chemical reactions) to create chemical changes. Rather, it is factors such as cell genetic makeup, oxygen transfer rates, and metabolic rates that determine bioreactor and fermentor yields, and while advances are rapidly occurring to quantify these factors, much research and development is required to achieve reliable paradigms.
- While it would greatly improve design and scale-up efficiency, the fact is there is little exact correlation data existing for actual scale-up and design of manufacturing fermentors and bioreactors. Much of the published information is a result of academic studies that most often obviate actual manufacturing conditions such as power consumption, cleaning and validating, maintenance schedules, and parts repair and replacement. Commercial bioreactor and fermentor suppliers are probably hesitant to supply reliable information since it could compromise their products if such information was made available.
- Verify the design criteria selected is correct. When the key factors required for successful scale-up are often identified, it is important to incorporate them properly in the design. For instance, while the oxygen transfer is most important, failure to properly provide the required (at a minimum) flow for the full size design could result in lower yield. Also, it is most important

that a careful analysis has been performed to accurately identify the correct variables required for scale-up procedures. Perhaps the most direct and effective technique for augmenting the task is through the implementation of a design of experiments (DOE) program, a technique that is described in Chapter 11.

- Intrinsic variables should be consistent in the scale-up operation. Items such as broth composition (is there a tendency for the cell media to accumulate during processing?), viscosity variation during fermentation, pH during the operation, and processing temperature must be considered during the scale-up activity.

- Geometric similarity should be maintained as an integral part of the scale-up. Often, if the scale-up commences on a lab-scale basis using laboratory containers such as a large magnetically stirred Erlenmeyer flask, the fact that the actual reactor will be a straight wall tank may be obscured because attention is given to the stirring mechanism and other batch parameters.

- Maintain a fixed design basis during the scale-up design phase. Yield, cell concentration, cell activity are variables that are germane to the design. The proposed installation should focus specifically on what is required of the fermentor or bioreactor and scale-up and design should address the specific intent. Consequently, it is critical the specific objectives of the task be clearly defined since variable (yield, cell concentration, and cell activity) can affect the ultimate design. For instance, what is the effect of a shorter operating period on the cell concentration when product yield is the objective? Will this adversely affect temperature or broth viscosity? These parameters may have to be considered if, for instance, the design basis focuses on yield without consideration of the affect of other process factors.

- Be certain that numerical values obtained from using dimensionless groups such as the Reynolds number and power number are identical during scale-up activities.

- During scale-up, impeller sizing and design is very important. Variables such as pitch, number of impellers on the shaft, impeller location, impeller diameter, impeller length, and speed (i.e. rpm) are some of the factors that must be evaluated for geometric and dynamic similitude. Additionally, vessel variables such as total broth height, inside and outside vessel diameters and nozzles and sampling port locations and sizing must also be considered as well as the design factors and operating information located in Section 4.1.5.

- The bioreactor or fermentor system should include the necessary parts or parts availability to provide adequate interchangeable parts replacement for the planned lifetime of the system. Further, responsible facility personnel should clearly understand the concept of "replacement in kind" and the applicability of generic components to this concept.

- One important maintenance and operating issue is selection of cleaning solutions. Cleaning solutions should not impede cell growth when bioreactions or fermentation is the production mode. Additionally, the cleaning agents employed should not be flammable. Further, if the vessel is fabricated of stainless steel, potential harmful effects on the vessel must also be evaluated. One universal taboo when considering cleaning solutions is considering cleaning solutions containing chlorides. Failure to properly remove chloride-containing cleaning solutions residue could result in production runs where the chlorides, over time, adversely affect production. Chlorides in an acidic environment are also prone to exhibit a phenomenon known as chloride stress corrosion, details of which are presented in Chapter 8. For these reasons, cleaning solutions that are intended for fermentation and biopharma processing applications should operate in a higher caustic environment (Typically, pH > 10), which would minimize the potential to form free chloride ions and potentially reduce the solution surface tension and, in effect, making the cleaning agent "wetter." In addition to effective use of cleaning solutions, removal of vessel residue is also quite important. If approved procedures indicate that, subsequent to cleaning, a record of residue accumulation exists, it might be beneficial to consult with the cleaning unit vendor and possibly a firm specializing in equipment cleaning and validation consultants. Also, as explained earlier, if surfactants are employed in vessel cleaning operations, it is most important the surfactant employed is a USP/NF approved compound and is compatible with the broth components in all stages of manufacturing. An applicable technique to evaluate the efficacy of a cleaning operation where residue, as opposed to microbial buildup, is determining the impact on spray cleaning operations. The level of impact needed to clean a vessel is primarily a function of the type of residue formed, cleaning chemicals employed and the water temperature. An empirical relationship has been developed to evaluate the impact of a bioreactor and or vessel cleaning operation. The relationship is given as $I = KQP$ [11] where K is a constant, Q is the flow rate, and P is the liquid pressure. If determination of the constant presents a problem, the factors Q and P can be evaluated alone, preferably with other vessel cleaning systems to use as a comparison. In all cases, consistent units should be employed when performing an impact calculation or comparing systems. Regardless of the relationship used (i.e. using the three dependent variables or the two independent variable formula), calculated values serve as a general numerical relationship to estimate the operational impact, I, on a spray cleaning system.

- Be certain that vessels employing spray nozzles or other cleaning devices are adequately sized to cover the internal surfaces for effective cleaning. It is not uncommon to employ two or more units to effectively encompass the critical cleaning surfaces. Since multiple spray nozzles are often mounted off a common cleaning solution shaft supply shaft and are located at different points (there are usually no more than two nozzles or spray balls on the shaft), this design should be capable of ensuring that the required spray area is covered upon installation. For fermentors or bioreactors that have two or more spray nozzles, or spay balls, on separate cleaning solution supply shafts (no more than two separate supply shafts), it would be reasonable to expect a degree of flexibility in the vertical positioning of the separate supply lines. For single shaft, dual nozzle installations, such an adjustment may be more problematic since modifying the height of one of the nozzles will also affect the other nozzle located above or below the nozzle of concern.

- These considerations identified above should also be evaluated when dealing with probes such as dissolved oxygen sensors, carbon dioxide sensors, pH sensors, and other installed probes. As bioreactors and to lesser extent fermentors, being supplied with more sophisticated instrumentation, the need for easily removable and rugged probes is increasing in importance.

- Part of the pre-operating procedures should detail the steps involved in how the bioreactor or fermentor is purged of residual oxygen. This is most often accomplished by introducing an inert gas such as nitrogen throughout the vessel for a specified period of time. Failure to address this item can result in incorrect oxygen levels that can compromise production due to incorrect oxygen concentration instrument readings.

In addition to scale-up design and operating considerations, such as those highlighted above, basic equipment criteria should also be considered; some of these issues include:

- Required utilities should be verified. This item is normally a standard requirement in most equipment installations. For bioreactors and fermentors, it is important to evaluate this factor. For instance, it is advantageous to have a common voltage supply for process equipment. Employing small pumps using 110 VAC and simultaneously having agitators relying on 220 VAC can cause a significant additional cost in money and time.

- Another utility that must be addressed is the air supply to the bioreactor or fermentor. Since the unit is regulated by cGMP criteria such as 21 CFR 211.42 (ventilation, air filtration, air heating and cooling), it is critical that the air, a source of oxygen, is filtered to a required purity level so the broth constituents will not be contaminated. Ideally, the bioreactor or fermentor air supply could be incorporated with an existing filtered air supply. Such an action could be cost effective and also shorten qualification activities, specifically the validation activities. Thus, preparing a comprehensive utilities program focusing on commonality and interchangeability could be quite beneficial.

- Replacement parts and components should be readily available. Certain control components such as specialty valves should be easily replaceable either by maintaining adequate spare parts or possessing a reliable supply chain. Also, the installed unit, fermentor or bioreactor, should be designed such that all components (valves, pumps, agitators, spargers, etc.) can be removed and replaced without compromising the vessel (e.g. replacement parts or components requiring maintenance should be in easily accessible locations). Also, be certain that vendor supplied items such as sanitary piping connections are not long lead time items that can adversely affect production schedules rather than maintaining a stored quantity of items. A large spare parts and maintenance inventory is undesirable since this inventory ties up the capital that could be applied to other facility applications. A spare parts inventory should be determined by employing the facility CMMS as well as mathematical relationships such as those found in Chapter 5. These particular logistics and provisioning considerations should be continuously monitored throughout the useable lifetime of the bioreactor or fermentor system. Ideally, this task should commence upon approval of the user requirements specification (URS) and the functional requirements specification (FRS), details of which are found in Sections 10.1–10.6.
- The design should emphasize cleaning and sanitizing operations. Ideally, cleaning and sanitizing operations should be designed such that a minimum of turnaround time is required to have the system return to operation. To achieve this goal, a combination of effective reliability planning and implementation coupled with efficient maintainability activity should be part of the operational procedures. Such programs can be based on the material provided in Chapter 5.
- As has been noted previously, it is important to reemphasize the need to have a design that deletes or minimizes angular joints and crevices. If possible, baffles should be eliminated or designed with tapered joints and not right angles in order to minimize the potential for locations to accumulate a buildup from previous manufacturing runs. By incorporating smooth gradient free surfaces in the design, minimization of residue accumulation is an achievable goal; an objective that would please cleaning validation personnel. For large bioreactors and fermentors, even the most effective CIP unit cannot guarantee a totally clean vessel.
- Verify that materials of construction are compatible. While it may be cost effective to use less costly materials for certain connections, the durability of the less expensive items should also be evaluated. For instance, if a probe casing is fabricated with a polymeric body, be certain the body will not be affected by temperature changes during the fermentation process or the bioreaction. If the coefficient of the probe casing differs significantly from the coefficient of the fermentor or bioreactor shell, the vessel integrity may be compromised due to the different rates of expansion and contraction due to temperature changes during processing due to gaps forming between the probe casing and the vessel body. A quick evaluation can be performed by comparing the coefficient of linear expansion for the vessel and the probe casing materials. The coefficients of expansion should be of the same order of magnitude.
- Associated vessel piping should contain a minimum of elbows, branches, and tees. Such configurations could make effective cleaning problematic due to buildup of residue and

unflushed product. While the piece parts may have sanitary connections, the difficulty may be in the internal piping lengths between the sanitary joints. Also, if elbows must be employed, wide radius pieces should be considered to minimize the potential for buildup.

- Discharge lines and drain lines should be sized to minimize the potential for clogging. If discharge valves are automatically actuated, manual overrides should be considered should a condition upset occur. The additional cost would be minimal compared to the risk presented, particularly if the product being produced has an identifiable health hazard.
- Confirm internal diameters, (bore) of piping components are identical and have no gaps between pieces being joined.
- When evaluating the fermentor or bioreactor vessel diameter, it is advisable to design for a minimum allowable diameter. As the design diameter increases, the fabrication costs also increase, usually not in a linearly. Also, for larger complex fabrications with multiple nozzles and very tight tolerances, the number of qualified and experienced fabricators qualified to bid on the fermentors or bioreactors will diminish. This could prevent a competitive bidding process for the required vessels and could cause the user to accept a design that was not originally planned. This, of course, introduces the possibility of an installation that may not perform as originally intended.
- The internal surfaces should be machined to the smoothest finish possible. Typically, this translates to a Ra of less than 20, as defined in ANSI B46.1-2009 (surface texture, surface roughness, and lay), or a grit size of 180 (a basic sheet finish designation no. 4). Internal surface finish is another factor contributing to the desirability of using the smallest diameter that can reasonably be specified. As the diameter increases, obtaining the smooth finishes desired becomes more expensive.

While cleaning operations must be specified in accordance to cGMP criteria such as 21 CFR 211.63 and 21 CFR 211.67, it is also important to optimize the vessel and bioreactor cleaning procedures for both cost effectiveness and operational purposes. Several effective steps have been suggested to achieve this goal. These factors include [11]:

1) Minimizing the quantity of water used in spray cleaning operations. Because heating water, particularly treated purified water, is a costly operation, it could be economically justifiable to perform a small study to determine an optimum water temperature that can be used for cleaning operations. An efficient and meaningful study could be predicated on using the procedures explained in the DOE section of Chapter 11. Also, evaluating the ability to effectively use recycled cleaning water at a lower temperature may also minimize the hot water temperature normally used and still provide effective cleaning. By recirculating and filtering the cleaning solution prior to discharge, it is conceivable the volumetric bioburden to the waste treatment system can be reduced significantly.

2) Decrease the number of cleaning cycles. Minor adjustments to the flow rate and spay pressures could produce a shortened cleaning cycle and consequently reduce the system downtime. As a result, a shorter cycle time would lead to an improved operational availability, as discussed in Chapter 5.

3) Be certain the nozzle design allows for 360° coverage and extends to the inner wall with sufficient pressure to be effective.

4) The impact of new designs incorporating the rapid integration of single use and disposable reactors and components should be evaluated. Chapter 9 can serve as a basic information source for this topic.

4.2   Safety Relief Valves and Rupture Discs

The United States, as is the case with virtually every country in the world, has its own codes and regulations regarding the design, specification, and sizing of safety relief devices. The regulations governing aspects of these types of devices are contained in the provisions of the ASME Pressure Vessel Code, specifically ASME Section VIII, Division 1. Specific parts of the code covering relief devices extend from UG-125 to UG-136. Additionally, Appendix 11 provides useful information for determining flow capacities of relief devices as well as associated information. Appendix M of the code (ASME VIII, Division 1) also contains relevant information concerning relief device installation and pressure vessel code markings.

Knowledge of relief devices is desirable since once these devices are installed, it is often the responsibility of the organizational engineering and maintenance operations to verify proper relief device operation. Also, unlike water systems or HVAC installations, the owner of the installed relief devices is, most often, responsible for all aspects of relief devices at a location. For this reason, it is important for personnel involved in manufacturing, engineering, maintenance, and quality to have a reasonable knowledge of relief devices. Since relief devices will always be encountered in facility operations, it is important to understand, as a minimum, safety, design, specification, installation, maintenance, and testing requirements for relief valves. Additionally, these devices represent significant capital investments, and proper design, installation, and maintenance can be significant cost savings if properly implemented.

Figure 4.7 displays a schematic of a conventional safety relief valve. One variation to the device shown in Figure 4.7 would be a valve intended for steam, air, and hot water service (a temperature >140 °F). For this design, the valve would be equipped with a lever, as required by the ASME Code.

4.2.1   Safety Relief Devices, Definition of Terms

While Section 4.1.1 presents terms applicable to code pressure vessels, the definitions below are specific to installed relief devices associated with pressure vessels.

Safety Relief Valve This is a general term and includes all types of spring controlled, automatic pressure-relieving devices. The purpose of his safety relief valve is to protect personnel and equipment by preventing an extensive accumulation. Figure 4.7 illustrates the components and assembly of a typical safety relief valve.

Relief Valve An automatic pressure-relieving device actuated by static pressure upstream from the valve. A relief valve is intended for liquid service.

Safety Valve This spring-loaded pressure relief device is intended for steam, vapor, and gas service. As is the case with all ASME Code pressure-relieving devices, it is actuated by a static pressure upstream of the safety valve.

Rupture Disc A rupture disc consists of a thin diaphragm held between two flanges. Each can be made of metal or graphite. The thickness of the diaphragm is such that it will rupture completely at a specified pressure typically within a ±5 to 15% tolerance, dictated by manufacturers' specifications and procedures. Figure 4.8 displays a typical rupture disc.

Maximum Allowable Working Pressure (MAWP) The MAWP depends on the type of material, its thickness, and the service conditions set as the basis for design. A vessel may not be operated above this pressure or its equivalent, at any metal temperature other than that used in its design; consequently, for that metal temperature, it is the highest pressure at which the safety or relief valve may be set to open. The MAWP for specific pressure

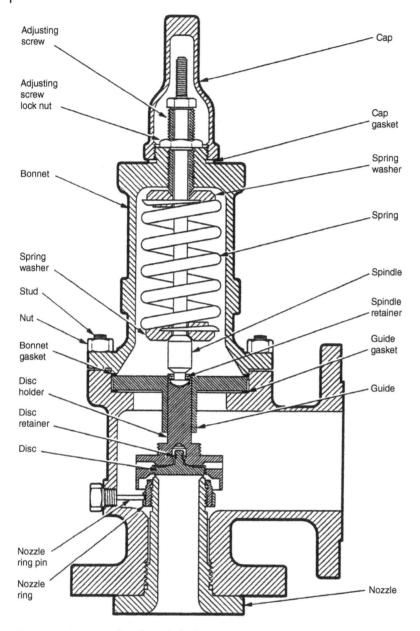

Adjusting screw

Adjusting screw lock nut

Bonnet

Spring washer

Stud

Nut

Bonnet gasket

Disc holder

Disc retainer

Disc

Nozzle ring pin

Nozzle ring

Cap

Cap gasket

Spring washer

Spring

Spindle

Spindle retainer

Guide gasket

Guide

Nozzle

Figure 4.7 Diagram of a safety relief valve.

vessels should be available from approved drawings or, if the pressure vessel is installed, from the visible and required data plate installed on the pressure vessel. It should be noted that vessels encapsulated with insulation should also have the data plate visible for inspection at all times.

Set Pressure The set (or relieving) pressure is the pressure at which a safety relief valve starts to open and discharge (liquid service or pop on steam, gas, or vapor service) against atmospheric back pressure. The MAWP is usually identical to the set pressure.

Figure 4.8 Standard rupture disc.
Source: From Fike Corporation [12].

For pharmaceutical applications, the set pressure is given in units of pounds per square inch. For a safety or safety relief valve in gas or vapor service, the set pressure is the pressure at which the valve pops under service conditions.

Accumulation The pressure increases over the relieving pressure for safety relief valve during discharge through the valve, expressed as a percentage of the relieving pressure. The valve will be wide open at the relieving pressure plus the accumulation allowed. Normally, the accumulated pressure for piping should be the same as for unfired pressure vessels.

Back Pressure Pressure developed on the discharge side of the safety relief valve has two different forms:

1) Superimposed the back pressure – This is the pressure in the discharges line before the safety relief valve opens either constant pressure (e.g. at the suction of a reciprocating pump or the pressure to the discharge of other valves) and is referred to as superimposed back pressure. A superimposed back pressure is usually not present in pharmaceutical applications. A common source of superimposed back pressure is often associated with flares that are often visible from petroleum refineries. The flare is a result of waste gases generated from operations. These gases have a constant pressure in addition to the normal pressure produced from normal operations.

2) Built up back pressure (BBP) – The pressure in the discharge header, which develops as a result of flow rate in to the valve, is called built up back pressure. It is the superimposed back pressure plus the pressure drop caused by flow from the discharging valve through the relief header. Just as the release of fluid from the relief device has a pressure, air, most often contained in the relief piping, also pushes in the opposite direction, creating a pressure in the opposite direction and must be accounted for in pressure drop calculations. The back pressure, or ΔP, is actually the set pressure minus the discharge pressure to the environment or a drain. The discharge pressure is merely the atmospheric pressure. Thus, the final discharge pressure is

the relief device set pressure minus the atmospheric pressure ($14.7\,lb/in^2$). Further, the relationship for determining the numerical value for the back pressure is given by Equation (2.12). This relationship is:

$$\Delta P = \frac{f\rho L v^2}{1442\,gD} \tag{2.12}$$

Here, ΔP is, again represented as the set pressure (lb/in^2), minus the discharge pressure to the environment that is $14.7\,lb/in^2$ at sea level.

It is most important to note the BBP cannot exceed 10% of the set pressure for a non-fire scenario, 16% for a vessel with multiple relief devices (e.g. a relief valve and a rupture disc protecting the same pressure vessel), or 21% for a pressure vessel exposed to fire.

Just as there are criteria for discharge pressure drops, there is a general guideline for inlet pressure drops to the relief device. This guideline is intended to minimize the entrance effects that may be present in the flow regime. To account for entrance effects, the BBP of a relief device inlet line is limited to 3% of the relief device set pressure. For relief device settings in the 50–$60\,lb/in^2$ range and lower, a fixed BBP inlet pressure drop in the range of 1–$3\,lb/in^2$ should suffice.

Conventional safety relief valve A conventional safety relief valve has a housing that is designed to vent on the discharge direction of the safety relief valve.

Balanced relief valves Valves of this type are designed so that back pressure has little influence on the relieving pressure. The design can be of either the bellows type or the piston type. The bellows type is the design most often employed.

Bellows valve For the bellows design, the bellows area is identical to the nozzle area (relief valve inlet area). Additionally, a bellows containing relief valve has no eductor tube and does contain an open vent on the valve bonnet that vents directly to the atmosphere. Bellows are added to conventional valves for several reasons. They are:

- There is superimposed back pressure exceeding 10% of the relieving pressure.
- The fluid is highly viscous or a slurry type composition.
- The fluid is corrosive to components of the relief device.

If any of the above conditions exist, use of a bellows should be evaluated. Because of the availability of retrofit kits, converting from a conventional valve to a bellows configuration is easily accomplished at a minimal cost.

Piston valve For a piston valve, the guide is vented such that the back pressure on the opposing faces of the disc cancels each other such that the venting occurs at atmospheric pressure through the bonnet vent.

For pharmaceutical cGMP operations, the preferred device is the conventional type of design. A major reason is the cost difference between conventional valves and balanced valves. On average, a balanced valve is, at least 30% more than a conventional type. For operations where a large number of relief devices are employed, this is a significant cost. Because of this, a careful audit should be performed on new and existing relief valves when a program involving such devices is encountered. If vendor supplied installations are employed, it is, again, important to verify if devices match the corporate engineering specifications relating to relief devices.

4.2.2 Relief Valve Design and Specifications

When considering safety relief device design and specification, an understanding and examination of the system/installation that requires protection is most important. A significant aspect of relief device knowledge is a basic ability to perform and verify the calculations and causes of overpressures in an easily understood and documented format. This is best achieved by initialing reviewing the most current applicable P&ID for the installation. The P&ID should clearly define the section or parts that are covered by the relief device(s) of concern. Also, it should identify the set pressure at which the relief device will act for vendor installed systems with relief device installations, data should be available with supporting calculations and information. If the installation involves in-house process design, the data must also be available as part of a verified relief device design record.

4.2.3 Requirements and Capacity

This section is applicable to unfired pressure vessels (U-stamped vessels) regulated within the scope of ASME Section VIII, Division 1.

Relief valves are specified to prevent pressure in the vessel from raising more than a certain percentage of the MAWP, depending on the application. For the following relief situations, the allowable pressure accumulations, as identified previously, in percent, are specified as:

	Accumulation (%)
• Fired boiler (ASME Section I)	6
• Unfired pressure vessel (single device)	10
• Unfired pressure vessel (multiple devices)	16
• Unfired pressure vessel exposed to fire	21

By combining the applicable accumulated pressure value with the MAWP values given above, a maximum allowable set pressure for the relief device can be obtained by the following relationship:

$$P_A = (MAWP) \times \frac{(100 + \%Accumulation)}{100} \tag{4.35}$$

P_A = Maximum allowable accumulated pressure, psig
MAWP = Maximum allowable working pressure, psig

While unusual for pharmaceutical applications, the pressure vessel may be one of a series of vessels and each vessel in the series must be evaluated as an individual component.

The above relationship is the maximum pressure a relief device can be set at for safe operation.

Example 4.8 A pressure vessel containing a solvent is used in a blending operation. It is an unfired vessel (a "U" stamp). The MAWP designated on the vessel identification plate is 100 lb/in^2 and is equipped with a single relief device. An examination of the P&ID indicates no fire scenario is envisioned. Based on the information given, at what pressure should the relief device be set to actuate due to a failure of the discharge valve to open?

If a fire scenario is envisioned, what should the set pressure be for the vessel if exposed to a fire?

Solution

For these situations, Equation (4.35) will be used for both parts of this example. For the first scenario (i.e. a non-fire failure), the solution is:

$$P_A = (100) \times \frac{(100+10)}{100} = 110 \, \text{lb/in}^2 \leftarrow \textbf{Solution} \left(\text{MAWP for non-fire hazard}\right)$$

For a fire event where the pressure vessel is exposed to fire, a 21% accumulation is required:

$$P_A = (100) \times \frac{(100+21)}{100} = \textbf{121lb/in}^2 \leftarrow \textbf{Solution} \left(\text{MAWP for vessel exposed to a fire hazard}\right)$$

The two values 110 and 121 lb/in^2 represent the maximum values at which the respective relief devices should be set. In practice, the values could be set to 1 or 2 lb/in^2 lower than the calculated values.

Although a fire condition is a common cause of relief device actuation, there are several possible causes of relief device overpressure. The need for a risk analysis detailing the possible overpressure for the vessels, including heat exchangers, should be performed as part of the engineering program. It is at this point system that P&ID's are of value. If the system is an as-built unit, this information should be supplied as available documentation. This documentation should include the flow rates into and exiting the pressure vessel as well as other covered equipment such as heat exchangers. This information should be noted on the P&ID or related project documents and should also be clearly documented for easy identification and accessibility. This may also be required should validation or commissioning operations be required. In addition to being required as part of the project phase, this information may be required for commissioning and validation operations, depending on the detail involved in these programs.

As indicated previously, there are several potential causes of overpressure. These causes include:

- Heat from external fire.
- Thermal expansion of liquids.
- Utility failure (failure to control inlet steam supply, cooling water failure, electricity, etc.).
- Instrument failure.
- Heat exchanger tube failure.
- Pump failure (malfunction related to inlet/outlet failure or exchanger tube failure).
- Pressure vessel cooling system failure.
- Closed vessel outlets (feed continues or outlets closed with heat source continuing).
- Compressor failure (e.g. high input).
- Unexpected exothermic reaction or decomposition.
- Backflow from downstream equipment.
- Control valve failure.
- Introduction of a more volatile fluid in the vessel.
- Unexpectedly high fluid input.
- Backflow from downstream equipment.
- Failure of downstream check valve.

While not a standard design element, when an operation incorporates a centrifugal pump, there are several scenarios when installation of a relief device is a practical decision. In the following situations, a hazard and operability study (HazOp) should be performed to determine the hazard potential due to operating overpressure:

- Closed suction and discharge valves.
- Heating raises liquid vapor pressure above the specified operating pressure.
- Sufficient fluid flow to result in significant heating of fluid being pumped.

For the latter two failure possibilities, it is possible to reasonably determine the heat generation caused by the pumping operation. The applicable relationship is given by:

$$Q = (2545)(P) \tag{4.36}$$

where

Q = heat input, Btu/h
P = pump power consumption, hp

Equation (4.36) is also applicable for determining the power consumption for other devices, such as mechanical stirrers powered by electric motors.

Once the failure mode has been identified, a quantitative determination of the discharge rate of the over pressured vessel involving the most probable failure modes is necessary.

While many failure causes have been identified, there are only a few types of actual failures. That is, a failure of a downstream check valve will have the same effects as a failure of a discharge valve to open. These failures will have essentially the same result as an unexpectedly high fluid input into the vessel. By employing this concept, it is possible to determine reasonably accurate values that can be used to size the area of a relief valve that is required to safely discharge contents of a pressure vessel (i.e. liquid, vapor, and/or steam) exposed to several events (e.g. vessel exposure to fire, thermal expansion).

There are several common calculations available for determining the discharge orifice area of pressure relief devices that are a part of a pressure vessel or similar equipment processing high pressure materials.

While the following methods for valve sizing are adequate for confirming vendor supplied relief device information, access to two American Petroleum Institute (API) documents would be helpful should a more in-depth analysis of relief devices be required or desired. The two publications are:

- *Sizing, Selection, and Installation of Pressure-Relieving Devices in Refineries, Part I – Sizing and Selection*, American Petroleum Institute, Washington, DC, 2005
- *Sizing, Selection, and Installation of Pressure Relieving Devices in Refineries, Part II – Installation*, American Petroleum Institute, Washington, DC, 2005

Another excellent source for obtaining required data needed for sizing calculations are the catalogues available from device manufacturers. Most have sections that clearly explain the steps required to perform relief device area calculations employing ASME and API sizing formulas as well as the supporting data required for the calculation.

A common process in pharmaceutical operations involves steam sanitization of pressure vessels as part of the cleaning operation. One possible cause of failure of the operation would be an unplanned failure of the steam discharge line or a steam trap malfunction. An overpressure of the pressure vessel would cause the relief valve to actuate. An accurate orifice area for the relief device is necessary. The ASME equation for sizing the area in a steam environment is given by the relationship

ASME sizing formula for steam

$$A = \frac{\text{lb/h}}{51.5 \times K \times P \times K_{sh}} \tag{4.37}$$

A = Relief valve orifice area, in^2
lb/h = Rate of steam generation
K = Effective coefficient of discharge = 0.97
P = Relieving pressure (set pressure) × 1.21 + 14.7 (for external fire)
P = Relieving pressure (set pressure, lb/in^2) × 1.1 + 14.7 (all other scenarios)
K_{sh} = Superheat factor (this value is = 1.0 for pharmaceutical steam)

The steam sizing equation above also includes a correction factor for backpressure correction. Since this calculation is intended to verify the vendor relief valve sizing and is not intended to be a rigorous calculation, a backpressure correction value of 1.0 is assumed. For pharmaceutical applications, this is a reasonable assumption since most cGMP processes discharge into the environment at a low ΔP (i.e. a low BBP).

Example 4.9   Steam is being fed into a pressurized storage tank at the rate of 15 000 lb/h. The pressure relief valve is set at 100 lb/in^2. Determine the discharge orifice area required for the relief valve.

Solution
The solution is achieved by employing the equation for steam.

$$A = \frac{lb/h}{51.5 \times K \times P \times K_{sh}}$$

$$A = \frac{15000}{51.5 \times 0.97 \times \left[\left(200 \times 1.1\right) + 14.7\right] \times 1.0}$$

$$A = \mathbf{1.279 in^2} \leftarrow \textbf{Solution} \ (\text{calculated relief area})$$

The calculated relief area is the minimum area required to achieve safe discharge should a steam overpressure occur. To specify a unique orifice area for each steam relief situation would create the need for many available relief valves, each with a given required area. This situation would be not only uneconomical but also very untenable. To simplify the situation, manufacturers have created a standardized series of orifice areas. There are two series of valve sizes; one is the responsibility of the ASME, while the other is specified by the API. Regardless of the orifice capacity series employed, all approved relief devices must bear an ASME "U" stamping on the device.

For liquid relief device orifice areas, the ASME sizing designations are normally used. Conversely, for extensive vapor releases, the API sizing designations are often employed, particularly in petrochemical operations. For pharmaceutical applications, employing ASME sizing designations are primarily employed. For sizing steam relief orifice areas, vendor data should be employed regarding size designations. Relief device sizing designations range from a designation of D (0.1279 in.) to size T (28.62 in.).

Virtually all relief device vendors employ the ASME or API sizing tables and are identified in their respective catalogues and/or design manuals. Table 4.5 lists relief device designations and their associated orifice areas for both ASME and API designations. By observation, it can be seen that, on average, an ASME designation has about a 14% higher orifice area. It is also important to note that individual manufacturers have varying capacity factors for orifice designations. Careful evaluation should be employed when relief device orifice sizing is considered. For instance, capacity for a vessel or line containing pressurized air at a set pressure of 50 lb/in^2 and a size D designation has a relieving capacity of 139 SCFM, whereas the same orifice designation for natural gas at the same pressure setting (50 lb/in^2) is 181 SCFM (data from Consolidated® Relief Valve Tables).

Table 4.5 Relief device designations and corresponding orifice areas.

ASME orifice designation	ASME orifice area (in^2)	API orifice designation	API orifice area (in^2)
D	0.1279	D	0.110
E	0.2279	E	0.196
F	0.3568	F	0.307
G	0.5849	G	0.503
H	0.9127	H	0.785
J	1.496	J	1.287
K	2.138	K	1.838
L	3.317	L	2.853
M	4.186	M	3.60
N	5.047	N	4.34
P	7.417	P	6.38
Q	12.85	Q	11.05
R	18.60	R	16.0
T	28.62	T	26.0

The above example (Example 4.2) produced a calculated orifice area of 1.279 in^2. By referring to data in a vendor manual specified for steam discharge at a set pressure of 200 lb/in^2 plus 10% overpressure (non-fire), a set pressure of 220 lb/in^2 is shown on the table (in this instance, the Consolidated® relief valve capacity table for steam is referenced). While the calculated value of 1.279 in^2 is not shown, there is a value of 1.496 in^2. This value corresponds to an orifice size "J" designation. By referring to a vendor relief valve manual or Table 4.1, a *J* orifice designation for steam service indicates the discharge capacity for this designation (*J*) is 15 460 lb/h (steam). Since the discharge rate for the vessel in the example is 15 000 lb/h, the selected orifice is designation J, with an excess capacity of

$$\frac{15460 \text{ lbs/hr}}{15000 \text{ lbs/hr}} \times 100\% = 103.1\%$$

This value, 103.1%, indicates the selected orifice area has an excess capacity of 1.03%. Under no condition should the calculated discharge be greater than the capacity of the valve size selected. Also, the discharge capacity selected should not exceed 300% of the required capacity. If the valve size exceeds 300%, for this case, a smaller orifice size should be employed. This rule-of-thumb value is desired to prevent phenomena known as "valve chatter" from occurring. Valve chatter is caused by the valve disc opening and closing rapidly and repetitively, striking against the seat many times a second resulting in improper seating of the disc. If improperly seated, the disc will not actuate properly when an overpressure occurs.

For pressure vessels containing liquids, the relief device area for discharge can be determined by the relationship:

ASME sizing relationship for liquids

$$A = \frac{Q(G)^{1/2}}{K_u K_d (P - P_d)^{1/2} K_v K_w} \tag{4.38}$$

A = Relief valve orifice area, in^2

G = Specific gravity of liquid contained in the vessel, a dimensionless value (as opposed to density that has dimensions of lb/ft^3)

Q = Flow rate, if Q = gal/min, K_u = 38

if Q = ft^3/h, K_u = 304.798 = 304.8

K_d = 0.744

K_w = Correction for a bellows type relief valve (no correction is required for conventional relief valves)

K_v = A dimensionless value employed to correct for viscosity if the liquid is not water. The correction is employed subsequent to initial orifice area calculation and can be determined from Figure 32 of API 520, Part I, Sizing and Selection.

P = Set pressure of the relief device, lb/in^2

P_d = Discharge pressure of the relief device (if discharging to the atmosphere or drain through a piping configuration, P_d = 14.696 = 14.7 lb/in^2 (at sea level)

A typical failure mode causing the relief valve to actuate would be a situation where an unplanned excess liquid flow rapidly enters the vessel and the fluid accumulates. Another overpressure mechanism is the failure of a spray ball cleaning system to automatically stop as programmed and no alarm actuation available or functioning.

Example 4.10   A pressure vessel containing citric acid (G = 1.665) solution is designed to discharge at a rate of 80 gal/min. The calculated BBP for the discharge line is 4.5 lb/in^2. The relief device will be set at 70 lb/in^2. Determine the orifice area of the relief device required for this pressure vessel.

Solution

The required area will be determined using the ASME sizing for liquid relief given above.

$$A = \frac{80(1.665)^{1/2}}{(38)(0.744)(70-14.7)^{1/2} K_v}$$

$$A = \frac{103.22}{208.31 K_v}$$

$$A = \frac{0.495}{K_v} \text{in}^2 \leftarrow \text{Solutions (calculated relief area)}$$

As indicated, K_v is the dimensionless value used to correct for the reduction in the safety relief valve capacity. Determination of this correction is obtained by employing Figure 32 of API 520 Part I-Sizing and Selection. This material is also available in most vendor supplied relief valve catalogues.

The first step in obtaining viscosity correction factor is calculating the Reynolds number (see Chapter 2) for the citric acid solution. The Reynolds number for this correction is given by the relationship

$$N_{RE} = \frac{Q(2800)G}{\mu(A)^{1/2}} \tag{4.39}$$

N_{RE} = Reynolds number

Q = Volumetric flow rate = 80 gal/min

G = Specific gravity of the citric acid solution = 1.665
μ = Viscosity of citric acid solution = 7 cP
A = Orifice relief area prior to viscosity correction

$$N_{RE} = \frac{80(2800)(1.665)}{7(0.704)} = 75682$$

Referring to Figure 32 of API 520, for a Reynolds number of 75 682, the viscosity correction factor, K_v, is 1. The preliminary relief device area of 0.495 in^2 need not be altered. Thus, for this determination, it is not necessary to employ a correction factor. If the calculated Reynolds number for liquid relief sizing is greater than 40 000, performing a viscosity correction calculation is not required. If, on the other hand, the Reynolds number is less than 40 000, a K_v calculation should be performed. In the event the corrected area exceeds the calculated standard orifice area, the calculations should be repeated using the next larger standard orifice size.

Referring to the ASME relief device capacity tables, for a set pressure of 70 lb/in^2, an orifice letter designation of **G** with an orifice area of 0.5849 in^2 appears suitable. The pressure vessel containing citric acid is discharging at the rate of 80 gal/min. A size G orifice discharges water at the rate of 130 gal/min. The excess capacity of the selected orifice is:

$$\frac{130 \text{ gal/min}}{80 \text{ gal/min}} = 1.625 \times 100\% = 162.5\% \textbf{ excess capacity}$$

It is most critical that the orifice size selected always exceed the calculated orifice area. As a general rule, the next higher orifice area designation should be selected for the specified orifice area.

While the capacity tables (ASME or API) are based on water as the relieving liquid, the tables are valid regardless of the liquid employed. This is the purpose of the K_v viscosity correction factor, which, in this example, was not required.

When evaluating relief device sizing for vapors or gases such as nitrogen, oxygen, or air, the following ASME sizing equation is employed for relief device discharge area sizing:

ASME Sizing Relationship for Vapors and Gases

$$A = \frac{Q(T)^{1/2}(Z)^{1/2}}{C K_d P(M)^{1/2} K_d} \tag{4.40}$$

Q = Volumetric flow rate at ambient temperature and pressure
T = Discharge temperature, °F + 460 °F
Z = Compressibility factor, if unknown use a value of 1.0
C = Constant based on ratio of specific heats C_p/C_v (table 9, API 520, Part I), for an ideal gas (Z = 1.0), use the value C = 315
K_d = Discharge coefficient = 0.825
P = Set pressure, lb/in^2 × 1.2 + 14.7 (for fire scenario)
Set pressure, lb/in^2 × 1.1 + 14.7 (for non-fire scenario)
M = Molecular weight of gas or vapor
K_b = Back pressure correction factor (Figure 27 API 520, Part 1)

For a more specific determination, refer to the relief device vendor catalogue.

Example 4.11 Nitrogen is used as a blanketing gas for a reactor. The set pressure of the pressure vessel reactor is 70 lb/in². Nitrogen at 50 lb/in² and ambient conditions is entering the reactor at the flow rate of 5000 ft³/min. The relief device is set at 90 lb/in². The BBP is 10% and the discharge line is 2 in schedule 40 steel piping. The reaction poses no exothermic or fire hazard. The failure mode considered is a failure of the nitrogen inlet line. The discharge piping has yet to be specified.

Determine:

A. The orifice area required.
B. The maximum equivalent length of discharge piping required (assuming new schedule 40 nominal 2 in. new steel pipe).

Solution

A. For sizing the relief area, the ASME sizing formula for vapors and gases is used.

$$A = \frac{Q(T)^{1/2}(Z)^{1/2}}{CK_d P(M)^{1/2} K_d}$$

$Q = 500\,\text{ft}^3/\text{min}$
$T = 68\,°F + 460\,°F = 528°, (T)^{1/2} = 22.98$
$Z = 1$ (assuming ideal gas behavior)
$C = 319$ (table 8, API 520, table 9, API 520)
$K_d = 0.825$
$P = 90(1.1) + 14.7 = 113.7\,\text{psia}$
$M =$ Molecular weight of nitrogen gas $= 28, (28)^{1/2} = 5.29$
$K_b = 1.0$ (Figure 27, API 520)

$$A = \frac{5000(22.98)(1)}{319(0.825)(5.29)(113.7)(1)} = 0.730\,\text{in}^2 \leftarrow \textbf{Solution A}(\text{Calculated Area})$$

The set pressure is 90 lb/in² with a flow rate of 5000 ft³/min with a calculated orifice relief area of 0.730 in². To meet the requirements of this operation, an ASME orifice of designation *L* is required. The relief area for a **designation L is 3.317 in²** with a relieving capacity of 5918 ft³/min. The excess relieving capacity is:

$$\frac{5918\,\text{ft}^3/\text{min}}{5000\,\text{ft}^3/\text{min}} = 1.184 \times 100\% = \textbf{118.4\% excess capacity}$$

B. Maximum length of discharge piping

This calculation employs, as identified above, Equation (2.11).

1. *Reynolds number determination*
   Cross-sectional area of 2 in. pipe – 2 in. nominal diameter, Sch 40 pipe = 2.067 in.

$$\frac{1\,\text{ft}}{12\,\text{in}} \times \frac{2.067\,\text{in}}{1} = 0.17225\,\text{ft} = \text{Diameter of 2 in. pipe}$$

   Cross-sectional area of pipe, A

$$A = \frac{\Pi D^2}{4} = \frac{(3.14159\,\text{ft})(0.029\,67\,\text{ft})}{4} = 0.02330\,\text{ft}^2 = \text{Cross sectional area}$$

   Velocity of nitrogen gas at ambient conditions $= \dfrac{5000\,\text{ft}^3/\text{min}}{0.02330\,\text{ft}^2} = 214592\,\text{ft}/\text{min}$

Density, ρ, of nitrogen gas at ambient conditions (calculation based on ideal gas law)

$$\rho = \frac{PM}{RTZ} \tag{4.41}$$

P = Ambient pressure (absolute) = 14.7 lb/in^2
M = Molecular weight of nitrogen gas = 28 g/g mol
R = Gas constant = 10.72 ft^3 / (lb mol lbin^2 °R)
T = Absolute temperature, °R = 68 °F (1 °R/°F) + 460 °R = 528 °R
Z = Compressibility (dimensionless correction factor) = 1.0

$$\rho = \frac{\left(14.7 \text{ lbs/in}^2\right)\left(28 \text{lb} / \text{lb mole}\right)}{\left(10.72 \text{ft}^3 / \text{lbmolelbin}^2 \text{°R}\right)\left(528 \text{°R}\right)\left(1.0\right)} = 0.0727 \text{lb} / \text{ft}^3 \text{Density of nitrogen}$$

μ = Viscosity of nitrogen, = 0.000 702 lb/ft min
2. *Reynolds number calculation*

$$\text{Re} = \frac{Dv\rho}{\mu} = \frac{\left(0.17225 \text{ft}\right)\left(214592 \text{ft} / \text{min}\right)\left(0.0727 \text{lbs} / \text{ft}^3\right)}{0.000702 \text{lb} / \text{ft min}} = 3827983 \leftarrow N_{\text{Re}}$$

Referring to the Moody chart in Figure 2.1, the friction factor for a Reynolds number of 3 827 983 is 0.0091 (f = 0.0091).

The variables required to obtain the required equivalent length of the discharge piping are now known. Accordingly, rearrangement of Equation (2.10) to determine the required equivalent length yields:

$$L = \frac{\Delta P(144)(2)(g)(D)}{f(\rho)\left(v^2\right)}$$

Recalling that g, the gravitational force, is 32.2 ft/s^2 and ΔP = (113.7 lb/in^2) – (14.7 lb/in^2), **the pressure drop (ΔP) through the discharge piping is 99 lb/in^2.** For the dimensions to cancel, the velocity, (214 592 ft/min) must be converted to ft/sc. Thus,

$$\frac{214592 \text{ft}}{\text{min}} \times \frac{1 \text{min}}{60s} = 3577 \text{ft/s}$$

$$L = \frac{\left(99 \text{lb/in}^2\right)\left(144 \text{in/ft}^2\right)(2)\left(32.2 \text{ft/s}^2\right)\left(0.17225 \text{ft}\right)}{\left(0.0091\right)\left(0.0727 \text{lb/ft}^3\right)\left(12794929 \text{ft}^2 / \text{s}^2\right)} = 18.7 \text{ft} \leftarrow \text{Solution B}$$

The required equivalent length for this design is 18.7 ft. This equivalent length includes elbows, tees, and other pipe components needed to assure proper discharge. This example represents the need for analysis of discharge piping in addition to relief valve orifice sizing. Again, unless in-house relief valve sizing is being performed, the facility owner/operator should have access to these calculations as part of the total relief device package.

The discharge piping design should evaluate materials of construction, thermal expansion, adiabatic cooling, and discharge piping orientation.

The type of piping should take into consideration wall thickness. While the relieving pressure is 14.7 lb/in^2, at the beginning of the discharge (at the relief valve orifice, at discharge), the pressure can be significantly higher upstream of the discharge point. While Schedule 10 piping may be more inexpensive than Schedule 40 pipe, an evaluation should consider the pressure, relieving constituent temperature and the yield strength of the piping.

Another very important consideration when dealing with discharge piping is also an important consideration for the inlet piping. It is critical that there are no dead legs in both the inlet and outlet piping configurations.

Thermal expansion can be a significant factor if a temperature increase occurs as a result of the relieving fluid. While less likely to occur, significant cooling can also affect discharge operations if there is a significant decrease in relieving fluid pressure. This phenomenon is referred to as adiabatic cooling.

The composition of discharge piping should also be evaluated. A corrosion evaluation of the discharge piping should include potential of corrosion if the relieving material has the possibility of forming a galvanic cell.

Discharge lines should either be oriented in an upward or downward direction. For vapors and discharging gases, a vertical direction with a minimum of ten feet of discharge piping above the discharge valve is reasonable. Liquid discharge distance should be such that the discharge terminates at an approved draining system. For both liquid draining and vapor discharge, there should be no valve or flow restricting device between the discharge of the relief device and the termination of the drain line. And, of course, there should be no dead legs in the venting or liquid discharge piping.

Should discharge fluids pose either an environmental or a safety hazard, an evaluation should be performed by responsible facility organizations such as engineering, maintenance, or the facility process safety operation.

While relief valves are generally provided in both a threaded and a flanged availability, the flange option is often preferred. This is preferable if a scheduled valve testing and inspection program is standard procedure for maintenance operations. With flanged relief valves, removal of the relief device is easier, even if unbolting, as opposed to unscrewing, is required. This operation is required if bench testing is part of the maintenance procedures. Also, in addition to removal of threaded valves, standard valve reattachment usually includes the addition of Teflon pipe tape on the threaded valve. This action is advised since thread contamination of threading is possible if simmering or valve activation occurs.

If threaded relief devices are employed, it would be advisable to use threaded flanges as an addition to the threaded relief valve. For pharmaceutical applications, 150 lb rated flanges are typically employed. Flange selection is determined by the pressure rating of the relief device pressure setting and most often employs either a 150 lb rating or a 300 lb rating for pharmaceutical cGMP operations.

The fact that sanitary connections are not employed in relief devices is somewhat moot. If a relief device discharges, it will cause the batch to most likely be discarded since the product will be an incomplete or rejected batch.

OSHA 29 CFR 1910 Part 119, known as process safety management (PSM), includes identification and allowable quantities of flammable materials that can be present on manufacturing sites. While most pharmaceutical manufacturing operations have implemented programs to minimize the quantities of flammable materials, most typically flammable solvents, smaller operations such as API manufacturers must maintain amounts of these materials as well as a surplus quantity to cover manufacturing irregularities. Often, these flammables are stored in above ground ASME Code pressure vessels. In all cases, the allowable limit for flammable materials is 10 000 lb.

Relief device sizing for a fire scenario is more detailed than the other sizing procedures thus far encountered. In addition to acting as storage containers for planned operations, pressure vessels, when designed appropriately, are also used as reactors in pharmaceutical operations. While the advent of bioreactors has significantly reduced the size of many bioreactors, there are still many operations employing large reactors. Reactors ranging in size from 12 000 to 25 000 gal are still employed in API operations, generics as well as small molecule cGMP manufacturing processes. Similarly, while many operations are eliminating or minimizing the amounts of flammables on their respective facility locations, the threat of a fire exists in many operations with pressure vessels and/or reactors.

Sizing relief device orifice areas is predicated on the scenario that the flammable liquid in the vessel will vaporize due to heating of the vessel during a fire.

The initial step in sizing relief valve orifice sizing for a fire scenario involves determining the vessel or reactor dimensions (required for surface area calculations) and mounting configurations (i.e. horizontal, vertical, and above or below surface location). Additionally, the percent of normal tank volume is also required. Another value, the environment factor is required and explained in this section.

When determining the surface area for fire sizing, the surface area in contact with the flammable liquid in the vessel is an important variable. A vessel that is 80% filled with a flammable liquid presents more of a potential hazard than the same vessel that is 20% full. The correlation factor that relates the wetted surface area of a vessel containing flammable liquids is referred to as the wetted perimeter factor, F_{wp}. Figure 4.9 (Courtesy of Consolidated ® Relief Valves) displays the wetted perimeter factor F_{wp} as a function of the volume of liquid in a vessel.

In addition to requiring the wetted wall perimeter factor, F_{wp}, it is necessary to know the surface area of vessels containing liquids. Table 4.6 provides common relationships for calculating

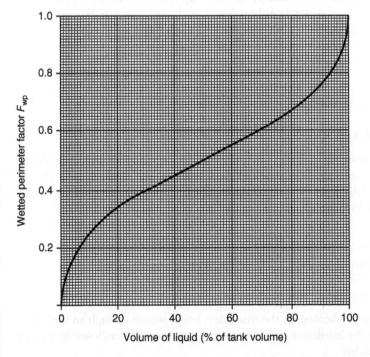

Figure 4.9 Wetted wall perimeter factor as a function of vessel liquid level.

Table 4.6 Surface area formulas for pressure vessel heads.

Configuration	Vessel head description	Surface area formula, A^a
Vertical cylinder	Flat ends	$A = \pi(DL + D^2/2)$
Vertical cylinder	Elliptical	$A = DL + 2.61D^2$
Vertical cylinder	Hemispherical	$A = \pi(DL + D^2)$
Horizontal cylinder	Flat ends	$A = \pi(DL + D^2/2)$
Horizontal cylinder	Elliptical	$A = DL + 1.084D^2$
Horizontal cylinder	Hemispherical	$A = \pi(DL + D^2)$

^{a} A, is the wetted surface area.

vessel surface area for most ASME pressure vessel designs. For this table, D = diameter in feet and L = vessel length.

With this information, it is possible to determine the total wetted surface area of a vessel. The total wetted surface area is given by the relationship:

$$A_{wsa} = F_{wp} \times \text{Total vessel surface area} \tag{4.42}$$

A_{wsa} = wetted surface area of a vessel
F_{wp} = wetted perimeter factor

Also required for fire sizing of relief devices is the latent heat of vaporization in Btu/lb. While this data is readily available for flammable liquids employed for pharmaceutical applications, a reasonably accurate estimate is available for this determination. Although intended for petroleum products, it serves as a reasonable calculation for verification or estimating purposes when actual heat of vaporization data is not available, although with such easy access to the internet and DIERS, obtaining latent heat data should not be difficult. The relationship is given by:

$$\lambda_v = \frac{(110.9 - 0.09t)}{SG} \tag{4.43}$$

λ_v = latent heat of vaporization, Btu/(lb°F)
t = temperature, °F
SG = specific gravity of liquid at ambient conditions

Another required variable needed for sizing orifice areas in a fire scenario are values identified as environment or environmental factors and designated as F. These factors (F) are intended to account for insulation that may be present on pressure vessels or storage tanks. Environmental factors, F, are as follows:

F = 1.0 for a bare vessel
F = 0.3 for vessels covered with 1 in. of insulation
F = 0.15 for vessels covered with 2 in. of insulation
F = 0.75 for vessels with 4 in. of insulation
F = 0 for vessels that are stored underground

These insulation factors are predicated on the insulation being secure enough so that a fire hose stream will not dislodge the insulation. It is also assumed the vessel is enclosed by a sump with sufficient drainage capability.

Before a calculation can be performed to size for a fire event, an estimate on the rate of discharge (lb/h) is required. For this determination, the following relationship applies:

$$Q = \frac{21000F\left(A_{\text{wsa}}\right)^{0.82}}{\lambda} \tag{4.44}$$

For this relationship, Q is the heat flow entering the vessel through the wetted surface area, A_{wsa}. The heat flow input is predicated on fluid being vented having physical properties similar to hexane ($\lambda = 157$ Btu/lb, boiling point $= 140.4\,°F$). For pharmaceutical applications, this is a reasonable assumption since many, if not most, liquids that are stored for pharmaceutical processes possess similar physical properties.

The numerator for Equation (4.16) is sized for a vessel with a wetted surface area greater than $2800\,\text{ft}^2$. While this relationship (the numerator) is generally used for all tanks and is valid for a quick verification of design adequacy, the numerator of Equation (4.16) also has similar relationships if vessel heat absorption is particularly relevant. A typical case in point would be two or more smaller vessels containing flammable liquids and in close proximity to one another. For this layout, the following relationships can be considered when designing or verifying potential fire situations:

- For situations where sufficient fire suppression and drainage are not installed, use

$$Q = \frac{34000F\left(A_{\text{wsa}}\right)^{0.82}}{\lambda}$$

- For vessels with a wetted wall surface area, A_{wsa}, ranging from 20 to $200\,\text{ft}^2$, consider using the calculation

$$Q = \frac{21000F\left(A_{\text{wsa}}\right)^{0.82}}{\lambda}$$

- For vessels with a A_{wsa} from 200 to $1000\,\text{ft}^2$,

$$Q = \frac{199300F\left(A_{\text{wsa}}\right)^{0.566}}{\lambda} \tag{4.45}$$

- For vessels with a A_{wsa} from 1000 to $2800\,\text{ft}^2$

$$Q = \frac{963400F\left(A_{\text{wsa}}\right)^{0.338}}{\lambda} \tag{4.46}$$

- For vessels with a A_{wsa} in excess of $2800\,\text{ft}^2$,

$$Q = \frac{21000F\left(A_{\text{wsa}}\right)^{0.82}}{\lambda} \tag{4.44}$$

Most of the flammable liquids stored in pharmaceutical, biotech, and API operations usually have boiling points below $212\,°F$. As such, when these liquids approach boiling, the liquids will start discharging from the relief device orifice more as a result of liquid expansion than vapor formation since significant heats of vaporization must be absorbed prior to significant vaporization of the liquid in the vessel. While a minimal amount of liquid may vaporize, most relief devices intended for liquid service can adequately process up to 20% vapor phase discharges with no need to evaluate for two phase flow.

Example 4.12 A vertical pressure vessel with flat ends and an inside diameter of 7 ft and a height of 14 ft contains acetone (SG = 0.8 ρ = 49.92 lb/ft^3 = 5.98 gal/ft^3), the tank is normally 90% full. The vessel has 1 in. of secured insulation. Acetone has a latent heat of vaporization of 233 Btu/lb. The relief device is a conventional device set at 75 psig (assuming a 21% overpressure for a fire scenario). At ambient conditions, the viscosity of acetone is 0.32 cP.

Determine the acetone discharge in lb/h of acetone if a fire occurs in the vicinity of the tank.

Solution

Calculation of the surface area is required. The surface area relationships from Table 4.6 are used, specifically, $A = \pi(DL + D^2/2)$, the formula for the configuration described.

$$A = 3.14159\left(7\,\text{ft} \times 14\,\text{ft} + \frac{(7\,\text{ft})^2}{2}\right) = 3.14159\left(98 + \frac{49}{2}\right) = 385\,\text{ft}^2 \text{ surface area}$$

Referring to Figure 4.9, for a liquid volume of 90%, the wetted perimeter factor F_{wp} is 0.77. Equation (4.44) is needed to determine the wetted wall surface area of the vessel. Thus,

$$A_{wsa} = 0.77 \times 385\,\text{ft}^2 = 296.45\,\text{ft}^2 = 297\,\text{ft}^2 \text{ wetted wall surface area, } A_{wsa}$$

The hourly discharge from the vessel in lb/h can now be calculated using Equation (4.44) and recalling the data given to determine the weight of fluid discharged.

$$Q = \frac{21000F\left(A_{wsa}\right)^{0.82}}{\lambda}$$

$$Q = \frac{21000(0.3)(297)^{0.82}}{223} = 3010.89 = 3.011\,\text{Btu/h, hourly discharge from vessel} \leftarrow \text{Solutions}$$

For purposes of comparison, it may be of value to examine the results employing the relationship for vessels between 200 and 2000 ft^2 of wetted surface area employing the same data input $[Q = 199300(F)(A_{wsa})^{0.566}/\lambda]$. The calculation for this is

$$Q = \left[\frac{199300(0.3)(297)^{0.566}}{223}\right] = 6728\,\text{Btu/h alternative heat absorption method}$$

While the two values differ, they are both of the same order of magnitude. For vendor design calculations, Equation (4.16) is sufficient. However, if the calculation is required as a process engineering component, the relationship for vessels with a wetted wall surface area between 200 and 2000 ft^2 should be considered.

While the solution to Example 4.16 is acceptable, it is not in a format for relief device orifice sizing. For liquids, the area for liquid devices, as per ASME sizing criteria, is predicated on dimensions of gal/min and not Btu/h. Thus, to achieve the desired results, then a dimensional change is required.

$$\left(3011\,\text{lb/h}\right)\left(1\,\text{h}/60\,\text{min}\right)\left(1\,\text{ft}^3/49.92\,\text{lb}\right)\left(5.98\,\text{gal/ft}^3\right) = 0.168\,\text{gal/min discharge}$$

With correct dimensions, the liquid relief device orifice area can now be determined.

For this calculation, the previously employed relationship (Equation 4.38) for sizing liquid relief device orifice sizing is selected. Hence, relationship

$$A = \frac{Q(G)^{1/2}}{K_u K_d (P - P_d)^{1/2} K_v K_w}$$

is applicable.

For this calculation,

$Q = 0.168 \, \text{gal/min}$
$G = 0.8$
$K_u = 38$
$K_d = 0.744$
$P = 75 \, \text{psig}$
$P_d = 14.7 \, \text{psig}$ (the vessel discharges to the atmosphere)
$K_v = $ Viscosity correction factor applied subsequent to determination of discharge area
$K_w = 1$

$$A = \frac{0.168(0.8)^{1/2}}{(38)(0.744)(60.3)(1)} = 0.0000881 \, \text{in}^2 \leftarrow \textbf{Uncorrected relief device discharge}$$

To determine the viscosity correction factor, the Reynolds number must be calculated.

$$N_{RE} = \frac{Q(2800)(G)}{C_p(A)^{1/2}} = \frac{(0.168)(2800)(0.8)}{0.32(0.00939)} = 125240 \leftarrow N_{RE}$$

Using Figure 2, API 520, Part I, the correction factor is 1.0. The corrected orifice discharge area is $0.0000881 \, \text{in}^2$. At this low discharge, an ASME designation D with a relief area of $0.1279 \, \text{in}^2$ is applicable.

References

1 Livingston, E. and Scavuzzo, R.J. (2000). Pressure vessels. In: *The Engineering Handbook* (ed. R.C. Dorf). Boca Raton, FL: CRC Press.

2 Carlson, B. (2011). Pipeline bodes well for biologics growth. *Genetic Engineering & Biotechnology News* 31 (12): 16–17.

3 Morrison, R.T. and Boyd, R.N. (1959). *Organic Chemistry*, 45–46, 77, 128–130, 284–285. Boston, MA: Allyn & Bacon, Inc.

4 Kittsley, S.L. (1955). *Physical Chemistry*, 99–111. New York: Barnes and Noble.

5 Robinson, R.N. (1987). *Chemical Engineering Reference Manual*, 4e, 11-1–11-21. Belmont, CA: Professional Publications.

6 Holland, C.D. and Anthony, R.G. (1979). *Fundamentals of Chemical Reaction Engineering*, 2e, 50–99. Englewood Cliffs, NJ: Prentice Hall.

7 Charles, M., Petrone, J., and LeBlanc, J. (1993). *American Institute of Chemical Engineers, North Jersey Section, 33rd Annual Spring Symposium* (18 May 1993).

8 ZAIN Technologies. *Mixing Guide*. Bloomingdale, IL: ZAIN Technologies. http://www.zmixtech.com/mixing-guide/ (accessed 19 December 2017).

9 Oldshue, J.Y., Herbst, N.R., and Post, T.A. (1995). *A Guide to Fluid Mixing*, 57–67. Rochester, NY: SPX Flow Technology.

10 Johnstone, R.E. and Thring, M.W. (1957). *Pilot Plants, Models, and Scale-up Methods in Chemical Engineering*, 12–26, 71–73, 173–181. New York: McGraw-Hill.

11 Wood, A. (2011). Effectively clean tanks and reactors. *Chemical Processing* 74 (3): 30–35.

12 Fike Corporation (1997). *Process Safety Protection*. Blue Springs, MO: Fike Corporation. www.fike.com (accessed 19 December 2017).

Further Reading

American Institute of Chemical Engineers (1976). *Practical Aspects of Heat Transfer*, 11, 152–167. New York: American Institute of Chemical Engineers.

Castellan, G.W. (1964). *Physical Chemistry*, 3e, 813, 848. Reading, MA: Addison Wesley Publishing Company.

Crandall, S.H., Dahl, N.C., and Lardner, T.J. (1959). *An Introduction to the Mechanics of Solids*, 163–164. New York: McGraw-Hill.

Farr, J.R. and Mann, J.H. (2001). *Guidebook for the Design of ASME Section VIII Pressure Vessels*. New York: American Society for Mechanical Engineers.

Gerald, M.C. (1981). *Pharmacology: An Introduction to Drugs*, 2e, 27, 274, 542. Englewood Cliffs, NJ: Prentice Hall.

Glaser, V. (2011). Finding a bioreactor that's right for you. *Genetic Engineering & Biotechnology News* 31 (14): 44, 46, 47.

Harvey, J.F. (1985). *Theory and Design of Pressure Vessels*. New York: Van Nostrand Reinhold.

Kachelhofer, K. (2010). Decoding pressure vessel design. *Chemical Engineering* 117 (6) (June).

Levenspiel, O. (1962). *Chemical Reaction Engineering*, 53, 99–135. New York: John Wiley & Sons, Inc.

Morrow, J.K., Jr. (2006). Disposable bioreactors gaining favor. *Genetic Engineering & Biotechnology News* 26 (12): 42–45.

Perry, R.H. and Kirkpatrick, D. (1984). *Perry's Chemical Engineers' Handbook*, 6e, 6–114, 6–115. New York: McGraw-Hill.

Sanders, R.E. (2010). Treat tanks with care. *Chemical Processing* 73 (11): 29–32.

Schaepe, S., Lubbert, A., Pohlsceidt, M., et al. (2012). Bioreactor performance: Insights into the transport properties of aerated stirred tanks. *American Pharmaceutical Review* 15 (6): 52–59.

Treybal, R.E. (1955). *Mass Transfer Operations*, 78–103. New York: McGraw-Hill.

U.S. Occupational Safety and Health Administration (1989). *Publication 8-1.5: Guidelines for Pressure Vessel Safety*. Washington, DC: U.S. Occupational Safety and Health Administration (14 August).

5

Reliability, Availability, and Maintainability

5.1 Introduction to RAM

An often unaddressed cost of operating pharmaceutical, biotech, and device operations is the need to maintain these systems subsequent to start-up and operation. As the installed system ages through use, maintenance costs will also rise until a time is achieved when operation of the system is no longer efficacious.

The intent of this chapter is to define, explain, and describe the procedures necessary to successfully evaluate and enhance operations relevant to pharmaceuticals and related products. Areas such as reliability, availability, and maintainability (RAM) can now be viewed as cost-effective techniques to enhance pharmaceutical operations, in particular, as well as the evolving biotech and medical device operations. While accuracy in RAM-related problem solving is most important, an ability to decrease the timeline can also be very effective. Hence, performing mundane calculations and shortening project cycle time are desirable. To assist in achieving this objective, items such as nomographs, tables, and charts are very helpful in adding a practical aspect to this area of pharmaceutical engineering.

While RAM operations are somewhat common in aircraft, automotive, electronics, defense, and logistics operations, RAM implementation has, for the most part, not advanced as quickly as pharmaceutical and biotech manufacturing.

With many traditional small molecule pharmaceutical operations facing patent expirations and a reduced product pipeline, procedures, operations, and methods not considered in the past are now viewed as viable techniques. It is this rationale that allows, or perhaps should require, RAM to be included in pharmaceutical as well as medical device and biotech engineering programs.

Practical Pharmaceutical Engineering, First Edition. Gary Prager.
© 2019 John Wiley & Sons, Inc. Published 2019 by John Wiley & Sons, Inc.

To fully benefit from the efficiencies and cost savings that can be derived from a RAM program, a commitment to the RAM concept should be implemented at the earliest phases of a new system concept or configuration. Just as there are project managers, project engineers, quality representatives, and vendor personnel representing specific interests of the new project, a reliability/maintainability presence should be an integral part of the team. Typically, the reliability/maintainability presence would be a part of the engineering operation (site or corporate engineering), but a qualified reliability/maintainability specialist could come from virtually any department in the project matrix. However, because a successful RAM program is often difficult to quantify financially, the reliability/maintainability personnel usually originate from an operation that is chargeable to overhead costs such as engineering, or the facility maintenance operation. On occasion, quality assurance engineers with sufficient background and experience may have a very effective involvement in the RAM planning and project implementation. This may, of course, vary with the site and how costs are distributed.

The methods described in this chapter are separate from cGMP requirements for system, equipment, and/or software validation as defined in 21 CFR. As noted previously in this text, the term system is defined as a configuration of components designed to perform a specific pharmaceutical, biotech, or medical device task (e.g. a water manufacturing system or a packaging line). Systems are, for the most part, composed of components obtained from a myriad of original equipment manufacturers (OEM) whose components are incorporated into the configuration referred to as a system. Thus, while one company is responsible for design, installation, and operation, several different OEM vendors often supply pumps, motors, reverse osmosis systems, filters, and control systems. The selected vendor responsible for the overall system should maintain reliability and maintainability data for the OEM components incorporated in the system of concern.

Since the equipment and components employed are already conforming to FDA criteria, subsequent to validation, additional qualification is not required when considering existing cGMP components for spare parts. This chapter focuses on efficient and cost-effective deployment of certain spares and components. In general, the material contained in this chapter is not regulated by the quality provisions of 21 CFR but, rather, is in accordance with the pharmaceutical operation's corporate quality standards.

The techniques employed to accomplish this task are identified with the general terminology of RAM. Since these three factors are interrelated, an understanding of all these variables is required.

5.2   The Role of Reliability

Reliability is a term that implies different meanings to different people. A most common definition is, reliability is the probability of an item or event performing successfully an intended function for a specified period of time under specified time under specified conditions.

Achieving a successful reliability operation is dependent on several major variables:

- Estimation of reliability goals
- Redundancy
- Components conformance (tolerances and specifications)
- Environmental factors
- Worst-case analysis
- Failure modes and effects analysis (FMEA)
- Design reviews

- Failure analysis
- Corrective and preventive actions (CAPA)

The global process industries (including pharmaceuticals) lose $20 billion or 5% annual production due to unscheduled downtime and poor quality. The Alberta Reliability Committee (ARC) estimates that almost 80% of these losses are preventable. In fact, 40% of these losses are largely attributable to human error [1].

Reliability, in addition to equipment, also encompasses procedures, software, and personnel that use the equipment. Further, it is dependent on requirements related to reliability, validation, and effective management through the lifecycle of the equipment/system, commencing with approval and system and functional requirements.

With the increasing procurement and installation of "as-built" systems, it is not uncommon for one or more system elements/components to be sacrificed due to factors such as cost, schedule, reliability, and/or performance. These, of course, are management decisions and are often made without a demonstrated verification of the efficacy of element(s)/components that are sacrificed. Consequently, it could be worthwhile to consider reliability modifications that can enhance the proposed installation without a significant sacrifice. It is important to recall that reliability is the responsibility of the vendor, and once the equipment/installation is complete, the vendor, saved for a consulting role, is not involved. Thus, it is important to address design and operating variables as early as possible in the design/procurement phase.

Some of the reliability modifications to be considered include [2]:

1) Components – Using high reliability parts or lower cost parts and the effect on lifecycle costs
2) Materials – Evaluation of stainless steel, high performance polymers (e.g. PVDF) or alloys for cGMP and non-cGMP applications (a more comprehensive analysis of pharmaceutical and biotech materials is addressed in Chapters 8 and 9).
3) Operations and procedures – The vendor should provide, as part of the Engineering Turnover Package (ETOP), all the required documentation (e.g. spare parts list, operating manuals, maintenance procedures, drawing packages, etc.). This is most important since many standard operating procedures (SOPs) and related facility documentation is based on the ETOP information.
4) Definition of the project – The expected performance requirements should be clearly defined. For instance, a facility chilled water system should detail operating parameters, allowable variances, and operating times at the contracted performance levels.
5) Amount of attention to detail – Tasks such as subcontractor qualifications should be detailed by the construction manager or prime contractor and furnished to the user.

In the design and procurement phases, typical elements that may improve the system's reliability and are addressed by system vendors and associated contractors should include:

1) Cost of the design effort – If an as-built system is selected, areas where additional component details should be evaluated so the user can be aware of costs that may be added as a result of required as well as unanticipated change orders.
2) Costs of parts – The vendor should provide data defining the cost trade-offs relating to reliability versus component price.
3) Scheduling.
4) Availability of purchased components and spare parts – These constituents should be available from several vendors and OEM to assure a capability to select items from a variety of suppliers. Additionally, spare parts provisioning requirements should be provided as part of the ETOP.

5) Information relating to equipment failure modes – While operator error (human factors) can be a large contributor to failures, other modes (e.g. motor failures, hose assemblies, alarms and indicators, etc.) should be detailed and submitted to the user for incorporation in the user computerized maintenance management system (CMMS).
6) In addition to employing reliable components, the vendor should confirm standard components, and parts will be used as often as possible.
7) Interchangeability of parts is also an important consideration. The design phase should focus on the system lifecycle by providing parts and components that are conveniently available or capable of being manufactured at a competitive cost. Unique components or parts often cost significantly more if, due to system age or design uniqueness, replacement parts must be custom manufactured. Also, replacement in kind criteria should be evaluated so that it is possible for suppliers other than the OEM to qualify as approved suppliers. Often, the misconception that replacement in kind requires one supplier is uniquely qualified to meet a specific need.
8) Redundancy – As part of the design phase, critical operating equipment should be evaluated for feasibility of using a parallel "backup" unit (see Section 5.6).
9) Safety – Always an overriding concern, design safety cannot be overemphasized and should be considered in all phases of the lifecycle of the system as a continuous part of the reliability and maintainability process.
10) Identification – In addition to being a critical cGMP element (21 CFR, parts 211.63, 211.65, 211.68 as examples), unique equipment identification can significantly enhance reliability and maintainability procedures and activities.

While reliability factors are primarily the responsibility of the vendor and the user has limited options regarding design input (other than addressing the reliability issues noted above), the user can significantly affect the system maintainability (i.e. mean time to repair (MTTR) – addressed more comprehensively in Section 5.2 and, consequently, efficiency, productivity, and cost savings. Some significant guidelines that could enhance maintainability of an installation, and thus lengthen the lifecycle of the installation, include the following:

- Develop and implement a maintenance plan for the entire lifecycle of the system. Typically, this would include anticipated or planned spare parts requirements for the useful life of the system or installation (calculations for spare parts requirements are provided further in this chapter).
- Reduce operational downtime by enhancing maintainability through simplification of test and repair procedures to minimize trouble shooting and scheduled maintenance checks and correction time. An example would be to verify that the installed system has easy access for standard maintenance procedures such as minor adjustments and instrument calibration tags.
- In addition to using standard parts and components, procedures should be in place for replacing interchangeable parts, components, and assemblies at the appropriate repair level (i.e. while the facility may not recalibrate instruments in house, maintenance procedures should detail device and instrument calibration expiration dates and routine calibration tests).
- Plan maintenance requirements so that standard maintenance tasks can be performed by facility personnel, thereby minimizing the cost for outside contractor assistance.

A synopsis of the material thus far addressed can be observed in Figure 5.1, which displays the classic "bathtub curve." Figure 5.1 conveniently displays failure types (i.e. early failures, random failures, and wearout failures). Additionally, it also identifies the maintenance phases of the lifecycle of a system, component, part, or item with the end-of-life region being the part

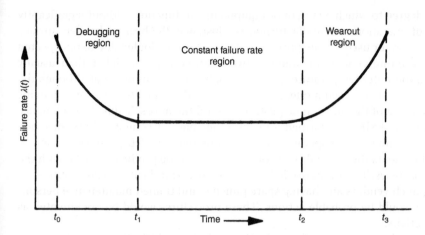

Figure 5.1 Bathtub curve. *Source:* AMCP 706-132 Ref. [3].

of the curve where failures are significantly increased at a shorter time. Also, the failure rate initially decreases and then remains reasonably constant until the end-of-life region, hence the bathtub-like shape. While the bathtub curve is common in maintainability operations, it is important to note that not all items, components, and systems exhibit this behavior [4].

The plot of the failure rate in Figure 5.1 of an item (whether repairable or not) is plotted versus time.

As can be seen, the bathtub curve is divided into three distinct regions, the debugging region, the constant failure rate region, and the wearout region. The debugging region, extending from time t_0 to time t_1, represents the "start-up" time of the system. This period is the time required to correct installation faults, defective parts and components, required procedural corrections, and operator on the job time. This start-up period is also identified as "burn-in" time.

The constant failure rate region (t_1 to t_2) represents the normal operational period of the system or equipment. As indicated, during this time period, parts replacement and repairs occur at specified intervals that can be expressed mathematically as a function of time and a term defined as a failure rate constant, which is defined by Equation (5.3) and covered in depth in Section 5.7. While Figure 5.1 may not be a description of actual failure modes in the useful operating life range, it is a suitable model to describe the reliability in this region.

The final zone, the interval from t_2 to t_3, is the wearout period. As Figure 5.1 illustrates, the failure rate increases significantly in this zone. During this time frame, the system or equipment fails primarily as a result of deterioration, wear, and aging. It is in the wearout region that maintenance, repair time, and parts replacement costs render the system or equipment too inefficient or costly to continue operations.

An effective tool for evaluating the reliability of a design is by addressing potential failures and their effect on the system. To accomplish this task, a procedure for determining the cause of failures and minimizing their effects has been developed and is known as failure modes and effects analysis (FMEA). An FMEA can be implemented at any level ranging from a complete system to individual parts or components. The basic mode of a FMEA is to identify the failure modes of a component or part and detail the cause or causes of the item failure. An analysis is performed by itemizing the failure modes of failure as well as identifying the effects of the failure of each item. Subsequent to the tabulation of the item failure modes, changes can be implemented to correct, reduce, or eliminate the cause of the failure. The course of action is determined by the criticality or seriousness of the effect of the failure.

Criticality is the degree to which a system or equipment can function without significantly affecting the level of operation or manufacturing. For instance, the heat exchanger (surface condenser) of an evaporator used to cool purified water has become fouled due to unexpected tube fouling. The resulting condensate water is two or three degrees higher than usually recorded but within the water temperature specification. It was determined that although the condenser is operating in a degraded mode, the performance does not affect the productivity. As a result, scale removal of the unit can be postponed until the next scheduled maintenance shutdown, based on the FMEA performed. On the other hand, frequent failures of tablet punches will cause the unit to cease operation since the number of tablets produced is significantly reduced and presents the possibility of contamination of the product (tablets) with the metal tablet punch particles. This event would be considered a critical failure, and actions such as using alternate punch vendors and having spare punches and trained maintenance personnel readily available would be probable actions. These procedures would be documented as part of the FMEA activity.

An FMEA can be used as a technique for evaluation, redesign, substitution, or replacement, depending on the project phase (i.e. design, installation, start-up, operation). As a general rule, an FMEA should be performed as early as possible in the project phase.

An FMEA is commonly divided into two phases. The first phase (phase 1) of the task is performed in parallel with the start of the detail design of the project. If the project is an in-house effort, staff and contract personnel are generally responsible. If the installation is an as-built project, the effort should be the responsibility of the prime contractor with assistance from the construction manager. The user representative should be closely involved since the user ultimately approves the FMEA. Phase 2 should be performed simultaneously with the issuance of the detail drawings.

Typically, phase 1 consists of several steps:

1) Documentation including a flow diagram identifying the critical components with the probable failures identified
2) An FMEA identifying failure modes and consequences (effects) of components such as:
   - Open circuits
   - Short circuits
   - Wear
   - Documented system and component identification
   - A documented and approved critical component and parts list

Phase 2 involves performing the following tasks:

1) Updating the phase 1 flow diagram by specifically providing unique identification of the critical items and documented on a critical items list.
2) The criticality level is assigned a numerical value with the analysis performed as described in Chapter 10.

Extensive data has been compiled to identify failure modes and failure occurrence of many standard parts and components encountered in various operations, including pharmaceutical applications. Table 5.1 presents failure modes and percentage of occurrence for parts and components often encountered in pharmaceutical operations. While the information contained in the table is not an actual FMEA, it can serve as a basis and can be modified as the FMEA implementation process matures. Also, while several of the parts identified may be constituents of larger systems whose disposition is predetermined when the failure occurs, often these decisions based on the availability of the component and the failed part may be replaced and operable for a brief period until the replacement equipment is available and installed.

Table 5.1 Part failure modes [5, 6].

Part	Failure mode	Approximate percentage of occurrence (%)
Blowers	Winding failure	35
	Bearing failure	50
	Slip rings, brushes, and commutators	5
Bearings	Loss or deterioration of lubrication	45
	Contamination	30
	Misalignment	5
	Brinelling (excessive stress)	5
	Corrosion	5
Capacitors – fixed ceramic dielectric	Short circuits	50
	Change of values	40
	Open circuits	5
Capacitors – fixed aluminum	Open circuits	40
	Short circuits	30
	Excessive leakage current	15
	Decrease in capacitance	5
Capacitors – fixed, mica, or glass dielectric	Short circuits	70
	Open circuits	15
	Change of value	10
Capacitors – fixed metalized paper or film	Open circuits	65
	Short circuits	30
Capacitors – fixed paper dielectric	Short circuits	90
	Open circuits	5
Capacitors – fixed, electrolytic, tantalum	Open circuits	35
	Short circuits	35
	Excessive leakage current	10
	Decrease in capacitance	5
Circuit breakers	Mechanical failure of tripping device	70
Clutches – magnetic	Bearing wear	45
	Loss of torque due to internal mechanical degradation	30
	Loss of torque due to coil failure	15
Coils	Insulation deterioration	75
	Open winding	25
Connectors, interstage	Shorts (poor sealing)	30
	Mechanical failure of solder joints	25
	Degradation of insulation resistance	20
	Poor contact resistance	10
	Miscellaneous contact resistance	15
Connectors, standard	Contact failure	30
	Material deterioration	30
	Mechanical failure of solder joints	15

(*Continued*)

Table 5.1 (Continued)

Part	Failure mode	Approximate percentage of occurrence (%)
Diodes, silicon, and germanium	Short circuits	76
	Intermittent circuits	18
	Open circuits	6
Hose assemblies	Material deterioration	85
	End fitting mechanical failure	10
Indicator lights	Catastrophic (opens)	75
	Degradation (corrosion, solderability)	25
Insulators	Mechanical breakage	50
	Deterioration of plastic material	50
Meters, mounted gauges	Catastrophic (opens, glass breakage, open seals)	75
	Degradation (accuracy, friction, damping)	25
Motors and generators	Winding failures	20
	Bearing failures	20
	Slip ring, brushes, and commutators	5
Oil seals (rubber)	Material deterioration	85
O-rings (rubber)	Material deterioration	90
Pump seals	Seal face	45
	Dynamic component	40
	Static seal seat	5
Relays	Contact failures	75
	Open coils	5
Resistors – fixed, carbon, or metal film	Open circuits	80
	Change of value	20
Resistors – fixed, composition	Change of value	95
Resistors – variable, composition	Erratic operation	95
	Insulation failure	5
Resistors – variable, wirewound	Erratic operation	55
	Open circuits	40
	Change of value	5
Resistors – variable, wirewound, precision	Open circuits	70
	Excessive noise	25
Switches, rotary	Intermittent contact	90
Switches, toggle	Spring breakage (fatigue)	40
	Intermittent contacts	50
Transformers	Shorted turns	80
	Open circuits	5
Valves – check and relief	Poppets sticking (open or closed)	40
	Valve seat deterioration	50

It is important to include in the FMEA only those systems, components, and parts that are the responsibility of the facility maintenance activity. Systems such as programmable logic controllers are usually maintained by outside contractors (commonly the system supplier) rather than facility maintenance.

Table 5.1 identifies that a significant number of failures can be attributed to connector and connection failures. Often, the type of connection selected represents a trade-off between reliability and maintainability. Admittedly, it could be time consuming and not cost effective to analyze for optimal connection technique to be used in the many component or parts replacement typically employed in maintenance actions. However, a ranking of expected reliability of various connections is available as a reasonably accurate evaluation technique. An approximate idea of relative connector failure rates is indicated by an index shown against each type (below). The most reliable is given as one (1) or unity [7]:

Wrapped joint	1
Welded connection	3
Machine-soldered joint	7
Crimped joint	8
Hand-soldered joint	10
Edge connector (per pin)	30

As a general rule, the smaller the connection, the less reliable it will be. Similarly, a more reliable connection will generally require a longer time to complete the activity.

When installations are in the design phase and reliability and maintainability criteria can be addressed as design issues, here, again, is an issue to be considered. However, it is also possible to focus on connections in a mature system, particularly if a failure rate for a particular type of connector method is significant (Figure 5.1). Often, this failure rate increase can serve as an indicator of problems that can occur with related components or parts that may relate the connector failure rate (e.g. resistors, capacitors, and lighting).

For installations where FMEA implementation cannot be fully performed, the information provided in Table 5.1 can serve as a basis or aid for facility preventive maintenance (PM) programs, particularly when scheduled maintenance operations are considered.

5.3 The Role of Maintainability

Maintainability can be defined as the probability that a device or operation that has failed can be restored to operational effectiveness within a specific period of time upon completion of the maintenance action in accordance with approved procedures.

In dealing with RAM, it is most important to differentiate maintenance activities with maintainability. Maintainability is a characteristic of design and installation of equipment [2] that is expressed as the probability that an item will be retained in or restored to a specified condition within a given period of time, when the maintenance is performed in accordance with prescribed procedures and resources. A continuous maintenance operation is a regulatory requirement for cGMP operations as defined by 21 CFR (211.58 Maintenance), which states, "Any building used in the manufacture, processing, packing, or holding of a drug product shall be maintained in a good state of repair."

Maintainability can be defined as the probability that an item will be retained at a specified condition or restored to that position within a given time period, when maintenance is performed in accordance with approved procedures and employing approved resources [4]. As is the case with reliability, maintainability must also be designed into the equipment or system.

Basically, maintenance is composed of activities that allow for specified item operation throughout the lifecycle of the installed equipment. Maintenance involves actions required to retain an item, or system, or restoring the item, or system, to a specified condition [4].

Maintainability engineering, while similar to other engineering disciplines, is unique in that its emphasis is on recovery of equipment subsequent to failure and with the goal of minimizing upkeep costs. While this may be a novel concept in many pharmaceutical operations, maintainability engineering is a common operation in many chemical, petrochemical, and petroleum refining operations. Hence, by incorporating some basic concepts in pharmaceutical operations, it is conceivable that with a basic reliability and maintainability background, significant efficiencies and cost savings can be achieved in the early phases of design and installation by possessing a familiarity with basic concepts. In fact, significant benefits can be achieved throughout the useful life of the equipment if an effective program can be developed and implemented. Such programs are often available as part of an online CMMS. However, typically, pharmaceutical operations do not take advantage of this option since it requires extra training, input, and documentation.

Also, as noted, any element of the project team can contribute to a maintainability program at any point in the design, installation, and operation of the system. If the project is an in-house design or an "as-built" installation, maintainability (and reliability) concerns and criteria can be addressed starting with the basic plan elements (i.e. user requirements specification [URS] and the functional requirements specification [FRS]). Another area where maintainability and reliability can be highlighted is during the scheduled design reviews and vendor/contractor presentations.

Maintainability goals are divided and allocated in three major categories: (i) equipment design, (ii) personnel, and (iii) support (i.e. test equipment, spare parts, etc.). These objectives are accomplished by defining the operations provided in the PM and scheduled maintenance procedures for the facility. Another important component of maintenance planning and implementation would be incorporating the mathematical concepts such as those presented in Section 5.2. If such data is not available, previous experience with similar or identical units can serve as a basis for maintenance procedures until quantitative data specific to the planned installation becomes available.

When identifying system maintainability goals, it would be relevant to identify design guidelines that could affect the system maintainability. Some relevant design aspects are [8] the following:

1) Consider reducing maintenance needs by designing reliability into equipment to ensure desired performance throughout the anticipated performance lifecycle.
2) Incorporate reliability improvements to minimize maintenance time and manpower by reducing PM requirements and, consequently, provide a longer operating time. An example of this element would be scheduled testing of relief devices. Standard procedures require relief valves to be tested at twelve (12) month intervals. While an annual performance is standard for a pharmaceutical reactor, this is not the case for a relief device intended to protect an ion exchange tank where the significant damage would be water and resin being released.
3) Evaluate the possibility of reducing downtime by improving maintainability through simplification of test and repair procedures to reduce troubleshooting and correction time; as an example,

other than providing easy access to monitoring and recording devices, required adjustments should be simple. Interfacing with maintenance personnel to ascertain possible modifications and improvements of proposed and existing procedures should be a standard event.

4) Decrease the logistics requirements by using available standard parts, spares, tools, and test equipment.

5) Operations and procedures should be capable of being performed by facility personnel where feasible and minimizing the need for outside contractors to perform maintenance tasks that could be performed in house. Often, this is a decision that requires a trade-off between replacing the component and repairing the part of the component that is defective. For instance, if a motor fails due to worn windings, if is more cost effective to replace the motor than to have in-house maintenance, redo the windings.

A more detailed examination of the general principles involved in preparing a comprehensive maintainability program should encompass the general principles of maintainability. While, ideally, these principles would be addressed during the design and planning phases of the project, the principles can be applied at any point in the equipment/system lifecycle and still present economic, efficiency, and safety benefits. Since maintainability is the responsibility of the system owner, maintainability principles should also be addressed to the vendors, construction managers and subcontractors, and project managers prior to bid award and also subsequent to contract award. Further, vendors should understand what contributions to the maintainability program they are responsible for concerning design maintainability provisions (e.g. access, tools, spare parts needs, and equipment safety provisions) and related items that can affect cGMP and non-cGMP maintenance operations.

A description of the elements describing the general principles of maintainability, which are identified in the previous paragraph, is presented below [3]:

1) Reduce or eliminate the need for maintenance.
2) Reduce the amount, frequency, and complexity of required maintenance tasks.
3) Provide for reduction of lifecycle maintenance costs.
4) Provide for single-use cGMP components.
5) Reduce the required levels of maintenance skills and training required.
6) Establish maximum frequency and extent of PM to be performed.
7) Improve procedures for maintenance.
8) Reduce the volume and complexity of maintenance procedures.
9) Provide components that can be adjusted for wear, and provide adjustment to preclude breakdown, when practical.
10) Provide characteristics of the equipment and its components that result in minimum downtime.
11) Ensure simple, adequate, and satisfactory maintenance technical data are available with the equipment when delivered.
12) Provide for repair times of components; reduce the MTTR.
13) Provide optimum accessibility to equipment and components requiring frequent maintenance, inspection, or replacement.
14) Provide for rapid and positive identification of equipment malfunction or marginal performance. Typically, this includes troubleshooting charts in logic diagram form, potential failures, and corrective procedures.
15) Ensure the human factor aspects (Section 5.4 are satisfactory and operability of controls and physical and manual limitations are satisfactory.
16) Provide human engineering elements for access to maintenance points such as electrical, pneumatic, hydraulic, and lubrication.
17) Provide optimum capability to verify performance, anticipate and locate a malfunction, and perform calibration.

18) Provide for adequate, clear, and rapid identification of parts and components that need repair or replacement.

19) Provide current drawings, P&IDs, and material safety data sheets (MSDS) for API, and finished products have been reviewed, and procedural requirements for dealing with these materials, if applicable, are in place.

20) Reduce the quantities of tools, tool sets, and equipment necessary to maintain the system or equipment. Eliminate, wherever possible, the need for special tools.

21) Reduce to a minimum the number and types of replacement parts and components required to support maintenance activities.

22) Use standard parts and components to the fullest extent possible.

23) Provide for maximum interchangeability.

24) Provide maximum safety features for both equipment and personnel in the performance of maintenance tasks (Section 5.4).

25) Provide adequate lifting and hoisting equipment.

26) Provide for maximum storage life with minimum maintenance rehabilitation.

27) Ensure required environmental compatibility (e.g. corrosion, temperature control, humidity) for the equipment, components, and parts.

28) Provide bearings and seals of sizes and types that require a minimum replacement and servicing on a lifecycle basis. Select adjustable parts and components to take care of wear.

29) Provide gears of adequate size to satisfy overload requirements, which can be modified for lifecycle modifications (e.g. specify centrifugal pump impellers below the maximum size; specify electrical components that can accommodate possible overloads).

30) Ensure accessibility procedures are being implemented, including rapid access to components and parts by incorporating rapid fasteners, covers, and doors and minimizing the amount of bolts, fasteners, etc., with the goal of reducing the MTTR.

31) Ensure components requiring frequent maintenance are located to preclude the need to remove other components to gain access to the specific component.

32) Ensure adjustment controls are readily accessible.

33) Provide adjustment control locking devices.

34) Ensure all required test equipment and calibration equipment are available.

35) Ensure sufficient storage for spare parts and components are properly stored and identified.

36) Ensure adequate and sufficient guards are installed over dangerous moving mechanisms.

37) Ensure adequate protection from electrical shock is provided for maintenance and operating personnel.

38) Ensure that fire extinguishing equipment is installed and adequate.

39) Ensure sufficient and readily available drains are properly located.

40) Verify cleaning procedures conform to acceptance criteria.

41) Ensure maintenance of the system or equipment is performed by personnel wearing the required coverings and footwear when operating in ISO classified areas.

42) Ensure piping and critical equipment are visible and properly labeled.

43) Ensure maintenance procedures, maintenance manuals, operational procedures, and safety information such as the MSDS are readily available.

44) Consider advantages of component replacement versus part repair versus "throwaway" design.

With relevant maintainability principles identified (Table 5.1), it would be helpful to identify specific maintainability principles that should be considered as design properties of a system or equipment installation. Table 5.2 identifies several maintainability design principles that should be addressed as the project evolves.

Table 5.2 Maintainability design characteristics [8].

If all or part of the installed equipment/system is under a warranty, it is critical that written procedures clearly define what components are under the warranty and require vendor support (e.g. calibration, adjustments, and component replacement)

Manuals, checklists, charts, aids

Labeling and coding

Test equipment

Tools

Test points

Controls

Adjustments and calibrations

Displays

Test hookups

Test adapters

Handles and handling

Cases, covers, doors

Openings

Accessibility

Mountings and fasteners

Connectors

Installation

Standardization

Lubrication

Fuses and circuit breakers

Interconnecting wires and cables

No maintenance-induced faults

Components

Interchangeability

Size and shape

Modular design

Cabling and wiring

Ease of removal and replacement

Operability

Personnel numbers

Personnel skills

Safety

Work environment

Illumination

Training requirements and procedures

Failure indicators and alarms

Emergency shutoffs

A convenient method to review critical maintenance items for planning PM servicing and inspection needs is through the preparation of a maintenance checklist. Such a checklist (normally scheduled during downtime periods) could be very helpful, particularly when routine maintenance inspections involve complex installations.

When preparing and documenting a maintenance checklist (see Table 5.2), or a PM Plan, checklist items should include and address the following maintenance considerations [2]:

1) Have maintenance and inspection intervals been selected to assure PM and inspection procedures are consistent with production schedules?
2) Have adequate considerations been prepared for inspection (e.g. convenient access door, inspection windows and/or ports, test points and displays, comprehensive inspection procedures located at identified locations and accessible to operators and technicians, sufficient lighting, and safety)?
3) Have provisions made to facilitate rapid fault detection?
4) Are critical operating parameters automatically monitored (e.g. are human–machine interface [HMI] devices required for the operation)?
5) Do maintenance procedures monitor for degradation and failure prediction?
6) Can critical equipment or component failures be isolated to the affected unit?
7) Are indicators and alarms located and positioned in a location that will assist in locating and isolating the defective or failed component?
8) Are defective or failed components readily repairable or replaceable?
9) Are approved procedures in place for dealing with repair, replacement, and discarding of failed components, including a documented component replacement and disposition record?
10) Are spare and parts locations documented to provide for access and maintenance? (This topic is addressed in further detail in Section 5.9.)
11) Are approved test and inspection procedures in place to verify the maintenance action?

The factors described above, although intended for the design phase of a program, can be effectively utilized at any phase in the system lifecycle. Maintainability is a dynamic task that is subject to improvement whenever data or intuition can effectively demonstrate a change or modification to an operation or system improves efficiency, cost, or safety.

5.4 The Preventive Maintenance Program

A significant portion of pharmaceutical maintenance involves PM operations. Preparation and implementation of PM programs is the responsibility of the operation owner once the installation has been completed and is operational.

For cGMP activities, the need for a documented preventive program is specifically addressed in 21 CFR 211.67 and 211.68. For non-cGMP installations, an effective PM plan is a key contributor to a successful system lifecycle. An effective scheduled maintenance program consists of identification, scheduling, and determination of tasks and installation equipment and requirements for PM activities. Generally, the steps involved in this planning effort involve the following [3]:

- Compile data – The data compilation should include engineering drawings, parts lists, repair manuals, approved procedures, failure data, and PM tasks.
- Determine the criticality of repairable and replaceable parts.

- Mean time between failure (MTBF) data for critical equipment.
- Determination of the average operating cycle of the equipment per day. The operating cycle is the amount of time, in hours, the equipment operates in 1 day. The average operating period, h_{av}, is obtained by dividing the number of operating days into the total hours the equipment is operated while in service. The applicable relationship in a mathematical format is given by

$$h_{av} = \frac{h_1 + h_2 + h_3 + \cdots + h_n}{d_t} \tag{5.1}$$

where

h_{av} = average operating cycle per day, h/day
$h_1 + h_2 + h_3 + \cdots + h_n$ = the sum of the individual operating hours per day, h
d_t = total number of operating days.

A brief application of Equation (5.1) is a liquid level control that operates for 300 days and operates for 2910 h. The average operating cycle per day is

$$h_{av} = \frac{2910}{300} = 9.7 \, \text{h/day}$$

A more comprehensive coverage of operating times and its relationship to reliability and maintainability is detailed in Section 5.7.

- Specify the frequency of PM tasks and monitor the frequency as the system progresses in the lifecycle.
- Schedule lengthy PM tasks during scheduled shutdown periods; this action will minimize overtime expenses. For instance, bench testing of relief valves should be scheduled during periods where the protected equipment is not in use during normal operating periods. Typical periods would be equipment cleaning operations between batch runs.

It is understood that implementation of many of the programs described is time consuming and, for those not familiar with reliability and maintenance planning and implementation, not effective, but a well-planned and implemented program ultimately results in a more efficient, safer, and ultimately more efficient manufacturing operation.

5.4.1 System Replacement Considerations

Ultimately, even well-maintained pharmaceutical systems require replacement. In some cases, the system is obsolete due to the increasing need for higher output and not cost effective due to the increasing need for maintenance (Figure 5.1) and the increasing cost of spare and repair parts over time. While many manufacturing operations recognize the need to install new systems to remain competitive, there is, often, little corporate guidance on an approach to new system installation. As a result, guidelines for new system installation can be useful.

No project can succeed without adequate preparation. Proper planning is critical for success. That planning must include identifying the needs as well as limitations in available staffing, funding, and other resources. One of the most important parts of a new system installation preparation is to understand common misconceptions about modernization and system replacement. The issues and common misconceptions described below should be considered prior to implementing a new or significant upgrade or system installation [9].

The first misconception is the assumption that the current system vendor is the best since he/she is knowledgeable in the current system. The problem with this line of thought is that a new system is being installed and the evaluation should be based on knowledge and requirements of a new system, not the existing system. To avoid this potential problem, evaluate all qualified suppliers employing identical standards.

Another myth is the guarantee of system support of an existing upgraded installation (legacy system) for an extended period, often 10–15 years often; these guarantees are contingent on mandatory software and parts upgrades. In many cases, particularly process control systems, the upgrade typically does not affect the original wiring, I/O mechanism, or basic configuration. Very often, upgrading an existing system merely extends the existing life of the legacy system; an upgrade is not equivalent to a new system.

A system upgrade is most applicable as a short-term "fix" when planning for a new installation. By upgrading an installation, the upfront costs may be reduced; the basic rationale for a new system (e.g. outdated technology, maintenance costs, repair costs, software inadequacy) still remains.

Another common fallacy is the thought that changing vendors entails the need to support two systems at the same time during the changeover. Regardless of vendor selection, two systems will require system support until the legacy system is decommissioned and the new system is validated. It is problematic to assume one vendor (new or current) can effectively decommission the existing system and install, commission, and validate the new system in the same time frame with the same personnel.

Another misconception is the assumption that every vendor provides the same new system installation options. In actuality, vendors differ in relevant aspects specific to system changeover. For instance, one vendor may possess in-depth expertise in control system software installation and validation, whereas another vendor may be proficient in system hardware installation such as control valves and feed systems. It is most important that a selected vendor has a demonstrated proficiency in the total system selected. Failure to adequately address this issue will often lead to cost overruns, delays, and, ultimately, costly change orders as a result of improper contractor/vendor inadequacies.

5.5 Human Factors

One of the most important aspects to be considered in equipment design and operation is that it should be capable of being operated and maintained by humans, the variable upon which human factors engineering is based (also referred to as ergonomics). Human factors are involved in designing, operating, and maintaining equipment and systems that require an interface between humans and machines. Ideally, machines operate with no requirement for human involvement. However, because maintenance, repair actions, and operation require human involvement to some degree, it is important to view HMI as a tool that can promote reliability, maintainability, efficiency, and safety.

It would be helpful, when dealing with human factors, if a distinction could be made between helpful properties of humans and machines. Table 5.3 emphasizes some of those characteristics.

One critical characteristic of maintainability and reliability that is commonly encountered in human factors is accessibility. If a particular maintenance activity cannot be performed by clearly seeing, reaching, and manipulating the item requiring maintenance, there can be a tendency to delay or omit the maintenance action, make mistakes, and accidentally damage equipment. Poor accessibility to routine service and inspection points and equipment parts

Table 5.3 Characteristics of humans and machines [8].

Characteristics tending to favor humans	Characteristics tending to favor machines
1) Ability to detect certain forms of energy	1) Monitoring of men or other machines
2) Sensitivity to a wide variety of stimuli within a wide variety of stimuli	2) Performance of routine, repetitive, precise tasks
3) Ability to perceive patterns and generalize about them	3) Responding quickly to control signals
4) Ability to detect signals (including patterns) in high noise environments	4) Exerting large amounts of force smoothly and precisely
5) Ability to store large amounts of information for long periods and to remember relevant facts at the appropriate time	5) Storing and recalling large amounts of data
6) Ability to use judgment	6) Computing ability
7) Ability to improvise and adopt flexible procedures	7) Range of sensitivity to stimuli
8) Ability to handle low probability alternatives (i.e. unexpected events)	8) Handling of highly complex operations (i.e. performing many tasks simultaneously)
9) Ability to arrive at new and completely different solutions to problems	9) Deductive reasoning ability
10) Ability to profit from experience	10) Insensitivity to extraneous factors
11) Ability to track in a wide variety of situations	N/A
12) Ability to perform fine manipulations	N/A
13) Ability to perform when overloaded	N/A
14) Ability to reason inductively	N/A

N/A = not applicable.

reduces efficiency by lengthening the maintenance and inspection time. If it is necessary to significantly dismantle a component to obtain access to a specific component or part, the availability of the equipment is decreased and the maintenance costs are increased.

Accessibility, when evaluated as a "stand-alone" activity, is not part of maintainability. Merely having access to a component or part does not indicate it can be maintained. Accessibility requirements are, ideally, determined by the planned maintenance action, which may require visual and physical activity, depending on the defined task. Consequently, accessibility is primarily a function of two factors: access to the component or part requiring inspection or maintenance and convenient space to perform the defined activity such as adjustment, repair, or replacement [2].

Well-designed equipment access is desired for ease of maintenance and should be integral to maintenance procedures requiring removal, inspection, opening, fitting, or dismantling equipments, components, or parts. Thus, the design should address equipment access with attention to parts likely to encounter more frequent maintenance and cleaning (i.e. tablet presses encountering powder buildup during manufacture and pan coater spray nozzles becoming clogged during production runs) actions than typically planned or expected.

While many accessibility design features, including human factors, are incorporated in the design, there are factors that can affect accessibility and can be incorporated in maintainability operations as part of the planned maintenance actions. Some of the accessibility factors to be considered with documented procedures include [2]:

1) Location of the unit
2) Frequency with which access must be entered
3) Classification of the access area (i.e. cGMP location, ISO classification)
4) Maintenance functions to be performed
5) Time requirements identified
6) Tools and accessories required
7) Work clearances and space required to perform a task
8) Type of clothing required to perform a task in the specified location
9) Illumination required to perform a task requiring access
10) Identified hazards related to access and maintenance

With accessibility concerns considered, maintenance design guidelines are to be evaluated. While it is desirable to incorporate accessibility factors in the design phase, where user input can influence design, such modifications are more problematic when dealing with a mature pharmaceutical installation. However, with the transition of older small molecule pharmaceutical operations to newer large molecule biotech facilities, it is reasonable to expect more attention will be given to design phase criteria, including accessibility.

Nevertheless, for older operating pharmaceutical facilities, there can be nonexpensive maintenance modifications that can, in general, enhance operations and improve specific factors, such as accessibility (confirming illumination requirements, for example). The 10 items identified above can serve as a pre-tasking list that can be a basis when routine, scheduled maintenance activities are to be performed.

Once access to the area requiring maintenance, repair, replacement, or inspection is accounted for, convenient access to the component part or assembly must be considered. To assist in this function, design guidelines that improve ease of the maintenance activity would be helpful. Such design guidelines would typically include, but not limited to, the following maintenance evaluation elements:

1) Locate parts and components such that other assemblies do not need removal to service the unit requiring maintenance attention.
2) If the design requires specific components, parts, or assemblies be placed behind another, consider positioning the item requiring less maintenance activity in the front of the access location.
3) Do not locate a component, part, or assembly under pipes, hoses, or other parts of the equipment that present removal difficulties.
4) Access to the assembly, component, or part should require only single access by the maintenance personnel.
5) Access should be at the front and not through the side or back of the installation.
6) Removal of units, parts, or components should be designed to be linear motions, without the need to "juggle" around corners or position in awkward locations.
7) Assemblies, components, and parts should provide sufficient work area such that maintenance tasks can be adequately performed.
8) When feasible, plug-in units should be employed.

The design guidelines described above indicate that human factors involve an operation in which men and machine (i.e. units, components, and parts) must interact to successfully perform a given task or operation. When dealing with human factors, meeting the defined objectives of performance and reliability is a measure of the system success. Successful completion of the defined task involves speed, accuracy, reliability, and maintainability. Other important factors are operator and maintenance personnel skill level and safety.

A tablet coater is an example of a typical human factors operation. The coater operator is responsible for monitoring operations such as coater pressure, coater temperature, and the other variables identified on the HMI at the coater location. The operator is responsible in reporting sudden excursions in the operating conditions in a timely manner. This function represents the machine component of the man–machine human factors paradigm. The reaction of the operator and the response of maintenance staff is the human element of the man–machine relationship. The machine element of the relationship is primarily determined during the design phase. Maintainability is to a large part determined by the human aspect of the man–machine system. As has been pointed out, a large fraction of the success, or failure, of the operation is the reaction to the required maintenance to correct the operational failure. This time to recovery action has been identified as the MTTR. Once installed, there is little that can be done to increase the reliability of the system, the machine element of the human factors system. However, it is possible to enhance the MTTR by modifying some machine factors, discussed previously, and some human factors.

One action is on the area of approved procedures. Few maintenance procedures actually describe step-by-step actions that specifically detail the steps required to remove, replace, or repair installed systems, components, or parts. This is often due to the fact that many facility maintenance procedures are little more than the vendor maintenance manuals, usually provided as part of the ETOP that is provided to the user upon completion of the installation. The user rarely focuses on how the manuals are specific to the installation at the location provided. Items such as distance from the equipment requiring maintenance to the facility maintenance office are rarely addressed. For a large facility composed of many buildings, the transit time could significantly affect the MTTR. For instance, would there be a benefit if maintenance and repair personnel assigned to a packaging line were permanently located near the packaging line, or is the skill level such that their packaging line responsibilities are just a part of their overall responsibilities and not dedicated solely for packaging line activities? Also, it might be beneficial if specific parts and components that require more frequent repair and replace actions were conveniently located near the packaging line, space permitting.

The crucial determinant of man–machine performance interface is the complexity of human tasks associated with the system and an analysis of the need for specific trade-offs between man–machine operations. A system that requires frequent and precise maintenance action or adjustments by an operator or maintenance technician can create reliability difficulties related to wearout, incomplete tasks, or maintainability difficulties resulting from repetitive replacement and/or repair of the same part or component. Conversely, a system equipped with automatic adjustment devices can cause problems of cost, reliability, maintainability, and safety as a result of the design and operational complexity when maintenance or repair is required. For the same level of effectiveness, an installed system, because of access difficulty, repair difficulty, or safety considerations, has to exhibit greater reliability than a system with a less complex HMI (e.g. is a panel mounted digital flow meter actually more reliable and efficient than a mounted flow meter, particularly for non-complex process monitoring such as measuring ambient water flow). Accordingly, HMI can enhance as well as degrade performance, depending on the trade-offs and interactions provided in the design and maintenance and repair procedures employed. While the original design can be influenced by evaluating human factors in the early phases of the program, continuous monitoring by operators, technicians, and the engineering staff can considerably enhance the human factors aspects of the system lifecycle.

It is also most important to implement and modify procedures through additional training and comprehensive change control documentation. Effective and documented training is the most crucial element of human factors considerations as well as virtually all operations involving reliability and maintainability. Comprehensive training is a key contributor to minimizing human errors, which are often a cause of system failures. While training and well-documented procedures can significantly enhance equipment maintainability, there are human factors that can have a detrimental effect. These include the following [10]:

• Forgetting – Often, scheduled periodic reviews of the maintenance procedures can reduce the failure rate caused by this element.
• Murphy's law – When dealing with human factors, Murphy's law can be stated as, "If it is possible for someone to do it wrong, someone will." By preparing and continuously modifying procedures as required and maintaining updated training reviews, maintainability and reliability can be improved.
• Human error – Again, effective procedures clearly written, easily understood, and reflecting current modifications, coupled with training updates, are the most effective techniques to minimize human error.
• Motivation – Current procedures, proper tools, availability of components and parts, and effective training allow for a shorter repair time, all of which contribute to higher motivation.
• Maintenance management – A significant element of maintenance management is clearly defining the functions of the operator and the maintainer. In a production mode, operators may perform an action they may not be trained to undertake in an effort to maintain production. Similarly, maintenance personnel may perform a task that, because of its apparent simplicity, will be performed to speed the maintenance activity, thus introducing an opportunity for greater malfunctioning. It is important that tasks are clearly delineated and defined to avoid difficulties when dealing with the operator maintenance interface.

As it has been shown, human factors is an ongoing and continuous process that is required throughout the lifecycle of the system. What often makes a human factors program difficult is the lack of a defined goal in the program. While it is possible to quantify actions such as reliability improvement and maintainability by employing factors such as MTBF and MTTR (details of both terms are covered in Section 5.6 human factors improvement evaluations cannot be easily measured. However, failure of an effective human factors program can often, unfortunately, be identified in qualitative measures. Accidents, production breakdowns, and equipment failures due to poor maintenance or repair activity often are the result of failure to adequately address aforementioned items such as procedures, training, and safety considerations. To strive for better outcomes, these qualities and behaviors need cultivating through both inspired leadership and motivated employees. When these human factors converge, more effective pharmaceutical operations will result and, ideally, the benefits of a human factors program will be recognized [11].

Perhaps the most salient element of a human factors program involves not only the reliability and maintainability of a system but also the safety aspects of the operation. Just as a pharmaceutical operation has to evaluate work orders based on which cGMP regulations apply, if any, there should also be an element requiring which OSHA regulations are applicable to the task. Since task specificity is a detailed individual requirement for each system, part, or component, it would be advantageous to have a checklist that could address human factors safety elements for review and analysis of installation. Ideally, the checklist would focus on the design phase of the system. Of course, this could introduce severe difficulties when dealing with older operations. Benjamin S. Blanchard and others (Reference [12]) have prepared a safety checklist specific to the design and development phase of a new installation [13]. However, this checklist is also applicable to existing systems requiring maintenance actions at frequent intervals. The checklist should require a "Yes" or "No" response. A No response requires an analysis of the checklist responses specific to the safety aspects of the installation. Safety considerations include [12, 13]:

1) Have fail safe provisions been incorporated in the design?
2) Are protruding devices addressed and protected?
3) Have provisions been incorporated for protection against high voltages? Are all external parts adequately grounded?
4) Are sharp metal edges, access openings, and corners protected with rubber fillets, fiber, or plastic coating?
5) Are electrical circuit interlocks employed and identified?
6) Are standoffs or handles provided to protect equipment from damage during the performance of bench maintenance?
7) Are tools used near the high voltage locations adequately insulated at the handle or other parts of the tool that is likely to be touched?
8) Are the environments such as personnel safety addressed? Are noise levels within the safe range (Section 3.2.8 contains details concerning noise criteria)? Is illumination adequate?
9) Have personal protective equipment (including footwear) requirements and procedures been identified for areas where the work environment can be detrimental to safety?
10) Has a system hazard analysis (HAZOPS) been completed where required?
11) Does the facility have a continuing safety training program?
12) Are facility housekeeping procedures in place for non-cGMP activities?
13) Are written and verbal communication programs established?

While several items in the checklist appear to be repetitious, it should be noted that the above checklist is specific to safety, whereas other checklists are applicable to reliability and maintainability evaluations. When addressing specific safety concerns, these issues should be addressed in a detailed format using the FMEA format or a hazard and operability (HAZOP) examination. Consequently, it is important to focus on the safety aspects of the checklist, presented above, with more of an in-depth analysis than the reliability and maintainability checklists presented herein.

Stopping at the human error level of accident causation is not enough for true accident prevention. Companies that are serious about keeping their facilities safe must dig deep and go further in their investigations of incidents and near misses. The human machinery of any complex system is apt to fail as much as any piece of complex equipment. Companies should focus not only on the symptom of human error but instead on remedying the underlying weaknesses in the systems.

5.6   The Role of Availability

As will be seen shortly, the term availability is the dependent variable that brings reliability and maintainability together to create a general relationship capable of evaluating the two variables (reliability and maintainability) as a systematic relationship that can be applied to effectively resolve many RAM issues. Availability can be defined as the proportion of time during which a system is available and ready for use. The role of availability is addressed in more detail in Section 5.7

5.7   Basic Mathematics for Reliability, Availability, and Maintainability

Chapter 11, specifically, Section 11.1.1, is a helpful source to augment the topics addressed in this section.

Prior to delving into this section, two terms that are critical to RAM should be further defined. While the MTTR and the MTBF have been alluded to in the previous sections, a more formal definition would be helpful when dealing with this section. While it will be seen that MTTR and MTBF are most often applicable in mathematical operations, qualitative definitions are also important. Thus these terms are identified as

- MTBF – The average or mean time a system operates between failures. A system can be a single piece of equipment or a unit composed of several pieces and components. The unit of measurement is most often given in hours: minutes and seconds can also be used, depending on the system.
- MTTR – The average or mean time to restore a system to an operating mode. The measurement units should be consistent with those of the MTBF.

Reliability can be defined simply as the probability that a system will perform in a satisfactory manner for a given period of time when functioning under specified operating conditions. This is expressed in terms of *failure rate*, λ, or in terms of the MTBF, both of which are parameters of the reliability. It can be shown that if failures occur at random, the failure rate, λ, is then constant, λ, and the reliability (R) is equal to

$$R = e^{-\lambda t d} \tag{5.2}$$

where t is the period of time referred to in the definition of reliability and d, the duty cycle factor, is the ratio of the operating time to total elapsed time of operation.

Typically, pharmaceutical systems such as tablet coaters, tablet presses, and fluidized bed dryers operate in a batch mode using production schedules. However, most water systems operate continuously. Hence, for pharmaceutical water systems, the duty cycle ratio is assumed to be 1 and is not germane to Equation (5.2). The equation for water reliability of water manufacturing systems can then be modified as

$$R = e^{-\lambda t} = e^{-t/\text{MTBF}} \tag{5.3}$$

The failure rate, λ, can also be expressed as

$$\lambda = \frac{\text{Number of failures}}{\text{Total operating hours}} \tag{5.4}$$

where λ is given in reciprocal units of time (i.e. h^{-1}, min^{-1}).

A closer examination of Equation (5.4) reveals the greater the number of failures per operating hours, the more reliable the value of λ will be. For this reason, recording reliability data should be a continuous operation throughout the planned lifecycle of the component or system.

One commonly occurring difficulty when dealing with reliability calculations is obtaining dependable values for λ. It is not uncommon for manufacturers, designers, and users to withhold this data for a variety of reasons. Often, a reliable source of this information is the government. When dealing with federal procurements, reliability and failure data is often required to conform to contractual requirements. Such information can often be obtained from the Department of Defense programs involving integrated logistics support (ILS) where RAM data is an integral component of the development program. While not specific to pharmaceutical operations, through due diligence, many common components and equipments used in pharmaceutical systems may be located in the ILS package.

Another valuable source of RAM data that can be applied to pharmaceutical operations is available through the Nuclear Regulatory Commission (NRC). One most valuable source is a document published by the NRC commonly referred to as WASH-1400. This publication (available for no charge online) contains many reliable values for λ not typically available from other sources.

Table 5.4 presents values for the failure rate, λ, obtained from WASH-1400 that is common to many pharmaceutical operating systems. It displays some components and equipment that are common to both nuclear operations and pharmaceutical operations accompanied by their failure rates, λ.

Rather than using the statistical mean (average), Table 5.4 uses the median (Section 11.1.1). Recalling the median is the midpoint of a set of values formatted in ascending order, where the middle value is the median value. If the values are composed of an even number of values, the median is then the calculated mean of the middle two values. For example, a set of five values is 3, 8, 9, and 10. The two middle values have a mean value of 8.5 ($[8+9]/2 = 8.5$).

Thus, the median is 8.5 for this number set. For most analyses of this type, the median is accurate enough to substitute for a mean.

Table 5.4 is useful to permit quick and reliable data that assists in obtaining needed information for establishing a RAM database.

The failure rate may be expressed or recorded in several formats such as

- Failure/hour
- Percent failure/1000 h
- Failures/1 000 000 h

Table 5.4 Failure data for operating components [14].

Component	Failure mode	Median failure rate, λ/h
Motors, electric	Failure to run	1.0×10^{-5}
	Failure to run, given start	1.0×10^{-5}
	Failure to run, given start (excessive environment)	1.0×10^{-3}
Relays	Failure of NO (normally open) contacts to close	1×10^{-7}
	Short across NO/NC contact	1×10^{-8}
	Coil open	1×10^{-7}
Pumps (includes drivers)	Failure to run, given start	1×10^{-5}
	Failure to run, given extreme environment	10×10^{-3}
Valves		
Motor operated	Failure to remain open (plugged)	3×10^{-7}
Motor operated	Rupture	1×10^{-8}
Solenoid operated	Rupture	1×10^{-8}
Air-fluid operated	Failure to remain open (plugged)	1×10^{-8}
Check	Severe internal leak	3×10^{-7}
Check	Rupture	1×10^{-8}
Vacuum	Rupture	1×10^{-8}
Manual	Rupture	1×10^{-8}
Relief	Premature open	1×10^{-5}
Pipes		
<3 in. diameter per section	Rupture/plugged	1×10^{-9}
>3 in. diameter per section	Rupture/plugged	1×10^{-10}

The term reliability can be interchanged with the word probability, denoted by P, and Equation (5.3) can also be expressed as

$$P = e^{-\lambda t} \tag{5.5}$$

The term P can also be interpreted as the probability that a component will successfully operate over the time period desired. In this context, P is also understood to be defined as the reliability of the component to successfully function for the desired time period required.

Another important RAM term is a variable, Q_x:

$$Q_x = \text{Unreliability of the component} = 1 - P \tag{5.6}$$

This relationship, Equation (5.6), provides a method of determining the probability that the component will not successfully operate for the time required.

Example 5.1

A. Calculate the probability of survival of a pump that is to operate for 500 h and consists of four primary components (e.g. seals, packing, impeller, shaft) having the following MTBFs:
Component A (seals): MTBF = 5000 h
Component B (packing): MTBF = 3000 h
Component C (impeller): MTBF = 15 000 h
Component D (shaft): MTBF = 15 000 h

A random failure rate is assumed.

B. Calculate the probability that the pump will not survive 500 h of operation.

Solution

A. The failure rate of the components are computed as

$\lambda_A = 1/5000$ failure/h
$\lambda_B = 1/3000$ failure/h
$\lambda_C = 1/15\,000$ failure/h
$\lambda_D = 1/15\,000$ failure/h

The total failure rate for this (series) system is

$$\text{Total} = \frac{1}{5000} + \frac{1}{3000} + \frac{1}{15\,000} + \frac{1}{15\,000} = 0.000\,66 \text{ failure/h}$$

The probability of the system performing for 500 h is

$$P = e^{-(0.000\,66)(500)} = 0.719 \leftarrow \textbf{Solution}$$

B. $Q_x = 1.000 - P$ (Equation 5.6)

$$Q_x = 1.000 - 0.719 = 0.281 \leftarrow \textbf{Solution}$$

The solution to this problem states that the system has a 71.9% probability if operating for at least 500 h. Also, as solution B states, the pump can be expected to fail to operate for the required 500 h 28.1% of the time it is in operation.

The failure rates given in the above example are intended for problem solving purposes. In reality, the failure rate for a properly installed centrifugal pump is much higher, typically on the order of several years, if properly maintained with a focus on packing and seal performance.

The term unreliability is actually another word for failure or, in this instance, the probability of a failure. Often, a great deal of RAM information can be obtained from failure data. Thus, when a high unreliability is encountered, it is of value to analyze the system or component performance for possible failure causes. Perhaps the best methodology to analyze failure modes is to initiate failure analysis program upon issuance of the URS and the FRS.

For pharmaceutical operations, the most common causes of system or component unreliability (failure) are as follows:

- Incorrect design basis – Salient design characteristics were incorrectly specified, incorrectly designed, or overlooked completely.
- Improper materials of construction – Formation of galvanic or corrosion cells because of improper material selection (as will be seen in Chapter 8, even stainless steels can, when operating in certain environments, undergo corrosion). Also, components that were not designed to properly withstand mechanical forces and operating stresses resulting in metal fatigue are another common failure mode.
- Poor manufacturing documentation – In addition to nonadherence to the provisions of 21 CFR 210 and 211 (e.g. 21 CFR 211.100 & 211.186), SOPs that contain errors can be very costly to manufacturing operations.
- Improper or inadequate commissioning documents – In addition to commissioning documents (a detailed presentation of commissioning requirements is found in Chapter 10), site acceptance tests (SAT) and factory acceptance tests (FAT) should be carefully reviewed to obviate assembly and inspection errors.
- Improper or poorly prepared protocols – Preparation of the IQ, OQ, and PQ of an instrument component or equipment involves a significant investment in time and **cost.** Thus, it is critical that qualification testing be designed to evaluate all relevant aspects of the subject configuration. From a RAM viewpoint, successful qualification and requalification should be an integral aspect of the installed operation. Improper OQ and PQ testing can result in reliability failures long after the qualification protocols have been approved.
- Operational abuse – A common cause of this failure is attributed to operator error. This failure is often a result of improperly prepared SOPs that may fail to identify problems not identified by operating personnel for a variety of reasons. It is important to interface with personnel involved in the operational aspects of the system of concern.
- Improper utilization – Even with proper installation and system implementation, misapplication of a unit, equipment, or system can occur. A typical scenario can involve HVAC units. Units that have been in use for many years often undergo alterations such as adding cooling ducts to areas not originally intended for cooling. If the additional ductwork is in a non-cGMP location, change control procedures are often given a cursory evaluation. Further, one ductwork modification can often create a replication of the original modification (i.e. adding additional ducting subsequent to the original modification). Ultimately, the result of this expansion will be a unit that is underperforming because associated HVAC components such as the compressor were not evaluated for the increased load requirement. Often, for older installations, evaluation of the modified installation is difficult since the original drawings and specifications are not available because the location was non-cGMP and not considered critical.

The failure is often identified only after the cooling operation fails to perform properly. A situation such as this can occur even with an operating building management system (BMS).

- Damaged components – If the installation is being outsourced to an engineering design firm and installation is the responsibility of a construction management firm, it is most important to have approved material and equipment receiving and inspection documentation procedures in place. It is this documentation that will identify material that has been improperly packaged or shipped.
- Improper system transition – The trend for the last few years has been to contract pharmaceutical projects to outside vendors. Tasks such as design, installation, and start-up are usually the responsibility of the contractors selected to perform the task. Generally, system start-up is the responsibility of the equipment vendor and/or the construction manager. It is critical to the successful operation of a system that thorough training be given to the user personnel who will be using and maintaining the new equipment instrument, equipment, or system. Proper start-up training, documentation, and initial operation are not only required for cGMP conformance, but they are also important for efficient and economic facility operation, whether cGMP or non-cGMP components are involved.

The term total operating hours is a predetermined period of time during which equipment is evaluated. For instance, in a manufacturing environment, under normal operating conditions, data is often collected on operating equipment at quarterly or semiannual intervals. For computational purposes, a quarterly operating interval is usually identified as 1975 h. Similarly, a semiannual operating interval is defined as 3950 operating hours. When considering an annual operating period, 7900 h are employed.

For normal random failures, the reciprocal of λ is the MTBF. Thus,

$$MTBF = \frac{1}{\lambda} \tag{5.7}$$

Example 5.2 Five identical centrifugal pump components were operated under identical conditions. The components (which are not repairable) failed as follows:

Component 1 failed after 100 h
Component 2 failed after 125 h
Component 3 failed after 150 h
Component 4 failed after 275 h
Component 5 failed after 400 h

There were five failures and the total operating time was 7900 h (1 year).
Based on the data given,

A. Calculate the failure rate, λ, of the pump components.
B. Determine the mean time between failure (MTBF) for the pump components.

Solution
A. Employing Equation (5.4), the calculated failure rate per hour (λ) is

$$\lambda = \frac{5\,\text{failures}}{7900\,\text{h}} = \frac{0.000633\,\text{failures}}{\text{h}} \leftarrow \textbf{Solution}$$

B. Using Equation (5.7), the mean time between failures (MTBF) is

$$\frac{1}{0.000633\,\text{failures}\,\text{h}^{-1}} = 1580\,\text{h between failure} = \textbf{MTBF} \leftarrow \textbf{Solution}$$

Basically, solution A states that after 1 h of operation, there will be 0.001 26 failures. The result is more discernable if 1000 h of operation is considered. If the failure rate is 0.001 266 failures/h, then, after 1000 h, the failure rate would be

$$\frac{0.000633\,\text{failures}}{\cancel{h}} \times 1000\,\cancel{h} = \mathbf{0.633\,failures} \leftarrow \textbf{Solution}$$

In this format, it can be seen that after 1000 h of operation, a failure of about one pump component (0.633) can be expected. This data is required when planning a preventive maintenance program in a pharmaceutical operation, although the approach is valid for virtually all systems (specifically aerospace, defense, and nuclear operations). This approach differs from standard maintenance operations because maintenance activities are typically performed at arbitrary scheduling such as 3- or 6-month intervals with little or no concern for when failures actually occur. A more cost-effective procedure would be to perform a pump component maintenance/inspection activity between 1000 and 1500 operational hour intervals. If the calculation is performed for 7900 operating hours, it can be seen that after 1 year of operation, about five pump component replacements can be expected. As will be seen, this procedure will be useful when determining the number of spares that will be required for the pump component on an annual basis.

It is also noted that the MTBF of the components identified in Example 5.2 are not germane to the example solution and are only used as convenient values.

Examples 5.1 and 5.2 present some interesting details about λ, reliability, and MTBF. We note that reliability is, as indicated in the solution of problem 5.2, a probability of successful operation, not a guarantee of successful operation.

The common perception is that calculation of the MTBF is achieved by recording estimated run times until failure by manually inputting and recording estimated failure times and run times and subsequently recording the data in a computerized maintenance management system (CMMS). While this technique is widely accepted as the norm, automated and highly reliable instruments are available to record and analyze equipment failure times such as pumps or motors. A device capable of performing these tasks is an electronic instrument known as a multichannel pulse input data logger [15]. These data loggers can continuously record and measure energy and power use for many types of equipment such as motors and pumps. Once installed, the unit is capable of simultaneously measuring and recording pulse signals (when a motor starts and stops), events (unexpected motor trip), state changes (when a motor is turned on or off), and recorded run times. Additionally, these data loggers are capable of automatically calculating and entering the MTBF in a database as well as recording the result for entry in a CMMS. The data loggers are also helpful in performing analyses for building control systems (Section 3.2.9) where they have the ability to record data such as current, voltage, amp hours, kilowatt hours, kilowatts, temperature, and humidity.

In 2012 dollars, the installed cost of a typical four-channel hardware installation would be about $1500, exclusive of installation costs.

Failures occur in a completely random fashion. This assumption implies that the component has no memory of its previous use. Thus, the probability of failure at any point in time is equivalent to the probability of failure at any other time. This is generally the case with electronic and detailed mechanical components. Less complex mechanical items, such as manual valves, and chemical failures, may not be totally random. Systems conforming to the randomness criteria (e.g. HVAC, mechanical components of water purification units, or tableting operations) can be assumed to conform to a Poisson distribution.

This definition of reliability and maintainability stresses the elements of probability, unsatisfactory performance, time, and specified operating conditions. These four elements are extremely important, since each plays a significant role in determining the system reliability. Reliability, as an engineering discipline, experienced rapid development shortly after World War II as an outgrowth of requirements of missile and space technology. Within recent years, the realization that, in many cases, a more cost effective can be obtained by trading off some reliability for the availability to maintain a system easily has led to a considerable research and development effort to describe a relatively new engineering discipline, maintainability. This discipline is not in this new basic concept, but rather in the concentration given to its attributes, its relationship to other system parameters, the quantitative prediction and evaluation of maintainability during design, and its management. Maintainability is a characteristic of the system and equipment design. It is concerned with system attributes such as accessibility, testing, controls, tools, maintenance manuals (SOP), checklists, check out, and safety. Maintainability engineering is a discipline that is concerned with the design, development, installation, validation, and operation of a specified system such as a USP water system, a tablet coating operation, or a pump installation.

Maintainability may be defined as a characteristic of design and installation that imparts to a system or an item a greater inherent ability to be maintained, so as to lower the required maintenance man-hours, skill levels, tools, facilities, and logistic costs, and to achieve greater availability (significant cost savings). Maintainability is usually stated in terms of its own parameter, mean time to repair or MTTR (also identified by the letters MTR), which can be expressed as the repair rate, μ, and also equivalent to the reciprocal of the mean time to repair (1/MTTR). The word "repair" implies that the concern is with the time to perform corrective maintenance actions only, whereas the time taken to carry out preventive maintenance is also of interest. The fact is giving rise to an improvement in corrective repair time will reduce preventive maintenance time in the majority of cases.

Maintenance and maintainability have different meanings. Maintenance is concerned with those actions taken by a system user to retain an existing system/equipment in, or restore it to, an operable condition. Maintainability is concerned with those actions taken by a system/equipment designer, during development, to incorporate those design features that will enhance ease of maintenance. Its function is to ensure that – when produced, installed, and operated – the system/equipment can be maintained at minimum lifecycle support cost and with a minimum downtime.

The lifecycle support (owner) aspect is the responsibility of the maintenance engineering organization, although the name may vary with the pharmaceutical firm or organizational entity.

Availability (*A*) or the measure of the degree of a system is in the operable and committable state at the start of an operation (i.e. in pharmaceutical terminology, a batch run). This definition is more detailed than the more basic definition of availability presented in Section 5.6. Availability is often called "operational readiness." It is a function of operating time (reliability) and downtime (maintainability/supportability). Availability can be understood as the percent or probability that a component or system is operable or ready to operate when required. The availability term is also the reliability of the system part or equipment expressed as a fraction. As will be seen, both maintainability and availability affect each other and influence the costs of routine maintenance, repair, and system outage. The parameter *availability*, which combines maintainability and reliability, is therefore of interest. Another definition of availability, known as the steady-state definition, is that the proportion of time during which a system is available for use (i.e. not in a failed state). As we will see, by determining the relationship between reliability, availability, and maintainability (RAM), a powerful tool will be available for developing and employing a method of pharmaceutical system and equipment facilities management that will save both time and money. These techniques will be comprised of non-complex mathematical relationships. However, it is necessary to first become familiar with the terms RAM in a quantitative format. As such, the following relationship can be defined:

$$A_i = \frac{\text{MTBF}}{\text{MTBF} + \text{MTTR}} \tag{5.8}$$

where A_i is the *inherent availability* and is composed of the MTBF and the MTTR. The relationship excludes idle time (nonoperating time), logistic time (the time required to transport the component or part to the maintenance location), waiting time (the time required to prepare for the maintenance operation), and preventive maintenance time (usually the time required to perform routine inspections as a scheduled activity).

Figure 5.2 presents a nomograph that relates the inherent availability (A_i) with the two other variables of Equation (5.8).

Another availability term is the expression *operational availability*, A_o. The operational availability, A_o, can be defined mathematically as

$$A_o = \frac{\text{MTBM}}{\text{MTBM} + \text{MDT}} \tag{5.9}$$

For this relationship, MTBM is identified as the mean or average time between maintenance. Similarly, MDT is the mean maintenance downtime. The MDT incorporates factors such as idle time and logistic time. As is the case with MTBF, MDT and MTBM are recorded in hours, which is the case for all pharmaceutical components and equipment.

Equation (5.9) is shown as an example of the several types of availability determinations that are employed. While these various availability definitions are particularly relevant in the aerospace and defense industries, such in-depth analysis is not required in the pharmaceutical, biotech, or medical device industries. Also, since most pharmaceutical manufacturing operations usually maintain a parts inventory on-site or have ready access to a parts supplier or vendor, terms such as idle time and logistic time become relatively minor fractions of the availability equation. Consequently, these operations (or lack of operation) can be ignored. Thus, for most pharmaceutical, biotech, and device operations, the most effective relationship for our purposes is Equation (5.8), which is, again,

$$A_i = \frac{\text{MTBF}}{\text{MTBF} + \text{MTTR}} \tag{5.8}$$

For situations where a rapid determination of A_i, MTBF, or MTTR is desired, the nomograph presented in Figure 5.2 can be useful if two of the three variables are known.

When dealing with pharmaceutical, biotech, and medical device system designs, installation, maintenance, and operation, A and A_i are used interchangeably.

A closer look at the relationship reveals some useful information. For instance, when numerical variables are used in lieu of symbols (MTBF and MTTR are the symbols), the result will be a fractional value that can vary from 0 to 1. For instance, if a component exhibits no required repair time during the evaluation period (e.g. 7900 h, assuming a scheduled component replacement on an annual basis), then A_i would be equivalent to 1.0. As a percentage, this would be a 100% availability. This implies that the pump component will always be in operating condition and can successfully perform its intended function. It is also observed that the shorter the MTTR, the higher the availability. For RAM applications, a higher availability is most desirable.

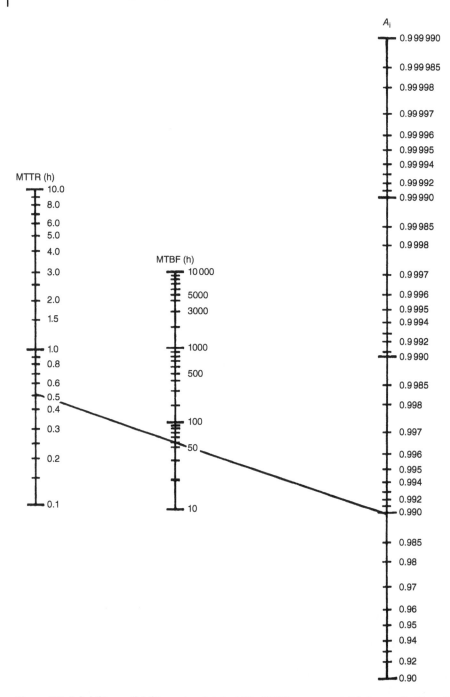

Figure 5.2 Reliability, availability, and maintainability (RAM) nomograph [7]. *Source:* Smith and Babb [7]. Reproduced with permission of John Wiley & Sons, Inc.

In addition to Table 5.2, another useful source for obtaining information on maintenance and repair times for many components involved in pharmaceutical/biotech operations is a document prepared by the US Department of Energy for the Idaho National Laboratory (INL) [16]. Table 5.3 contains repair times predicated on the material contained in the INL publication and,

Table 5.5 Component repair times.

Component	Median repair time (h)
Motor pump	10.0
Flow control valve	4.0
Level control valve	5.0
Gate valve	1.0
Butterfly valve	4.0
Ball valve	2.0
Check valve	2.0
Backflow preventer	4.0
Safety valve	3.0
Y strainer	2.0
Screen mesh filter	0.5
Filter separator	5.0
Air eliminator	1.5
Pole transformer	3.0
Breaker	0.75
Fuse	0.75
Bus	4.0
Cable	2.5
Contactor	1.0
Pipe	4.0
Flow indicator	1.0
Tank	0.03
Human error	0.1

as will be seen, can serve as a practical and speedy resolution to several aspects of RAM needs. While Table 5.5 is predicated on components employed for a specific application (i.e. jet fuel and airport fire equipment), it is a reliable source for pharmaceutical/biotech applications.

With the information provided in Tables 5.2, 5.3 and Figure 5.2, it is possible to determine basic RAM solutions rather quickly and in a practical approach.

Example 5.3   A small contract manufacturer specializing in processing high solids content fluids and slurries is preparing an annual cost estimate to determine maintenance and operations expenditures. The facility uses older processing equipment and seeks to use the existing equipment for another year of manufacturing before purchasing new components. The facility has a number of older motors (eight) of the same design and specification. It would be helpful for budgeting purposes to have an approximate knowledge of the time period the motors will be operable during the year. For budget planning, it would also be helpful to know the number of spare safety valves needed to support the eight rotary pumps in the plant operation. The eight pumps operate an average of 2400 h/year. All of the rotary pumps are equipped with a safety valve in the discharge lines. The most common failure mode of the pumps is the pump motors.

While the facility product line has become more diverse, failure to adequately update the pumps and motors is a concern. Thus, it would be helpful to know what fraction of the time the pump motors will be available for maintenance planning purposes. Similarly, a knowledge of the number of motor spare parts would be helpful to approximate funding needs.

To achieve these objectives, the following information is required:

1) The inherent availability of the motors for the year. (There is a small parts warehouse in the facility making for minimal administrative delays. Maintenance and repair actions will be performed by the in-house maintenance staff.)

Solution
Equation (5.8) and Tables 5.4 and 5.5 are the basis of calculation.
For Equation (5.8), the required variables from Tables 5.4 and 5.5 are

Motors, electric (Table 5.4) $\text{MTBF} = 1/\lambda = 1/1 \times 10^{-5\,\text{h}} = 100\,000\,\text{h}$
Motor pump (Table 5.5) MTTR (median) $= 10\,\text{h}$

Inserting the required variables into Equation (5.8) $(A_i = (\text{MTBF/MTBF} + \text{MTTR}))$

$$\frac{100000\text{h}}{100000\text{h} + 10\text{h}} = 0.9999 \leftarrow \textbf{Solution}$$

The resulting calculation implies that the motors should be operational 99.99% of the processing period (2400 h) and should present no maintenance action.

While Example 5.3 was quite straightforward, it serves to illustrate how available information and a basic relationship can be quite useful for predicting the projected status of common process components. Additionally, as these relationships are used and documented in real-time operations, the RAM characteristics of actual components in a specific facility will lead to a custom database for a particular pharmaceutical/biotech operation without a significant effort. This is even more effective and efficient if available software specific to CMMS applications are incorporated in facility use and training.

The MTBF of a system or component is the responsibility of the quality element of an engineering or maintenance organization. Ideally, the RAM provisions for a component or system should be incorporated at the user requirements specification (URS) and functional requirements specification (FRS) phases of system procurement. Vendors should be capable of providing reliable MTBF information for typical operating systems under the required operating conditions. The MTTR criteria is the responsibility of the assigned engineering or facility maintenance activity since corporate quality assurance organizations are often not equipped to facilitate a technical activity at this level. Rather, it is the responsibility of quality assurance to verify equipment or component acceptability conforms to the cGMP provisions of 21 CFR parts 210 and 211 and part 11.

Reliability is an inherent feature of the system/component that has been procured by the company. Once the unit is installed, any reliability obligation of the pharmaceutical, biotech, or device is, basically, moot. Factors such as MTBF should be addressed prior to system vendor selection. It is difficult, if not impossible, to retrofit reliability into an installed unit. Such an ex post facto operation would create a morass of documentation revision, change control, revalidation issues, and significant cost considerations. For this reason, when considering the addition of a RAM capability, the preferred option is to incorporate reliability requirements and specifications at the beginning of a system proposal (i.e. as part of the FRS and URS).

Effective reliability programs should be designed to accommodate the system needs as well as constraints such as lifecycle costs (utility costs, maintenance and preventive maintenance cost, parts replacement, calibration, amortization, etc.).

As previously noted, reliability is a concept whose responsibility lies with the vendor or vendors selected to design and install the equipment or system. However, the system user or owner should have a basic knowledge of RAM before evaluating proposals incorporating RAM technology. Ideally, the system integrator (the vendor) should have a database with reliability and maintainability information relating to the OEM, as well as vendor-manufactured components in the system. Often, if an ISO 9000 firm is the vendor, the desired reliability information is available. In the bid selection process, incorporating this data as a contract deliverable should be considered.

5.8 Series and Parallel Configurations

Thus far, our analysis of RAM has been limited to single-component examples such as pumps. However, pharmaceutical systems are generally composed of several individual units configured to create an operating system. For example, a tablet packaging operation usually consists of bottle labelers, bottle fillers, bottle cappers, bottle unscramblers, and other units as required by regulation and/or design criteria.

As a rule, the units performing operations (i.e. bottle cappers and bottle fillers) operate in a sequential mode with one operation following another on what, in this case, is identified as a packaging line. Successful operation of the packaging line is contingent on continuous operation of all components in the line. When one component, such as the bottle capper unit, fails, the entire tablet filling operation is inoperable due to a downtime.

This scenario creates a significant expense in lost operating time. In this scenario, while all but one component operates successfully and one unit malfunctions (such as the device that verifies the correct size bottle cap is being employed), production ceases for this packaging line. While the possibility exists that there may be a similar unused packaging line available in the facility, this is a very remote possibility. Also, the time required to start an alternate packaging line is longer than the time required to remediate the inoperable component of the original packaging line (provisioning and locating inventory is often referred to as asset allocation).

5.9 Spares and Replacement Parts

Most large small molecule pharmaceutical manufacturing operations have a building or facility that maintains an inventory of spare parts, replacement parts, or components. Often, these parts operations are maintained by an outside vendor who keeps track of the parts contained in the inventory. Often, the main task of the parts inventory operation is to replace spares and components as they are consumed with little concern for whether or not there are too many parts or too few parts in the parts inventory (another element of asset allocation). If an uncomplicated mathematical relationship can be offered to calculate the required quantity of spares, the relationship could be beneficial in terms of economy and efficiency. In fact, such a relationship, based on known variables, can be derived and conveniently applied.

Recalling Equation (5.3) where λ (also defined as 1/MTBF) is the failure rate constant, a relationship can be defined for a required part in a given time period as

$$F = \lambda t \tag{5.10}$$

where

F = time elapsed until a specific part fails and requires replacement spare part, h
t = annual operating hours of the part, h.

If more than one component is performing the same task (e.g. two identical centrifugal pumps feeding coating solution to a tablet coater), Equation (5.10) can be modified to

$$N_s = n\lambda t \tag{5.11}$$

where

N_s = number of spare parts required, dimensionless
n = number of identical parts performing the same service.

With another slight modification, Equation (5.11) can be modified to a more convenient format by a substitution ($\lambda = 1/\text{MTBF}$) to yield the relationship

$$N_s = \left(\frac{1}{\text{MTBF}}\right)(t)(n) \tag{5.12}$$

Example 5.4 Referring to Example 5.3, determine the quantity of rotary pump relief valves needed for the facility to perform operations for the 1 year period prior to the facility improvements.

Solution
For this example, Equation (5.12) serves as the basis of calculation for determining the number of spare parts required (N_s)

$$N_s = n\left(\frac{1}{\text{MTBF}}\right)(t)$$

N = number of pumps performing the process operations = 8
$\lambda = 1/\text{MTBF} = 1 \times 10^{-5}\,\text{h}$
t = 2400 h of annual operation

$$N_s = 8(0.00001)\text{h}^{-1}(2400\text{h}) = 0.192 \leftarrow \textbf{Solution}$$

The result indicates no spare relief valve would be needed since the number of replacement relief valves is less than one. While the result indicates no need for a spare relief valve, a brief analysis might be performed to consider obtaining a spare relief device if the spare relief device cost is low, and it could be kept if new pump purchases will be compatible with the spare relief valve.

Example 5.5 is another slightly modified version of the previous example. Determining spare parts requirements is an important contribution to providing a cost effective method of eliminating the need for maintaining excess and costly spare parts facilities in pharmaceutical/biotech manufacturing and research facilities.

Example 5.5 Two exhaust fans are employed to remove ether vapors from two separate ether storage facilities operated by an API solvent supplier. The fans discharge the vapors in the storage areas through activated carbon beds prior to atmospheric discharge. Each of the fans is actuated by individual motor starter units, which operate for about 10 min prior to workers entering the storage areas to remove the ether containers for shipping to customers. On average, the motor starters operate 1100 h annually. On average, the motor starters fail after 2000 h of operation. Calculate the number of spare motor starters that are required to maintain in the spare parts inventory. Because motor starters for this application are inexpensive, unit repair is not a consideration.

Solution

Equation (5.12) is the basis of solution for this example. Since the MTBF of the motor starters are 2000 h, λ, the failure rate is $1/2000\,h = 0.0005\,h^{-1}$.

The annual operating hours for the individual motor starters both have the same operating times of 1100 h. The motor starters therefore are assumed to be identical units since the service and service times are the same; the units are identical with the same vendor identification numbers. The variables required for Equation (5.12) are

$1/\text{MTBF} = 0.0005\,h^{-1}$

$t = 1100\,h$

$n = 2$

$$N_s = \left(0.0005\,h^{-1}\right)\left(1100h\right)(2) = 1.1 \approx \textbf{2 spare motor starters required} \leftarrow \textbf{Solution}$$

While the completed calculation indicates one spare motor would be required, it is common to err on the side of caution and round off the fractional value to the next highest value. Since this motor starter is inexpensive and does not occupy a large volume, there is no significant risk in having the extra part as part of the spare parts inventory.

In addition to presenting an introduction to RAM, as well as an ability to perform some of the basic calculations involved, there is also the goal to understand the economic advantages that can be achieved by incorporating an effective reliability and maintainability programs. The "if it ain't broke, don't fix it" approach to maintenance is obsolete. Because of the increased complexity and higher quantity of automated operations in pharmaceutical and biotech operations, a higher level approach to reliability and maintenance is required. For most manufacturing operations, including pharmaceuticals and biotech, 15–60% of all operating costs involve maintenance activities. To address this significant business expense, concepts such as asset allocation and spare parts requirements calculations should be actively integrated in the corporate mindset.

Merely having the capability to determine spare parts/replacement requirements is only a partial solution. To effectively control and manage assets and assure efficient management of maintenance activities requiring monitoring, repair, replacement, and related reliability operations requires a comprehensive strategic asset management plan, which often requires increasing responsibilities for facility maintenance departments. The intent is to maximize the return on assets (RoA) and extend the benefits of the existing asset base. Internal and external pressures – reduced operational and capital budgets, rising material costs, and aging assets – are driving the need to initiate improvements [15]. By planning and implementing a strategic asset management program, the salient variables of a successful operation can be achieved. By integrating a coordinated production, maintenance, and storeroom management program (i.e. asset management), operations, while often remaining in the "putting out the fire" mode, can nevertheless proceed in a more organized and efficient manner.

Three action areas (strategies) have been identified as the relevant components of a successful asset management program. While other sectors can be identified as germane, these three strategies can be initiated in a relatively short time frame and can produce reasonably rapid results [17].

Of course, a successful program requires input from the key operations involved in creating or improving the process. Typically, these basic organizations are maintenance, production, IT, procurement, quality assurance, and health and safety operations.

The first strategy for consideration is the implementation of a criticality-based maintenance activity. Since reliability (Section 5.2 and spare parts requirements (Section 5.9) are the keys to effective asset management, reliable equipment and processes typically result in more efficient operations and often with fewer health and safety issues.

Consequently, by focusing on reliability and identifying equipment criticality as a vital element, effective maintenance is a very germane consideration. And since half of all planned maintenance activities are considered unnecessary, according to a study by the ARC Advisory Group [18], in these cases, prioritizing and identifying critical equipment and provisioning for related spares can significantly reduce downtime and maintenance costs.

A criticality-based maintenance program relies, in large part, on careful economic analysis of lifecycle costs. The decision to repair or replace is generally a function of the maintenance frequency, downtime, and parts replacement costs. Also integral to the evaluation are possible health and safety issues associated with the repair/replace evaluation.

An effective criticality-based maintenance program can be an effective driver for improving the return on investment (ROI) by enhancing safety, maintaining high reliability, reducing maintenance costs, reducing spare parts requirements, and eliminate potential collateral equipment and parts damage.

The second asset management strategy involves effective storeroom management. While appearing to be a rather mundane topic, an effective reliability operation cannot be performed with a high degree of efficiency without effective storeroom management. Section 5.9 has provided a paradigm that can be applied to identifying a spare parts float. However, knowing the spares requirements in quantitative terms offers no assurances that the components will be available when required in a timely manner. Further, effective planning and inventory maintenance is also critical. Storerooms that do not have correct spares when needed and having the correct spares unavailable is not an uncommon occurrence. In fact, according to Rockwell Automation customer data, 50% of downtime can be attributed to insufficient spares, while more than 60% of spares are classified as inactive, excess, or obsolete. In more organizations, maintenance, repair, and operations (MRO), spares range from 15 to 25% of carrying costs. Consequently, stocking inactive excessive or obsolete spares creates unnecessary costs for maintenance operations [17].

Often, a poorly managed spares inventory is a significant contributor to defective storeroom operation. A key to improved storeroom efficiency is the reduction of excess inventory. This is commonly accomplished by writing off obsolete assets. However, this action can result in the failure to replenish the discarded assets. This action can have a detrimental effect on critical parts inventories if a careful analysis is not performed to identify the required critical parts float.

Inadequate provisioning has a ripple effect – e.g. spares for critical equipment aren't available when needed and must be rush ordered, causing downtime to lengthen profitability to suffer as a result. The result is a constant "firefight" to maintain uptime. Not only does this impact a maintenance operation's ability to focus on larger issues such as resource requirements, it ultimately erodes profit margins [17].

Perhaps the most important element in provisioning for critical parts and required spares is to have an updated inventory of parts and spares that are actually available in the storeroom or maintenance facility. An updated and validated CMMS is the best tool to implement this need. With the wide availability of this software at competitive pricing, even the smallest facility can maintain a cost effective CMMS.

A properly provisioned inventory is also an important component of inventory management. While an inventory lacking in the adequate quantity of spares is certainly not desirable, an inventory with too high a spares count is also a burden. As noted, in addition to tying up capital, an excessive parts inventory also consumes facility space. As a result, a parts overload can create configurations where components may be moved from identified locations to other areas. While the intent may be to only temporarily relocate items until a rearrangement repositions the components, it is not uncommon to overlook this "temporary solution."

Having the required spares available at a critical time is also an issue to be considered. If spares and components are not conveniently located to permit timely requisition, valuable downtime is expended. Locating critical parts in convenient locations, other than the prime maintenance location, may be beneficial. For instance, maintaining repair parts and spare parts for a pharmaceutical packaging line in the vicinity of the line, rather than the maintenance facility, could allow for a faster response time. Rather than all packaging operation cease until the parts are available, when both maintenance personnel and repair parts are convenient to the line, lost time is minimized.

Also contributing to storeroom operations is the efficiency of the operation. Components that are clearly identified (i.e. labeled or tagged components that are clearly visible) in defined locations (e.g. an updated floor plan showing the location of, at a minimum, critical parts/spares) in the storeroom area can affect downtime when an operation is halted due to a part failure. In addition to identifying the failed component, speedy identification and access to the replacement part reduces the downtime (MTTR), which can enhance the operational availability (A_o).

The third strategy element involves converging databases or data systems. The main cause between production, maintenance, and storerooms is lack of data continuity. Many companies, including pharmaceutical operations, in the past 10 years have implemented software-driven enterprise resource planning (ERP), which results in often eliminating obsolete components, storeroom management, and maintenance systems in the process.

For many pharmaceutical manufacturing facilities, an ERP capability emphasizes planning, manufacturing, and inventory maintenance as the important elements.

Unfortunately, these systems do not always fulfill the intended objective. As a result, maintenance operations often resort to spreadsheets for tracking, or not tracking at all. By merging ERP systems with production systems, including CMMS, significantly higher productivity and efficiency can be realized from the pharmaceutical plant manufacturing operation (i.e. diagnostics, reliability, asset management, process control) to the storeroom and maintenance operations.

Pharmaceutical operations, as well as other manufacturing processes, using or planning to incorporate ERP with production systems might verify some key items that are evaluated:

- Material masters (inventories) are current to the critical subcomponent level (21 CFR 211.58, 21 CFR 211.105).
- Lead times for indirect materials are identified and up to date.
- Equipment records are accurately maintained (21 CFR 211.105) and are current.

It is most important to understand that pharmaceutical companies be cognizant of the regulatory criteria imposed on ERP software. As part of the selection, the user should be knowledgeable of the impact of cGMP regulations of the validation requirements such as defined in 21 CFR part 11 and presented in more detail in Chapter 10 [17].

When dealing with an ERP option for pharmaceutical applications, the potential vendor/installer selected to install or validate the system should be competent in at least three important implementation criteria. The vendor must provide comprehensive services that extend beyond the standard consulting and recommendations and should be capable of assisting in validation and implementation as well. Such in-depth post-installation support is generally not common in pharmaceutical and biotech vendor installed software-based operation.

The selected ERP system vendor should also have a knowledge of the intended application to a degree such that recommendations, based on the defined applications, are technically sound and acceptable to the user. The vendor should also be capable of clearly defining the benefits, in detail, of the asset management components of the installed system. Specifically, the custom-tailored aspects of the operation must be identified to confirm successful post-installation functioning. This may require extended support subsequent to successful validation and initial operating success. As a result, in-depth post-installation support from IT and management should be integral to the vendor support.

As has been identified above, the user also has the responsibility to verify that equipment records are accurately maintained [19].

A survey question, given to processing operations management personnel inquired, "Do you have a strategy and plan for sustaining reliability at your plant?" The responses recorded that 41.6% of the plants surveyed had a documented plan as part of the business strategy. Another 12% of the respondents indicated that while they did have such a plan, it was not documented. Further, 16.6% of the respondents had no plans to proceed with an ongoing reliability program. While the survey presents a positive approach, the 16.6% of the respondents implies that there is still a large sector of decision makers who are not committed to the need for an ongoing reliability and maintainability program. The survey initiators concluded that embarking on a reliability and maintainability program without evaluating the impact on affected areas of the plant is a recipe for failure [1].

References

1 Hollywood, P. (2012). Making the business case for reliability. *Hydrocarbon Processing* 91 (6): 15.
2 Schnell, K.R. (1984). Risk of modeling spares demand at a maintenance depot using constant arrivals. *Logistics* 19 (3): 20–23.
3 US Army Materiel Command (1975). *Engineering Design Handbook, Maintenance Engineering Techniques*, AMCP 706-132 (June). Alexandria, VA: AMC.
4 US Army Material Command (1976). *Engineering Design Handbook, Development Guide for Reliability, Development Guide for Reliability, Part Six, Mathematical Appendix and Glossary*, AMCP 706-200 (January). Alexandria, VA: AMC.
5 Department of Defense (1966). *Maintainability Program Requirements (For Systems and Equipment)*, MIL-STD-470 (March). Alexandria, VA: Department of Defense.
6 Penrose, H.W. (2012). Mechanical seal failure in pumps. *Maintenance Technology* 25 (6): 37.
7 Smith, D.J. and Babb, A.H. (1973). *Maintainability Engineering*. New York: John Wiley & Sons, Inc.
8 Department of the Army (1976). *Engineering Design Handbook, Maintainability Engineering Theory and Practice*, AMCP 706-133 (January). Alexandria, VA: Department of the Army.
9 Alsup, M. (2012). Don't get misled by modernization misconceptions. *Chemical Processing* 75 (9): 39–46.
10 Bureau of Ships (1965). *Reliability Design Handbook*, NAVSHIPS 94501. Washington, DC: Fleet Electronics Effectiveness Branch, Bureau of Ships.
11 Cunningham, C.E. and Cox, W. (1972). *Applied Maintainability Engineering*. New York, NY: John Wiley & Sons, Inc.
12 Atkins, R.L. (2012). Basic safety considerations. *Maintenance Technology* 25 (6): 32–35.

13 Blanchard, B.S. (1986). *Logistics Engineering and Management*. Englewood Cliffs, NJ: Prentice Hall.

14 United States Nuclear Regulatory Commission (1975). *Reactor Safety Study: An Assessment of Accident Risks in U.S. Commercial Nuclear Power Plants*, WASH-1400 (NUREG-75/014), Appendices III and IV (October). www.nrc.gov (accessed 24 December 2017).

15 Onset Computer Corporation (2012). *Onset®, HOBO® Data Loggers*, Product Catalogue. Bourne, MA: Onset Computer Corporation. www.onsetcomp.com (accessed 24 December 2017).

16 Cadwaller, L.C. (2012). *Review of Maintenance and Repair Times for Components in Technological Facilities*, INL/EXT-12-27734 (November). Idaho Falls, ID: Idaho National Laboratory. www.inl.gov (accessed 24 December 2017).

17 Hermans, M. (2013). Converging production, maintenance and storeroom management. *Maintenance Technology* 26 (11): 40–43.

18 *PdM User Strategy Survey 2012: Current Practices and Future Plans* (August).

19 *ABC Strategy Report, EAM–FSM User Survey 2013: Current Practices and Future Plans* (April).

Further Reading

Aberdeen Group (2013). *Aberdeen Group Analyst Insight Report, Asset Management Using Analytics to Drive Predictive Maintenance* (March). www.aberdeen.com (accessed 24 December 2017).

Carter, A. (1972). *Mechanical Reliability*. New York: John Wiley & Sons, Inc.

Department of the Army (1976). *Engineering Design Handbook, Design for Reliability*, AMCP 706-196 (January). Alexandria, VA: Department of the Army.

Hurlock, B. (2011). The call for zero defects everywhere. *Maintenance Technology* 24 (1): 14–17.

Lewis, E.E. (1987). *Introduction to Reliability Engineering*. New York: John Wiley & Sons, Inc.

Mackenzie, C. and Holmstrom, D. (2009). Investigating beyond the human machinery: A closer look at accident causation in high hazard industries. *Process Safety Progress* 28 (1): 84–89.

Robinson, R.N. (1987). *Chemical Engineering Reference Manual*, 4e. Belmont, CA: Professional Publications, Inc.

Shooman, M.L. (1968). *Probabilistic Reliability: An Engineering Approach*. New York: McGraw-Hill.

6

Parenteral Operations

CHAPTER MENU
Introduction, 279 Parenteral Definitions, Regulations, and Guidelines, 280 Lyophilization, 282 Lyophilizer Maintenance Issues, 294 References, 296 Further Reading, 296

6.1 Introduction

Parenteral drug delivery is the second largest segment of the pharmaceutical market following solid oral dose delivery (see Chapter 7) and accounts for nearly 30% of the market share. In 2015, the parenteral market value was $27 billion, with an expected value of $51 billion in the near future.

Many pharmaceutical companies do not have the resources necessary to manage the increasing complexity of producing and filing parenteral substances. Further, stringent requirements must be followed, including maintaining compliance with US and international regulatory requirements, including current good manufacturing practices (cGMP-21 CFR 210 and 21 CFR 211) to protect product safety, identification, strength, purity, and quality [1].

Parenteral articles are preparations intended for injection through the skin or other boundary tissue, rather than through alimentary canal so that the active substances they contain are administered, using gravity or force, directly into a blood vessel, organ, tissue, or legion. They are produced scrupulously by methods designed to ensure they conform to US Pharmacopeia (USP) requirements for sterility, pyrogens, particulates, and other contaminants and, where appropriate, contain inhibitors of the growth of microorganisms. An injection is a preparation intended for parenteral administration and/or for constituting or diluting a parenteral article prior to administration.

Administration for parenteral products into the body is accomplished by the following specific routes:

1) Intravenous
2) Intra-arterial
3) Intraosseous injection (into the bone marrow)
4) Intramuscular
5) Intracerebral (into brain parenchyma – the areas of the brain containing neurons and myelin producing glial cells)

Practical Pharmaceutical Engineering, First Edition. Gary Prager.
© 2019 John Wiley & Sons, Inc. Published 2019 by John Wiley & Sons, Inc.

6) Intracerebroventricular (into cerebral ventricular system)
7) Intrathecal (into spinal canal)
8) Subcutaneous (under the skin)

6.2 Parenteral Definitions, Regulations, and Guidelines

Parenteral articles are preparations intended for injection through the skin or other external boundary tissue, rather than through the alimentary canal, so that the active substances they contain are administered, using gravity or force directly into a blood vessel, organ, tissue, or lesion, as defined previously. These are prepared scrupulously by methods designed to ensure they conform to pharmacopeial requirements for sterility, pyrogens, particulate matter, and other contaminants. Parenterals may also contain inhibitors to prevent the growth of microorganisms. An injection is the preparation form designed for parenteral administration and/or for constituting, or diluting, a parenteral item prior to administration.

Parenterals have very rigorous nomenclature and definitions, and it is critical to understand and use these terms properly when dealing with parenteral articles.

6.2.1 Nomenclature and Definitions

6.2.1.1 Nomenclature

The following nomenclature, as defined by the U.S. Pharmacopeia (USP 29, <1>), pertains to the five general types of preparations, all of which are suitable for and intended for parenteral administration [2]. They may contain buffers, preservatives, or other added substances:

1) *[DRUG] injection* – Liquid preparations that are drug substances or solutions thereof
2) *[DRUG] for injection* – Dry solids that, upon the addition of suitable vehicles, yield solutions conforming in all respects to the requirements for *injections*
3) *[DRUG] injectable emulsion* – Liquid preparations of drug substances dissolved or dispersed in a suitable emulsion medium
4) *[DRUG] injectable suspension* – Liquid preparations of solids suspended in a suitable liquid medium
5) *[DRUG] injectable suspension* – Dry solids that, upon the addition of suitable vehicles, yield preparations conforming in all respects to the requirements for *injectable suspensions*

6.2.1.2 Definitions

A *pharmacy bulk package* is a container of a sterile preparation for parenteral use that contains many single doses. The contents are intended for use in a pharmacy admixture program and are restricted to the preparation of admixtures for infusion or, through a sterile transfer device, for the filling of empty sterile syringes.

The closure shall be penetrated only one time after constitution with a suitable sterile transfer device or dispensing set, which allows measured dispensing of the contents. The *pharmacy bulk package* is to be used only in suitable work areas such as a laminar work hood (or an equivalent clean air compounding area).

Designation as a *pharmacy bulk package* is limited preparations from nomenclature categories 1, 2, or 3 as defined previously. *Pharmacy bulk packages*, although containing more than one single dose, are exempt from the multiple-dose container volume of 30 ml and the requirement that they contain a substance or suitable mixture of substances to prevent the growth of microorganisms.

Where a container offered as a *pharmacy bulk package*, the label shall (i) state prominently "Pharmacy Bulk Package – Not for direct infusion," (ii) contain or refer to information on proper techniques to help assure safe use of the product, and (iii) bear a statement limiting the time frame in which the container may be used once it has been entered, provided it is held under the labeled storage conditions.

6.2.1.2.1 *Large and Small Volume Injections*

The designation *large volume intravenous solution* applies to a single-dose injection that is intended for intravenous use and is packaged in containers labeled as containing more than 100 ml or less.

6.2.1.2.2 *Biologics*

The pharmacopeial definitions for preparations for parenteral use generally do not apply in the case of biologics because of their special nature and licensing requirements (see Biologics <1041>).

The pharmacopeial monographs involving biologics conform to the Food and Drug Administration (FDA) regulations in covering those aspects that are of particular interest to pharmacists and physicians responsible for the purchase, storage, and use of biologics. The FDA identifies a biological product as "...any virus, therapeutic serum, toxin, antitoxin, or analogous product applicable to the prevention, treatment or cure of diseases or injuries of man. A virus is interpreted to be a product containing the minute living cause of an infectious disease and includes but is not limited to filterable viruses, bacteria, rickettsia, fungi, and protozoa."

Another important FDA definition is the term toxin. The FDA identifies a toxin as a product containing a soluble substance poisonous to laboratory animals or to man in doses of 1 ml or less (or equivalent weight) of the product and have the property, following the injection of nonfatal doses into an animal, of causing to be produced therein another soluble substance that specifically neutralizes the poisonous substance that is demonstrable in the serum of the animal thus neutralized.

Biological products are regulated by FDA regulations contained in 21 CFR Part 600: Biological Products: General, 21 CFR Part 601: Licensing, and Part 21 CFR Part 610: General Biological Products Standards. It is most important to differentiate parenterals from lyophilization, which is addressed in Section 6.3 in further detail.

6.2.1.2.3 *Parenteral Glossary*

Chapter 2, specifically Section 2.1, has a more comprehensive and detailed definition of many of the terms defined in this glossary.

Adulterated: The presence of an impurity that lowers the quality of a preparation and may endanger the well-being of the patient or user.

Ampule: An all glass hermetically sealed container for a parenteral or other sterile container.

Bioburden: The number of microorganisms per unit of material or product that are present prior to sterilization.

Pyrogen: A fever-producing compound of microbial origin. The most potent and widely encountered type is a lipid associated with a polysaccharide and is an endotoxin produced by *Bacillus* microorganisms.

Sterile: The state of freedom from all viable microorganisms, recognized in terms of the probability that the item is free from virtually all microorganisms.

Compliance Policy Guides (CPC) were developed by the FDA as method of spreading

Table 6.1 Guidelines to parenteral drug products.

1) Guide to Inspections of Lyophilization of Parenterals, July 1993
2) Guideline for Submitting Documentation for the Manufacture of and Controls for Drug Products, February 1987
3) Guideline for Submitting Samples and Analytical Data for Methods Validation, February 1987
4) Guideline for Submitting Supporting Documentation in Drug Applications for the Manufacture of Drug Substances, February 1987
5) Guideline for Submitting Documentation for the Stability of Human Drugs and Biologics, February 1987
6) Guideline for Submitting Documentation for Packaging for Human Drugs and Biologics, February 1987
7) Guideline on Formatting, Assembling, and Submitting New Drug and Antibiotic Applications, February 1987
8) Submission in Microfiche of the Archival Copy of an Application, February 1987
9) Guideline for the Format and Content of the Microbiological Section of an Application, February 1987
10) Guideline for the Format and Content of the Nonclinical/Pharmacology/Toxicology Section of an Application, February 1987
11) Guideline for the Format and Content of the Chemistry, Manufacturing, and Controls Section of an Application, February 1987
12) Guideline for the Format and Content of the Summary for New Drug and Antibiotic Applications, February 1987
13) Guideline for the Format and Content of the Clinical and Statistical Sections of New Drug Applications, July 1988
14) Guide for Drug Master Files, September 1989

1) FDA policy to specific operations (i.e. FDA district offices). The purpose of the guidelines is to explain and clarify how the FDA will enforce various regulations specific to parenterals. Table 6.1 lists guidelines that were prepared to assist in understanding and implementing parenteral drug product issues [3]. Copies of these guidelines are available online through www.fda.gov.

6.3 Lyophilization

6.3.1 Background

Though recognized by the public since freeze-dried coffee hit the market 30 years ago, lyophilization has found an important niche in pharmaceutical, diagnostic, and biotech operations. The number of such products has been growing steadily: over 370 biotechnology products are currently in development, with more than 300 sterile product approvals expected and revenues in excess of $50 billion. Traditionally comprising blood products and vaccines, this group has now been supplemented by a panoply of different molecules expressed as the products of molecular engineering in cell culture: growth factors, recombinant vaccine components, and, of course, a profusion of monoclonal antibodies and antibody fragment fusion proteins. However, these biological molecules are by their nature, as well as function of their manufacture and extraction, susceptible to degradation and deterioration. For this reason, the stabilization of these molecules by lyophilization is an effective manufacturing technique [4]. Approximately 25% of these products will be lyophilized.

The advantages of lyophilization include [5]:

- Ease of processing a liquid composition that simplifies aseptic handling
- Enhanced stability of a dry powder
- Removal of water without excessive heating of the product, thus reducing energy costs

- Enhanced product stability in a dry state
- Rapid and easy dissolution of the reconstituted product
- Low particulate contamination
- Less in-process degradation when operating at a low temperature

Some disadvantages of lyophilization are:

- Increased handling and processing time.
- Need for sterile diluent during reconstitution operations.
- Difficult to produce crystalline material.
- Process is long and expensive.
- Cost and complexity of equipment: a new manufacturing lyophilizer system can cost in excess of $1 000 000, while a used or rebuilt system is commonly in excess of $200 000.
- A freeze-dryer can weigh in excess of 54 000 lb and may require significant installation space as well as detailed utilities specifications and diagrams.

Freeze-drying is usually the last option for drying drug products because of the cost and time required.

6.3.2 Lyophilization Glossary

Amorphous phase: A multicomponent phase consisting of uncrystallized solute and uncrystallized water.

Atmospheric pressure: The pressure exerted on the Earth's surface by the atmosphere. One atmosphere is the equivalent of 760 mm of mercury (the pressure exerted at ambient conditions at an elevation of sea level).

Backstreaming: An action that occurs at a low lyophilizer chamber pressure when hydrocarbon vapors from the vacuum system enter the lyophilizer chamber.

Circulation pump: A pump used for circulating heat transfer fluid primarily throughout the freeze-dryer chamber shelves.

Compressor: A lyophilizer component used in the refrigeration operation. The compressor (single stage or double stage) is used to control shelf temperature for both cooling and preventing the shelf temperature from overheating using a temperature controller.

Condenser (cold trap): A vessel that collects moisture on the plates (shelves) and holds it in the frozen state; it protects the vacuum pump from water vapor that could contaminate the vacuum pump oil.

Condenser/receiver: A component that condenses (i.e. changes phase from gas to liquid) and subsequently reverts to the gas phase as part of the refrigeration cycle.

Degree of crystallization: The ratio of energy released when freezing a solution to that of an equal volume of water.

Degree of supercooling: The number of degrees below the equilibrium freezing temperature at which ice first starts to form.

Desiccant: A drying agent used in the freeze–drying operation.

Equilibrium freezing temperature: The temperature at which ice will form in the absence of supercooling.

Eutectic point: On a phase diagram, the temperature and composition coordinate below which only the solid phase exists.

Expansion tank: A tank located in the circulation system used as holding and expansion tank for the heat transfer fluid (not to be confused with the refrigerant in the refrigeration operation of the lyophilizer).

Filter or filter/dryer: There are two systems that have their systems filtered or filter/dried. They are the circulation and refrigeration systems. In newer freeze-dryers, the filter or filter dryer is the same.

Heat exchanger: The heat exchanger is a component of the circulation and refrigeration system and transfers heat from the circulation system to the refrigeration unit.

Heat transfer fluid: A liquid with a specified vapor pressure and viscosity range used to transfer heat to or from a lyophilizer component. Heat transfer fluid is contained in the chamber shelves and the condenser in the freeze-dryer.

Inert gas: A group of gases such as helium, neon, argon, and nitrogen that are chemically inactive. Nitrogen is used in freeze-dryers.

Matrix: A system of ice crystals and solids that is distributed throughout the freeze-dried product.

Melting temperature (melt-back): The temperature at which the mobility of water is detectable in a frozen system.

Nucleation: The formation of ice crystals on a surface or as a result of the growth of water clusters.

Reconstitute: The dissolving of dried lyophilization product (cake) into a solvent or diluent.

Shelf heat exchanger: The transfer of heat from the heat transfer fluid flowing within the shelves to the refrigeration system through tubes in the heat exchanger that, in turn, elevates the temperature of the compressor vacuum to raise the temperature during the freeze-drying operation.

Shelves: In terms of the freeze-drying process, the type of heat exchanger, within the chamber (Figure 6.2), which has a circuitous liquid flow through the shelves, entering at one end and flowing to the other side. They are part of the heat transfer circulation system.

Sterilization: The use of steam to destroy bacteria that may contaminate the lyophilizer or the product.

Torr: A measurement unit equivalent to a pressure of 1000 μm.

Vacuum: A volume where the total pressure is less than atmospheric (<760 mm of mercury).

Vacuum pump: A mechanical technique of reducing pressure in a vessel, such as a lyophilizer chamber, to below atmospheric pressure where sublimation can occur. There are three types of pumps: rotary vane, rotary piston, and mechanical booster.

Vapor baffle: A part inserted in the condenser to direct vapor flow in order to promote an even condensate distribution.

Vial: A small glass bottle with a flat bottom and flange intended for stoppering.

6.3.3 Lyophilizer Design and Operation

Lyophilization, as it is known in the pharmaceutical and related fields, and alternatively referred to as freeze-drying, consists of the freezing of a sterile solution in a sterile container, with the resultant frozen water removed aseptically by sublimation under a vacuum at predetermined and controlled shelf temperatures under vacuum level and times appropriate for the product being lyophilized. Sublimation is a process of removing water from the product after it is frozen and placed under a vacuum and allowing the ice to change directly from solid to vapor without passing through the liquid phase. The final product is a water-free cake that is then reconstituted in a form that is approved for parenteral administration. The basic operation consists of three independent processes: freezing, primary drying (sublimation), and secondary drying (desorption).

Sublimation can be illustrated by using the pressure/temperature phase diagram shown in Figure 6.1.

Figure 6.1 Pressure/temperature phase diagram.

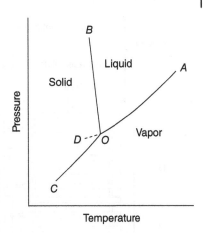

Figure 6.1 presents the equilibrium conditions existing in the three phases in the system (i.e. water, vapor, and ice). Since this is a one-component system, the concentration cannot vary, and Figure 6.1 is actually a plot of pressure versus temperature – in this case a water system. Figure 6.1 is for illustrative purposes and not to scale due to the range of pressures involved. Vapor, liquid, and solid phases are identified in Figure 6.1.

The line OA is the vapor pressure curve for water. Above OA, only water in liquid form is present. Similarly, below the line, only water vapor is present. An equilibrium state exists along the OA line.

The line OC is the sublimation curve of ice. Above OC, only ice is present and below OC, only water vapor exists. Along OC, both ice and water vapor exist in equilibrium.

The line OB is the effect of pressure on the melting point of ice. Since the line slopes to the left, it is observed that the melting point of ice decreases as the pressure increases. Along OB, ice and water, in liquid form, are in equilibrium.

The dotted line OD is an extension of OA and represents the vapor pressure of supercooled water. Dotted lines are used in phase diagrams to indicate less stable equilibria. Such lines are identified as *adjustable equilibria*.

At point O, solid, liquid, and water vapor are in equilibrium with each other. The point at which the three phases are in equilibrium is known as the *triple point*.

As can be observed, the line OA stops at the critical temperature. The theoretical limit of OC is $0\,K$ [6].

More specifically, the freeze-drying operation generally encompasses the following operations:

- Dissolving active pharmaceutical ingredient (API) and excipients in an approved solvent, most commonly water for injection (WFI), which is comprehensively covered in Chapter 2
- Sterilizing the solution by passing the mixture through a $0.22\,\mu m$ bacteria-retentive filter
- Filling the individual sterile containers (also referred to as vials or bottles) and partially stoppering the containers under sterile conditions
- Transporting the partially stoppered containers to the lyophilizer and placing the containers into the chamber under aseptic conditions (Figure 6.2)
- Freezing the solution by placing the partially stoppered containers on cooled shelves in the chamber

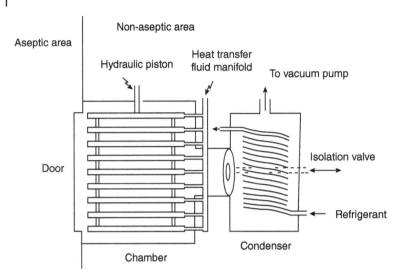

Figure 6.2 Schematic diagram of a freeze-dryer.

- Pulling a vacuum in the chamber and heating the shelves to evaporate the water from the frozen ice
- Completion of the stoppering operation by using a hydraulic (common technique) or a screw rod stoppering mechanism

At process completion, the product will maintain its original properties specific to volume, crystalline.

6.3.3.1 Lyophilization Chamber

The freeze-drying process starts in a lyophilizer unit called the chamber. The chamber filled with sterile nitrogen and the containers are automatically stoppered, providing a sterile product. Figure 6.2 illustrates a common lyophilizer schematic configuration.

Typically, lyophilizers are independently controlled employing computer system consisting of a programmable logic controller (PLC) often integrated with a supervisory control and data acquisition (SCADA) capability for automated processing. In addition to automated operation, units can also be operated manually or semiautomatically though the preparation (steam sterilization) and lyophilization cycles, product loading and unloading, are commonly a manual operation. The loading and unloading doors are under vertical laminar flow to provide ISO 14644-1 Class 5($<3520\,$particles/m^3) conditions.

Freeze-drying identifies three relevant concepts as critical to effective lyophilization design and operation:

1) The sublimation properties of the product are, in large part, dependent on the frozen (crystalline) structure.
2) The structure cannot be altered or modified during operation.
3) Product temperature plays a significant role in the last three of the four freeze-drying procedures, and successful processing is predicated on the selection of operating parameters such as vacuum level, heating/cooling rates, crystal growth rates (nucleation), etc.

A lyophilizer, or freeze-dryer, is composed of the following components:

- A drying chamber, consisting of a vacuum tight unit containing a loading door, inspection windows, a set of jacketed shelves for indirect heating/cooling, a hydraulic stop burning unit and a gas bleed system.
- A heat transfer system capable of producing shelf temperatures commonly ranging from −20 to +75 °C.
- A vacuum system capable of achieving 20 μm (0.02 torr) vacuum. Pressure monitoring of the vacuum is accomplished by using a capacitance manometer type gage or a Pirani device. A Pirani gage uses the electrical resistance of a gas to measure vacuum. For lyophilizers, the effective measurement range varies from 0.5 to 10^{-4} torr. A capacitance manometer uses the dielectric constant of the gas as a basis for vacuum measurement. The capacitance manometer has a range from 0.1 to 10 torr.
- A condenser and associated refrigeration system.
- An internal stoppering system that provides a sterile seal for the product bottle or vial.
- A control unit and installed instrumentation.

The lyophilization process involves the eight primary procedures:

1) Product preparation. This step involves formulating the material in a manner that will allow for start-up of the freeze-drying operation. This step includes equipment checkout, a steam sterilization procedure, product introduction at the correct location, and other procedures performed in accordance with approved operating procedures (SOP), normally supplied by the system vendor.
2) An initial freezing process. It is during this stage the formulation is cooled. At the completion of the initial freeze-drying operation, the product should exhibit the desired crystalline structure. Critical to crystalline structure is the requirement to verify that the composition is frozen below its eutectic temperature (the lowest temperature at which a liquid phase exists in the product).
3) A primary drying or sublimation phase (in specific temperature and pressure above which only ice exists in the product and below, only water vapor is present) where the partial pressure of the vapor surrounding the product is lower than the pressure of the vapor from the ice while remaining at the same temperature. This step requires the energy provided as heat remains lower than the product's eutectic temperature. This thermodynamic phenomenon is basically a phase change that involves transitioning directly from a vapor phase to a solid phase with no liquid phase.
4) A secondary drying aimed at eliminating the final traces of water that remain due to absorption and where the partial pressure of the vapor emanating from the product will be at its lowest levels.
5) Product preparation. This step involves formulating the material in a manner that will allow for start-up of the freeze-drying operation. This step includes equipment checkout, a steam sterilization procedure, product introduction at the correct location, and other procedures performed in accordance with approved operating procedures, normally supplied by the system vendor.
6) An initial freezing process. It is during this stage the formulation is cooled. At the completion of the initial freeze-drying operation, the product should exhibit the desired crystalline structure. Critical to crystalline structure is the requirement to verify that the composition is frozen below its eutectic temperature (the lowest temperature at which a liquid phase exists in the product).
7) A primary drying or sublimation phase (in specific temperature and pressure above which only ice exists in the product and below, only water vapor is present) where the partial pressure of the vapor surrounding the product is lower than the pressure of the vapor from

the ice, while remaining at the same temperature. This step requires the energy provided as heat remains lower than the product's eutectic temperature. This thermodynamic phenomenon is basically a phase change that involves transitioning directly from a vapor phase to a solid phase with no liquid phase.

8) A secondary drying aimed at eliminating the final traces of water that remain due to absorption and where the partial pressure of the vapor emanating from the product will be at its lowest levels.

More specifically, the freeze-drying operation generally encompasses the following operations:

- Dissolving API and excipients in an approved solvent, most commonly WFI, which is comprehensively covered in Chapter 2
- Sterilizing the solution by passing the mixture through a 0.22 μm bacteria-retentive filter
- Filling the individual sterile containers (also referred to as vials or bottles) and partially stoppering the containers under sterile conditions
- Transporting the partially stoppered containers to the lyophilizer and placing the containers into the chamber under aseptic conditions (Figure 6.2)
- Freezing the solution by placing the partially stoppered containers on cooled shelves in the chamber
- Pulling a vacuum in the chamber and heating the shelves to evaporate the water from the frozen ice
- Completion of the stoppering operation by using a hydraulic (common technique) or a screw rod stoppering mechanism

The material of construction for the chambers, shelves, and condenser is type 304 or 316 stainless steel and certified (ASME Section VIII, Division 1, as addressed in Chapter 4) to withstand a design pressure of 20 psig and a full vacuum. This pressure is required for sterilization operations that are performed at a temperature of 121 °C (250 °F). The stainless steel components should also be passivated using 12.5 N nitric acid, as addressed in Chapter 8, Section 8.5.1 ("Passivation").

6.3.3.2 Shelves

There are a specified number of shelves in the dryer chamber. It is common design practice to include an extra shelf, in addition to the number of shelves specified in the design (typically, the extra shelf area varies from 12 to 20 ft^2).

As a general design consideration, provisions for ease of maintenance should be available for all components in all design phases (maintenance activities are addressed in Chapter 5).

The chamber shelves are also constructed of stainless steel equipped with side and rear shelf guards. The shelves contain internal baffles (not to be confused with vapor baffles, defined previously) to optimize the flow of recirculated liquid heat transfer fluid. The shelves are designed to operate under a pressure of about 50 psig. It is critical that shelves are machined to a very flat level surface (approximately +0.01 in., with the shelves machined to a 320 grit or no. 8 finish designation) for effective heat transfer between the shelves and the vials containing the product. The bottom of each shelf is machined flat to the above tolerance, and the surface should be parallel to the top face to assure a smooth surface for the stoppering vials. Also, shelves should be installed parallel to each other so that an even stoppering pressure is maintained.

Shelves are connected by means of stainless steel flexible bellows hoses.

Lyophilizer shelves are hydraulically moveable to allow placement of product vial placement on the shelves. Shelf clearance between the top of one shelf and the underside of the shelf above is typically a minimum of 4 in. determined, in large part, by the specified vial size.

The chamber is also equipped with necessary openings for sterilization, bypass valves, drain, pressure relief, and vacuum break valves. These components are required to provide a break in the drain line to preclude the drain line contents from backing up or clogging the chamber or condenser.

As might be expected, internal fittings in the chamber are fabricated of stainless steel.

The chamber bottom is normally sloped toward the front of the chamber with a "dam" and door lip shield intended to channel condensate from the dryer chamber door to the front screened drain. These components are used during sterilization operations as well as permitting easier cleaning of the chamber. Additionally to facilitate cleaning and maintenance of the bottom of the chamber area, the bottom section of the shelf assembly is capable of being lifted approximately 16 in.

6.3.3.3 Condenser

Also identified as a key component of a freeze-dryer is the condenser. As is the case with the chamber design, the condenser is designed and ASME code stamped to withstand a minimum pressure of 20 psig and can withstand a full vacuum. The condenser components (e.g. condenser plates and internal piping) are fabricated, as noted, of 304 or 316 stainless steel. For certain components, such as weld neck flanges or butt weld spool pieces, welding grades 304L or 316L should be considered.

The condenser is also equipped with integral drying units to minimize space requirements (the "footprint").

A condenser has connections for the vacuum outlet, a steam inlet, a defrost discharge, a water outlet, a relief valve, and thermocouple connections. Typically, T-type thermocouples are used to measure lyophilizer product temperatures. For type T, the operating temperature range is −328 °F (−200 °C) to 662 °F (350 °C). Additional information on thermocouples and similar temperature measuring equipment is located in Chapter 11, specifically Section 11.3.

The condenser is also equipped with baffles to prevent "ice bridging" between condenser plates.

When considering condenser design, the choice is between an internal design and an external configuration. The internal design has condenser plates or coils that are mounted on the side walls of the unit or in the bottom of the chamber. The external design features a separate condenser unit also fabricated with either refrigerated plates or coils and connected to the dryer chamber via a short large diameter stainless pipe equipped with an isolation valve that isolates the dryer chamber from the condenser.

While there are some advantages to the internal condenser design (e.g. lower cost, minimal footprint or floor space, a less detailed sterilization procedure), most manufacturing lyophilizers employ the external condenser design. The primary rationale for an external condenser design is that when it is located between the product (shelves) and the vacuum pump, more efficient collection of water vapor and condensate occurs. By trapping water vapor efficiently, the volume of vapor reaching the vacuum pump is minimized. Similarly, an external condenser design minimizes backstreaming of hydrocarbons (defined in Section 6.3.2) originating at the vacuum pump and terminating at the freeze-dryer. Additionally, the external condenser configuration produces an even distribution of ice on the external condenser surface area, resulting in a more efficient operation. Further, the internal condenser installation has a proclivity to form ice on the side of the plate nearest the product shelf. Another potential advantage of an external condenser is the fact that a batch of product may be processed prior to the condenser being completely defrosted. Of course, a successful operation is contingent on several factors, including confirmation that the freeze-dryer is adequately sized.

6.3.3.4 Stoppering System

The stoppering system is an internal hydraulic system that uses the product shelves as stoppering plates. In addition to moveable stainless steel shelves, the stoppering installation contains:

- A mounted stop/start switch and pilot light
- An adjustable stoppering pressure control that allows the operator to adjust and lock the stoppering system at the specified stoppering pressure
- A safety relief valve or similar approved mechanism

The stoppering system should also be capable of moving shelves from their full clearance to a minimum of 1¼ in. under full stoppering.

Vial selection can be a very important variable in efficient lyophilizer selection and operation. Failure to properly specify the optimal vial size can affect both productivity and performance. For instance, by not verifying the correct bottle size for individual shelf loading, production quantities can be significantly affected.

Sizing vials to accommodate shelf parameters is standard in freeze-dryers when purchased as new equipment. However, some contract manufacturing operations (CMO) may choose to purchase a rebuilt lyophilizer for economic and/or other criteria. In this instance, it is critical that the lyophilizer vials are suitable for the shelves of the purchased unit.

6.3.3.5 Refrigeration System

The freeze-dryer refrigeration system is commonly equipped with the following components:

- Two-stage (2) compressors with installed piping. The compressor is designed for direct expansion into the condenser plates. (The four phases of the refrigeration cycle are evaporation, compression, condensation, and metering.) The compressors are interconnected such that either can function as a backup capability in the event of one compressor failure. The interconnection is configured to permit one or both of the compressors to assist the shelf heat transfer compressor during the initial chilldown of the transfer medium and pre-freezing of the product.
- An expansion valve
- A heat exchanger
- Non-elastomeric couplings
- Refrigerant

The refrigeration system provides for rapid cooling of the chamber and condenser, subsequent to sterilization through shelf cooling. The refrigeration unit should be capable of providing a condenser temperature of at least −70 °C.

The refrigeration unit is also equipped with a dryer strainer or a trap and an indicating glass, installed ahead of the expansion valve.

For many years, the refrigerant most commonly used was known as R-22 (chlorodifluoromethane). However, R-22 is being phased out as a refrigerant in developed countries due to the compound's ozone depletion potential. R-22 has also been identified as a greenhouse gas. As a result, R-22 is, as noted, being phased out in new equipment (including lyophilizers) and replaced by other refrigerants such as propane and other blends (two refrigerants, R-424A and R-422D, are used for low temperature applications). The US EPA estimates R-22 inventories will be depleted by 2020. As a result, design consideration should be given to existing freeze-drying systems. New refrigerants do not possess the same physical and thermodynamic properties as R-22. As a result, the refrigeration units may experience a decrease in operating efficiency. This is a design consideration that may influence present and future existing refrigeration system operations. Consequently, consideration should be given to the possible need for refrigeration unit modification or replacement, if operation is seriously affected.

6.3.3.6 Vacuum System

The vacuum system typically consists of the following: two vacuum pumps complete with electric motors, stainless steel piping with gaskets suitable for operation to −70°C (silicon rubber or equivalent), remotely operated high vacuum valve installed between the chamber and the condenser, remotely operated valves located at the condenser outlet and on the chamber side of the high vacuum, and condensate pumps, primarily required for process and testing requirements.

The vacuum pumps are sized for normal operation of both units. The pumps are of two designs – a rotary vane belt drive oil pump (wet pump), the most common type, or an oil-free belt drive dry pump. Lyophilizers are usually the oil type. Oil-type pumps have the disadvantage of producing vapors from the water, and the condensate mixes with the oil that can significantly affect pump performance without an installed treatment unit. However, rotary oil pumps are more commonly used since they are less expensive than equivalent oil-free designs.

Because maintaining the required vacuum is critical to effective lyophilizer operation, the extra (spare) pump is a common component. In the event of the failure of one unit, the remaining operating vacuum pump can continue to function at the specified vacuum level. While normal operations in the degraded mode occur, additional time may be required to achieve the specified operating vacuum.

6.3.3.7 Shelf Heat Transfer System

The purpose of the shelf heat transfer system is for recirculation of the heat transfer fluid to the product shelves. Until recently, a common heat transfer fluid was trichloroethylene (C_2HCl_3). However, trichloroethylene has been phased out due primarily to environmental concerns. Currently, silicone oil and a commercial fluid (Lexol) are approved alternatives.

The shelf heat transfer system includes the following components:

- A compressor with an installed oil separator
- Heat exchangers
- Expansion valves
- Electric immersion heaters fitted with temperature controls
- Moisture-proof shelf covers
- A control panel with a manually operated switch
- Interconnecting stainless steel piping
- A heat transfer fluid circulation pump
- A heat exchanger
- Desiccant-type dryer strainer
- Pressure gages
- Isolation valves installed for maintenance of the dryer strainer desiccant element, thus obviating the need for draining the system

Cycle times for a freeze-drying operation are crucial for efficient product manufacturing. The time required to remove water from the product is a major factor of the cycle time. The time required to produce an acceptable batch is a key to manufacturing profitability. Sublimation, the phase change from ice to water vapor, is the key to producing an acceptable moisture-free dry product. While it would appear that this phase change would be the rate-limiting variable in the cycle, other factors are at play. The phase change is a speedy process. However, since sublimation is a thermodynamic phenomenon and independent of time, drying time is very much a function of the rate of heat transfer from the shelf to the product in the vial. A great deal of this heat transfer mechanism is accomplished by conduction of heat, a slow heat transfer when compared with heat transfer via convection (both topics are comprehensively detailed in Chapter 2, specifically Sections 2.3 and 2.4). The heating of the product on the shelf is primarily a conduction process. Also, cooling the product in the condenser is an operation requiring additional time for completion. While convective heat transfer also occurs in the chamber, during processing, the convective heat transfer primarily involves molecular heat transfer in a gas phase, a very slow operation.

It would be helpful to evaluate the cycle time required to obtain a dry cake when performing freeze-drying operations and compare it with actual cycle times. A relationship is available to calculate the evaporation rate (sublimation) for a substance undergoing lyophilization. The equation is [3]

$$W = 5.83 \times 10^{-5} P_\mu \left(\frac{M}{T}\right)^{1/2} \tag{6.1}$$

where

P_μ = vapor pressure in μm Hg
T = temperature (K)
M = molecular weight of subliming compound (in most cases, the substance is water)

Example 6.1   A freeze-drying vial with an internal diameter of 1.8 cm and contains 2.0 g of ice at −18 °C. Determine the rate of evaporation and the theoretical time required for complete drying, i.e. cake formation.

Solution
Equation (6.1) is the basis of calculation for this example:

$P_\mu = 939 \, \mu m$ Hg
$M = 18$
$T = 255°K$

Inserting the given variables for Equation (6.1) yields the evaporation rate

$$W = 5.83 \times 10^{-5} (939) \left(\frac{18}{255}\right)^{1/2} = 0.0146 g/cm^2 \, s$$

Evaporation predominantly occurs at the surface of the vial contents, i.e. the top area of the vial. The top surface area of the vial is given by

$$A = \frac{\pi D^2}{4}$$

where

A = area in cm
D = vial diameter = 1.8 cm

$$A = \frac{3.14159 \left(1.8 cm\right)^2}{4} = 2.544 cm^2$$

The time required to dry 2 g of product is

$$\left[\left(2g\right) \times \left(1 cm^2 s\right)\right] / \left[\left(0.0145g\right)\left(2.544 cm^2\right)\right] = 54.34 \, \textbf{seconds} \leftarrow \textbf{Solution}$$

As indicated by the answer to Example 6.1, the theoretical drying time for 2 g of water removal via lyophilization is quite rapid. However, the actual drying time for 2 g of product is in the 4–8 h range. As a rule, completion of a production lot of freeze-dried product requires at least one shift (8 h). However, it is not uncommon for a lyophilization, depending on the product, to require several shifts to complete the process.

6.3.3.8 Automated Loading and Unloading Systems

Autoloader systems can reduce the risk of product contamination during freeze-drying operations. Operators who load a lyophilizer in an ISO Class 5 cleanroom are a major source of contamination [7]. Just as an aseptic product is required for freeze-drying products, the vial loading operation is also required to be free of contaminants prior to lyophilizer operations. To minimize product contamination of freeze-dry drug products, automated loading products (autoloader) have found wide use in the freeze-drying process.

There are three types of autoloaders. The first is the batch loader, which constitutes a stack of shelves of equal capacity to the shelf stack in the lyophilizer. Vials discharge from the filling line and accumulate on the shelves of the autoloader. Once full, a vehicle travels on tracks to the front of the freeze-dryer. When the chamber door is fully opened, the loader clocks and pushes the vials into the shelves. To remove the vials, the vehicle removes them, one shelf at a time. When full, the vehicle proceeds to the next finishing operation.

A more common type of autoloader is the shelf loader. This unit is a vehicle that can move on the floor or a track. This system is most suitable for placing the product on a cold shelf through a sub-door of the freeze-dryer. The vials are then transferred on the chamber shelf, at the rate of one shelf per cycle. The vehicle then returns to the main vial supply station where the shelf loading process is repeated until all shelves are filled with vials. Upon completion of the lyophilization cycle, the vials are retrieved via a device that transfers the vials through the sub-door and on the vehicle. Again, the process is repeated until all the shelves are cleared of vials.

The third autoloader design is intended for operation in small locations and is known as a row loader. This unit is also for loading product on a cold shelf through a sub-door. For this design, a conveyor is used to transport vials from the filling line. The unit is placed in front of the freeze-dryer. Using an indexed counting device, the vials are sorted in a row equivalent to the width of a shelf. A transfer platen then extends through the sub-door and positions next to the shelf. A pushing device then moves the row toward the shelf, thereby allowing for enough space for the next row to accumulate. The operation is continued until the shelf is full. The operation is repeated until all shelves are full.

Row loaders use an extraction device to remove the vials from the shelf, also using an indexing, or counting mechanism. The operation continues until all shelves are empty.

It is important to evaluate the efficacy of purchasing an autoloader. Since an autoloader is expensive, acquiring an autoloader may not be practical for many companies, particularly CMO. Most autoloaders are used by the largest drug companies, primarily due to the cost factor. However, for operations where cGMP compliance is an important factor when minimizing human involvement in the process is a concern, an autoloader may be worthy of consideration [7].

6.3.3.9 Control Panel and Instrumentation

The freeze-dryer control panel is control center for the lyophilization process. The control panel is remotely mounted with required switches, gages, and pilot lights for the rotary compressor, the vacuum pump, electric heaters, remotely operated vacuum valves, hydraulic pumps, the heat transfer system circulation pumps, and a piping schematic displaying the controls and pilot lights in an ergonomically usable configuration that is easily accessible to manually operated controls. In addition to the identified controls and pilot lights, the following should also, as a minimum, contain:

- Shelf and condenser temperature controller
- Chamber and condenser vacuum controller to allow for stepwise temperature and vacuum control installed with a manometer capable of measuring the vacuum in the chamber
- Vacuum indication equipped with elapsed hours operating meters for refrigeration and vacuum pump motor
- Hour meters to monitor refrigeration and vacuum pump motors
- A multichannel recorder (a minimum of 30 channels) to record temperatures and pressure of various probes and components
- Panel-mounted motor starting equipment
- Process gages
- System malfunction indicator and alarm with terminal strips connected to the installation malfunction alarm and the control center or SCADA unit
- Ammeter to monitor electric heaters and power consumption (volts × amps = watts)
- Pressure gages for oil and refrigeration metering pressure for the compressor
- Condenser evaporator pressure gage mounted on or adjacent to the refrigeration system

6.3.3.10 Utilities

The lyophilizer basic electric power characteristics are 3 phase, 60 Hz, and 460 VAC. The freeze-dryer is also equipped with step-down transformers, or another source, capable of supplying 1 phase, 60 Hz, and 115 VAC. This lower power requirement is principally for controls and instrumentation applications. Electrical equipment such as motor starters, circuit breakers, and transformers are prewired and housed in NEMA 12 enclosures (NEMA classifications for motors are presented in Chapter 3, specifically Section 3.1.2).

Other required freeze-dryer utilities include:

Clean saturated steam supply, nominally 60 psig (151.7 °C = 305 °F)
Sterile air, 100 psig
Instrument air, 80 psig
Sterile nitrogen, 75 psig
Cooling water, ≈85 °F (29 °C)

6.4 Lyophilizer Maintenance Issues

6.4.1 Maintenance Systems Analysis

Maintaining freeze-dryer components in proper working order is a key to assure the equipment is not a significant cause of process failures. Establishing an effective maintenance program (Chapter 5, Section 5.3) as required in 21 CFR Part 211.67 (Equipment cleaning and maintenance) is critical to assuring satisfactory lyophilizer operation. Scheduled maintenance checks and testing procedures for cGMP systems can provide reliable, uninterrupted, and compliant processing. Thus, addressing specific lyophilizer components and some germane component maintenance issues could be beneficial [8].

An often overlooked item of freeze-dryer maintenance is the computer control system, the PLC/SCADA system. To maintain peak efficiency, backup or archived files should be periodically updated as per the facility SOP. Additionally, files should be backed up at least once per month, preferably, more frequently. This action will permit recovery of data in the event of a computer system malfunction. Further, computer data processing, such as defragment hard disk operations, should not be performed to coincide with freeze-drying operations.

The computer system should be a dedicated operation, and it is not a recommended practice to use for other applications. Computer system maintenance activities must also conform to the provisions of 21 CFR Part 1, "Electronic Records; Electronic Signatures," when performing computer system maintenance activities.

The instrumentation installation consists primarily of the temperature and pressure sensors in the lyophilizer containing 2 or 3 types of instrumentation that require scheduled calibration, primarily of the temperature sensors and the pressure sensors. It is most important to verify that correct instruments are used for specific operations. For instance, T-type thermocouples monitor product temperatures and shelf heat transfer fluid temperature. Should replacements be required, the component should conform to the criteria set forth in 21 CFR Part 211.68, "Automatic, mechanical, and electronic equipment." Most lyophilizers use several instrumentation installations requiring scheduled calibrations (usually at a 6-month interval). The calibrations are performed in accordance with approved procedures. Vacuum sensors should also be calibrated in accordance with an approved procedure at regular intervals.

Calibration procedures are generally performed using an in situ non-complicated check. As an example, thermocouple calibration can be accomplished by using crushed ice (produced from USP grade water). The basis is a temperature of $0\,°C$. If sensors other than thermocouples are used, it may be necessary to use an electronic simulator for calibration operations.

A common calibration method for both the capacitance manometer and a Pirani gage is running the freeze-dry unit on empty.

Vacuum pump design and operation has been highlighted in Section 6.3.3.6. For oil-type vacuum pumps, the oil change frequency is a key maintenance parameter. Most wet vacuum pump manufacturers recommend changing the pump oil between 2000 and 3000 operating hours [8]. However, in some freeze-drying applications, the time between oil changes should be more frequent. Should the freeze-drying operation employ solvent other than water, a more frequent oil change may be required. When performing routine maintenance tasks on the lyophilizer, a simple evaluation is to observe the color of the subsequent to a production run. If the color is visibly darker, an oil change should be performed at the completion of the processing cycle. The color change is a possible indication that solvent is in the vacuum unit.

It would also be helpful to check the vacuum system trap for acceptable operation. In addition to capturing spurious solvents, the trap also serves to protect the vacuum pump and should, for most effective operation, be placed between the freeze-dryer and the vacuum pump. Maintenance normally involves visual inspection of the configuration [9].

Routine maintenance of the lyophilizer chamber door should routinely include visual inspection. This action includes verifying the manifold, and chamber door silicon rubber gaskets show no visual signs of dirt or spurious contaminants. These gaskets should be free of grease or oil since they do not require lubrication, although, on occasion, small amounts of vacuum grease may be used. If such is the case, the grease should be removed at specified time intervals, or unless visual contamination is observed, with new grease being applied to obviate dirt or contaminant buildup over time.

Vacuum system leak tests are a critical maintenance procedure. A typical vacuum leak test procedure involves evacuating a clean dry freeze-dry system from 1 atm pressure to a vacuum level of about $100\,\mu m$ and maintain that vacuum level for a period of time (typically, about 15 min). The system is then evacuated to about $20\,\mu m\,Hg$. The leak rate should not exceed about $3\,\mu m/min$, commencing at $25\,\mu m\,Hg$.

During the maintenance procedure of the vacuum system, the malfunction alarm should be tested to verify whether the unit is properly functioning. Similarly, check valves should be visually inspected for integrity and functioning.

Maintenance of the refrigeration system is dependent on the test mode employed. For newer units, equipped with an automated built-in test (BIT) capability, the automated functioning test should be performed on a scheduled basis, typically every 6 months, or quarterly.

Should the freeze-dryer not be equipped for automated functioning, an approved specified maintenance/preventive maintenance procedure should be employed incorporating information detailed in Chapter 5. Of particular relevance, Table 5.1 and Section 5.4 may be of value. While performing refrigeration system maintenance operations may appear to be cost effective, the manufacturer often provides contract assistance to provide some or all maintenance tasks.

The chamber also has some maintenance considerations of note. Cleanliness is a germane consideration. It is very important to stress the need to maintain clean shelf surfaces in particular and a clean chamber in general. An approved cGMP procedure should be in place for chamber and shelf cleaning subsequent to each production run. Part of the operation should include inspecting for broken glass, stoppers, and empty vials. It would also be prudent to inspect the chamber unit prior to starting freeze-drying operation [8].

References

1 Husain, S.T. (2015). Bolstering capabilities for parenteral drug development and manufacturing. *Pharma's Almanac* (October 2015), pp. 34–36 (a supplement to the October 2015 issue of *Pharmaceutical Technology*).
2 U.S. Pharmacopeia (2008). USP 29, Chapter <1>, Injections. Rockville, MD: U.S. Pharmacopeia, p. 2455. www.usp.org.
3 Avis, K.E., Liberman, H.A., and Lachman, L. (1993). *Pharmaceutical Dosage Forms; Parenteral Medications*. New York: Marcel-Dekker.
4 Matejtschuk, P., Malik, K., Duru, C., and Bristow, A. (2009). Freeze drying of biologicals: Process development to ensure biostability. *American Pharmaceutical Review* 12 (2): 12–18.
5 U.S. Food and Drug Administration (1993). Lyophilization of Parenteral (7/93) (July). Silver Spring, MD: U.S. Food and Drug Administration. www.fda.gov (accessed 24 December 2017).
6 Kittsley, S.L. (1955). *Physical Chemistry*. New York: Barnes & Noble, Inc.
7 Roedel, W. (2004). Automated loading and unloading for lyophilizers. *Pharmaceutical Technology Supplement – Lyophilization*, 28–35.
8 Dellinger, M. (2012). Basic lyophilizer maintenance helps assure good results. SP Scientific (December) www.spscientific.com (accessed 24 December 2017).
9 Sprung, J. (2014). Top 5 mistakes made in the lyophilization process. *Labconco News* (6 March).

Further Reading

Bucur, B. and Smith, T. (2009). Low risk lyophilization. *Pharmaceutical Manufacturing* 8 (3): 33–36.
Challener, C.A. (2014). Predicting lyophilization performance. *Biopharm International* 27 (3).
Gitig, D. (2011). Vaccine development and production. *Genetic Engineering & Biotechnology News* 31(7): 37–39.

Ho, K., Tchessalov, S., Kantor, A., and Wayne, N. (2006). Development of freeze and thaw processes for bulk biologics in disposable bags. *American Pharmaceutical Review* 11 (5): 64–70.

Jennings, T.A. (2009). Back to the future of lyophilization for 2009. *American Pharmaceutical Review* 12 (3): 20–25.

Johnson, R.E., Teagarden, D. L., Lewis, L.M., and Gieseler, H. (2009). Analytical accessories for formulation and process development in freeze – drying. *American Pharmaceutical Review* 12 (5): 54–60.

Nail, S. and Searles, J.A. (2008). Elements of quality by design and scale-up of freeze dried parenterals. *Biopharm International* 21 (1): 44–52.

Riordan, W. (2012). Principles of freeze drying. Yardley, PA: American Lyophilizer, Inc. www.freezedrying.com (accessed 24 December 2017).

Sprung, J. (2014). Maintaining your freeze dryer and vacuum pump. *Labconco Lab & Science News* (30 October). www.labonco.com/news.

Teixeira, J. (2005). Lyophilization: Growing with biotechnology. *Genetic Engineering News* 25 (16): 52–54.

7

Tableting Technology

CHAPTER MENU

7.1   Introduction

In addition to acquiring a background in tablet manufacturing and related operations, this chapter is intended to provide an overview of the role of the Food and Drug Administration (FDA) in the manufacturing and storage of drug products. Some of the material has been noted in previous chapters in a general reference mode. This chapter, as well as Chapter 6, should provide a more specific insight in the two of the most common methods of pharmaceutical dosage, including biotech, preparations.

Tableting operations, as will be seen shortly, encompass pharmaceutical powder flow, an important variable in tablet and capsule manufacture. Pharmaceutical products are particularly important since approximately 80% of manufactured pharmaceutical items are powder originated products.

As more attention is given to FDA initiatives such as process analytical technology (PAT) and quality by design (QbD), these topics (further addressed in Chapter 11) will clarify why emphasis on evaluating pharmaceutical powder flow evaluation techniques that allow for continuous analysis of in-process manufacturing is important. Powder flow properties are critical to obtaining rapid, real-time process information during critical steps in tableting operations.

In this section, emphasis is given to the equipment design, operating procedures, and evaluation methods involved in performing tableting operations since, in addition to being costly, an error in equipment selection can seriously affect the profitability of a tableting operation by reducing the yield and product quality.

Practical Pharmaceutical Engineering, First Edition. Gary Prager.
© 2019 John Wiley & Sons, Inc. Published 2019 by John Wiley & Sons, Inc.

7.2 The Role of the FDA in the Manufacturing, Processing, Packing, and Holding of Drugs: The Relationship Between Regulations and Pharmaceutical Engineering

Tablets are, arguably, the most recognizable forms of medication. Typically, little thought is given to how these dosage forms are prepared in such large, reliable, and consistent quantities. Yet, tablet manufacturing often employs pharmaceutical-like equipment also used in other diverse processes such as manufacture of certain confectionary and fertilizer products.

Unlike other manufacturing applications that may use equipment similar to pharmaceutical manufacturing, drug manufacturing operations are subject to Federal Regulations that are defined in 21 CFR, which, among other responsibilities, regulates drug manufacturing via the FDA regulations. In addition to regulating design and operations of drug manufacturers, the FDA is also responsible for assuring the products adhere to requirements for safety, product identity, strength, and quality and purity, as defined in the requirements presented in 21 CFR 210.1, status of current good manufacturing practice (cGMP) regulations.

The cGMP regulations presented in 21 CFR Parts 210 and 211 and ICH Q7A (International Council on Harmonization) are, in large part, the differentiation between standard engineering practice and what is commonly referred to as pharmaceutical engineering. The cGMP regulations can serve as a basis for defining the functional requirements specification (FRS) for many drug manufacturing operations and are requirements that, if not properly implemented, can result in FDA-administered penalties such as a 483 warning or the more serious consent decree, with significant financial penalties and oversight actions implemented until the error or errors are corrected.

Additional cGMP regulations, specific to drug manufacturing operations, identified in 21 CFR Part 211 include criteria such as:

- 21 CFR 211.42 (Subpart C): Design and construction features – This part of the regulations addresses the buildings (e.g. size, location, cleaning, and maintenance) processing, packing, and storage of drug products. Also covered by this subpart are additional regulated details such as floors, walls and ceiling construction and materials, temperature and humidity controls, air supply, airflow, and filtration requirements. (Many of the issues described herein are presented in more detail in other chapters, including Chapters 2–4. They are again referenced to highlight the cGMP criteria for specific drug manufacturing operations rather than perceiving 21 CFR as general nonspecific criteria.)
- 21 CFR 211.46 (Subpart C): Ventilation, air filtration, air heating and cooling – this element of Part 211 is specific to HVAC systems and their applicability to drug manufacturing.
- 21 CFR 211.56 (Subpart C): Sanitation – the germane criteria to ensure and confirm buildings involved in the manufacture, processing, packing, or holding of a drug shall be maintained in a clean and sanitary condition. Additionally, these criteria must be addressed in approved written procedures. This includes disposal of trash, organic wastes, and critical components that are involved in the manufacturing process; further details of waste disposal are addressed in Chapter 9. The need for specifics such as rodenticides, insecticides, fungicides, fumigating agents, and cleaning and sanitizing agents are specified in paragraph c of 211.56.
- 21 CFR 211.58 (Subpart C): Maintenance – specifically denotes the need for any building involved in the manufacture, processing, packing, or holding of a drug product to establish and implement a maintenance operation as an integral activity of the building.
- 211.65 (Subpart D): Equipment construction – While Subpart C is focused on cGMP activities for buildings, Subpart D addresses the equipment involved in drug manufacturing, processing, packing, or holding of drug products. The germane requirement of 21 CFR 211.65 is paragraph (a) that states "Equipment shall be constructed so that surfaces that

contact components, in-process materials, or drug products shall not be reactive, additive, or absorptive so as to alter safety, identity, strength, quality, or purity of the drug product beyond the official or other established requirements."

While many types of equipment exist to perform the particular drug processing operation required, they must be validated and approved in accordance with cGMP regulations prior to performing operations related to pharmaceutical processing. As a result, while equipment such as mixers may be widely used in industry, if they don't comply with the cGMP criteria, applications for ethical drug operations will be problematic, at best.

- 21 CFR 211.80 (Subpart E) – this subpart, titled "Components and Drug Product containers and Closures: 211.80 General Requirements" focuses primarily on the need to verify drug product containers and closures "…shall at all times be handled and stored in a manner to prevent contamination."
- 21 CFR 211.84 (Subpart E) – Testing and approval or rejection of components, drug product containers, and closures. This part of 21 CFR 211 paragraph (a) identifies that the need for each lot of components, drug product containers, and closures will be withheld from use until the manufactured lot has been sampled, tested, or examined, as appropriate, and released for use by the quality control unit.
- 21 CFR 211.113 (Subpart F – Production and Process Controls) – Control of microbiological contamination. This section of 21 CFR 211 is also referenced in Section 3.3.1, which introduces the criteria involved in aseptic operations, a critical element of drug manufacturing. In Chapter 3, 21 CFR 211.113 focused on paragraph (b) that is specific to control and prevention of microbiological contamination by requiring approved design and validation procedures by requiring an aseptic procedures and designs. Paragraph (a) is more general by addressing drug products not required to adhere sterilization criteria but nevertheless clearly defining the need for approved written procedures designed to prevent objectionable organisms in drug products.
- 21 CFR 211.130 (Subpart G) – Packaging and labeling operations – Once manufactured, the pharmaceuticals must be packaged and labeled to preclude mix-ups, cross contamination, or distribution errors. Part 211.130 states that the need to provide written procedures here shall be written procedures designed to assure correct labels, labeling, and packaging materials are used for drug products. Also required are approved written procedures that shall be followed. To further prevent mix-ups and cross contamination, physical or spatial separations are required to separate individual drug operations. An additional requirement identified in 21 CFR 211.130 includes the need to identify unlabeled products that will be packaged and labeled at a future date (21 CFR 211.139(b)). Additional design and operational regulations specific to 21 CFR 211.130 address manufactured drug lot identification (c) (including a unique identifier) as well as a traceable documented history of the lot. The importance of packaging materials examination labeling and materials, coupled with specific documentation, is also specified in 21 CFR 211.130(d). The need to inspect packaging and labeling facilities, prior to operation to verify drug products from previous operations, has been removed. Additionally, the need for an inspection to assure packaging and labeling materials that are not suitable for subsequent operations has been removed (e).

The above regulations are specific required design criteria for drug manufacturing operations in general and not solely for solid dosage products. However, these design considerations are, in this chapter, focused on tableting operations and related solid dosage operations such as capsule filling. Chapter 6 will address these regulations as applicable to liquid manufacturing and lyophilization operations in further detail.

Drug products administered in the form of creams, ointments, ophthalmic (drops), otic (ear drops), suppositories, transdermal patches, and inhalers are, if identified as ethical pharmaceuticals, also subject to FDA regulations, including 21 CFR Part 211.

As pointed out, many of the provisions of 21 CFR Part 211 are directed at pharmaceutical equipment design and construction. While noted, the design and construction features identified should be incorporated in design criteria and specifications. There are also many elements though not specifically identified that are also important considerations when specifying equipment for solid dosage manufacturing operations. Solid dosage processing is somewhat more problematic since, unlike liquid operations, continuous movement of liquids is generally much easier and cost effective since, unlike bulk solids, individual unit operations do not require stepwise transfer of materials from one secure operation to another requiring manual operations and equipment. Most often, piping, pumping and perhaps filtration are not labor or significantly labor intensive. Also, as biotech products are increasing at a rapid rate in the marketplace, relative to small molecule drugs, the manufacture of small molecule products decreases in importance. As a result, the incentive to implement significant modifications in tableting operations requires increased scrutiny. In several cases, it is more cost effective to transfer older established drug products to offshore facilities where labor and construction costs are significantly lower.

Tableting, and other solid dose operations, presents a problem because of the dust generated during manufacture. This dust is composed of the ingredients that constitute the tablet being manufactured. The accumulation of this product dust impedes production since it collects on critical operating components of the tablet press, often causing cessation of operations in order to clean the affected parts of the press. Additionally, the generated dust affects the product yield since it represents a loss of product.

Also associated with the dust accumulation is the need to protect operating personnel from ingesting the dust. The most common method of personnel protection is worker protective devices and clothing such as masks, gowns, and foot coverings. These items also assist in mitigating outside impurities from entering solid dosage manufacturing areas.

In addition to particulate accumulation, simplification of the tableting operations is also an item requiring additional emphasis. Solid dosage preparation as previously alluded to typically involves transfer of bulk materials from one operation to another through the use of bulk containers that include fiber drums and steel containers of various sizes. Since the transfer operations are not closed processes, a degree of product loss and contamination is possible, thus creating extra potential costs and product safety concerns. Also, these material transfers often involve transfer of these intermediate products through either manual or powered vehicle transfer cars. These material movers are usually powered by DC batteries, although, depending on the safety and size of the operation, gas-powered transfer are occasionally employed to move unfinished bulk powder and tablets between production steps. These transfer operations are, if not manual operations, can be quite energy intensive as well as requiring both scheduled and unscheduled maintenance operations. These activities become more costly as solid dosage equipment ages. Adding to the costs is the fact that many replacement parts are often more expensive as aging occurs since the demand for original spare parts declines with age and use. Often, this is more costly since the need for replacement parts declines as the fewer units are used in manufacturing. As a result, costs of replacement parts generally increase due to lower demand and the often misunderstood concept of "replacement in kind" for parts replacement.

With the persistence of dust and particulate accumulation often impeding tableting operations and the reliance on container bulk transfer, there is an evolving need to more specifically address equipment and processing operations, such as tablet manufacture.

When evaluating tableting and related manufacturing operations such as capsule filling, there are several equipment purchase considerations stakeholder and stakeholder representatives should evaluate prior to procurement plans. Equipment evaluations for tableting and related operations include:

- Capability to interchange solid dosage equipment and components with minimal disconnections – Automated liquid operations can be varied by adjustments to valves and control systems with a minimum of personnel input, powders and similar granulations usually rely on a high degree of bulk transfers involving labor-intensive container transportation from location to location. Also, valves that are used for solids transfer, such as transferring tablets from a tablet coater, are often merely manually opened gate-type valves that empty coated tablets to wheeled bins located beneath the coater for transfer to the next operation, usually to the packaging line.
- The capability to process solids of different flow characteristics – Particle size distribution has always been considered a critical material attribute in pharmaceutical tableting quality and process performance. Dynamic flow properties (including basic flowability energy, specific energy, aerated energy, and flow rate index) are bulk powder properties that require evaluation when tableting equipment procurement is planned. These variables require analysis to optimize the flow regime in the tableting unit and die filling operations. These analyses are required to minimize occurrences of attrition, segregation, and agglomeration, common causes of tableting production failures. While dynamic flow properties of solids are very important, criteria such as shear flow of the material during processing, the impact of compression on the powder, the metal content of the powder, and the airflow rates in the powder are also elements that are important [1]. In many solid dosage drug manufacturing operations, one tablet press is often used to manufacture one specific tablet. A more rigorous analysis of the variable noted above could, with minor modifications, qualify one tablet press for more than one dedicated tableting operation. By eliminating the need for a tablet manufacturing line, the cost savings could be quite significant. Consequently, with a modest upfront analysis, it is possible to significantly reduce overall tableting costs.
- Preliminary planning and provisions for future tableting, and related processes, employing yet to be identified constituents for use in various operating scenario operations (i.e. optimal and worst-case operating conditions).
- Ability to perform automated wash down and removal, as well as disposal, of effluent waste containing pharmacologically active wastes – Manual cleaning of tableting equipment, unlike many liquid drug process equipment is, typically, a manual operation that, in addition to being labor intensive, also consumes a great deal of time. And since tableting operations consist of individual discrete operations, there is seldom a system view of tableting production. As a result, procedures such as container cleaning and tablet press cleaning are performed at intervals that often vary, resulting in a need to have spare equipment available, which, in many instances, can be a superfluous, and often overlooked, capital cost.
- Capability to conform to dynamic regulatory changes – In addition to 21 CFR criteria set forth by the FDA, OSHA (29 CFR) also has regulations that involve the operation of certain pharmaceutical equipment. Careful consideration of existing and future changes to FDA and OSHA regulations warrants continuous attention to the direction these government administrations appear to be focusing on regarding updated regulations.
- Concerted effort to minimize processing times – While material transfer between solid dosage manufacturing steps is commonly practiced and most likely performed in as efficient a

mode, it is also relevant to evaluate equipment modifications, such as two or more tablet coaters operating in series, when considering options to minimize processing time. An example of this would be multiple equipment operating at approximately the same time. An example would be using two tablet coaters operating sequentially, one configuration identified as continuous coaters. Here, for this continuous coater configuration, two coaters can be installed in series with the tablets transferred by a conveyor. Basically, one tablet batch could be transferred to an uncoated batch to a front coater. Once the first batch is coated, a second uncoated batch can be fed to the rear coater and both batches can be subjected to the coating operation. While a larger manufacturing area may be necessary, being able to coat two batches in approximately the same time required to complete one batch offers a significant time reduction. Also, with enhanced control system, a reduction in operating personnel can be realized.

As noted, unlike liquid processes, tablet production is composed of discrete independent steps. At a minimum, six manufacturing operations can be identified:

Blending and mixing the ingredients required for tablet manufacturing – These ingredients consist of the active pharmaceutical ingredients (API), the critical component or components of the tablet, excipients or inert ingredients used to enhance manufacturing properties and tablet characteristics such as enhanced powder flow, improved powder bonding, tablet disintegration time, taste, and color.
Transfer of the blended powder mix into a hopper and processed in a tablet press where the feed powder goes into a die filling and compression operation.
Transfer of the tablets to a tablet coating operation.
Fill and finish operations, including packaging and labeling operations [1].

7.3 Tablet Blending Operations

Approved API used in solid oral dosage forms are comprised of particles matching critical quality attributes that enable a patient to realize the therapeutic advantage from the formulated products. In cases where the desired product attributes cannot be achieved, formulators may look for modifications of the active ingredient to achieve the desired outcome. A classic example is to micronize the API to increase the surface area and, potentially, the dissolution rate. As will be seen, many other examples of the need for particle control exist throughout tableting operations. The pharmaceutical industry has designed processes to grow particles for ease of filtration and drying, followed by granulation (milling) of the dried material. The granulation techniques required for tablet manufacturing are critical to tablet manufacturing [2].

With the basic tableting operations identified, a more in-depth analysis of these operations can be undertaken, commencing with the basic tablet components, the API and excipients. While the API is the key component or components, the remaining ingredients are required because a suitable tablet cannot be composed of just the active ingredient. In addition to the variables noted above (bonding, color, taste, etc.), processing properties such as enhanced compressibility and bulk density are very important and also require the addition of excipients to satisfy tablet product requirements.

A very key design element that must be considered in excipient implementation is the finished tablet or capsule size. In addition to other factors, the tablet must be easily swallowed with a pleasant taste. Thus, it is most important to control the quantity of excipients that are incorporated in the finished tablet or capsule.

There are some definitions that are germane to understanding the role of excipients in tablet and capsule manufacturing. Some of the relevant terms involved in excipient operations include:

- Antiadherent – A tablet powder constituent intended to reduce the adherence of a tablet powder blend from metal components that contact the tablet powder mix.
- Binders – Compounds that maintain the integrity of the manufactured tablet. Binders are, basically, adhesives to keep the powder particles from separating from the tablet bulk.
- Coating agent – A substance intended to increase the stability, enhance the visual appearance, and promote the disintegration of a tablet product.
- Coloring agents – In addition to enhancing the appearance of tablets, coloring excipients allow for identification of the tablets.
- Controlled release – One form of manufactured tablet is known as a controlled release (CR) tablet, also known as an extended release (ER) tablet or an enteric tablet. This type of tablet often has what is referred to as a bilayer coating or multilayered tablet if more than two layers are involved in the tablet preparation. The intent of these tablets is to target specific tissues or cells in the body by delaying the onset of pharmacologic effects until the desired location in the system is reached. One technique used to achieve this goal includes coating the tablet with a polymer layer. Two common polymer coatings are cellulose acetate and cellulose butyrate. While the bilayer tablet involves other manufacturing steps, these tablet coatings are important to ensure successful medication distribution.
- Diluent – A tablet or capsule manufacturing compound used to add bulk to the dosage form. Diluents are also referred to as bulking agents or fillers.
- Disintegrant – A compound that facilitates the disintegration of a tablet, capsule, or other dosage forming contact with water and/or stomach fluids. Often referred to as a solubilizing agent. The dissolution time for a tablet varies from 2 or 3 min to 4 h.
- Direct compression – The process of manufacturing a tablet without granulation. For this process, the excipients usually consist of a filler, a disintegrant, and a lubricant. The blend is fed directly to a tablet press to undergo compression.
- Flavoring agents – Since many active ingredients have unpleasant tastes, flavor additives to disguise the taste are added to the tablet mix. For instance, a sour taste is overcome by the addition of a raspberry- or licorice-flavored excipient or a citrus-based product. Citrus-flavored excipients are also used to overcome a sour taste.
- Glidant – An excipient that is employed to enhance the powder flow for tableting by reducing the friction of powder particles.
- Lubricants – The purpose of lubricants is to reduce the friction (the coefficient of friction) between components during the blending, tableting, or capsule manufacturing process. A lubricant also lowers the friction between the powder and the well wall of the dye and the punch powder. This permits easier tablet formation and reduces the tendency of the granulation to adhere to the punch or accumulate in the dye cavity.

While there are many compounds that are approved as excipients, excluding those used for coloring and coating appearance, less than 100 are commonly employed in tableting operations. Table 7.1 identifies many common excipients and their function as tablet excipients.

7.3.1 Dry Granulation

While the most commonly recognized application for pharmaceutical powders are tablets and capsules, pharmaceutical powders are also active ingredients in injectables, nasal powders, oral powders, and topical applications (skin). While these administration forms are gaining market

Table 7.1 Common pharmaceutical tablet and capsule excipients and their function.

Excipient	Common applications
Cellulose acetate	Bilayer tablet coating
Cellulose butyrate	Bilayer tablet coating
Croscarmellose sodium	Tablet and capsule and disintegrant, subsequent to ingestion
Silicon dioxide	Adsorbent, anticaking agent, stabilizer, viscosity additive
Cellulose (several forms)	Adsorbent, suspension agent, diluent
Carnauba wax	Binder
Corn starch	Binder
Crospovidone	Disintegrant
Dextrin	Binder
Ethyl cellulose	Binder
Gelatin	Higher viscosity additive, suspension enhancement, coating agent, tablet binder
Hydroxypropyl methyl cellulose (HPMC)	Coating agent, thickening agent, tablet lubricant, tablet binder, viscosity additive, wet granulation agent, coating agent
Ferric oxide	Coloring agent
Lactose	Filler (bulking agent), tablet filler, diluent, binding agent (diluent)
Magnesium carbonate	Glidant
Magnesium stearate	Tablet lubricant (antiadherent)
Mannitol	Ingredient in chewable tablets
Microcrystalline cellulose	Binder, adsorbent, suspension agent, diluent
Polyvinylpyrrolidone (PVP) or providone	Tablet binding agent, disintegrant, suspension agent, wet granulation agent
Polyethylene glycol (PEG)	Binding agent, tablet lubricant, dispersant/surfactant
Polysorbate 80	Solubilizing additive
Starch	Diluent, binder, disintegrant
Silicon dioxide	Anticaking agent, disintegrant, adsorbent, viscosity elevating agent, glidant
Stearic acid	Lubricant
Titanium dioxide	Coloring pigment, tablet coating compound
Talc	Bulking agent, anticaking agent, lubricant, glidant

share, the primary products employing pharmaceutical powders are still tablets and capsules. Consequently, tablet and capsule manufacturing warrant particular emphasis for pharmaceutical powder operations.

Having the ingredients needed to manufacture tablets or capsules is a preliminary operation in the tableting operation. Once the ingredients are identified, they must be blended using an operation that will provide, among other powder properties, a consistent uniform powder composition resulting in the production of acceptable and repeatable tablet product. Other reasons for granulation include providing a free-flowing powder, creating a free-flowing powder, reduction of dust generation associated with powder processing, providing a more uniform blend, enhancing the compressibility of the blend, and providing a higher powder density, uniform powder density, and increased particle hardness. To achieve these objectives, tableting compositions are often prepared using two formulation processes, wet granulations, and dry granulations.

Dry granulation is more often used to provide specific properties required for required tablet or capsule production. Dry granulation is desirable when the granulation is composed of moisture-sensitive materials such as the ability to undergo hydrolysis reactions when moisture is present. Another element that favors the use of dry granulation is the ability of dry granulation to process heat-sensitive active ingredients. Dry granulation is also preferred since disintegration or dissolution is not a concern. Other advantages of dry granulating are a general lack of color change resulting from the presence of moisture or a drying operation.

Dry granulation is accomplished by two processing procedures, the slugging operation (also referred to as rotary compaction) and roller compaction.

Slugging is a basic operation where a larger tablet or slug is produced by using a larger diameter punch and dye (>3/4 in.) to produce a large diameter product with the slug undergoing further size reduction by using a hammer mill or an oscillating granulator. These smaller particles are in turn fed to a tablet press to produce tablets.

The slugging operation has several disadvantages. The major drawback is a low yield subsequent to slugging. The milling operations create large amounts of scrap product in the form of fines and larger particulates. The larger particles are commonly recycled back to the feed unit for reuse in the operation. The smaller particles (dust) usually require dust collection equipment such as air filtering or a dust collection unit. Another disadvantage is the inability of the slugging procedure to operate in a continuous mode. But perhaps the greatest disadvantage when performing a slugging operation is the difficulty in obtaining a uniform, consistent product. An additional milling step is often problematic since the milling procedure cannot produce a granulation with consistent features because of the variation in the slug produced. For these reasons, roller compaction is the preferred blending technique for dry processing of pharmaceutical products for tableting and capsule production.

The other more common technique for dry granulation is roller compaction. Roller compaction offers several advantages over the slugging process. One advantage is roller compaction is used to provide a uniform particle or powder for further processing in a tablet press. Also, roller compaction offers a lower disintegration time if a binder excipient is not required. An improved powder flow and a consistent powder or particle uniformity is a further advantage of roller compaction. The operation basically consists of the powder being fed from a hopper, usually a circular feed opening, which is often equipped with a motor-driven shaft (auger) in the middle of the hopper bin to promote flow. From the hopper, the powder flows between two roller cylinders rotating circumferentially in opposite directions. The powder is fed between the compaction rollers with a resulting uniform product of varying shapes. The roller surfaces often have serrated surface configurations that in addition to being smooth can produce small pellet products or a continuous ribbon of product, depending on the roller surface design.

An important zone of the powder flowing between the rollers is region where the feed particles deform due to the pressure imparted by the compaction rollers. The operation commences at a point known as the nip angle or nip region. This angle, between the roller centerline and the narrowest opening at the feed hopper, is the angle where there is no longer powder slippage and the powder adheres to the outer roller walls, allowing for the formation of the desired finished material in the desired shape (i.e. ribbon, pellets, ribbons, strip, or a more defined powder granulation).

Once passing through the space in the rollers, the particles (powder, pellets, or continuous strips, also called a ribbon product) is often screened with out of specification (OOS) materials fed back into the feed hopper. Figure 7.1 exhibits a diagram of a typical roller compaction

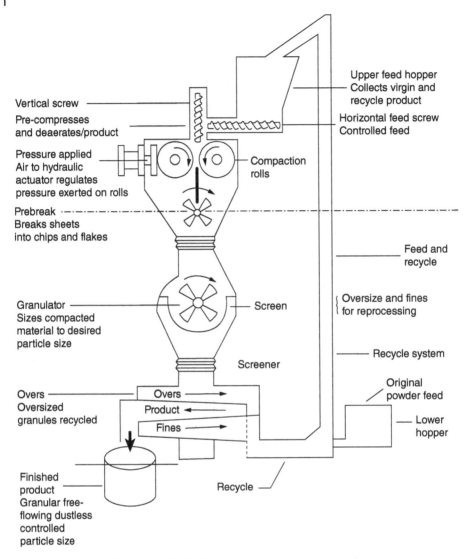

Figure 7.1 Typical roller compaction feed system.

system with the hopper above the rollers. Other roller systems may have different hopper and feed configurations, such as a horizontal feed and hopper where the powder is horizontally introduced in the rollers via a variable speed auger screw, but, nevertheless, perform the same function.

Once the finished form exits the compaction roller operation, it is typically fed to a device known as a granulator or mill. Here, the roller compression product, strip, ribbon, or pellet is introduced into the mill where it is converted to a fine powder for introduction into a tablet press.

Certain characteristics, or design features, of roller compression systems require detailed evaluation when addressing equipment design, procurement, specification, and maintenance. Perhaps the most critical element of a dry roller compaction unit is the feed unit to the rollers. While most feed hoppers are equipped with a motor-driven auger, the configuration of the feed unit is important. The decision to use a horizontal feed or a vertical feed should be based on the properties of the feed material. Depending on the design, a horizontal feed is often employed when the unit is equipped with a variable feed rate, while a vertical unit inputs hopper feed material at a constant rate. Vertical hopper and feed configurations are, typically, not equipped with the variable speed feed option since they are specified for one product designed to feed material at a fixed input rate.

Another parameter in the roller compaction operation is the pressure exerted on the rollers. The pressure typically varies, depending on the unit, between 4 and 140 tons.

While it would seem the pressure exerted on the rollers is very critical, actually, the roll pressure is not as critical as rotational speed of the feed screw (auger). The roller pressure is intended to accommodate a range of granule properties ranging from soft powders to harder particles. The feed screw operation is specified to powder to the compression operation at a consistent rate and consequently deliver a consistent product such as pellets and ribbons.

The diameters of the rollers vary from 4 in. diameters to a 16 in. diameter. The powder processing rate may vary, throughput rates between 33 and 22 000 lb/h, depending whether the unit is a pilot unit or a production system and if the product is a powder, pellets, ribbon, or strip.

The roller rotation speed is a critical design element in unit design and specification.

Power requirements range from 3 kW to as high as 80 kW for large production units, such as the example given in Figure 7.2.

The rotation speed is an independent variable of the roller compaction system. That is, the speed of rotation (most often specified in rpm) is part of the specifications of the unit and provided by the end user or end user representative. Other independent variables that are provided by the user include:

- Horizontal feed speed – Most often specified for variable feed screw feed.
- Vertical feed speed – Most often specified for fixed feed input.
- Compaction roller speed – Also specified by the user based on product specifications (pellet, strip, powder, etc.) and intended product yield.
- Roller pressure – Specified by user in terms of oil pressure where a pneumatic drive filled with hydraulic oil exerts pressure on an actuator that maintains a defined roller pressure.

Identification and implementation of the input design, specification, and operating requirements (independent variables) produce a product with features and characteristics that are integral to the product. These features are the dependent variable properties of the finished items such as pellets or ribbons. For dry roller compaction, the two important dependent variables for further tableting operations are compressibility and particle size. Compressibility can be defined as:

A material property which is constant for a given moisture content and temperature at time of compaction [3].

The measure of compressibility is given by the dimensionless factor, K, known as the compressibility factor. The compressibility factor is determined through a test configuration where the log of increasing pressure is applied on materials of known bulk density (the bulk density is the initial mass of free-flowing particles divided by the volume the particles occupy). The abscissa or x coordinate is plotted against the increasing bulk density, the ordinate or y coordinate. At higher pressures, the inverse of the slope is determined and defined as K or the compressibility factor.

Figure 7.2 Production scale roller compaction unit. *Source:* Powtec Machinen und Engineering GmbH.

For pharmaceutical powder and/or particles composed of an API and excipients, the K value is typically in the range of 8–14. For K values in the range of 40, the material is, by pharmaceutical criteria, noncompressible and not amenable to tableting operations. Similarly, materials with a K value in the range of 5–8 are compressible, and their applicability to tableting operations must be evaluated using other criteria.

Operations, such as roller compaction, involving powder flow depend in large part the ability to accurately characterize powder flow. Unlike liquid pharmaceutical operations, powder flow properties are not easily defined. Powder flow is important if accurate design properties are not specified. An incorrect equipment selection can significantly alter the economics of an operation. While determinations such as the K value are helpful in characterizing a powder property, many variables determine the critical properties of powder flow required to ensure viable and economic powder processing operations. As is the situation with the K value, more than addition of glidant additives is required to ameliorate powder flow in pharmaceutical operations such as roller compression, tableting, and gelatin capsule filling operations. To effectively address effective powder flow, there are a plethora of tests and procedures intended to define powder flow characteristics and related properties. While no single specific test or standard can comprehensively define the relevant flow characteristics of a specific powder flow, employing a small number of straightforward, reproducible, and meaningful tests are desirable. While there exist several testing and evaluation techniques, the methods of testing can vary widely ranging from visual observations to analytical instrumentation, depending on the degree of accuracy desired. It is important to note that current techniques for powder flow are empirical and have little theoretical basis and serve more as a guideline for powder flow evaluation.

When considering the type of testing and evaluations required, the operating environment should be reasonably similar to the actual powder processing conditions. Factors such as humidity and actual powder moisture content at the time of testing and evaluation have a significant influence on the test results. Often overlooked details such as storage time and storage conditions, prior to testing versus the actual manufacturing conditions, can affect the results and explain differences between test results and the manufactured product.

For pharmaceutical powders, the US Pharmacopeia has identified four frequently used methods for testing and evaluating powder flow (USP 30 Monograph <1174>) that can serve as initial standards for tests and evaluation. Further, the methods are popular since the test equipment is inexpensive and the test procedures are easily performed. The four powder flow test methods are (i) angle of repose, (ii) compressibility index/Hausner ratio, (iii) flow rate through an orifice, and (iv) shear cell.

7.3.1.1 Angle of Repose
The angle of repose is a powder flow property specific to friction or resistance of individual particle flow. The angle of repose is basically the result of allowing powder to flow down a feed funnel or hopper and forming a three-dimensional cone on a horizontal surface. The angle of repose, represented by the Greek letter alpha, α, is calculated by the following relationship (Equation 7.1):

$$\tan(\alpha) = \frac{\text{Height}}{0.5\text{base}} \tag{7.1}$$

Equation (7.1) can also be formatted in the form of a more usable inverse trigonometric function as

$$\tan^{-1} = 2\frac{h}{d} \tag{7.2}$$

where the height, h, is the vertical distance from the base of the powder cone to the tip of the cone and the base is the linear length (diameter) of a fixed horizontal base, given as d, or diameter of the collection ring if equipped with a circular "lip" intended to retain the powder that flows

Table 7.2 Common test values for the angle of repose for pharmaceutical powders.

Angle of repose (degrees)	Performance rating	Comment
25–30	Excellent	
31–35	Good	No mechanical aid required
36–40	Passable	Powder may hang up during testing
41–45	Poor	Requires vibration or agitation to complete tests
46–55	Very poor	Vibration or agitation usually required to complete tests
≥65	Fail	Not recommended

from the funnel on the surface enclosed by the raised lip. The angle of repose is basically the perpendicular height of the powder cone divided by the diameter of the collection ring. If the angle of response is determined in a dynamic mode (i.e. the height and base are measured at discrete time intervals, or continuously, if so equipped, until the powder sample is depleted), the distance from the tip of the powder cone to the funnel exit should be maintained at a distance in the range of 2–4 cm. The midpoint of the funnel exit should be approximately in line with the midpoint of the diameter of the circular base surrounded by the lip. As a guide, the funnel diameter can be about 1.42 in. (3.2 cm) and the diameter of the area enclosed by the raised lip on the collection surface about 3.6 in. (20 cm). It is important to note the difference between the aforementioned nip angle and the angle of repose. The nip angle is a parameter of the dry compression roller, an equipment variable, whereas the angle of repose is a property of powder flow.

For pharmaceutical powders, general values for the angle of repose are given in Table 7.2 [4].

In most evaluations involving pharmaceutical powders, an angle of repose greater than 50° is considered to be unacceptable for tableting operations.

Example 7.1   A specification is being prepared for procurement of a dry roller compaction unit. Prior to procurement some preliminary evaluations are to be performed on the powder to be processed. One evaluation is the determination of the angle of repose. The test unit consists of a fabricated circular dish measuring 20 cm in diameter surrounded by a boundary wall with a height of 1.5 cm. The powder will pour through a funnel with a diameter of 3.2 cm. A preliminary calculation positions the midpoint of the funnel discharge 10 cm above the midpoint of the measuring dish. After three powder flow tests, the average vertical height of the powder cone is 7.80 cm (the second decimal is an estimate). Determine the angle of repose and the initial applicability of the powder for roller compression operations.

Solution
The basis of calculation for this example is predicated on Equation (7.2), the inverse tangent calculation. Inserting the given values for Equation (7.2) yields

$$\tan^{-1} = \frac{2h}{d} = \text{arc tan} \ \frac{2(7.85 \text{ cm})}{20 \text{ cm}} = 38.0$$

Referring to a table of Trigonometric functions or a calculator,

the angle of repose is found to be 38°. ← **Solution**

Referring to Table 7.2, a 38° angle of repose is classified as an acceptable value for this particular test. For further verification, an additional procedure, such as the compressibility Index or the Hausner ratio, would be helpful for specification preparation and post-installation start-up.

7.3.1.2 Compressibility Index or Hausner Ratio

While the compressibility index (also referred to as the Carr index) and the Hausner ratio are identified using one measurement technique, there is a differentiation between the two variables, an initial volume and a final volume. The compressibility index (C) is given by

$$C = 100 \times \frac{(V_0 - V_f)}{V_0} \tag{7.3}$$

V_0 represents an untreated or unsettled volume, and V_f is a final volume subjected to compression until the volume no longer varies under an applied load and maintains a constant volume after two or three compression cycles. Another powder flow evaluation parameter, which by one definition is closely related to the compressibility ratio, is the Hausner ratio. When density is used as a measurement basis, the Hausner ratio is also given as

$$\text{Hausner ratio} = \frac{\rho_{\text{tapped}}}{\rho_{\text{untapped}}} \tag{7.4}$$

Here, the Hausner ratio (defined by H) is identified as the ratio the final compressed density (ρ_{tapped}) to the bulk density (ρ_{untapped}) in lieu of the compressibility index as an evaluation method.

The Hausner ratio can also be defined by using the initial observed volume of a powder sample (V_0) and the final tapped volume (V_f) of the same powder sample in lieu of the density. In this simplified form, the Hausner ratio is given as

$$H = \frac{V_0}{V_f} \tag{7.5}$$

Often, these values are determined by using as little as a 100 g sample and a 250 graduate cylinder. While more sophisticated analytical techniques can, as noted, be employed, it is again emphasized parameters such as those defined for the compressibility index (C) and Hausner ratio (H) are useful empirical powder flow parameters with no theoretical basis. It is possible, however, to correlate the compressibility index (C) with the Hausner ratio (H) by employing Equation (7.6), given as

$$C = 100 \left(1 - \frac{1}{H}\right) \tag{7.6}$$

As a relative measure of flow capability, or flowability, available experimental data has been summarized in Table 7.3.

For flow characterizations classified as poor, very poor, and very, very poor, retesting and/or other another technique (e.g. the angle of repose) should be considered.

Example 7.2   Three 100 g samples are funneled into three individual 250 ml graduate cylinders and the unsettled volume recorded in each cylinder. The unsettled average volume of powder in the three individual cylinders was recorded as having a volume of 125.5 ml. A rubber stopper was used to tap the samples until each sample exhibited a final constant tapped volume. The average final tapped volume was 118.0. Determine the Hausner ratio, H, and the compressibility index (C).

Solution
The determination of the Hausner ratio is based on applying the simplified Equation (7.5) as a solution. Inserting the values obtained in Equation (7.5),

Table 7.3 Relative flowability [5].

Compressibility index (%)	Hausner ratio	Flow characterization
≤10	1.00–1.11	Excellent
11–15	1.12–1.18	Good
16–20	1.19–1.25	Fair
21–25	1.26–1.34	Passable
26–31	1.35–1.45	Poor
32–37	1.46–1.59	Very poor
>38	>1.60	Very, very poor

Source: Data from Carr [5].

$$H = \frac{125.5\text{ml}}{118.0\text{ml}} = 1.06 \leftarrow \textbf{Solution}$$

The compressibility index can be obtained from Equation (7.6):

$$C = 100\left(1 - \frac{1}{H}\right) = 100\left(1 - \frac{1}{1.06}\right) = 5.7 \leftarrow \textbf{Solution}$$

The calculated results indicate that the pharmaceutical powder exhibits excellent flow properties and should satisfactorily perform in manufacturing operations such as roller compaction and tablet presses. Nevertheless, it would be recommended other tests also be performed to further verify the results described above.

While there are more rigorous methods of determining the angle of repose, the compressibility index (Carr index) and Hausner atio are available in publications such as ASTM D 6393-08, "Standard Test Methods for Bulk Characterization by Carr Indices," the calculations contained in these documents are identical to those defined above. The major variances are highlighted in operator experience and the number of tests and calculations required.

One valuable aspect of ASTM D-6393-08 is the ranges in which these tests are designed to function. Ideally, powder particle diameters should not exceed, on average, a 2.0 mm (0.0787 in.) diameter. The funnel outlet should be free flowing with an outlet funnel diameter in the 6.0 mm (0.2362 in.) to 8.0 mm (0.3150 in.) range. While these particle sizes are not regulations or universally accepted criteria, they serve as reasonable reference values for planned powder flow tests and measurements.

When recording values such as V_0, V_f, ρ_{tapped}, and $\rho_{untapped}$, the recorded readings should be based on the average of about three determinations and not a single measurement since the readings are often observational measurements and subject to human error.

7.3.1.3 Flow Rate Through an Orifice

As is the case with determining the angle of repose, there are several techniques employed to measure powder flow through an orifice. By approximating actual flow conditions, such as monitoring powder flow through specific types of orifices (e.g. funnels and hoppers), the experimental results can serve as a basis for a production unit requiring little or no post-installation changes. Hoppers from actual production units are also used for evaluating design tests involving flow through an orifice. While considering actual powder flow feed orifices is an effective method, the important flow property for flow through an orifice is that the powder is a free-flowing material that is not subject to "lumping," agglomeration or cohesion exhibited by the powder.

Just as there is no specific test apparatus for evaluating orifice flow, there are no set standards for measuring orifice flow rates. There is, however, a widely used empirical relationship that is frequently employed to evaluate, in a quantitative form, the relationship between powder flow and orifice size. The formula is known as the Beverloo formula or Beverloo's law and is given by

$$\bar{Q} = C\rho_b \sqrt{g} \left(D_0 - kd \right)^{5/2} \tag{7.7}$$

where

Q is the mass flow rate through the orifice, g/s
C is a dimensionless constant in the range of 0.55–0.65 (0.58 is a common value)
ρ_b is the bulk density of the powder, g/cm^3
g is the gravitational acceleration, 980 cm/s^2 at sea level
D_0 is the outside diameter of the hopper orifice, cm
k is a dimensionless value commonly given as 1.5 for a circular orifice and 2.4 for a rectangular discharge orifice
d is the mean diameter of a powder particle, cm.

While it has been emphasized for pharmaceutical laboratory operations, the SI system is used; the Beverloo formula is applicable only to SI units. The primarily empirical equation was formulated using SI units with the constants derived in SI units.

While equation is a reasonable prognosticator of powder flow through an orifice, some guidelines are available to assist in analysis of the results. As a general observation, powder flow through an orifice should be, at a minimum, of 100 g/s. A higher powder flow in excess of 200 g/s is not uncommon. A powder flow of less than 10 g/s should be reevaluated to verify powder flow accuracy or investigate changes in particle sizing and powder properties. Also, the transition range for clogging during powder flow appears to be when the ratio D_0/d is about 4.5 for many particles and powders, including regulated excipients and API [6].

The Beverloo equation (Equation 7.7) is applicable for free-flowing powders, specifically pharmaceutical compositions, in the range where $D_0/6 > d > 400\,\mu m$ (0.004 cm). In addition to the diameter of the flow orifice being greater than 6 times the average particle diameter, the holding container, if cylindrical, should be at least twice the outside diameter of the flow orifice.

Example 7.3   As part of a preliminary specification for a tablet press intended for production of a generic tablet to be manufactured by a small API manufacturer, it was concluded that a used vertical hopper roller compaction unit would be the most cost-effective procurement decision. The toll manufacturer is to produce the tablets, and a contract packaging operation will complete the packaging, labeling, and shipping requirements. The API powder composition has the following properties:

a) Bulk density $= 0.75\,g/cm^3$
b) Average particle diameter, $d = 0.141\,cm$
c) $D_0 =$ outside diameter of hopper orifice $= 3.2\,cm$
d) $k = 1.5$, a circular orifice
e) $C = 0.58$
f) $g = 980\,cm/s^2$

Based on the data supplied, calculate the flow rate for the dry roller compaction unit and its applicability for use in this application. This result, coupled with the other tests and evaluations described, should serve as accurate predictors regarding the performance of this particular equipment.

Solution
Equation (7.7), the Beverloo equation, is the basis of calculation for this example. Inserting the given values for Example 7.3 yields

$$\bar{Q} = C\rho_b \sqrt{g} \left(D_0 - kd \right)^{5/2}$$

$$Q = (0.58)(0.75\text{g/cm})(31.30\text{ft/s}^2)\left(3.2\text{cm} - 1.5\left[0.141\text{cm} \right] \right)^{5/2}$$

$$= 210\text{g/s} \leftarrow \textbf{Solution}$$

As the computational result indicates, the compaction unit appears to satisfy the flow requirements for the selected equipment.

As has been pointed out on several occasions, a single series of tests or a single computation should not serve as a basis for equipment procurement.

7.3.1.4 Shear Cell

Being solids, powders possess yield strength and elastic deformation. As such, powders do not act as fluids. Powder movement is due to the slippage of flow in the planes of the individual powder particles in the bulk material while undergoing a shear force. Thus, unlike fluids, powders cannot generally be described in terms of a gradient as a paradigm. Powders move due to slippage of the planes the particles are part of when undergoing an applied stress. Powder flow occurs when the applied stress exceeds the yield strength of the powder particles. As a result of this "overstressing," the powder accumulates to a point where excess compaction of the powder occurs. To predict when the overstressing phenomena will occur and to evaluate other properties of shear powder flow, procedures known as shear cell methods has been developed. Shear cells are devices that measure the force required to initiate powder flow. These devices are further augmented with the introduction of sophisticated software that can estimate mass discharge rates, feed rates, powder caking, and other associated variables involved in tableting manufacturing operations. In fact, shear cell technology has evolved to the point where ASTM Standards are available to measure and evaluate the properties of bulk powders, including pharmaceutical powders, which are applicable for use in the design of hoppers and bins when evaluated during steady-state measurement and testing. One such publication is ASTM D 6128-06, *Standard Shear Test Method for Testing of Bulk Solids Using the Jenike Shear Cell*. The Jenike cell is essentially a device that measures the ratio of the yield (σ_1) of a confined powder to that of free-flowing powder with no constraint (σ_c) in the test cell or shear cell when subject to a horizontal force [7]. This ratio, σ_1/σ_c, is defined as the flowability of a powder and given by ff$_c$. Thus, the flowability, ff$_c$, can be expressed as

$$\text{ff}_c = \frac{\sigma_1}{\sigma_c} \tag{7.8}$$

One difficulty common to shear cell testing and measuring devices is that the further the measuring device is from the location of the applied stress, the lower the value of the applied stress. Thus, powder particles flow with more force at the initial stress area and diminish as the distance from the applied stress is increased. This phenomenon is analogous to a shear stress (horizontal to the material or powder) applied to a solid material such as a metal rod or rigid polymer; the greater the applied stress on the rod, or bar, the greater the deformation or proclivity to break when the maximum yield strength is exceeded [8]. As is the case with powders, the phenomena are more pronounced as the applied stress is further from the stress locus.

Studies using the flowability ratio (ff_c) have provided a range of values that can be employed for evaluating designs and certain powder formulations as candidates for consideration of a particular design prototype or production element (e.g. a hopper or powder composition). The ff_c ratios are given as [9]

$1 < ff_c < 2$ – Very cohesive, little, or no flowability potential
$2 < ff_c < 4$ – Cohesive, low flowability potential
$4 < ff_c < 10$ – Easy flowing
$10 < ff_c$ – Free flowing

Referring to Table 7.3, it is interesting to note the similarity between the optimum flowability for the compressibility index (%) and the ff_c; both values exhibit the same integral value, although the variables differ.

Also, since powders do not exhibit uniform behavior when exposed to a shear stress, phenomena such as varying powder density, lumping, ratholing (inability of powder to flow near the wall of the bin or hopper), and arching (formation of a bridge of nonmobile powder formed at the exit of the bin or funnel and causing a cessation of powder flow) will occur unless corrective/preventive actions are implemented.

Another type of shear cell test and measurement device is the Schulze ring shear tester. This device, whose description, operation, and procedures are found in ASTM D 6773-02, *Standard Shear Test Method for Bulk Solids Using the Schulze Ring Shear Tester*, is also an effective tool for obtaining design data for hoppers and bins. A major difference between the Jenike device and the Schulze shear cell is that the Schulze cell can obtain data in a non-flow operation without reliance on a constraint measurement.

An advantage in shear cells is the ability to automatically record test and measurement powder data in a consistent reproducible format.

With the four common techniques common to powder flow testing identified above, focus can now be given to the critical variable of powder particle sizing and distribution.

As noted, another critical independent variable integral to dry roller compaction is the particle size distribution. If the particle size is too large or too small, significant production problems will occur when the roller compression product is fed to the tablet press. While particle size is an important variable in tableting operations, it is gaining increasing importance in the area of nasal drug administration where aerosols are the focus of effective nasal administration. The introduction of pharmaceuticals for treating diabetes is an example of this evolving therapy.

Determining an optimal particle size for pharmaceutical powders is often problematic. Generalizing on a common diameter or common range is difficult since each powder is subject to individual product composition performance. Factors such as dissolution rate in water and gastric fluids are greatly affected by particle surface area and diameter. Traditionally, particle diameter was analyzed by segregating powder samples through a series of stacked screens, with different screen diameters nested vertically atop one another. This screening is the technique used for separating pharmaceutical powders, and other non-drug commodities, into two or more size fractions where the oversized powder particles are trapped above the screen and the smaller diameter powder particles pass through the particular screen. The screens are then vibrated, horizontally, and tapped simultaneously, vertically, with an oscillation in several directions causing the powder sample to separate by particle diameter, commencing with the

Figure 7.3 Test sieve shaker device. *Source:* W.S. Tyler® Industrial Group.

largest screen diameter on the top and terminating with the smallest screen diameter at the bottom of the arrangement. The common device designed to do horizontal shaking, as well as simultaneously tapping, from the top is a machine referred to as a Ro-Tap® Sieve Shaker, such as the type shown in Figure 7.3, a device that has been used for particle sizing since the early twentieth century.

The arm atop the screens is the mechanism that oscillates with a fixed up and down action, vertically, when the equipment is in operation. Similarly, when operating, the screens oscillate, horizontally, creating the motion in that direction.

The widely used set of screens through which the powder passes is the W.S. Tyler® Test sieves. The sieve screen openings range in size from a 2 in. diameter (50 mm) opening to 0.0015 in. (0.037 mm) diameter, depending on the screen diameter selected (the standard sieve screen diameters are 8 and 12 in.). For a maximum particle diameter, the ASTM D 6393-08 particle diameter of 2.0 mm (0.0787 in.) can serve as basis, while actual particle size can be larger or smaller, depending on the product specification. This correlates to a Tyler screen mesh size of 9 Mesh, while another commonly used sieve sizing system, US Sieve Size, identifies a No. 10 Mesh as the equivalent of a Tyler 9 Mesh. Table 7.4 is a summary of common pharmaceutical powder sizes often encountered in particle distribution testing.

The mesh numbering system is a measure of openings per linear inch of mesh or screen.

The sizing varies by $\sqrt{2}$ (1.414). The US Screen sizes differ from the Tyler sizing system as they are arbitrary identifiers.

Particle sizing by sieve analysis for pharmaceutical applications commonly uses the SI units, as is common practice in most pharmaceutical laboratory operations.

Once the powder distribution is obtained, obtaining the effective powder particle size range is the critical procedure. The particle size is ascertained by studies performed in the early phases of drug development that have determined the effect on particle size on factors such as solubility, dosage requirements, dosage frequency, and tablet compression requirements.

Table 7.4 Common pharmaceutical powder particle sizes by diameter.

US sieve size	Tyler sieve size	Mesh diameter	
		mm	in.
—	2½ Mesh	8.00	0.312
—	3 Mesh	6.73	0.265
No. 3½	3½ Mesh	5.66	0.233
No. 4	4 Mesh	4.76	0.187
No. 5	5 Mesh	4.00	0.157
No. 6	6 Mesh	3.36	0.132
No. 7	7 Mesh	2.83	0.111
No. 8	8 Mesh	2.38	0.0937
No. 10	9 Mesh	2.00	0.0787
No. 12	10 Mesh	1.68	0.661
No. 14	12 Mesh	1.41	0.0555
No. 16	14 Mesh	1.18	0.0469
No. 18	16 Mesh	1.00	0.0394
No. 20	20 Mesh	0.841	0.0331
No. 25	24 Mesh	0.707	0.0278
No. 30	28 Mesh	0.595	0.0234
No. 35	32 Mesh	0.500	0.0197
No. 40	35 Mesh	0.420	0.0165
No. 45	42 Mesh	0.354	0.0139
No. 50	48 Mesh	0.297	0.0117
No. 60	60 Mesh	0.250	0.0098
No. 70	65 Mesh	0.210	0.0083
No. 80	80 Mesh	0.177	0.0070
No. 100	100 Mesh	0.149	0.0059
No. 120	115 Mesh	0.125	0.0049
No. 140	150 Mesh	0.105	0.0041
No. 170	170 Mesh	0.088	0.0035
No. 200	200 Mesh	0.074	0.0029
No. 230	250 Mesh	0.063	0.0025
No. 270	270 Mesh	0.053	0.0021
No. 325	325 Mesh	0.044	0.0017
No. 400	400 Mesh	0.037	0.0015

While most hopper designs are furnished with standard-sized hopper, it is sometimes desirable to have an idea of the mass contained in a hopper during normal operating periods. This information is useful to determine if difficulties such as arching, bridging, ratholing, or powder flooding are occurring during operations. In this case, visual observation can provide a quick analysis of the operating problem encountered. To assist in this quick analysis, if needed, an approximate design relationship is provided. The relationship correlates hopper design height with the specified powder flow. Thus, operating problems can be addressed by the design hopper height and comparing it with the observed height (volume) during operation. The approximate value is typically applicable to larger storage bins intended to hold large masses of stored bulk quantities but can be helpful in smaller powder flow configurations. The relationship is given by [10]

$$H = \frac{m}{\rho A} \tag{7.9}$$

where

H is the cylinder height, m
M is the mass to be stored in the hopper during processing, kg;
ρ is the bulk density, kg/m^3
A is the cross section area of the cylindrical hopper, m^2.

Example 7.4 It is planned to maintain about 10 kg of powder in the hopper employed in Example 7.3 during the continuous processing of the tablet press powder through the roller compaction unit. As a visual check, a line will be painted on the hopper feed wall to indicate the approximate height of the feed volume to the roller compaction unit. At what height above the hopper feed entrance to the roller compaction unit should the line be drawn?

Solution
For this example, the basis of calculation is Equation (7.9). The required variables, referring to Example 7.3 and converted to applicable dimensions, are

$\rho = 750 \, \text{kg/m}^3$
$m = 100 \, \text{kg}$
The diameter of the hopper orifice is 3.2 cm = 0.032 m.
The orifice cross-sectional area, A, is given by

$$A = \frac{\pi D}{4} = \frac{(3.14159)(0.032 \text{m})^2}{4} = 8.04 \times 10^{-4} \, \text{m}^2$$

Inserting these values in Equation (7.9) yields

$$H = \frac{10 \text{kg}}{(750 \text{kg/m}^3)(0.000804 \text{m}^2)} = 16.58 \text{cm} \leftarrow \text{Solution}$$

The line should be placed 16.58 cm or about 7 in. above the feed hopper nozzle and should remain at that approximate level during equipment operation to indicate proper operation.

7.3.2 Wet Granulation

As noted in Section 7.3.1, dry granulation is the most common approach to powder granulation operations. However, wet granulation operations are still employed for specific blending requirements where dry granulating is not the preferred route. There are several reasons why a wet granulation operation is preferred. They include:

- Greater compressibility – This operation allows for a greater mass of active ingredients to be administered in a given volume. Increased compressibility is typically accomplished by adding a solvent such as water to the powder and mixing the composition for a given time at a predetermined mixing rate. The operation is referred to as wet massing.

- Dissolution – Often, ingredients in solution will dissolve more readily than dry granulations.
- Dust reduction – Wet blending significantly minimizes the presence of dust during blending operations. In addition to being a cost factor, powder dust presents a safety concern as a physical hazard (dust explosion, often requiring venting and protective devices) and a health hazard (inhalation and physical contact with the dust particles, often requiring personal protective equipment).
- More uniform particle distribution – Particles in solution tend to clump or bulk as can occur in dry powder form.
- More effective mixing – Particles in solution can mix more homogeneously than dry blending.

An important objective of wet granulation is to achieve a more continuous granulation, similar to dry granulation, thus improving the efficiency and thereby reducing processing costs.

Wet granulation also presents disadvantages not common in dry granulation. Among the disadvantages are:

- More costly manufacturing – Because wet granulation requires additional manufacturing and equipment costs, careful evaluation is required when evaluating this procedure in lieu of dry granulation. Also, wet granulation may require additional safety concerns if hazardous solvents are used in the wet granulation operation. Issues such as solvent vapor exposure limits and flammability must be assessed.
- Additional processing operations – Because of the need to add a solvent, the need for additional components such as mixers, drying units, and agitators are frequently required to produce a product that meets the specifications set forth. Additionally, the additional equipment must conform to cGMP criteria for details such as materials of construction and cleaning that require longer run times and often more detailed maintenance activity than dry granulation.

The wet granulation unit operation is a relatively straightforward procedure consisting of the following steps [3]:

1) The dry constituents, including drug excipients, are mixed together.
2) Subsequent to the addition of water and cornstarch (Table 7.1), the blend then undergoes granulation in a wet granulator.
3) The wet granulation is then dried and sized as per Table 7.4 criteria.
4) After the particle sizing operation, the acceptable mix is then blended with lubricants such as PEG or magnesium stearate (Table 7.1).
5) The finished mix is then ready for tablet compression operations.

Selecting liquid and specific granulating agents for the wet granulation process is the most critical element of the wet granulation processing. Several of the most commonly employed liquid excipients for wet granulation include alcohols (ethanol, isopropanol), USP grade water, alcohol/water blends, and other organic solvents such as acetone. Solvent selection should also include an evaluation of finished product odor and taste, often a by-product of solvent residue or wet component interaction. For these reasons, the use of an inert blanket gas such as nitrogen should be considered if oxidation is a cause of the undesirable odor/taste.

In addition to wet granulation liquid options, granulation agents are often used in wet granulation. The preferred compound is water; however, materials such as starches and sugars are also used. Polymers such as hydroxypropyl methyl cellulose (HPMC) and polyvinylpyrrolidone (PVP) are also incorporated in wet granulation blending, as noted in Table 7.1.

If liquid organic granulation options are to be employed, certain regulatory considerations must be evaluated prior to using these materials. For instance, US EPA criteria regarding storage and discharge of the liquids must be assessed. Additionally, safety considerations such as the explosive limits of the solvent vapors used in the wet granulation procedure must be addressed. Included in the safety program, relevant requirements of 29 CFR 1910 (OSHA) must also be evaluated.

7.4 Tableting Operations

Ensuring a uniform mixture of constituents that can be compressed in tableting devices is the most critical unit operation of tablet manufacture. To accomplish this task, several equipment types have been developed and are widely employed, depending on variables such as tablet quantity and mixing properties.

Blending operations are often used even after ribbon blenders and similar processes have been employed. Often, additional blending is needed to manufacture products that have several product concentrations and require additional blending to conform to various tablet strengths.

The tableting operation typically commences by loading premeasured ingredients into a blending device. The blender can be a large rotating drum, a conical blender, or for large tableting quantities a horizontally designed tank with a shaft mounted in the middle of the drum to which large mixing blades are attached. As the shaft rotates, the mixing blades blend the tableting constituents for a predetermined period of time.

A common type of blender is the twin shell blender designed for dry powder blending, although some are fitted with liquid addition capability equipment. Figure 7.4 illustrates a typical small twin shell blender. This design is also referred to as a double cone blender or a V blender. It is referred to as a V blender because of the V-shaped vessel's twin legs. This batch blender rotates, tumbling the ingredients to mix them. This type of blender may be equipped with internal baffles, internal breaker bars, or other components to break up agglomerates and enhance mixing. While other blender designs are available (e.g. horizontal drum blenders and conical designs), the twin shell design is arguably the most common design.

For established tableting operations, the blender volume is determined by experimentally determined variables and equipment design factors and scale-up factors. Table 7.5 presents some relevant values for various sizes of twin shell blenders, a common type of blender. The values shown in Table 7.5 can serve as rough guidelines for existing operations as well as a basis for scale-up parameters subsequent to ascertaining needed experimental scale-up data. It is worthwhile noting that for most manufacturing blender units, the size rarely exceeds 500 ft^3 (14 158 l) in volumetric capacity.

While it is generally acknowledged, laboratory scale-up studies are of questionable value when considering larger blender units; it can be helpful in obtaining an idea of what is involved to obtain reliable design data, a task that is generally performed by vendor.

Table 7.5 identifies typical operating and design parameter guidelines for blenders in general and a twin shell blender in particular (the twin shell blender is a most commonly used unit for pharmaceutical blending operations).

A very critical variable, mixing time, cannot be easily determined. Again, experimental studies, coupled with scale-up procedures, are the standard techniques to resolve this item. But, as a practical "rule of thumb," for actual manufacturing operations, mixing times typically range

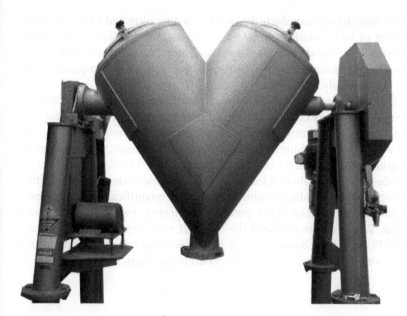

Figure 7.4 Patterson Kelly 50 ft³ twin shell blender. *Source:* Courtesy of HealthStar, Inc.

Table 7.5 Relevant design/operating parameters for twin shell blenders.

Working capacity (ft³)	Working capacity (l)	Mass (kg)	Scale-up factor	RPM range
20–50	566.3–1415.8	8–30	2	25–30
250	7079.2	80–150	3.5	23–28
500	14158	200–300	4	12–18

from 12 to 20 min. As more mass of mix is placed in the mixer, the effective mixing time increases. Consequently, the amount of material to be placed in the shells should not exceed 50–70% of the fixed volume. For V-type blenders, equal amounts of masses should be used for each shell. It is also more effective to have the discharge valve of the blender located at the bottom of the mixer.

When performing blender scale-up studies, for laboratory studies, a scale-up factor of 1 can be used for smaller laboratory blender units in the 1 l range. The scale-up factors are guidelines for studies involving masses and not volumes.

Determining effective RPM speeds is not necessarily transferrable since properties of the material to be blended (e.g. particle adhesion, particle density, moisture content, etc.) can vary with the volumes.

Another blending consideration is the order in which blending components are introduced in the blender. Physical properties, such as a tendency of components to agglomerate when in contact with one another, should be considered prior to randomly adding constituents. To preclude this possibility, a device known as a de-lumper is often introduced into the process prior to the powder flow into the blender.

The blending manufacturing operation is relatively straightforward; the constituents are fed into the top of the blender and the twin shells rotate around a full or partial shaft affixed to the perimeter of the blender body for a predetermined period, generally up to 30 min of operation.

Lending time is important since tablet manufacturing is a batch operation and longer blending periods affects the number of lots manufactured. Upon completion of blending, the mixed ingredients are typically stored in sealed containers until ready for use. The mass of the batch is recorded in kilograms. The containers are kept in controlled facilities where temperature and humidity conditions are continuously monitored.

Double cone twin shell blenders vary in size from about a $0.5\,ft^3$ capacity to about a $2000\,ft^3$ volume. The smaller capacity devices are commonly used for pilot plant and laboratory studies as a basis of scale-up for the larger volume units. Typically, the scale-up factors range from 11 for the smallest blender to 6 for a $500\,ft^3$ (14 158 l) unit. As a preliminary design basis, a $432\,ft^3$ capacity twin shell unit would require a scale-up factor of 3. Similarly, an $864\,ft^3$ blender, 24 468 l (rarely used), would employ a scale-up factor of about 5 for a preliminary sizing/scale-up design factor.

While obtaining scale-up data for blending is quite problematic, quantification of additional scale-up information can also be difficult. One potentially effective method is employing dimensionless groups (e.g. Reynolds number and the Schmidt numbers) and incorporating them in the Buckingham Pi paradigm. Some scale-up tablet blending considerations, which should be considered yet, are difficult to describe by an accurate paradigm include:

- Process variables
- Product variables
- Equipment variables
- Raw materials
- End point detection
- Humidity

It is important to reemphasize that the powder blending process, at this stage of manufacture, is a dry process devoid of significant moisture content. While, as noted, it may be difficult to quantify the impact of moisture content on a tableting blend, it is a most important consideration in scale-up design and planning.

7.4.1 Tablet Manufacturing

Subsequent to blending and controlled powder storage, the dry mix is then ready for tableting. This operation is accomplished by feeding the powder blend to a tablet compression press. The mass of powder feed is greater than the calculated amount for a calculated number of tablets. A larger actual mass is used to correct for powder losses that may occur during tablet manufacture. The mass of excess powder introduced is between 10 and 15% of the calculated theoretical quantity. The compression press is commonly referred to as a tablet press, as well as a tablet compressor. Figure 7.5 displays a profile view of a typical tablet high speed tablet press unit.

A production scale tablet press operates using powder compression pressures ranging from $300\,kg/cm^2$ ($4267\,lb/in^2$) to $3000\,kg/cm^2$ ($42\,670\,lb/in^2$). It is also common to state compression pressures in tons. For instance, $300\,kg/cm^2$ is commonly identified as a 2.1 ton compression force and $3000\,kg/cm^2$ is identified as 21.3 ton capability. It is important to design the punch and applied force with careful analysis. While the applied force on the punch may be several tons, the tablet surface area will be less than a square inch. This results in a pressure significantly greater than the pressure assumed. Thus, the nominal pressure (tons) is significantly greater, when addressing the pressure applied on the tablet. For instance, a ¼ in. diameter punch would be $\pi\,(1/8)^2$ (recalling the area of the tablet is $A = \pi r^2$, where r is the radius) or $(\pi) \times (1/64) \approx 0.05\,in^2$ or $40\,000\,lb/in^2$ or $20\,tons/in^2$. Such a great pressure would be a significant pressure overload. Thus, a careful analysis of punch forces must be performed when preparing operating specifications and procedures.

Figure 7.5 Typical manufacturing tablet press. *Source:* Courtesy of HealtStar, Inc.

For typical high speed tablet presses (>4000 tablets/min), the power required is in the range of 5–6 hp.

A basic high speed tablet press has a central component that is a circular stainless steel plate with evenly spaced cavities or dies located circumferentially in the plate. Located a few inches above the die is a steel alloy shaft known as the punch. Several types of steel are used for the punch and associated components (a more comprehensive coverage of alloy steels is located in Chapter 9, Section 9.1). The steel used in the punch is predicated on design factors such as tablet shape, powder abrasiveness, corrosion properties of the powder, and ductility of the punch alloy.

A high speed press has several punches affixed above the die location. In this configuration, the combination of punch and die is called a station. Typically, a high speed tablet press has about 40–47 stations operating during tableting operations. Another shorter length punch is located directly below the die. In steady-state operation, the upper punch will exert its maximum force downward simultaneously when the lower, shorter punch is pushing upward. The gap between the upper die and the lower die in maximum operating mode is about 20 mm (0.787 in.). This gap is where the blended powder is fed into the press to form a tablet. In addition to a specific size and shape, a typical tablet has a certain identification marking or logo. These tablet characteristics are determined by the tooling set used in the tablet press. A tooling set is composed of the components mentioned above, the die and upper and lower punches. The most common tool set is identified as BB tooling. The standard BB tooling is 5.25 in. long

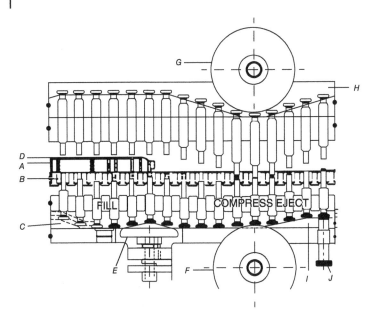

Figure 7.6 Schematic profile of a tablet press. (A) Powder feed mechanism – The blended mix is fed to the tablet press by gravity feed from the top of the tablet press. (B) Die. (C) Lower cam – The lower cam pushes the upper punch further into the die. (D) Wipe off mechanism – This blade removes excess powder from the die/punch surface. (E) Mass control cam – Fills the powder to the required die volume. (F) Lower compression roll – Moves lower punches vertically upward to the die cavity. (G) Upper compression roll – Moves upper punches downward to the die cavity. (H) Raising action – Upward movement of punch subsequent to passing over the upper compression roll (G). (I) Lower cam action – Upper movement of lower punch subsequent to lower punch passing over lower compression roll. This action causes the tablet to be flush with the upper die. (J) Tablet ejector device – The function of this device is to eject the tablet from the die with the completed tablet collecting in a container outside the tablet press enclosure. *Source:* Courtesy of Healthstar, Inc.

with a nominal diameter of 0.75 in. For larger tablets, a 1 and 1.25 in. diameter punch head is used to produce either a round-shaped tablet or the elliptical surface-shaped product.

The characteristics of the logo or other tablet markings also have certain tool design requirements. Important tooling design characteristics include:

- Uncoated tablets have an engraving cut of ~30° from the horizontal.
- For film-coated tablets, the engraving cut is at least 35–40°.
- The depth, width, and powder properties of the cut must also be evaluated.
- The tooling must be free of defects.

Figure 7.6 exhibits a profile view, with major component identification (A–J), of a common high speed tablet press.

While not identified in Figure 7.6, included in the process is a metal detector, whose purpose is to detect metal particles in excess of a specific diameter that are impregnated within in the tablets. These metallic particles often originate from press components such a damaged punch or die.

An important variable of a manufactured tablet is the "hardness" of the tablet. At this stage of manufacture, the tablet undergoes actions that affect the integrity of the tablet. These actions include mechanical stresses as a result of additional manufacturing operations (tablet coating), packaging, and shipping.

Tablet hardness is currently measured by applying a force that causes the tablet to break. The applied force is measured in kilograms. The acceptable hardness range for tablets varies from 5 to 8 kg.

Another dimensional measurement unit that is less frequently employed is a unit known as the kilopond. On 31 December 1977, the European Economic Council discontinued use of the kilopond. While older tablet hardness measurement units may still be in use, older tablet or dietary operations are, or have, converted to the kilogram system. For practical purposes, a kilopond is considered the equivalent of a kilogram.

There are several devices that are employed to determine tablet hardness. The most commonly used tablet hardness testers include:

- Monsanto tester (Stokes)
- Pfizer tester
- Strong-Cobb tester
- Erweka tester
- Schleuniger tester
- Copley Tablet Hardness Tester TH3

Tablet hardness (5–8 kg) is determined by a standardized USP procedure. Apparatus and defined procedures are contained in USP <1217>, Tablet Breaking Force. This monograph, <1217>, details the requirements needed to determine acceptable tablet hardness and the procedures.

Another important property of tablets is tablet friability. Friability is the tendency of uncoated tablets to break into pieces when exposed to forces (stresses) or contact with other tablets. To evaluate tablet friability, a standardized procedure, USP <1216>, Tablet Friability, is used. This procedure is applicable to uncoated finished tablets.

The test procedure for tablet friability testing consists of randomly selecting 30 uncoated finished tablets, 10 of which are selected for testing in a plastic drum that is 38 ± 2 mm in thickness and a diameter of 302 ± 5 mm. The drum has a curved "shelf" with an inside radius of 80 ± 5 mm. Figure 7.7 illustrates the basic configuration. The test procedure commences by 10 of the 30 tablets 100 times in the drum. Subsequent to 100 revolutions, the tablets are then individually weighed. The weight loss of each tablet is then recorded. After cleaning the test drum, another 10 tablets are selected and the test is performed a second time. If the average weight loss of the two tests is greater than 1%, the test should be repeated a third time. The average weight loss of the three tests should be less than 1% for an acceptable result.

Another USP quality test is USP <701>, Disintegration. This paragraph defines the required procedures and acceptable results required to conform to finished uncoated tablets, coated tablets, and gelatin capsules. For this procedure, the apparatus used for this dissolution test is the same, regardless of the tablet or capsule type. A random sample of 18 uncoated tablets is selected for the first test. The basic assembly consists of a circular transparent plastic rack container unit with a transparent basket contained within the container or beaker. The outside cylinder is 138–160 mm in height with an inside diameter of 97–115 mm. The unit has a design volume of approximately 1000 ml. Contained in the cylinder is a submersible basket with six concentric holes on a disk located on the top of the basket. The basket has a diameter of 90 of 92 ± 2 mm and a height of 77.5 ± 2 mm. Six removable transparent vials are placed in the holes prior to testing. The inner basket is attached to a vertical rod that moves up and down at a rate of 29–32 cycles/min. The vials contain five small concentric holes at the vial bottom. For uncoated tablets, the test beaker medium is distilled or USP grade water, which is stabilized to a temperature of 37 ± 2 °C. The testing commences by placing one dose of uncoated tablets (typically, one tablet in each vial). The tablets are submersed in the beaker for 1 h. After 1 h of

Figure 7.7 Typical tablet pan coater.

submersion in the water, the tablets are inspected. The tablets are inspected to verify they have all dissolved in the water. The procedure is acceptable if all six tablets have dissolved in the water. If one or two tablets have not dissolved, the test is repeated using the remaining uncoated tablets. The requirement for the repeat test is met if no fewer than 16 of the 18 tablets are disintegrated. When the tablets have disintegrated, any remaining residue must pass through a 10 Mesh screen (Table 7.4).

The disintegration time for an uncoated tablet is 5–30 min. Similarly, for coated for a coated tablet, the disintegration time is 1–2 h.

For coated tablets, while the basic apparatus is used, a different liquid test media is used. For coated tablets, a simulated gastric fluid is used. Simulated gastric fluid is a composition pepsin, hydrochloric acid, and water. This solution is used to test the ability of coated tablets not to dissolve. For coated fluids, a simulated intestinal fluid is the test media. The simulated intestinal fluid is composed of a solution containing monobasic potassium phosphate, sodium hydroxide, hydrochloric acid, pancreatin mix, and water with the pH adjusted to 6.8.

Another important tablet quality control test is identified in USP <711> Dissolution [11]. This paragraph identifies and defines the apparatuses and procedures needed to successfully evaluate tablet dissolution testing. For this testing, four approved dissolution apparatuses are available. All of the four testing procedures are performed at a temperature of $37 \pm 0.5\,°C$. The four test apparatus designs are:

1) USP Dissolution Apparatus – Basket
2) USP Dissolution Apparatus – Paddle
3) USP Dissolution Apparatus – Reciprocating Cylinder
4) USP Dissolution Apparatus – Flow Through Cell

Of the four approved testing units identified, Apparatus 2, the Paddle configuration, is most commonly used. The equipment employed for Apparatus 1 is a basket design similar to the apparatus used in USP <701> Disintegration, with the exception that a paddle is used in lieu of a basket. The paddle design basically consists of a 75–75 mm long hollow closed rectangle with a width of 4.0 ± 1 mm with a $9.4–10.1 \pm 1$ mm diameter shaft attached to the center of the rectangular hollow paddle. The shaft is powered by a device that causes the paddle to rotate at a fixed rpm rate. The basic test consists of the paddle rotating submersed in a standard reagent solution with tablets contained in the rectangular paddle for a specified time. Further, the primary reagent for USP <711> is a standard solution composed primarily of pepsin, hydrochloric acid, and water.

Evaluation of tablet acceptability for USP <711> is predicated on a calculation technique based on a value identified by the letter Q. Q is the quantity of dissolved active ingredient expressed as a percentage of the labeled content.

The dissolution testing can be performed in three stages [12]:

1) Stage 1 – Six tablets are acceptable if all tablets are not than the monograph (<USP> 711) tolerance limit (Q) plus 5% if fail.
2) Stage 2 – Another six tablets are tested individually. The tablets are acceptable if the average weight of the 12 tablets is greater or equal to, but no 1 less than, (Q-15%).
3) Stage 3 – 12 more tablets are tested. The results are acceptable if average value of the 24 tablets is greater than or equal to Q and no more than 2 tablets are less than (Q-15%).

7.4.2 Tablet Press Maintenance

As might be assumed, tablet manufacturing is, arguably, the most germane aspect of the tableting operation. Maintenance activity is most important for two major reasons, the production equipment must conform to the regulations defined in 21 CFR 211.67 (Equipment Cleaning and Maintenance) and the need to have an efficient and profitable tablet press operation. High tablet manufacturing numbers, coupled with a time demand and a compressed maintenance and cleaning schedule, and operator involvement, in many cases, can create significant mechanical and operational stresses on many tablet press configurations. As a result, common tableting problems occur, regardless of the tablet press type or vendor. Some common tablet press issues have been identified along with recommended remediation solutions [13]. These issues include:

- Fill Cams (Figure 7.6I) – It is critical to verify the proper lower cam is installed. If the cam is too low, excess powder will create tablet weight variation. Also, lower cam positioning and vendor dimensions should be regularly verified as part of the preventive maintenance program.
- Feeding and flow problems – Bridging of powder in the feed hopper (<USP> 1174, Powder Flow and Section 7.3 contain a more detailed description of this phenomena). This action occurs when there is no flow in the powder feed hopper. The flow is halted by a buildup of powder at a location in the hopper. The cause can be primarily attributed to moisture (high humidity), particle shape (deformation during transit), or a formulation error. Installation of a sight window in the feed hopper is recommended as a resolution of the problem.
- Powder accumulation on the punch and die – This action is often a common occurrence when the tableting system undergoes start-up. The cause is often caused by a lack of lubricant on the punch face and die such as the excipients identified in Table 7.1. Prior to press operation, the punches and dies are thoroughly cleaned. As a result, there is no lubricant, initially, on these components. If powder is an issue, the expedient resolution is to have an

initial start-up where the operating maintenance issues are resolved by removing accumulated powder mix from the punch and die. By removing the powder mix only, a surface film of lubricant should remain on the die and punches.

- Adjusting tablet press parameters – Calibrating of rotation speed (i.e. RPM) and upper punch penetration in the die are dimensions and procedures that are crucial to effective press performance and should be performed as routine maintenance operations.
- Low tablet hardness – Low tablet hardness results in a higher tablet defects level. Tablet hardness should be continuously checked and monitored during tableting operations to verify tablet hardness is in the minimum range of 5–8 kg).
- Maintenance and training – A great deal of tablet manufacturing relies on personnel involvement in the operation. It is common for one individual to man a tablet press during operation. Often, because of the powder dust generated by the operation, operators wear protective garments and face protection to minimize inhalation of powder. This is particularly relevant when certain hazardous API compositions are used in the tableting operation. Further, since tablet manufacturing quantity is very important, training of personnel involved in tableting operation is very important. Consequently, operating personnel and maintenance personnel should be continuously updated with operations updates. This aspect of operations is implemented, primarily by updating involved personnel about change orders (updating of SOPs), equipment modifications, and maintenance alterations. These operational aspects are accomplished by continuous updating, via training. For maintenance personnel, an effective computerized maintenance management system (CMMS) is the most important element of maintenance and training. Chapter 5 addresses the role of RAM in equipment maintenance. Section 5.4 encompasses the importance of training and maintenance.
- Condition of punches – The tablet rejection rate is increased when worn tablet punches are used. Scheduled inspection of production punches is also a critical element for an acceptable product.

7.5   Coating

While this section addresses tablet coating unit operations for pharmaceutical and dietary operations, it is important to note that coating equipment is also used in the food (sugar coatings and glazes) and agricultural (insecticides and fungicides) industries and much of the designs employed for tablet coating are based on coating equipment derived from other processes.

7.5.1   Tablet Coating

Tablet coating is a process involving the application of thin (20–200 μm) polymer-based water soluble in coatings to an appropriate substrate (e.g. tablets, capsules, pellets, etc.) under conditions that permit an acceptable balance between liquid (the coating solution) addition rate and the drying rate to be achieved. The coating operation should ensure a uniform distribution of coating solution on the tablet surface and conform to visual and functional quality of the finished items in the batch. The coating is soluble in intestinal secretions but, typically, not those of the stomach (enteric coatings). These types of coatings are most commonly applied by devices known as pan coaters or fluidized bed devices. The most widely used coating system is the pan coater. Figure 7.7 displays a large tablet pan coater.

Tablet coating is performed to achieve several objectives. The capability to mask odor and the taste of a tablet is an important factor. Protecting the drug from degradation due to exposure to light, oxygen, and moisture (i.e. separating the ingredients from the environment) is a further rationale to coat tablets. Tablet coating also enhances ingestion of the drug in the gastrointestinal tract by timing the coating dissolution rate and thus modifying the drug release rate. Coated tablets also protect the stomach lining from certain drugs. Coated tablets can also improve the mechanical integrity and handling procedures during operations such as packaging.

The basic coating process involves placing a quantity of tablets in a cylindrical pan (depending on size, the pan can be identified as a bowl for smaller units or a pan for larger volumes), which rotates about an axis inclined at an angle of about 30° for smaller units and a horizontal direction for larger capacity coaters. The diameter of the pan varies from 12 in. for smaller units (bowls) to 48 in. for high capacity units (pans). For units with a 24 in. diameter, the pan rotates at a speed of about 12–13 rpm. Larger units with a diameter of about 48 in. (there are tablet pan coaters with a diameter of 65 in.) typically have a rotating speed of about 6 rpm. The length of the rotating cylindrical pan ranges between 10 and 14 ft for large production pan coaters. The *L/D* ratio for larger coaters is approximately 2.5–7. With a proper pan load (about 850 kg or 1875 lb), the coater supplies sufficient coating solution to effectively wet the tablet surfaces. Internal baffles (e.g. long stainless steel blades mounted perpendicularly from the back of the coater pan to the front of the pan and evenly spaced at angles within the pan inside wall; the baffle blade edges often are at a right angle) enhance the tumbling and mixing action while coating drying is simultaneously performed by a stream of hot dry air circulated throughout the pan. A typical tablet coater is equipped with 12–13 baffles in the pan. The pan has small perforations that allow the hot air to circulate within the pan for the drying operation. During the drying process, small quantities of an inert dusting powder may be added to reduce tablet cohesion and tackiness (sticky tablets). Located between 6 and 10 in. above the tablet bed and adjusted to an angle of between 50 and 60° are spray devices (spray guns) that spray a fixed amount of coating solution on the rotating tablet bed. For batch pan coaters, the tablet bed occupies about 10–15% of the pan coater volume, or as noted above, about 850 kg (1875 lb). Typically, the tablet bed depth during coating operations is about 20 in.

Tablet coating solutions are stored, mixed, and pumped to the spray guns from a tank located in a location convenient to the pan coater yet capable of providing enough space to perform filling, mixing, cleaning, and maintenance activities. The location should also conform to cGMP criteria.

Subsequent to approval, acceptance, and release of the production batch by the quality assurance operation, the tablets are then forwarded to the packaging operations. While each product may vary somewhat in packaging operations, the basic operations include filling a fixed number of tablets in a predetermined container size (the sequence of packaging varies), removal of any bulk materials from the packaging line (partial label rolls, rejected materials, and waste from other earlier packaging procedures), bottle unscramblers (alignment of bottles in the packaging line), a bottle capper that screws the cap to the plastic container with a predetermined torque, an automated metal detector, a capping system, the labeling operation and a heat sealing tunnel that seals a foil disc to the mouth of the tablet container, container labeling equipment that must conform to the requirements of 21 CFR Subpart E (Control of Components and Drug Product Containers and Closures), and devices known as cottoners that are often used to insert a small wad of cotton in the container prior to sealing.

7.5.2 Tablet Coater Maintenance

Maintenance operations are focused on four areas of pan coater performance:

1) Temperature – Temperature control instruments should be calibrated at a minimum of every 6 months.
2) Agitation/mixing during processing – Proper tumbling and mixing during processing are most important to assure uniform tablet coating. Maintenance for this procedure includes assuring baffle integrity inspection, deformed baffles, and confirming baffles are securely attached.

3) Airflow – The air circulating within the pan coater is generated from vents at the bottom of the pan, below the tablet bed. Hot air is passed through the perforations in the pan and into the pan volume above the tablet bed. Humidity in the air flow is also an important consideration. An acceptable coating is dependent on a proper aqueous concentration in the coating solution.

4) Spray gun pressure – Clogging of the spray gun nozzles is very common. When nozzle clogging occurs, it is common to halt the coating operations to clean clogged nozzles. The buildup of solids around the spray gun nozzle areas is known as "bearding." Minimization of bearding can often be accomplished by agitation of the coating solution in a location as near as possible to the spray gun. Ideally, an agitation mechanism located in the spray gun nozzle would be a significant operational enhancement and could significantly reduce the downtime due to spay gun nozzle cleaning.

To effectively monitor the coating operations, it is common to install a unit referred to as a human–machine interface (HMI). This unit provides real-time operating information for incoming air temperature, airflow, humidity, and coating pan rpm as basic information. Some units, in addition to monitoring, also have the capability of automatically adjusting process variables should an OOS operation occur during the coating process. These automated systems are often referred to as "SCADA" (supervisory, control, and data acquisition) systems.

In addition to employing an HMI system or an SCADA unit, there are other key components of a tablet coating system that should be identified as integral components of an effective tablet coating process. In addition to spray gun units and tablet coating pans, there are other components that are critical to operations and also require maintenance attention in varying degrees. These components are:

1) Solution tanks – Often, these tanks are equipped with baffles, installed vertically on the inner tank wall. The baffles serve to enhance mixing and agitation of the solution. One maintenance issue involving the mixing solution tanks is the accumulation of solids buildup at the crevice where the tank wall interfaces with the baffle. The buildup of solids during and subsequent to mixing operations can create a serious corrosion phenomenon if the tanks are fabricated of 300 Series stainless steel where chlorides are components of the coating solution. If not properly addressed, an action known as chloride stress corrosion can occur. (Detailed information concerning this corrosion activity can be found in Chapter 8 and, specifically, in Section 8.3.1, Electrochemical Action. Section 8.3.1 addresses maintenance issues as well as engineering controls when dealing with chloride stress corrosion.)

2) Mixers and blenders – An effective procedure for maintenance monitoring of mixer performance is to monitor the full-load amperage (FLA) of the mixer motor during operations. (A more detailed overview of the relationship of the FLA and motor performance is found in Chapter 3, specifically, Section 3.1.2.)

3) Pumps – Transporting coating solutions with a high solids content can create an operation and maintenance problem. Pumps, particularly when operating at low flow rates, can impede flow if solids accumulate in the pump casing/impeller. Similarly, low flow rates can cause low flow due to solids buildup in the line. Maintenance requires line flushing and/or pump cleanout, time-consuming maintenance activities. As a preventive maintenance action, if steel piping is employed, installation of wide radius elbows in the coating solution transfer line may reduce the proclivity for transfer line clogging. Another potentially effective and cost-effective action involves replacement of the traditional centrifugal pump or positive displacement pump with a peristaltic pump and associated flexible tubing, in lieu

of steel or high performance polymer piping. One economic consideration of the peristaltic pump and tubing installation is the solution being transported is a mixture of excipients and most often does not require components such as tubing disposal to be classified as a hazardous waste, thereby significantly reducing disposal costs and transportation cost (i.e. in lieu of a hazardous waste manifest, the wastes can be classified as non-hazardous waste, requiring only a bill of lading).

4) Steam supply – Monitoring for any change in the heat transfer rate. Over time, a reduction in normal operating temperature is an indication of scale buildup in the coater steam supply. If this scenario is a concern, a tube cleaning/steam system scale removal maintenance program should be considered. Also, the steam water system purity should be monitored on a continuous maintenance schedule.

5) Filters – An effective filter integrity and performance technique is to install pressure differential measuring devices. Often, filter replacement is routinely performed, whether or not filter replacement is necessary. By establishing a tablet coater pressure drop specification, a conceivable savings can be realized due to a lower requirement for filter replacement.

6) Alarms and warnings – As part of the maintenance operation, it is important to include specific alarms and warnings for the installed tablet coater. Critical alarms and warnings for effective maintenance and preventive maintenance should integrate the following *critical alarm* items: Emergency Stop, HMI/SCADA interfaces alarm, Control system power faults, and Main Air Fault Alarm. Preventive maintenance procedures can be formatted by fashioning a template similar to the example of Table 10.15.

When preparing a solution for the tablet coating blend, high purity water is the most common solvent. While water is commonly used, it can present some problems that may require careful evaluation. Some issues that may require evaluation include:

- Increased processing times – Because water has a higher boiling point than other organic solvents (e.g. ethanol), a longer residence time may be required due to the higher heat of vaporization. The higher energy cost of using an aqueous composition may significantly impact tablet coating costs. If an organic coating solution can be considered, savings may be realized by installing a solvent recovery unit to allow reuse of the solvent.
- Deleterious effect on specific tablet ingredients stability if water is not significantly removed during coating operations.
- Higher processing requirements may affect drug stability and dissolution properties.
- Aqueous-based solutions generally have a higher viscosity than organic solvent-based coating solutions. As a result, higher pumping costs may be encountered. Also, the increased viscosity of an aqueous-based coating solution can negatively impact the operation of the spray gun performance.

7.6 Capsules

In addition to tablets, another solid dosage product is in the form of capsules. While capsules are widely used for pharmaceutical and dietary product containment, neither form, hard or soft capsules, is not produced in the quantities as tablets. Tablets are often produced at a rate of 15 000/min, while high speed capsules have a production rate of about 3000 capsules/min. Capsules, however, do fill a niche requirement for certain solid dosage applications such as using capsules for clinical studies, where proper dosage can be quickly adjusted without the need to rely on a tableting line.

7.6.1 Capsule Fundamentals

Capsules offer an advantage over tablets because they are easily swallowed and quickly dissolved in the stomach. Other advantages of capsules are:

- Capsule shells inert and offer little or no interaction or digestive difficulty when administered.
- Presents no taste or odor problems when administered.
- Economical to produce when compared with tablets.
- Easily digested and offers little compatibility issues with other pharmaceutical products.
- By addition of excipients such as titanium dioxide, to the capsule formulation, it is possible to block out light, thereby protecting certain ingredients from light damage.
- Capsules can be easily opened, allowing for mixing the powder in an altered form such as adding the powder to water or other digestible substances, i.e. mashed potatoes, milk, for ingestion by individuals who encounter difficulty in swallowing tablets or capsules.
- Ease of handling and carrying, in ambient environment.

Capsules also have several disadvantages, including:

- Capsules are more expensive than tablets.
- Some drugs that may be hygroscopic and can absorb moisture from the capsule and making the capsule more brittle and resulting in an inability to use for capsule filling.
- Capsules are not air tight; the shelf life is shorter than tablets.
- Unsuitable for concentrated solutions that require dilution prior to encapsulation.
- Unlike tablets, capsules are filled with powder that is not significantly compressed and thereby occupies more volume than compressed tablets.
- Capsule filling volumes can limit the mass of product that can be filled, which can limit the potency of the capsule.

7.6.2 Capsule Materials and Manufacturing

Capsules can be divided in two basic categories, hard shell capsules and soft shell capsules, both of which are manufactured from the same materials. The major component of both capsules is gelatin. Gelatin is composed of the bones of crushed pig or calf followed by common unit operations. A major component of this crushed material is protein, which serves as the connective tissue for the gelatin. This abundant protein is known as collagen and accounts for about 30% of whole body protein mass.

Gelatin is a chemically modified (hydrolyzed) form of gelatin and serves as the "glue" for gelatin cohesion. The basic processing operations involved in capsule gelatin preparation are displayed in the flowchart shown in Figure 7.8 [14]. Since much of the gelatin manufacturing operations involve organic reactions, gelatin preparation can be a time consuming process and could require many weeks to complete the processing (in some cases, up to 30 weeks of processing is required).

While gelatin capsules were previously fabricated by pharmaceutical manufacturers, currently, capsule shell production is becoming an item that is increasingly being fabricated by third party suppliers.

Capsule production commences with solubilizing the dry gelatin with a mixture of plasticizer and purified water. The gelatin solution is then heated. At this point in the process, metal molds in the form of stainless steel pins are submersed (dipped) in the heated gelatin solution. The viscous gelatin is allowed to dry on the pins (about 150 pairs of pins per machine) and when dry are then removed from the pins. Capsule production is an automated operation, and

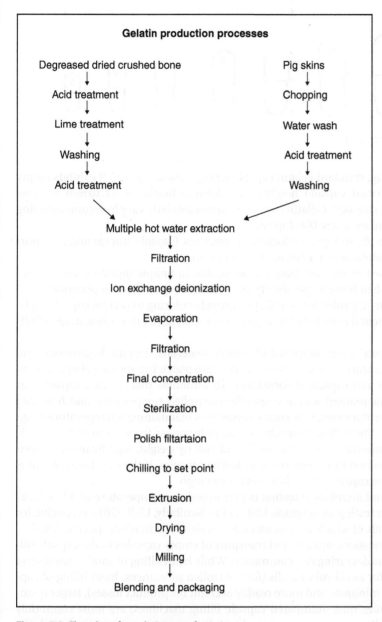

Figure 7.8 Flowchart for gelatin manufacturing.

in addition to dipping and drying, capsule finishing operations include automatic machining procedures such as spinning, cutting capsules to a fixed length, trimming, and joining the empty caps and bodies of the capsules. Blasts of cool air, which dry and solidify the capsules, are the next operation.

Once the capsules are dry, the capsules are separated from the pins by a set of mechanical "jaws" and then trimmed to the correct length by fixed blades while the capsules are being rotated on a fixed clamping device to keep the capsules in place. The operation is automated, and the manufacturing time for producing the 150 pairs (a total of about 300 capsules) is about 45 min.

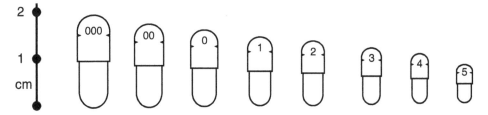

Figure 7.9 Empty gelatin capsule sizes.

To facilitate capsule filling, standard gelatin capsule sizing is used. Figure 7.9 exhibits empty gelatin capsule sizes. Finished capsule heights vary from a height of 11.10 mm for size 5–26.14 mm for the largest size, 000. Gelatin capsule volumes similarly vary in volumes ranging from 0.13 ml for 5 to 1.37 ml for a size 000 capsule.

Empty capsules have specific storage and logistics properties. Capsule storage and transport temperatures should be maintained at a constant temperature.

While pork are often used as the raw material, often, due to unique ingestion and dietary criteria, beef skin and crushed bone is specifically used for these specialized applications.

Another property of many capsules is the ability to provide coloring to certain capsules. The addition of coloring (excipients) is included to augment manufacturing in various stages of filling and finishing.

Since finished empty capsules are composed of organic materials, they are hygroscopic and contain a high degree of moisture. The water content acts as a plasticizer for the gelatin and can cause capsule degradation and capsule deformation. To minimize this activity, capsules, as noted above, should be maintained within a specific controlled temperature and humidity range. Typically, the temperature range for empty capsules is maintaining a temperature ranging between 15 and 25 °C. The relative humidity commonly ranges from 35 to 65%. Capsule moisture content varies, approximately, between 13 and 16% by weight. Significant deviations from the approved specification range can result in brittle capsules, if the moisture content is too low and softening of the capsules if the humidity is too high.

Capsule disintegration and dissolution testing is performed at a temperature of 37 ± 1 °C, as specified in the dissolution testing monograph USP <711> Similarly, USP <701> is specific for disintegration testing, details of which are discussed in Section 7.4.1 in more specific detail.

Upon completion of fabrication, storage, and transport of empty capsules to the capsule filling operation, the final manufacturing can commence. While hand filling of small quantities of capsules is still employed for small volume fills (for controlled substances, hand filling of capsules is sometimes used to minimize and more readily account for product losses), larger quantities, i.e. up to 3000 capsules/min, automated capsule filling machines, are most commonly used. While the capsule filling process may vary depending on the equipment selected, the filling operation uses some common operations in the filling operation. The filling process commences by aligning the empty capsules in a vertical position in the die so that the bottom half of the capsule is then oriented so that the exact mass of powder can be filled in the capsule bottom. Some filling machines use procedures such as applying a vacuum during the filling step. Other machines rely on an auger to implement the capsule filling step. Once filled the powder is then tamped by an overhead punch using a fixed force so as not to damage the capsule bottom. After the tamping step, the top of the capsule is then joined to the now filled capsule bottom using a device that will cap the filled capsule bottom, thereby creating a complete tablet that is then ready for surface cleaning, quality control analysis, and packaging.

A summary of the gelatin capsule filling process is presented in Figure 7.10.

Figure 7.10 Process of filling gelatin capsule.

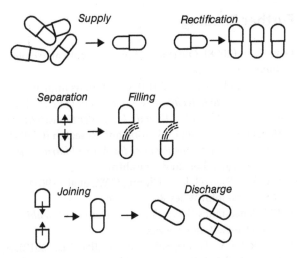

While other forms of capsules are manufactured (e.g. softgels, one piece capsules intended for liquid or oil based formulas), the manufacturing operations are basically akin to gelatin capsule operations.

References

1 Optimizing the tableting process with a quality by design approach (2012). *Pharmaceutical Technology* 36 (5): 48–53.
2 Eckrich, T.M. (2013). Emerging technologies create opportunities in the API outsourcing industry. *Pharmaceutical Outsourcing*, Special Edition 14 (5): 4–5.
3 *Tableting Technology 2000, Expediting the Formulation and Fulfilling Demanding Manufacturing Requirements*, Philadelphia, PA (15–16 June).
4 Solid dosage and excipients (2011). *Supplement to Pharmaceutical Technology* 2011 (2), Pharmaceutical Technology Editors, (April).
5 Carr, R.L. (1965), Evaluating flow properties of solids. *Chemical Engineering* 72: 163–168.
6 Thomas, C.C. and Durian, D.J. (2013). Geometry dependence of the clogging transition for tilted hoppers, University of Pennsylvania, Department of Physics (3 May).
7 ASTM International (2006). *Standard Test Method for Shear Testing of Bulk Solids Using the Jenike Shear Cell*, ASTM D-6128. West Conshohocken, PA: ASTM International.
8 Roark, R.J. and Young, W.C. (1975). *Formulas for Stress and Strain*, 5e, 10–11. New York: McGraw-Hill.
9 Jenike, A.W. (1964). Storage and Flow of Solids, Bulletin No. 123, the Engineering Experiment Station, the University of Utah, Salt Lake City.
10 Maynard, E. (2013). Ten steps to an effective bin design. *Chemical Engineering Progress* 109 (11): 25–32.
11 *United States Pharmacopeia* (2007), Edition 30, USP.ORG, Rockville, MD.
12 Sahoo, P.K. (2007). *Pharmaceutical Technology – Tablets*. New Delhi: Delhi Institute of Pharmaceutical Sciences and Research.
13 Kirsh, D. (2015). Fixing tableting problems. *Pharmaceutical Technology* 39 (5): 58–59.
14 Bhatt, B. (2007). *Pharmaceutical Technology – Capsules*. New Delhi: Delhi Institute of Pharmaceutical Sciences and Research.

Further Reading

Dave, R.H. (2008). Overview of pharmaceutical excipients used in tablets and capsules. *Drug Topics* (24 October).

Djune, D. and Hack, M. (2013). Continuous improvement in tablet coating and dry granulation. *Pharmaceutical Technology Supplement* (March) (2).

Ennis, B.J. (2006). *What Exactly Is Powder Flowability?* (May). www.powdernotes.com.

FDA, Center for Drug Evaluation and Research (CDER), ICH (2001). *Guidance for Industry, Q7A, Good Manufacturing Practice for Active Pharmaceutical Ingredients*, Rockville, MD (August). www.fda.gov/cber/guidelines.htm.

Foust, A.S., Wenzel, L.A., Clump, C.W., et al. (1962). *Principles of Unit Operations*, pp. 525–539. New York: John Wiley & Sons, Inc.

IDRG Consultancy Services (2009). *Detailed Feasibility Analysis of Drug Formulation*. Singapore: IDRG Consultancy Services.

Robertson, I.A., Tiwari, S.B., and Cabelka, T.D. (2012). Applying quality by design for extended release hydrophilic matrix tablets. *Pharmaceutical Technology* 36 (10): 106–116.

Singhai, S.K., Chopra, V.S., Nagar, M., et al. (2010). Scale up factor for determination of V blender: An overview. *Der Pharmacia Letter* 2 (2): 408–433.

Sutton, S. (2003). Formulation & solid dosage, a technology primer, *Supplement to Pharmaceutical Technology* (2).

8

Corrosion and Passivation in Pharmaceutical Operations

8.1 Corrosion

The intent of this chapter is to provide a basic understanding of the corrosion process and to provide fundamental information involving corrosion prevention, corrosion mitigation, and remediation of corrosive activity.

8.2 Corrosion and Corrosion Protection in Pharmaceutical Operations

As an indication of the cost of corrosion and corrosion-related maintenance in 2011, Air Force officials spend $1.5 billion annually, with the costs rising as the fleet ages, according to Carl Perazzola, a deputy director of the Research Laboratory at Wright-Patterson Air Force Base. Also, in the year 2006, a study headed by the National Association of Corrosion Engineers, the US Department of Transportation, and other involved organizations reported that the costs involved in corrosion damage to the US economy was $276–$333 billion annually. This cost represents about 3% of the gross national product [1]. The latter sum, $333 billion, represents the annual cost of all oil imports for the year 2007 [2]. A significant amount of this cost could be significantly obviated (estimates range to 30%) or reduced if effective corrosion control programs were implemented.

Practical Pharmaceutical Engineering, First Edition. Gary Prager.
© 2019 John Wiley & Sons, Inc. Published 2019 by John Wiley & Sons, Inc.

Another telling quote from Randall J. Brady [3],

> It has been estimated that corrosion costs the U.S. Department of Defense more than $20 billion annually and is the number one cost driver in military life cycle costs.

In addition to being a significant, often unnecessary, expense, failure to provide an effective corrosion control program could be a significant conflict with cGMP regulations. Three provisions that require evaluation include 21 CFR 211.63, 21 CFR 211.65, and 21 CFR 211.67. While not all operations in pharmaceutical facilities are covered by cGMP regulations, many contain manufacturing and clinical components and research equipment that is susceptible to corrosion or galvanic activity. And while not a factor in and of itself, failure to maintain an effective corrosion control program can be an indication of a pharmaceutical or biotech operations overall procurement, design, maintenance, and ultimately profitability.

This chapter is designed to provide the quantitative and qualitative information necessary to evaluate, define, and access the need for a corrosion prevention program at pharmaceutical and biotech operations. For the pharmaceutical and biotech industries, corrosion analysis is greatly simplified since the metal for cGMP applications is stainless steel, the preferred alloy for corrosion prevention. However, as this chapter will reveal, even stainless steel is susceptible to corrosion. Also covered in this chapter are metals, employed to a lesser extent, which require careful evaluation for pharmaceutical and biotech applications. While stainless steels are the predominate metal (stainless steel is actually an alloy and the difference between a metal and an alloy will be addressed shortly) employed in pharmaceutical and biotech critical operations, the general theory of corrosion is applicable to all metals, including stainless steels.

Effective corrosion protection is often a function of effective preventive maintenance and a comprehensive corrosion protection plan commencing at the beginning of the design and/or installation phase of the project. In some cases, ex post facto attention to corrosion prevention may be too late. Consequently, it is most desirable to include effective corrosion prevention planning at the earliest possible phase of a project.

Two terms that will be used in corrosion analysis are a metal and an alloy:

- Metal – Any class of elementary substances, all of which are crystalline when solid and many of which are characterized by ductility and conductivity, yielding positively charged ions in an aqueous solution.
- Alloy – A substance composed of two or more metals. A particular alloy of interest in the pharmaceutical and biotech industry is stainless steel, which is an alloy of iron (Fe), nickel (Ni), and chromium (Cr) of varying composition. Chromium is the key alloying element forming a resistant passive oxide film on the stainless surface. While stainless steel properties and information are used throughout this chapter, specific information relating to stainless steel composition and important properties are presented in Section 8.4.

For corrosion studies, the definition of a metal can also be defined as a substance that can become a cation (a positively charged ion such as Mg^{++} and Al^{++}) when exposed to aqueous solutions containing its salts. Anions are the negatively charged particles that form in aqueous solutions (e.g. Cl^{-1}, $(SO_4)^{-2}$, $(NO_3)^{-1}$). Ionization of metals in solution occurs when the solution is at a pH of below 7 (i.e. acidic solution).

Additionally, an alloy is a substance composed of two or more metals that have been intimately mixed by fusion, electrolytic deposition, or other methods. An alloy can often contain nonmetals such as sulfur and carbon. Stainless steels, in addition to metals such as chromium, molybdenum, manganese, and nickel, also contain small amounts (<1%) of sulfur and carbon.

Figure 8.1 Formation of zinc ions in dilute acid. *Source:* [4].

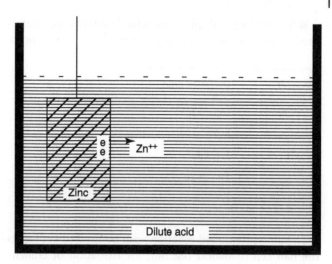

With this information, a brief overview of the chemical mechanisms relating to corrosion would be helpful in dealing with the information contained in this chapter with a particular focus on reduction and oxidation reactions, commonly referred to as redox reactions.

Two types of reactions that are central to the study of corrosion are oxidation and reduction reactions, generally referred to as redox reactions. Reduction is a gain in electrons, whereas oxidation is a loss of electrons. Also, two processes occur simultaneously. If one atom or ion gives up an electron, another atom or ion must absorb this electron. An oxidizing agent is an atom, ion, or molecule that takes up electrons from other materials that are present. A reducing agent gives up electrons to other substances. Redox reactions are those in which atoms undergo a change in valence. A simple redox reaction is given by $Zn + H_2SO_4 \rightarrow ZnSO_4 + H_2$, as shown in Figure 8.1 [4].

This is a redox reaction since the zinc atom changes valence (also referred to as the oxidation number) from +1 to 0. Further, the reaction $FeS + H_2SO_4 \rightarrow H_2S + FeSO_4$ is a double decomposition reaction since all the product atoms have the same valence as the starting components with the only alteration being a change in "partners."

Many redox reactions are reversible, or they can go in either direction. An example would be reacting the ferrous ion (Fe^{++}) with chlorine gas producing the reaction $2Fe^{++} + Cl_2 \rightarrow 2Fe^{++} + 2Cl^{-1}$. But by reacting the ferric ion (Fe^{+++}), the resulting reaction is $2Fe^{+++} + Cl_2\uparrow$. As can be seen, the first reaction, Fe^{++}, acts as the reducing agent, while the second reaction, Fe^{+++}, performs as an oxidizing agent.

It is convenient to divide redox reactions in two parts by focusing on the half cell or half reactions, the reduction and the oxidation reaction. The redox mechanisms can be identified as:

1) Reducing agent = oxidized species form + electrons – a higher valence (oxidation number)
2) Oxidizing agent + electrons = reduced form – a lower valence (oxidation number)

A complete redox reaction occurs when an oxidizing agent reacts with a reducing agent, which is described by combining or adding the two half reactions (i.e. an oxidation half cell reaction would be the chloride ion oxidized to chlorine gas, $Cl^- \rightarrow Cl_2\uparrow$, or a chloride ion reacting to form chlorine gas, $Cl^- \rightarrow Cl_2\uparrow$).

Redox reactions in an aqueous medium involving corrosion are ionic reactions. A simple example would be a solution containing $FeSO_4$, a reducing agent. But, since corrosion is specific to metals, the SO_4^{-2} ions are not altered and do not affect the reactions the metal is undergoing and are consequently not a focus of corrosion analysis.

When dealing with pharmaceutical systems and metals, several reducing agents are widely used. Some common reducing reagents and their associated oxidation products are:

Reagent	Oxidation product
HNO_3	$NO_3^- + H^+$
H_2SO_4	$SO_4^{-2} + 2H^+$
I^-, Cl^-	$I_2\uparrow, Cl_2\uparrow$
Fe^0	$Fe^{++} + 2e^-$
Fe^{++}	$Fe^{+++} + e^-$

The aforementioned reactions can be used to get half cell oxidation reactions. For instance, the half cell oxidation reaction for oxidation of chlorine is obtained by from the reaction $Cl^- \rightarrow Cl_2\uparrow$. A reaction more germane to metals would be the iron oxidation reaction $O_2 + H_2O + 4e^- \rightarrow 4(OH)^-$. From the previous list, an important iron oxidation reaction is

$$Fe + 2(OH)^- \rightarrow Fe(OH)_2$$
$$2Fe + 3H_2O \rightarrow Fe_2O_3 + 6H^+ + 6e^-$$

Also from the list, another important anodic oxidation reaction for steel is

$$Fe \rightarrow Fe^{++} + 2e^-$$

The associated cathodic reaction is

$$Fe^{++} + 2(OH) \rightarrow Fe(OH)_2$$

The iron oxidation half cell reactions are the reactions that cause corrosion. As a general rule, half cell reactions that create soluble ions in solution, such as hydroxides, are a cause of corrosion.

Oxidation caused by corrosion can, in some cases, produce an oxide layer that will protect the metal beneath the surface by creating a protective or passive oxide layer on the metal surface. This layer will significantly slow the rate of corrosion. An example of these passivation phenomena is aluminum oxide (Al_2O_3). This oxide protects aluminum from corrosion. Other metals that exhibit this passive surface oxide behavior are zinc and titanium oxides.

In addition to the oxidation reactions that play a role in corrosion, there are also important reduction reactions commonly encountered in corrosion and noncorrosion of metals. Some of these reduction reactions are

$$PbO_2 + H^+ (asid) \rightarrow Pb^{++} + H_2O$$
$$O_2\uparrow + H_2O + 4e^- \rightarrow 4(OH)^-$$
$$O_2\uparrow + 4H^+ + 4e^- \rightarrow 2H_2O$$
$$2H_2O + 2e^- \rightarrow H_2\uparrow + 2(OH)^-$$
$$2H^+ + 2e^- \rightarrow H_2\uparrow$$

The latter two hydrogen-forming reactions, unlike oxidation half cell reactions, are important in creating a barrier to galvanic corrosion and is a key component in the mechanism of

Pourbaix diagrams addressed in Section 8.7. It is important to differentiate the distinction between oxidation half cell reactions and reduction half cell reactions since this distinction is the important element in understanding corrosion.

8.2.1 Definition of Corrosion

Corrosion is the destructive attack of a metal by chemical or electrochemical reaction with its environment. The rate of corrosion is proportional to the electrical current discharged from the corroding metal into the electrolyte (any substance that dissociates into ions when dissolved in a suitable medium – usually water when evaluating corrosion – or when melted, creating a conductor of electricity). Deterioration by physical causes is not identified as corrosion, but is described as erosion, galling, or wear. In some instances, chemical attack accompanies physical deterioration as described by the terms corrosion/erosion, corrosive wear, or fretting corrosion. Nonmetals are not included in the definition. Plastics may swell or crack, wood may split or decay, granite may erode, and Portland cement may leach away, but the term corrosion is restricted to chemical attack of metals.

"Rusting" applies to the corrosion of iron or iron-based alloys with formation of corrosion products consisting largely of hydrous ferric oxides. Nonferrous metals, therefore, corrode but do not rust.

Corrosion in the pharmaceutical industry is somewhat unique since many of the materials and alloys employed are selected because of their resistance to corrosion. However, as this chapter will reveal, a great deal of preventive activities and maintenance activities are required to prevent and design out corrosion from pharmaceutical systems employing metals and alloys. If not performed properly, the results can be costly as well as compromising as cGMP intent and regulations.

Some definitions terms employed in this chapter and specific to metals and alloys are:

- Cold working – Strengthening of a metal by performing mechanical operations (e.g. drawing, pressing, extruding) below the crystallization point, often at room temperature
- Quenching – Rapid cooling of a hot metal
- Steel – Iron in a modified form containing carbon and other constituents

8.2.2 Corrosion Fundamentals

Corrosion occurs as deterioration of a metal by a chemical or electrochemical reaction with the environment. Essentially, all reactions of metals in a gaseous atmosphere (e.g. oxygen and water vapor) or in electrically conducting liquids or electrolytes (commonly aqueous solutions of salts, acids, or bases) are electrochemical in nature.

Corrosion may proceed at a rapid or a slow rate. It is very difficult to predict or prepare a paradigm that will accurately, rapidly, and reliably quantify corrosion rates. This is primarily due to the fact that corrosion is a thermodynamic phenomenon where time is not a criterion. However, several empirical models have been prepared to approximate corrosion rates for specific corrosion events and, depending on time and data collected, can, in some cases, be reasonably accurate. As is the case with all thermodynamic configurations, the intent is to verify a system dynamic (corrosion of a metal or alloy in this case) will occur rather than predicting the rate. The rate is controlled by the environment, the concentration of reactants, and the prevailing temperature (see Chapter 3 for pharmaceutical manufacturing operating temperatures).

Since a metal can vary from high purity to an alloy containing various other elements, a wide variety of corrosion behaviors is possible. Physical structure variability, because of heat treatment, quenching, or cold working, also will influence the susceptibility to corrosion. Also, the shape, form, or finish of the metal (e.g. concave shapes, surface discontinuities, sumps for the accumulation of corrosion-forming chemicals, cast or wrought forms, and grit-blasted or mechanical finishes) will influence the rate of reaction between the corrosive agents. Environmental conditions, such as moisture, active pharmaceutical ingredients (API), inactive ingredients (i.e. excipients), and temperature can accentuate or moderate corrosive reactions to a significant degree.

Environmental conditions also greatly affect the extent and nature and severity of damage to metals. While corrosion of a metal component may be observed as localized phenomenon, it may occur beyond the initial observation. Since a corrosive environment is a function of many variables (e.g. humidity, temperature, surface pH, cleaning solutions, excipients, and APIs in contact with the metal during manufacturing and cleaning operations), the various contributing factors should be carefully analyzed in order to formulate effective preventive and remedial procedures.

8.3   General Corrosion Protection in Pharmaceutical Operations

In almost all pharmaceutical application involving metals, some type of corrosion mechanism is possible. Ideally, pharmaceutical equipment fabricated of metal or containing metal (including stainless steels) is designed or provides prescribed measures that would preclude the onset of corrosion. If such is not the case, it is the responsibility of the owner/operator to take the lead in this activity.

In analyzing and correcting a potential or existing corrosion phenomenon, four principal evaluation steps are considered. These preliminary procedures are:

1) Considering the metal and determining whether the choice is acceptable or whether other materials may be better for the intended application (this is obviously an analysis/evaluation that should be performed prior to equipment selection)
2) Determining the environmental condition that prevails and, if warranted or if possible, altering the environment (for example, cleaning operations should be performed in an environment that is similar, or identical, to the location manufacturing operation was performed)
3) Analyzing the design in which metals are employed and considering modifications that could alleviate the cause of damage
4) Eliminating one, or more, of the factors that contribute to the corrosion dynamic (i.e. electrochemical reactions)

8.3.1   Electrochemical Action

Corrosion is an operation involving the alteration elemental metals to compounds of the metals through an electrochemical reaction with the surrounding environment. A flow of electrical current is associated with a corroding metal. The operating model or paradigm for the corrosion process applies to all metals and vary by degree. The conditions required for corrosion are:

- Electrolyte (a continuous liquid path). For pharmaceutical evaluation, this is, in virtually all cases, water containing acid salts such as chlorides, nitrates, sulfates, and/or phosphates. Electrolytes with a pH of less than 7 are more prone to exhibit corrosion activity.
- Pathway. A surface area or volume through which electrons can flow, most often due to the presence of an electrolyte on a surface.
- An anode (usually a conducting metal).
- A cathode.

Figure 8.2 Corrosion conditions. *Source:* [4].

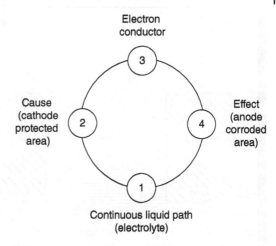

Electron
conductor

3

Cause
(cathode
protected
area)   2

4   Effect
(anode
corroded
area)

1

Continuous liquid path
(electrolyte)

- Two dissimilar metals.
- Oxidizer. Substances that readily gain or accept electrons, also referred to as oxidizing agents or oxidizers, details of which are covered in Section 8.2.

By eliminating one or more of these factors, the formation of a corrosion cell can be avoided.

An interesting symbolization of the factors contributing to corrosion is displayed in Figure 8.2. This illustration summarizes the conditions required for creating corrosion scenarios in the form of a "corrosion circle" [4].

Item 1 of the circle illustrated in Figure 8.2 represents the electron conductor, or the electrolyte. The electrolyte is the liquid solution that provides the pathway for electron transfer. While usually water with ions in solution, the conducting electrolyte can also be in the form of condensate or salt spray. Item number 2 is the source or cause of the corrosive electron flow, the cathode, the area where the electrons flow. Item number 3 represents the solid electron conductor where a structure (as opposed to an electrolyte solution) is most commonly a metal-to-metal contact such as bolts and spot welds. Item number 4 in the figure represents the effects of the corrosion cycle, the anode where the corrosive activity occurs.

Figure 8.3 exhibits an example of how the corrosion circle performs in a simple corrosion cell configuration. Note the evolution of hydrogen gas at the cathode where the reduction reaction takes place.

Figure 8.3 also illustrates the reduction reaction occurring at the zinc anode where the oxidation number increases from free zinc Zn to Zn^{++}.

When a metal is immersed in an aqueous solution, it has an electrical potential (commonly referred to as potential rather than electrical potential) in relation to the proclivity for the metal to enter the solution as metal ions and at the same time releasing electrons (e.g. in the case of iron, $Fe^0 \rightarrow Fe^{++}$).

The electromotive force series (Table 8.1) lists the metallic elements in order of potential when in contact with aqueous solutions containing one equivalent weight per liter[1] of metal ions at 25 °C (77 °F) at one atmosphere pressure (sea level).

The metals with tendencies to form ions in solution (magnesium, aluminum, manganese, zinc) are at the reactive or less noble end of the series. Thus, by examining Table 8.1, it can be observed that there is a relationship between the proclivity of a metal to corrode and its position in the series. As the electrode potential increases, the corrosion potential also increases.

---

1 The equivalent weight, in this case, is defined as the molecular weight of the metal divided by the valence (i.e. the number of hydrogen or hydroxide ions) and then blended with a liter of water.

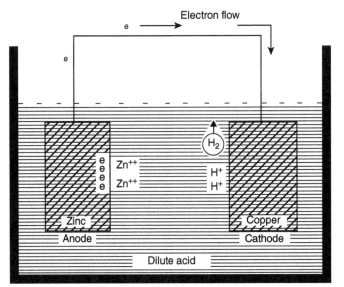

Electron flow

Figure 8.3 Corrosion at the anode and formation of hydrogen gas.

Table 8.1 Electromotive series of metals.

Least noble (anodic end/corrosive end)	Electrode reaction	Standard electrode potential ($E°$), volts at 25 °C
↑	$Mg^{++} + 2e^- = Mg$	−2.372
↑	$Be^{++} + 2e^- = Be$	−1.700
↑	$Al^{+++} + 3e^- = Al$	−1.63
↑	$Mn^{++} + 2e^- = Mn$	−1.050
↑	$Zn^{++} + 2e^- = Zn$	−0.762
↑	$Cr^{++} + 2e^- = Cr$	−0.91
↑	$Cr^{+++} + 3e^- = Cr$	−0.74
↑	$Fe^{++} + 2e = Fe$	−0.440
↑	$Cd^{++} + 2e^- = Cd$	−0.402
↑	$Co^{++} + 2e^- = Co$	−0.277
↑	$Ni^{++} + 2e^- = Ni$	−0.250
↑	$Sn^{++} + 2e^- = Sn$	−0.138
↑	$Pb^{++} + 2e^- = Pb$	−0.126
−	**Hydrogen**	0.000
	$2H^+ + 2e^- = H_2\uparrow$	
↓	$Cu^{++} + 2e^- = Cu$	+0.342
↓	$Cu^+ + e^- = Cu$	+0.521
↓	$Ag^+ + e^- = Ag$	+0.799
↓	$O_2 + 2H_2O + 4e^- = 4OH^-$	+0.401
↓	$Hg^{++} + 2e^- = Hg$	+0.854
Most noble (cathodic)↓ (least corrosive)	$Au^{+++} + 3e^- = Au$	+1.498
↓	$O_2 + 4H^+ + 4e^- = H_2O$ (in acidic solution)	+1.229
↓	$Pt^{++} + 2e^- = Pt$	+1.118

The corrosion potential is, in effect, the application of redox reactions, covered in Section 8.2, to an actual physical/chemical dynamic, an event that continuously occurs and, as noted, has a substantial cost to the economy.

A metal is capable of displacing from solutions any ions of another metal more noble than itself in the electromotive force series. For instance, when iron is immersed into a solution of copper salts and iron, this creates an ionic solution containing the metals, with metallic copper being precipitated. From an analysis of the electromotive force series, it can be concluded that metals on the more reactive side of the hydrogen electrode (less noble) will dissolve in acids with the formation of hydrogen gas.

The electrode potential of a metal is dependent on the concentration and type of ions in solution, which are usually very different from the conditions established for the electromotive series.

By employing the data displayed in Table 8.1, it is possible to ascertain if an electrochemical reaction (corrosion) can occur with the metals identified in the preceding table. By examining the half cell potentials in Table 8.1, it is possible to predict whether galvanic activity can occur if these metals possess the galvanic criteria for corrosive activity as defined and identified previously in this section. The determination can be made by evaluating the half cell reactions that occur should the conditions necessary to create a galvanic cell be present (Section 8.3.7.2). By identifying the voltage sign (+ or −) of the half cell reaction, the corrosion potential of the cell can be determined. If the half cell reactions have a positive electromotive force, the corrosion cell can form. If the half cell reaction is negative, corrosion of the material will not occur. (The + sign associated with the reaction indicates the voltage is being released, while a negative sign indicates voltage must be added.) The major variable creating the potential for corrosion is the voltage difference between two metals in the series; the greater the voltage difference, the higher the probability of corrosion. Accordingly, it would be advantageous to select metals closest to one another in the galvanic series.

The nomenclature used in Table 8.1 to identify the cell voltage is standard electrode potential (these values are compared to the hydrogen half cell reaction, $H^+ + 2e^- = H_2\uparrow$, which has an electrode potential of 0.000 V). However, other terms also used are standard potential and standard hydrogen electrode (SHE), and when dealing with graphs or diagrams, E (v), or a slight variation, is sometimes used. Again, all these symbols and terms refer to the standard electrode potential given in volts.

Example 8.1 The coupling of a pump and motor is not in alignment. The pump must have metal shims under it to align it with the motor. The pump carries phosphoric acid that is used as a cleaning agent that is used in a cheese whey evaporator heat exchanger cleaning operation. The pump has a slight leak. Two shim materials are available, strips of copper and strips of zinc. Is there a potential for a corrosive cell forming with the lead and copper strips?

Disregarding cathodic reactions (i.e. hydrogen formation) is acceptable since it is not germane to this example.

Solution
By examining the Table 8.2, we find the half cell reactions for lead (Pb) and copper, respectively:

$$Cu \rightarrow Cu^{++} + 2e^- \qquad E = -0.337V$$

(The sign is changed because the copper in contact with the phosphoric acid is oxidized and not reduced, as indicated in Table 8.1.)

$$Zn \rightarrow Zn^{++} + 2e^- \qquad E = +0.762V$$

(As noted earlier, the reverse half cell reaction is occurring; consequently, the sign is also reversed.)

Table 8.2 Factors influencing corrosion in solution.

Characteristics of the metal

Composition and homogeneity of the metal

Surface properties; inherent protective films

Corrosion cell formation potential; including the presence of variable identified in Section 8.3.2

Surface condition; physical homogeneity of metal and installed fittings

Protective deposits created as a result of contact with the solution

Characteristics of the solution

Hydrogen ion activity in the solution (pH of the solution)

Dissolved oxygen of the solution

Presence of oxidizing or reducing agents

Properties of other ions in the solution

Motion of solution in contact with the metal (e.g. piping, agitators, mixing vessels)

Temperature

Effect of corrosion products and corrosion progress

A metal oxide may inhibit corrosion reacting with other chemical elements in a solution to form a protective layer on the metal surface. Scale formation on boiler surfaces is an example of this activity

The formation of oxides may consume available oxygen where the supply of oxygen is limited. Such an action could obviate or retard further corrosion

---

Source: MIL-HDBK-721 (MR) [4].

For the copper shim, the resulting electromotive force is −0.337 V. The negative voltage indicates **the copper shim will not corrode.** ← **Solution**

The voltage for the lead shim cell is +0.762 V. Since the sign is positive, the zinc strip **can undergo corrosive activity.** ← **Solution**

Corrosion of a metal shim may appear to be minor; it should be recalled that the purpose of the shim was to align a pump and its associated motor. While the pump only operates when production is occurring, the corrosion of the shim is continuous. As time passes, the shim is corroding and losing alignment. Depending on the amortization time of the pump and motor, misalignment resulting from a failing shim may render the pump and/or motor inoperable before full write-off occurs.

Prepare the cell reaction and calculate the electromotive force for the following reaction at 25 °C to dissolve as the concentration of ions in solution increases. Similarly, the potential decreases as the ionic concentration decreases. Also, nonmetallic ions also affect the potential of the metal depending on whether or not the anions complex (combine) with the metal ions and promote dissolution of the metal. A typical case is the fact that cobalt is more reactive than iron in dilute citric or oxalic acid and tin is kept low by the complexing power of the acid. This example demonstrates that alterations of activity of the elements can occur by changes in the exposure environment in which the metals are exposed. Table 8.1 should be employed as a guideline for evaluating corrosion activities of differing metals in a particular environment.

8.3.2 Environmental Characteristics and Corrosion

There are several characteristics of an environment to which metals are exposed that have a significant effect on corrosion. As is the case with other chemical-related operations, pharmaceutical manufacturing, coupled with associated activities such as storage and transport, has exposure to activities capable of influencing corrosion scenarios. Some of the more relevant corrosion scenarios are discussed later in text.

As indicated previously, acidic environments (low pH) are generally quite corrosive to metals more active than hydrogen (Table 8.1) because of the tendency for the metal ions to displace hydrogen resulting in the continued dissolution of metal. Metals such as aluminum or sink may corrode in alkaline solution because soluble salts are formed at anodic areas and there is little tendency for polarization to occur, thus resulting in continued corrosion of the metal. In some corrosion environments employing acidic solutions, hydrogen is evolved as a gas. In this case, the cathodic reaction is

$$2H^+ + 2e \rightarrow H_2$$

Recall that the anodic reaction for the metal (M) in solution is

$$M - 2e \rightarrow M^{++}$$

The previous mechanism assumes the material is a divalent metal such as iron. Metals in a pharmaceutical environment are rarely used in environments that would produce hydrogen gas. This corrosion cell scenario is kept from forming because a thin film of hydrogen gas generated on the surface of the cathodic locales reduces the flow of electrons. This current reduction results in a lowered corrosion rate. However, if the hydrogen film comes in contact with air, the oxygen in the air can dissolve the hydrogen film. The hydrogen film is then destroyed and the metal is exposed to oxygen from the air, allowing the corrosion to proceed as such:

$$2H_2 + O_2 \rightarrow 2H_2O$$

Another possible mechanism is direct reduction at the cathode where water reacts with oxygen (from air or dissolved in water) to undergo a reaction such as

$$O_2 + 2H_2O + 4e^- \rightarrow 4OH^-$$

Slightly acidic solutions that are saturated with dissolved oxygen can undergo an increased rate of corrosion than situations where little or no oxygen is present.

Just as oxidizing agents can accelerate corrosion in some metals, they can slow attack on other metals by accelerating the production of protective films of oxides of adsorbed oxygen. Aluminum, chromium (a key constituent of stainless steels), and titanium are protected by oxidizing agents. Hence, by alloying these metals with other metals, such as iron, the resulting alloy often will possess an enhanced corrosion resistance.

Corrosion is, like most chemical reactions, usually accelerated by a rising temperature. An increase in temperature results in increased conductivity for the electrolyte as well as enhanced solubility and diffusion rates of many protective corrosion-resistant metals. Cathodic disassociation by dissolved oxygen is increased by an increased temperature. This phenomenon is the result of an increased diffusion of oxygen. Generally, when the temperature exceeds 80 °C (176 °F), the corrosion rate decreases. This is probably caused, in large part, by a decreasing solubility of oxygen in the surrounding solution.

8.3.3 Properties of Metals that Influence Corrosion

Most metals and alloys possess properties that affect corrosion. For instance, it is critical that metals and alloys are properly heat-treated in order to obtain optimum corrosion protection. An example would be a martensitic steel, which is normally quenched and tempered to provide required corrosion-resistant properties. Similarly, austenitic stainless steels are more resistant in an annealed phase and should be processed to resist intergranular corrosion. Aluminum alloys also require special processing procedures. For example, if aluminum alloys are quenched slowly, their corrosion resistance is decreased. Such alloys become vulnerable to intergranular attack and require an aging time (most often an artificial time) after slow clenching.

If an alloy is produced in a nonhomogenous process, it is possible, under certain conditions, to promote corrosion. If the extent of nonhomogeneity is manifested to a certain level, electrical potentials may form that are significantly different from the bulk metal causing anode and cathode formation.

A metal or alloy undergoing physical stress will corrode more quickly than the unstressed material when exposed to a specific environment. Residual or applied stresses in a metal or alloy can cause localized corrosion and cracking of the material.

In these instances, mill certification of the metals employed in fabrication, modifications, and maintenance activities should be available and implemented as part of an effective quality control program. Also, effective quality assurance procedures should be in place for evaluation operations. Pharmaceutical operations should not limit quality programs to only GMP- and cGMP-associated activities and systems. The corrosion process does not distinguish between regulated and nonregulated equipment and operations.

Another variable that can accelerate corrosion, is the velocity or agitation of the medium, such as water, that can expose the metal to potential corrosion in the solution (due to the increased interaction of the metal in the agitated media). Typical examples are water flowing through stainless piping and aqueous solutions being agitated/mixed in stainless vessels. Corrosive activity is often due to two specific phenomena. Specific corrosive elements, such as dissolved oxygen, are in surface contact with the metal at a higher rate as the flow or agitation increases. The second phenomenon is the removal of corrosion products, primarily metal oxides, at a higher rate that inhibits the products from accumulating and consequently inhibiting corrosive action. (Recall it is the oxide layer that prevents corrosion of stainless steels, aluminum, and other metals and alloys.) As the velocity and degree of turbulence in the fluid increases, the protective film on the surface of the metal tends to become thinner, rendering underlying metal surface vulnerable to corrosive activity. High turbulent flow (e.g. extremely high Reynolds number in excess of 50 000, typically) can also act to reduce or remove the protective film. As is the case with high velocity flow, high Reynolds number turbulent flow can also prevent and reduce formation of the protective film. A typical configuration creating this situation can be observed when small diameter tubes or pipes contain fluids with potentially corrosive components flowing at high velocity through elbows, tees, valves, and other flow-altering components. A similar action is also exhibited on agitators and centrifugal pumps when operating in this type of environment. In fact, high purity pharmaceutical water containing small quantities (5 ppm or greater) of chlorides are notorious for creating a form of corrosion identified as chloride stress corrosion, the details of which will be covered for further in this chapter.

8.3.4 Effects of Fabrication and Assembly on Corrosion

When performing tasks involving fabrication, assembly, metal working, and maintenance on both cGMP and non-cGMP, it is critical that these activities are performed in accordance with predetermined and approved procedures designed to avoid operations that may cause the product to be susceptible to corrosion. Without a detailed analysis, it is very difficult, if not impossible, to state that a certain task such as bolting, braising, casting, riveting, soldering, or welding can render a metal more vulnerable to corrosion. Primarily, the application, the environment, and most importantly the metal or alloying determine the material's resistance to the corrosion process. While corrosion resistance of installed vendor-supplied systems is most often the responsibility of the system or component supplier, the user should be familiar with the properties and standardized procedures including evaluating corrosion potential for various operations often employed in pharmaceutical installations and maintenance operations.

Corrosion resistance of bolted joints can be affected by the composition and electrode potential difference of the bolts as well as the joined parts. Here, trapped moisture, the type of exposure while the system/component is in service, undocumented stresses, and faulty installation can be causes or significant contributors to a corrosion-producing action.

While soldering is not commonly used in vendor-supplied stainless steel or carbon steel components, it may be employed for temporary quick fix, short-term maintenance operations, particularly if a nonstainless component needs immediate attention. This operation, soldering, may be employed since a soldering operation does not require metals to be heated to their melting points. A corrosion possibility exists if an electrolytic potential between the solder and the material to be joined exists. Various metals cannot be employed in corrosive environments if soldered. For instance, when joining aluminum, the corrosion resistance of welded or brazed joints is most often superior to soldered connections or joints. Again, soldering is a short-term, temporary fix and should not be used as a remedial activity.

Welding also requires careful evaluation and adherence to approved facility standard operating procedures (SOPs), vendor-supplied manuals, and related information. When welding corrosion-resistant steels such as 304L and 316L, it is critical to maintain high corrosion resistance in the vicinity of the welding operation. For this type of operation, it is imperative that welding grade stainless steel is employed (welding grade stainless steel is denoted by the letter "L" after the stainless type [e.g. 304L]). For stainless operations, certified TIG or MIG welders should be employed. It is also critical that the welding (or filler) rod and associated coating possess the same corrosion resistance as the metal to be joined via welding. A certified welder should be skilled in the proper technique that would obviate the formation of gas pockets, laps, undercuts, and excessive nonmetallic inclusions (slag) containing carbon, sulfur compounds, and other nonmetallic materials. The welding defects described previously may over time lead to localized corrosion. Associated variables such as proper welding temperatures, heat treatment subsequent to welding, inspection for defects in the vicinity of the weld, and removal of weld spatters are also important to ensure a maximum weld resistance.

Although design criteria are often beyond the scope of facility personnel because many installations are installed as an as-built configuration, design reviews should address installation features such as elbows (long radius vs. short radius), welding procedures, surface finishes and related components, and fabrication techniques that may be conducive to creating electrochemical cells. Often, short radius elbows, if not properly maintained and inspected, are susceptible to stress corrosion if deleterious substances such as chlorides invade locations that are exposed to greater stresses than the surrounding metal. This is particularly relevant when the fabrication material is stainless steel such as type 316.

The preferred evaluation technique for post-welding inspection is to employ approved quality control documentation, whether or not the operation is of a cGMP or non-cGMP component. Additionally, if a cGMP welding operation is proposed, it is desirable to have a knowledgeable welding/materials presence for management of change approval. Indeed, a quality inspection protocol should be in place for all of the operations identified in Section 8.6.

8.3.5 Protective Films and Corrosion

As noted in Section 8.6, the existence of protective films is, in many cases, a valuable component of many metals and alloys. The films or coatings that form on metallic surfaces are a result of chemical reaction with the environment and are actually the products of corrosion. As a result, these products are often oxides, other compounds, or adsorbed gases.

These films, or coatings, can vary in thickness. These coatings range from thin films to visible thickness. Invisible films that cause a metal to become more passive and exhibit more noble behavior are actually continuous nonporous films and protect the metal by barrier action and by minimizing the flow of current (amps) in the electrolyte of electrochemical systems (galvanic cells). Passive films are generally quite effective in protecting metals against corrosive attack when exposed to the medium in which they are formed. In addition to stainless steels, protective oxide coatings can form on metals such as titanium and chromium (the major component that imparts a protective coating for stainless steel). To maintain protective coating, an oxidizing environment (the pH > 7) is preferred. Conversely, a metal with a protective coating should not be exposed to a reducing environment (the pH is <7).

While thicker oxide films are effective, a thicker film is generally more porous, and this presents a potential shortcoming. More porous films can cause an electrochemical potential difference on a metal surface and can serve as an area for localized corrosion. For this reason, it is most important that effective surface treatment, such as passivation, be performed prior to implementation of the metal. Passivation is particularly relevant when dealing with stainless steel intended for pharmaceutical applications. This important topic will also be addressed in Section 8.5.1.

8.3.6 Corrosion Activity in Solutions

As described briefly in the previous sections, an electrolyte in solution (most often aqueous solutions) is the significant contributor to corrosion activity. For pharmaceutical applications, exposure to the atmosphere is often a significant environmental factor for corrosive activity. With an available source of moisture in the form of atmospheric humidity and a proclivity for moisture to condense on cooler surfaces, the formation of an electrolytic cell is quite realistic. Another scenario for corrosion cell activity is the failure to properly dry equipment subsequent to cleaning operations. While virtually all equipment cleaning procedures carefully define the cleaning criteria required, few address the need to ensure the removal of surface moisture subsequent to cleaning procedures. Most often, equipment and components that have been processed by employing carefully prepared cleaning procedures neglect to verify a drying operation. It is typical to allow components, particularly larger units such as blenders, mixing vessels, holding tanks, and agitators, to be allowed to air-dry once the cleaning operation is terminated. Similarly, for operations that have tubing and piping as components, little concern is generally given to water or water vapor contained in the piping when not in operation. Few cleaning operations require post-cleaning testing or inspection of piping or tubing employed in manufacturing operations. Yet, in many cases, corrosive chemicals, such as chlorides, can remain in the piping if inadequate pipe flushing is performed. Chlorides, in particular, when exposed to the variables present for corrosion cell formation (Section 8.3.7.3) are very likely to create the phenomenon identified as chloride stress corrosion.

Since water is the most plentiful and common component of a corrosion cell, a more detailed background regarding the corrosive action of various types of waters encountered in pharmaceutical operations could be helpful in dealing with the corrosion process.

While water in and of itself is generally a benign contributor to the corrosion process, it is the type and number of other components present in water that are of concern. Acids, basic chemicals, and dissolved oxygen can each determine a corrosion rate if the components required forming an electrochemical cell are present. It is most important to note that in many cases, the concentration of ionic species identified as corrosion cell-forming substances can, in many cases, be in a concentration range of single digit parts per million (ppm). For instance, oxygen dissolved in water varies in solubility from 14 ppm at 0 °C (32 °F) to 6 ppm at 45 °C (113 °F). At these low solubilities, it is possible to create an electrochemical cell. Further, water contained in pipes or closed containers may contain little dissolved oxygen content. While the dissolved oxygen present poses no general threat, dissolved materials such as bromides and chlorides do possess a corrosion threat, as is also the case with acidic aqueous solutions (e.g. pH < 7.0). Often acidic aqueous solutions are employed without the awareness of responsible personnel. For instance, cleaning solutions are often employed when stainless steel equipment or components undergo cleaning activity. Depending on the brand and type, cleaning solutions are most often either acidic or basic (often composed of a mineral acid such as phosphoric acid or a caustic solution typically containing potassium hydroxide). As identified in Section 8.3.7.3, an acidic electrolyte is a constituent of corrosive cell formation. Thus, when a repetitive activity such as equipment cleaning is performed using acidic cleaning agents, the probability of serious corrosive damage is quite real; this is particularly relevant when stainless steels such as type 304 are undergoing cleaning operations. Indeed, equipment failure because of chloride stress corrosion of stainless has cost at least one pharmaceutical firm in excess of $110 000 to replace blenders using chlorides in the mix and acidic cleaning solutions over a period of several years.

Temperature has also been found to play a role in chloride stress corrosion. Austenitic stainless steels, such as 304 and 304L, exhibit chloride stress corrosion in locations at a temperature of about 60 °C (140 °F), although there have been instances where chloride stress corrosion with temperatures as low as 30 °C (86 °F) to 40 °C (104 °F) and chloride levels as low as 10 ppm have occurred. As can be assumed, this is a very serious problem whose mechanism is not fully understood.

To preclude this type of damage, procedures should clearly identify the need to employ caustic cleaning agents and avoid acidic agents. (As will be described later in the chapter when Pourbaix diagrams are addressed, oxide protective coatings are stable at a pH of 7, or greater.)

Organizational procedures that fail to distinguish the difference in types of cleaning agents may experience similar catastrophic failures. Since many quality organizations possess a dearth of qualified personnel familiar with technical issues and quality assurance is the ultimate responsibility of organization when dealing with cGMP issues, it is important to have a knowledgeable presence to deal with corrosion issues.

When dealing with media in thermodynamic equilibrium, two fundamental characteristics apply when dealing with conductors in equilibrium with a solution:

> If a conducting substance is in thermodynamic equilibrium with a solution containing an oxidizing or reducing agent, its potential is equal to the potential of the solution;

And conversely:

> If a conducting substance has a potential equal to the redox potential of the solution in which it is immersed, the substance is in thermodynamic equilibrium with the solution, hence no reaction take place at the surface [5].

Table 8.2 summarizes many of the factors that can influence corrosion of metals in solution.

8.3.7 Types of Corrosion

Section 8.3.7 focused on a particular type of corrosion found in stainless steels. This section identifies and describes several other types of corrosion that is present when certain corrosion-contributing variables exist when dealing with metal process or manufacturing equipment and system components. It must, again, be emphasized that while stainless steel is the predominate material of construction when dealing with cGMP metals and noncritical systems such as certain HVAC applications, many pharmaceutical operations are not required to employ stainless steel. However, a failure to address corrosion phenomena when dealing with non-cGMP operations can result in significant facility cost and operating time losses.

In addition to chloride stress corrosion, other types of corrosion on a metal surface or in a localized area, such as a welded joint or a surface imperfection, include:

- Uniform corrosion
- Galvanic or dissimilar metal corrosion
- Concentration cell corrosion
- Stress corrosion cracking
- Fretting corrosion
- High temperature oxidation

Other terms that are used to describe specific modes of attack or effects include one or more of the processes identified earlier: (i) pitting corrosion, (ii) intergranular corrosion, (iii) erosion corrosion, (iv) impingement corrosion, (v) cavitation corrosion, (vi) chloride stress corrosion, and (vii) fatigue corrosion. While other types of corrosion mechanisms are identified, the forms of corrosion identified previously are most applicable to pharmaceutical materials and operations.

8.3.7.1 Uniform Corrosion

It is common for uniform corrosion to occur on metal surfaces with high composition homogeneity. Uniform corrosion is possible if access of the attacking species to the metal surface is generally unrestricted and uniform. If localized micro anode and cathodes are allowed to form at specific locations on the surface, uniform corrosion may be identified as localized corrosion that occurs consistently and uniformly over the surface of the metal and not at specific locations such as welds, surface imperfections, sanitary connections, piping elbows, and pipe system spool pieces.

Rusting of iron, fogging of nickel, and high temperature oxidation of iron or stainless steels are typical examples of uniform corrosion.

8.3.7.2 Galvanic Corrosion

Galvanic corrosion is encountered when two dissimilar metals are in contact with an electrolyte and a pathway is created (Section 8.3.7.2) to permit electrical current to flow from the active metal (anode) to the less active metal (cathode). As a result of this action, the less active metal, the cathode, is protected, and corrosive activity is observed at the more electrochemically active anode. While it would appear that the electromotive series (Table 8.1) would be useful in evaluating various metal couples for corrosion potential, the electromotive series is limited to metals and is not applicable to alloys such as stainless steels. Consequently, another relationship has evolved that includes alloys as well as certain theoretical considerations. This relationship has been identified as the galvanic series and identified in Table 8.3. The electrolyte employed in Table 8.3 is seawater. It should be observed that the metals in Table 8.3 are grouped together. While Table 8.3 resembles Table 8.1, its purpose is to identify metals that are more compatible for corrosion resistance. The metals in the same group can be joined (e.g. welded, bolted, etc.) and exhibit a minimum amount of corrosive activity. Similarly, metals in adjacent groups may

Table 8.3 The galvanic series of metals and alloys.

Anodic end

Magnesium

Magnesium alloys

Zinc

Alclad (high purity Al bonded to Al–Mg–Si alloy)

Aluminum 6053 (aluminum containing small quantities of Mg and Si)

Cadmium

Aluminum 2024 (aluminum containing small quantities of Cu and Mg)

Cast iron

Wrought iron

Mild steel

Stainless steel type 410 (active)

Stainless steel type 316

Stainless steel type 304

Tin

Lead

Lead tin solders

Naval brass

Manganese bronze

Muntz metal (brass alloy composed of 60% Cu, 40% Cu, and trace of Fe)

76 Ni-16 Cr-7 Fe alloy (active)

Nickel (active)

Silicon bronze

Copper

Red brass

Aluminum brass

Admiralty brass

Yellow brass

76 Ni-16 Cr-7 Fe alloy (passive)

Nickel (passive)

Silver solder

70–30 copper nickel

Monel

Titanium

13% chromium steel type 410 (passive)

Stainless steel type 316 (passive)

Stainless steel type 304 (passive)

Silver

Graphite

Gold

Platinum

Cathodic end

A more detailed source of information concerning the galvanic series can be found in the ASTM International (ASTM) document ASTM G98 (2009), *Standard Guide for Development and Use of a Galvanic Series for Predicting Galvanic Corrosion Performance.*

exhibit a slightly higher corrosive activity, whereas metals in groups that are distant can exhibit significant galvanic activity (corrosion). For instance, it would not be advisable to join magnesium, or a magnesium alloy, with titanium or a titanium alloy. However, joining magnesium, or a magnesium alloy, will not pose a corrosive situation when joined with zinc.

Variables that affect and control galvanic corrosion are:

- The pathway and electrolyte
- The degree of electrolyte polarization
- The ratio of anode and cathode surface areas

The most effective pathway for galvanic action is to have two metals or significantly dissimilar metals in contact with one another. If possible, gaskets, plastic films, or special coatings should be between the dissimilar metals. A configuration employing one or more of these materials will preclude the possibility of a galvanic cell forming. As noted in Section 8.3.7.2, elimination of any variable identified can prevent the formation of a galvanic cell.

Polarization occurs at the anode (less noble metal) and is often identified by deposition of corrosion products. It is also feasible for galvanic activity to occur at the cathode (more noble metal) with the evolution of hydrogen gas as the product. Both situations will eventually result in a decrease of electrochemical activity. The latter situation, referred to as cathodic polarization, is more relevant as it may serve as a barrier to further galvanic corrosion.

The areas of both the anode and cathode are also relevant to galvanic activity. If a large cathodic area is in direct contact with, or connected by, an electrolyte, a small anode is more concentrated and corrosion occurs at an accelerated rate.

One often overlooked item that may contribute to stainless steel corrosion involves cGMP cleaning with wire brushes. If this procedure is required as part of the equipment cleaning operations, it is most important that wire brushes are also fabricated from stainless steel. This precludes the possibility of dissimilar metals coming in contact and thus creating a corrosion cell potential.

8.3.7.3 Concentration Cell Corrosion

Concentration cell corrosion is basically an accelerated galvanic attack produced by a nonuniform electrolyte or a particular environmental variable. This electrochemical reaction occurs due to the difference in ion concentration or dissolved gases such as oxygen present in the electrolyte. This concentration difference causes a difference in the electrical potential on the surface of the same metal. It should be emphasized that a corrosion cell can form when two identical electrodes (the same metal) are in contact with one another. As indicated, differences in the environment may occur because of cracks, poor welds, faying (loose or poor connections), crevices, or imperfections on the metal surface. Also, confined areas, such as two metals joined together, may have a small localized area where an oxygen-deficient area acts as a cathode relative to a nearby location that has sufficient dissolved oxygen in the electrolyte. For this configuration, the oxygen-deficient location is the anode, which undergoes corrosion. Similarly, the location with the relatively rich oxygen concentration is, in effect, the cathode that does not exhibit corrosive effects [4].

While some of the factors contributing to concentration cell corrosion have been identified, it is often difficult to identify scenarios, design configurations, and environmental operating conditions that contribute to the phenomenon. Some protective measures that can be implemented to prevent or minimize concentration cell corrosion are as follows:

- If nonstainless metals are employed, using zinc coatings can be effective.
- Maintaining clean surfaces by employing corrosion-inhibiting cleaning (pH evaluation).

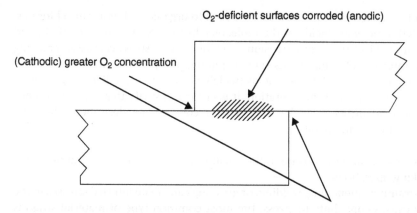

Figure 8.4 Concentration cell corrosion formation.

- Routine inspection of metals under faying surfaces such as riveted joints.
- Avoiding materials that can cause wicking (seepage) of moisture between faying surfaces.
- Verifying low carbon stainless is used in locations where corrosion cell formation may occur.
- Designs employing metals that have homogeneous surfaces, materials, and components (e.g. long radius elbows in the vicinity of dissimilar materials, welds, bolts).
- Employing approved gaskets for flanges, unions, and spool pieces such as elbows and branch tees, even when connections are of the same stainless steel composition.
- Using approved pipe tape for threaded connections even if the connections are the same stainless steel composition.
- Employing bleach (sodium hypochlorite) as a cleaning agent for stainless steel, particularly austenitic types, should be avoided. Using bleach should also be avoided in areas where stainless steel equipment is accessible (e.g. floors, walls, ceilings). The possibility of the presence of chloride ions should be minimized in all cases.
- Provide maintenance procedures that require inspection and cleaning of equipment locations where corrosion cell activity is possible.

A basic mechanism for concentration cell corrosion is displayed in Figure 8.4.

8.3.7.4 Stress Corrosion Cracking

As the title indicates, stress corrosion failure is the result of tensile stress and corrosion. Stress corrosion cracking is a common and dangerous form of corrosion. It is usually the result of three simultaneous events:

- Alloy grain boundary structure alteration – Often the result of metal treatment such as cold working, quenching, grinding, or welding that can produce stresses in the alloy.
- Alloy composition – Nickel-containing stainless steels are most prone to undergo stress corrosion. Alloys containing nickel in the 5–35% range are particularly vulnerable. Stainless steel alloys containing nickel in the 7–20% range are most prone to chloride stress corrosion. Chlorides in concentrations as low as 6–8 ppm are cause of this failure.

Ferritic stainless steels such as types 414, 430, and 440 containing little or no nickel will not undergo stress corrosion cracking. However, austenitic alloys such as types 304, 304L, 316, and 316L, because of their high nickel content (8–14% nickel), are very exposed to this corrosion technique.

- Alloy environment – Austenitic stainless steels exposed to organic and inorganic chlorides, sulfurous and polythionic acids (acids of the molecular form $H\,S_nO_6$, where $n = 0$–4), and caustic solutions are subject to stress corrosion cracking. Often, stress corrosion cracking takes place in the internals of components such as piping, pumps, and stainless vessels. If chlorides, caustic solutions, or polythionic acids (widely used as an ingredient in antibacterial topical ointments) are part of the operation, it is critical to verify that process components, such as those identified, are dry and lacking constituents described when the processing equipment is not in operation.

Tensile stress – As noted earlier this is often the result of a stress applied to an alloy subsequent to metal working operations.

Cold working or straining, quenching, welding, or grinding can create internal stresses in the metal. While many causes contribute to stress, the most common type of material stress is a localized and nonuniform stress. A typical example would be leaving a heavy tool placed on a nonstainless metal plate, adjacent to a bolted connection for an extended period of time (often a result of a maintenance action that failed to account for all components used in the maintenance action because of a time constraint). In such scenarios, the localized stressed location is a target of accelerated corrosion.

As explained in Section 8.3.7.3, metals that are subjected to applied stresses develop an anodic character when exposed to a corrosive environment. The adjoining unstressed areas, in this case a distance from the pipe connection, and exhibiting less stress, are less anodic or, in this case, more cathodic.

Two primary types of failure encountered with the combination of stress and corrosive activity are:

- Stress corrosion cracking
- Corrosion fatigue failure (addressed in Section 8.3.7.11)

While stress corrosion cracking does not occur with stainless steel or pure aluminum components, such action is possible when dealing with non-cGMP systems containing aluminum alloys and other metals. A typical example would be HVAC ducting in a cGMP location. Normally, stainless steel ductwork would not be employed in this installation because of the prohibitive costs involved. For this application, sheet metal (a thin steel plate easily bent and formed) would be the preferred material of construction.

Among the factors employed to control or prevent stress corrosion are:

- Control of temperature in the environment.
- Minimizing stresses on metals. An example would be to confirm hangers and supports for piping and ductwork installations are properly spaced to avoid building up a stress.
- Installed washers and gaskets are maintained.

It is important to note that as the degree of stress on a metal increases, the time to failure for the stressed metal component or equipment decreases.

8.3.7.5 Fretting Corrosion

When two metals, in surface contact, are subjected to a load and exhibit vibration, the configuration is a candidate for fretting corrosion. A pump shaft that is improperly aligned is subject to an abnormal stress and vibration at every revolution. Such can also be the case with improperly installed or worn bearings, couplings, frayed electrical contacts, defective bolting, pump mounts, and shims that are improperly aligned. As these components are continuously exposed to a stress during operation, metal fatigue can be encountered. This type of corrosion typically appears as a visible metal discoloration with visible pitting or depressions and often the visible presence of a corrosion product in the form of oxide debris. For ferrous alloys, this oxide, emanating from the metal interface, has a brown or reddish-brown color. Also, two metals in contact with one another and operating in a mode that causes constant mechanical contact in a cyclical or rapid vibrational mode can mechanically alter the surface properties of the metal. An example would be stainless steel parts contacting one another because a gasket has failed or is not present. As the stainless surfaces contact each other over time, the protective chromium oxide coating that gives stainless its unique properties gradually wears off and the constant metal contact renders it difficult for a new oxide coating to form. The introduction of surface oxygen and or moisture, in the form of humidity, can create a scenario similar to the concentration cell activity described in Section 8.3.7.3.

Since fretting corrosion is very common to pump installations, and without scheduled preventive maintenance for specific component inspection, fretting corrosion inspections should, in addition to the visible locations described earlier, occur in less obvious locations such as under:

- Grease seals, used for bearing protection
- Packing seals, commonly used to prevent liquid leaks
- Rubber seals with improper shaft connections
- Original equipment manufacturer (OEM) seals that are not designed for the intended pump application

As is emphasized in other sections of this chapter, an effective preventive maintenance program is the most effective technique to prevent and minimize galvanic phenomena in a pharmaceutical environment. Also, if monitored, the cost savings of effective corrosion prevention should easily be recognized.

8.3.7.6 High Temperature Oxidation

While not normally encountered in pharmaceutical operations, high temperature oxidation does present a corrosion topic worth evaluating since standard pharmaceutical or biotech operations rarely, if ever, have operations in the temperature range where high temperature oxidation occurs (250 °C or 482 °F). A direct reaction of an oxidizing agent such as oxygen or carbon dioxide with a metal at a high temperature is referred to as high temperature oxidation or dry corrosion (also referred to as oxidation reduction).

The major differences between high temperature oxidation and the other types of galvanic corrosion described are that, as indicated, the temperature of the former occurs at higher temperatures than other described corrosion mechanisms and there is no electrolyte (usually an ion-containing aqueous solution) required for high temperature oxidation.

Within specific temperature limits, corrosion products can form on the metal surface. With a continuous nonporous oxide film, the rate of oxidation (corrosion) will decrease over time.

Oxidation products (corrosion products) can form on metal surfaces if exposed to the higher temperature range described earlier. However, if the surface film is porous or exhibits spalling (extreme pitting), the oxidizing agent will have access to the surface and oxidation continues. In this case, the film is nonprotective and the metal will be attacked in the manner described previously.

8.3.7.7 Pitting

Pitting is a form of localized corrosion most often exhibited when high gage (thin) metal sheets or plate is a component of an installation. It can also occur when dissimilar metals are welded and, as is the case with other types of corrosion, when there are chlorides or similar halogens in the environment.

Pitting is believed to be the result of localized cell corrosion (Section 8.3.7.3) where a concentration cell (differences in oxygen concentration) can create corrosion products, most often accumulating in the pits in the metal. As is the case with other forms of corrosion, defective surface oxide coatings on stainless steel, aluminum, and other metals permit pitting by allowing corrosion-forming ions (in aqueous solution) to penetrate below the defective oxide layer. It is, again, important to emphasize that there is no minimal quantity of electrolyte solution required to create a corrosion (in this instance pitting) cell.

Three factors contribute to pitting corrosion:

1) Chloride concentration – As is the case with other forms of stainless steel corrosion, just a few ppm of chlorides, coupled with specific conditions such as pH and temperature, can initiate pitting corrosion.
2) pH – As a general guideline, the higher the pH, the lower the probability of pitting corrosion, and, conversely, the lower the pH, the higher the probability of pitting.
3) Temperature – Here, the general guideline is that the higher the temperature, the greater the probability of pitting corrosion.

While chromium alloyed stainless steels are moderately effective in controlling corrosion on surfaces, the alloys more resistant to localized corrosion, such as pitting, contain additional elements that enhance the resistance of the protective oxide layer from chloride-induced corrosion as well as other strongly oxidizing components. Other elements effective in decreasing pitting by increasing the content of the stainless are chromium, molybdenum, and nitrogen.

To assist in evaluating and predicting the pitting potential of stainless steels, an empirical relationship has been developed for this purpose. This relationship is known as the "pitting resistance equivalent number" or, more commonly, by its acronym PREN. There are several relationships associated with the PREN, but the most common is given by

$$PREN = \%Cr + 3.3(\%Mo) + 16(\%N) \tag{8.1}$$

As can be observed, Equation (8.1) states that molybdenum is 3.3 times more effective than chromium for preventing localized pitting corrosion. Equation (8.1) further notes that nitrogen is 16 times more effective than chromium.

While Equation (8.1) is an empirical relationship, another form of equation was actually evaluated in different applications where chlorides such as chlorine dioxide (based on pulp and paper mill operations) and chloride concentrations to 600 ppm chlorides as well as 300 ppm chlorates. All operations were performed at a pH of between 6 and 7 with the process operating maintained at 70 °C (158 °F).

The modified form of Equation (8.1) is specific to stainless steels that contain nickel, typically 304, 304L, 316, and 316L in the pharmaceutical industry. A type of steel that does not employ nickel is known as duplex stainless steel, a mixture of austenitic and ferrite alloys. The relationship that can more accurately account for the presence or lack of presence of nickel is given by the relationship:

$$PREN = \%Cr + 3(\%Mo) + 16(\%N) - 0.33(\%Ni) \tag{8.2}$$

As can be seen, the nickel content is detrimental to the PREN when taking into account the nickel content. Conversely, duplex steels due to their low nickel content (usually about 5% compared with 8–10% Ni for type 304 and 10–14% Cr for type 316) has a better resistance to pitting than nickel steels. However, there are several disadvantages when considering duplex stainless.

A major difficulty with duplex steel is that they are difficult to weld. Duplex alloys are also more brittle than higher nickel content stainless steels and also have a lower ductility. In addition, duplex stainless can separate into phases, usually over time that can cause one or more phases to deplete the chromium content of the duplex formulation and thus render it more susceptible to corrosion.

Example 8.2 Determine the pitting resistance equivalent numbers (PREN) for type 304 stainless steel, type 316L stainless, and a typical duplex stainless steel (type 2205) with evaluations for nickel content and not considering nickel content (recall the major difference between L and non-L grade stainless steels is the lower carbon content of the L grade).

Solution
For type 304 stainless steel, the relevant elements and contents by % are:

- Cr = 19%
- Mo = 0%
- N = 0.1%
- Ni = 10.5%

From Equation (8.1), PREN = 19 + 0 + 16(0.1) = **20.6** ← **Solution** (PREN for type 304)
For type 316L stainless steel, the relevant elements and contents by % are:

- Cr = 19%
- Mo = 2.5%
- N = 0.05%
- Ni = 0.5%

From Equation (8.1), PREN = 17 + 3.3(3) + 16(0.5) = **34.9** ← **Solution** (PREN for type 316L)
For type 2205 duplex stainless steel, the relevant elements and contents by % are:

- Cr = 22%
- Mo = 3%
- N = 0.5%
- Ni = 5.5%

From Equation (8.1), PREN = 17 + 3.3(3) + 16(0.5) = 22 + 3.3(3) + 16(0.5) = **39.9** ←**Solution** (PREN for type 2205 duplex)
For type 304 using Equation (8.2) = 19 + 0 + 16(0.1) − 0.33(10.5) = **17.13** ←**Solution** (PREN for type 304, employing Equation (8.2)).
For type 316L using Equation (8.2) = 17 + 3(2.5) + 16(0.10) − 0.33(12) = **22.14** ← **Solution** (for type 316L, employing Equation (8.2))
For type 2205 (duplex), the relevant elements and content by % are:

- Cr = 22%
- Mo = 3%
- N = 0.5%
- Ni = 5.5%

From Equation (8.2), PREN = 22 + 3(3) + 16(0.5) − 0.33(5.5) = **37.18** ← **Solution** (PREN for type 2205 duplex, employing Equation (8.2))

The lower PREN for type 2205 stainless steel indicates that this steel is not as effective as other types of stainless steel for protection against pitting corrosion but is more resistant to localized or cell corrosion. As can be seen from the PREN calculations employing Equations (8.1) and (8.2), 316 stainless steel has a higher corrosion resistance to pitting than 304 type stainless steel. Since pitting is often the precursor to other types of corrosion, it is important to recognize what needs to be done to preclude pitting of stainless steel from occurring. In addition to proper alloy selection, minimization and elimination of chlorides in an acidic environment (pH < 7) is the two most effective prevention measures.

While the PREN for duplex stainless indicates good pitting resistance, the metal must be carefully evaluated prior to use in pharmaceutical applications.

8.3.7.8 Intergranular Corrosion

Corrosive activity occurring at grain boundaries of metal phases is referred to as intergranular corrosion. Corrosion of this type is generally not encountered in pharmaceutical operations using stainless steel since this activity normally occurs when metals are exposed to temperatures in excess of 900 °F (482 °C) to a maximum temperature of about 1500 °F (816 °C). However, when there is a need for a stainless welding operation, this temperature range is not uncommon. For this reason, materials selection should include evaluation of components and locations (e.g. bin legs, structural supports) where welding maintenance can be performed. For these components and locations, it would be prudent to evaluate the efficacy of using low carbon stainless as an intergranular corrosion prevention measure [4].

Prolonged exposure to high heat operations, such as welding, can create a situation where the metal has undergone a transition in the metal known as sensitization. In this form, the heated metal is more exposed to intergranular corrosion.

In addition to careful materials selection when evaluating high temperature pharmaceutical applications such as welding, the proper graded welding rod for the type of stainless welding operation (i.e. TIG or MIG) be employed. A very critical part of the stainless operating procedure is to verify that the task is performed by American Welding Society (AWS) certified welders. Also, it is important to verify that proper filler wire is employed for stainless welding activities. Often, in-house maintenance operations rely on contract personnel to maintain spare parts and consumable materials in on-site maintenance operations. It is critical to monitor the type of welding material to be used for all welding operations. Constantly using the same type of welding rod/wire without analysis of the metals to be welded can create opportunities for intergranular corrosion as well as other types of corrosion.

If welding operations are performed by contractors, it is important to specify that welding operations are to be performed by certified welders employing approved materials.

8.3.7.9 Erosion Corrosion

This type of corrosion is also rarely encountered in pharmaceutical operations. This corrosion phenomenon appears when particles in a liquid stream impinge against the piping walls. In pharmaceutical applications, high fluid velocity in a high solids medium is the cause of this activity. While the stainless matrix is often minimally affected, removal of the protective oxide coating in a localized internal piping area is possible. The probability of erosion corrosion is enhanced if, subsequent to cleaning operations, the internal stainless internals are not dry and the equipment is placed in operation shortly after cleaning. Inactivity allows the nonprotected localized surface to be exposed to galvanic activity and susceptible to several types of corrosive action described earlier.

Chapter 2 provides guidelines for fluid velocities of pharmaceutical water transport during manufacturing operations. They also serve as sensible ranges for erosion corrosion prevention criteria.

8.3.7.10 Cavitation Corrosion

Cavitation corrosion can be viewed as a modified form of erosion corrosion. The major difference is that rather than corrosion occurring as a result of particle impingement on a metal, material loss is a result of a naturally occurring physical transformation. Cavitation is an event that occurs when a metal is physically transformed into a vapor phase without undergoing a liquid phase. This phase change is identified as sublimation. Sublimation, in pharmaceutical applications, occurs when considering one specific item for manufacturing and related operations, such as pilot plant scale-up. This item is a centrifugal pump. If the pressure of the fluid being pumped is below the vapor pressure of the fluid, vaporization of the fluid will occur. As a result, the vapor will form bubbles on the pump surface (in this case the pump impeller). The vapor bubbles most commonly form on the trailing side of the impeller surface. Because of the rapidity of the bubble formation, the protective oxide surface layer of the metal is rapidly worn away, and, as is the case with impingement corrosion (Section 8.3.7.9), the unprotected impeller surfaces are exposed to localized galvanic corrosion. Cavitation is usually accompanied by pump vibration, a drop in pump efficiency, and, upon pump inspection, visible pitting of the pump impeller.

To avoid cavitation corrosion, it is most important to verify that the pump is correctly sized. The primary design criterion, in this case, is to confirm that the pump has an adequate net positive suction head available (NPSHA) that is higher than the net positive suction head required (NPSHR) as well as a safety factor of at least 2.5 ft of liquid (most often water for pharmaceutical applications).

More detailed information relating to NPSHA and NPSHR is located in Section 2.2.1.

8.3.7.11 Corrosion Fatigue

Exposing a metal (including austenitic alloys such as stainless steel) to an extended cycle of stress weakens the metal by approaching the fatigue limit as the number of cycles and applied stress increases. While in a nonhostile environment, the number of stress cycles may be manageable (usually a function of the amortization time of the unit of concern), in a corrosive environment, the same cycling operation may prove to be a significant loss.

Typical examples of metals undergoing stress cycling include heat exchangers, such as surface condensers, undergoing continuous heating and cooling cycles and reactors that also serve as crystallizers. These actions, particularly if the materials being processed are prone to galvanic activity, can exhibit corrosion fatigue. For this reason, equipment that operates in a stress mode over time should be evaluated, particularly if the materials being processed are of a pH lower than 7 and/or chlorides are part of the operation.

Inspecting components subject to stress, particularly if corrosive environmental components are employed, should be incorporated in facility maintenance procedures.

8.3.7.12 Stress Corrosion Cracking

When a metal is under pressure, it is known to be in a state of compression. The opposite situation, when a metal is stretched or elongated, is known as stress. When a galvanic activity is associated with material stress, stress corrosion can occur. This scenario is more likely to occur if the stress is localized to a specific area of the metal. Stress corrosion failure is the result of three simultaneous events:

- Alloy grain boundary structure alteration – Often the result of metal treatment such as cold working, quenching, grinding, or welding that can produce stresses in the alloy.
- Alloy composition – Nickel-containing stainless steels are most prone to undergo stress corrosion. Alloys containing nickel in the 5–35% range are particularly vulnerable.

Ferritic stainless steels such as types 414, 430, and 440 containing little or no nickel will not undergo stress corrosion cracking. However, austenitic alloys such as types 304, 304L, 316, and 316L because of their high nickel content (8–14% nickel) are very exposed to this corrosion mechanism.

- Alloy environment – Austenitic stainless steels exposed to organic and inorganic chlorides, sulfurous and polythionic acids (acids of the molecular form H S_nO_6, where $n = 0$–4), and caustic solutions are subject to stress corrosion cracking. Often, stress corrosion cracking takes place in the internals of components such as piping, pumps, and stainless vessels. If chlorides, caustic solutions, or polythionic acids are present, it is critical to verify that process components, such as those identified, are dry and devoid of the chemical materials described when the components and equipment are not in operation.
- The presence of a tensile stress – As described previously, this is often the result of a stress imparted on an alloy subsequent to metal working operations.

While it is difficult to identify tensile stresses in metals, measures can be implemented to verify that process operations do not contribute to stress corrosion initiation or progression. Internals, such as piping, inside diameter heat exchanger tubing, and stainless vessel internals that process caustics, chlorides, and polythionic acids, should be devoid of these chemicals subsequent to use. Similarly, they should be dry and free of moisture. Procedures should be similar to cleaning validation protocols employed for cGMP components since the internals are critical components coming in contact with product or product components.

8.3.7.13 Microbiologically Induced Corrosion

A rarely encountered but existing form of corrosion is one that occurs from the interaction of bacteria with some metals and alloys. Several species of microorganisms, bacteria, yeasts, molds, and algae can be involved in electrochemical reactions that cause or influence corrosion. A process exhibiting this behavior is known as microbiologically induced corrosion (MIC). MIC is particularly relevant when dealing with bioreactor equipment fabricated of stainless steel and used in a cGMP environment. MIC scenarios in cGMP environments are most likely to occur in heat exchangers, liquid storage tanks, and possibly clean-in-place (CIP) spray heads that are not properly dried and sanitized subsequent to use.

Several species of microorganisms produce diverse effects depending on criteria such as dissolved oxygen concentration, pH, temperature, surface cleanliness, formation of localized electrochemical cells (Table 8.1), and specific organic and inorganic substances in the microenvironment [2]. For instance, certain forms of bacteria exhibit a metabolism that induces them to obtain nutrients and energy from the oxidation of inorganic compounds including iron, sulfur, hydrogen, nitrogen, and carbon monoxide. The chemical reactions that cause corrosion are part of the normal metabolism of these organisms. The mechanism of MIC can significantly increase the corrosion rate; the corrosion rate can vary from 10 to 1000 times the conventional corrosion activity. The MIC chemical reactions that cause this type of corrosion are part of the standard metabolism of these organisms. One form of MIC involves a sulfate-reducing bacterium that utilizes cathodic hydrogen to reduce soluble sulfates to sulfides (Figure 8.1). Further, some mechanisms involving sulfides can oxidize the metal ions at the anodic sites, resulting in a further increase in the corrosion rate [4]. Sulfate-reducing bacteria are classified as anaerobic bacteria. Anaerobic bacteria survive and multiply in an oxygen-free environment. Another bacterium, aerobic bacteria, requires oxygen for survival and multiplication. Aerobic bacteria can also function as a corrosion agent and accelerate the corrosion rate. Both aerobic and anaerobic bacteria can cause MIC.

One corrosion mechanism involves sulfur bacteria *Desulfovibrio*, a reducing agent forming hydrogen sulfide, and *Thiobacillus thiooxidans*, which oxidize sulfur to sulfuric acid. Similarly, certain bacteria known as nitrosifying bacteria convert ammonia to nitrites, and another species known as nitrifying bacteria further convert the nitrites to nitrates, which react with water to form nitric acid, lowering the pH in the area of the corrosion cell and consequently increasing the corrosion activity.

Another type of bacteria uses ferrous compounds (i.e. stainless steels) as a nutrient. The ferrous-containing alloy becomes vulnerable to concentration cell attack through the activity of the bacteria tubercles (a rod-shaped bacillus bacterium that is capable of causing some diseases). Corrosion due to tubercles is detected by formation of blisters of the corrosion product and is formed from biofilm deposits on iron-containing surfaces. The phenomena are also likely to occur in low flow velocity regimes. While not normally found in pharmaceutical operations, the rapid growth of biotech should be a consideration before discarding a consideration of tubercles. Also, bacteria and molds that live on organic matter in the immediate vicinity or in surface contact with certain metals are capable of forming biofilms, which adhere to the metal surface. This biofilm causes corrosion by depleting the oxygen supply of the localized corrosion cell. The by-product of these reactions is the release of compounds including organic acids, which can accelerate chemical attack (corrosion) of the host metal.

Materials selection can also affect the influence of MIC activity. For cGMP applications, stainless steel is the metal. One type of stainless that appears to be resistant to MIC is those stainless alloys that contain in excess of 6% molybdenum. The three grades containing the acceptable molybdenum level are:

- ASTM Grade S31254
- ASTM Grade N08926
- ASTM Grade S32654

As would be expected, these steels are quite expensive, and the need should be carefully evaluated before specifying this material for pharmaceutical cGMP applications.

While attention is normally focused on cGMP operations, the balance-of-plant (e.g. non-cGMP operations) operations using aluminum and copper and their alloys are also susceptible to MIC. While these metals and alloys pose little threat to cGMP operations, MIC, if uncovered, can result in an unplanned maintenance and potentially costly repair activity. Consequently, scheduled facility corrosion inspections and checks should include MIC.

While surface blistering due to tubercles is common to one type of bacteria, pitting is the most common form of microbiological corrosion. Further, corrosion products may be composed of various attacking microbes, and barring mitigation or remediation, the corrosion rate could increase.

While there is a paucity of MIC documented in cGMP facilities, basic preventive actions, such as routine documented visual observations, is a minor maintenance action that deserves attention in the evolving field of biotechnology.

8.4 Corrosion-Resistant Metals and Alloys

While relevant details of corrosion-resistant steels and other metals and alloys have been presented throughout this chapter, this section serves to present material often encountered by personnel involved in maintenance and project positions. The intent is to provide a brief introduction of the types and features of often used corrosion-resistant metals most often employed in pharmaceutical and related processes such as biotech and food processing. While 304 and 316 are most commonly used when critical components are involved (i.e. components that come in contact with pharmaceutical ingredients used in the manufacturing of a drug), a large percentage of pharmaceutical installations do not involve actual manufacturing processes.

8.4.1 Iron Alloys

Iron alloys containing in excess of 12% chromium are referred to as stainless steels. The range of chromium in stainless varies from 12 to 30% chromium with the most common types containing 16–22% chromium. For pharmaceutical applications, the chromium content of virtually all types of stainless steels contains chromium in the range of 16–20%. Depending on the type of stainless steel, other constituents are also contained in the stainless alloy. As noted previously in this chapter, chromium imparts a passive protective oxide film on the metal surface. While this film is extremely thin, 1–5 nm, it is very resilient and, when scratched, will quickly self-repair in standard atmospheres (>20% oxygen).

In strong oxidizing environments, stainless steels are often superior to more noble metals and alloys (Table 8.3). Stainless alloys are, however, more vulnerable to acidic environments and chloride-containing materials.

While stainless steels possess excellent corrosion resistance in most environments, stainless steel should not be used when holding or processing hot water above 50 °C (122 °F). This limiting temperature is primarily attributable to the possible presence of chlorides in the water.

Table 8.4 exhibits characteristics, including resistance information, of stainless steels typically used in pharmaceutical applications. While types 304, 304L, 316, and 316L are most commonly used, other types may also be used for specific non-cGMP applications.

While Tables 8.1, 8.3, and 8.4 may be useful in evaluating corrosion scenarios in given environments, the National Association of Corrosion Engineers (NACE) publishes very authoritative documents such as the *Corrosion Data Survey* and laboratory corrosion tests such as *NACE Standard TM0109* that can be implemented very effectively.

It is often difficult to ascertain if and how much corrosion actually occurs in stainless steel and aluminum equipment, parts or major components. Since visual inspections often hide corrosion formation in locations that are hidden from view (e.g. two dissimilar metals bolted or welded in an overlapping configuration), other options may be desired. An option would be to include iron content of specific pharmaceutical formulations as a part of the batch

Table 8.4 Properties of selected stainless steels.

Type	Grade	Corrosion resistance characteristics	Other characteristics
Austenitic	304, 304L, 316, 316L	Ordinarily, the most corrosion-resistant type of stainless steel; susceptible to chloride stress corrosion in certain environments	Nonmagnetic; can be welded and cannot be hardened by heat treatment. Major alloying additions are Fe/Ni/Cr
Ferritic	405, 409, 430, 439	Resistant to chloride stress corrosion cracking	Magnetic; subject to hydrogen embrittlement. Differs from 300 series since it contains 2–3% Mo offering greater acid resistance. Major alloying additions are Fe/Cr
Martensitic	410, 420, 440A, 440B and 440C	Less corrosion resistant than austenitic, ferritic, or duplex	Magnetic, difficult to weld, high strength and hardness. Major alloying additions are Fe/Cr/C
Duplex	2204, 2305	As name implies, a mixture of austenitic and ferritic alloys	Magnetic and highly resistant to chloride stress corrosion cracking; contains 2–3% Mo; difficult to weld. Major alloying additions are Fe/Cr/Ni

record of the formulation. This option may require an iron analysis detection level lower than the standard ppm. However, with the advent of process analytical technology (PAT), it may be beneficial to incorporate continuous online analysis of iron during pharmaceutical manufacturing.

Corrosion considerations should be incorporated as part of the design basis for new cGMP and non-cGMP pharmaceutical systems. Corrosion prevention should also be an integral component of pharmaceutical and biotech maintenance activities and incorporated as facility maintenance procedures.

An often unaddressed common pharmaceutical water system component is the heat exchanger. These units are common to WFI and USP systems employed as surface condensers and similar cooling applications. Heat exchangers are, of course, often used to heat process solutions. While attention is often given to piping systems, it is also important to note that heat exchanger tubing is also susceptible to the same corrosion mechanisms found in stainless steel piping. In fact, since many heat exchangers often employ a "U tube" design, special concern should be given to this exchanger design. Since many pharmaceutical operations are batch operations, it is not uncommon to neglect maintenance and inspection of heat transfer equipment, such as heat exchangers, when these units are not in use. If processing mixtures contain water and/or chlorides, stress corrosion cell formation is possible, particularly if there are gradients, such as bends or elbows, in the heat exchanger. For this reason, an effective maintenance program should pay particular note of heat exchanger internal tubing. One useful maintenance tool is the borescope. The use of a borescope for routine maintenance permits view of heat exchanger tubing internals without affecting the system configuration and should be considered as a valuable nondestructive inspection tool.

8.4.2 Aluminum and Aluminum Alloys

While aluminum is not used for cGMP equipment in the pharmaceutical industry, it is widely employed as sealing material tablet containers and blister packs and also as containers for ointments and consumer items. Aluminum containers are also used as containers for small quantities of analytical reagents used in quality control operations. For these applications, aluminum has several distinct advantages. In addition to being readily available and relatively inexpensive, aluminum is lightweight (approximately $173 \, lb/ft^3 \approx 0.1 \, lb/in^3$) and easily fabricated to desired configurations. Aluminum can be rolled to virtually any thickness, stamped, hammered, and extruded. Additionally, aluminum and its alloys are resistant to corrosion exposure in air. When exposed to oxygen in the atmosphere, oxide in protective film rapidly forms. Foreign matter such as moisture and dirt that may collect on the surface can initiate galvanic corrosion and can subsequently penetrate weak spots on the oxide film and create tiny pits. The reaction products of these pits become part of the oxide surface coating. As a result, the oxide surface coating may increase in thickness and serve to retard further attack. While there may be some discoloration, the aluminum will remain unaffected.

In general, aluminum and aluminum alloys exhibit little or no general attack when exposed to water. Water containing dissolved oxygen and most aqueous solutions also have no significant impact, but soft water such as USP and WFI and water carrying a high concentration of carbon dioxide are more corrosive. However, if aqueous solutions with high concentrations of acids or alkalis are present, corrosive activity can occur (Figure 8.6, Section 8.7).

For pharmaceutical applications such as tablet bottle seals, blister packaging, and ointment packaging, the aluminum usually has additional protective coatings applied prior to use. These protective coatings extend the shelf life of aluminum pharmaceutical packaging components.

While aluminum packaging components are resistant to corrosion, some additional procedures that may be of value for minimizing aluminum corrosion include the following:

- Avoid direct contact of aluminum with concrete. Even cured concrete contains about 3% moisture and aluminum products, such as foil-faced components, can be subject to corrosion if components are stored with little or no adequate ventilation. Aluminum components that are on the bottom of a large, nonpalletized stack in contact with a concrete floor would be an example.
- Avoid a condition where aluminum-based foils are in contact with dissimilar metals. The intent is to obviate the possibility of a galvanic cell forming. An evaluation of metal's compatibility and potential galvanic activity can be evaluated by referring to Table 8.1.
- If the aluminum-containing components are subject to storage, verify that the warehousing HVAC is operating within specified temperature and humidity specifications.

8.5 Passivation and Rouging

This section describes techniques intended to prevent damage to austenitic stainless steels from damage due to certain impurities often present in finished steels.

8.5.1 Passivation

For a stainless steel part to achieve maximum resistance to corrosion and in order to avoid discoloration, it is essential to remove all surface contaminants, including free iron that is present as a result of mill operations and stainless steel manufacture. A clean stainless steel surface will generate a protective oxide film in normal air. The steel is then said to be in "passive" condition. Foreign matters such as iron filings, steel particles, and embedded grit interfere with this film formation. Furthermore, the contaminants themselves may rust and course to stainless steel surface to appear streaked or otherwise discolored.

While nitric acid passivation is most often performed at the manufacturer's facility or a facility specializing in passivation activities, pharmaceutical operations should be aware of the passivation procedures employed for their new installation of new equipment. However, stainless steel components must be thoroughly degreased prior to any passivating treatment.

The passivation process can be speeded by immersing or otherwise contacting the part with a strong oxidizing agent; nitric acid solutions normally are being used for this purpose. Solutions of nitric acid in water provide a further advantage of dissolving hired or steel particles, thus removing areas of probable corrosive attack.

The most popular solution for passive 18 stainless steel consists of 20% nitric acid (by volume) operating at 120 °F. If immersed for a minimum of 30 min, the part should be thoroughly rinsed in clean hot water.

Considerable latitude exists in the choice of acid concentration, bath temperature, and immersion time. For an example, when passivating the 300 series stainless steels, those steels up to 400 series containing a 17% or more chromium, and except for the free machining grades (free machining grades are stainless steels manufactured with a composition and heat treatment intended to improve the machinability quality of the stainless steel), a solution containing one key to 14% nitric acid may be used at 130–160 °F for a time frame ranging from 30 to 60 min. The same bath may be used for the 400 series steels containing less than 17% chromium and the straight chromium free machining grades at a lower temperature range of 110–130 °F.

Also recommended for the free machining grades and high carbon heat treatable types of the 400 series is a solution containing 15–30% nitric acid (by volume) and 2% sodium dichromate (by weight) operating at 110–130 °F. This treatment is also applicable for use on material that shows evidence of damage such as etching [6].

Perhaps the best source for passivation operations information is ASTM Standard ASTM A967, *Standard Specification for Chemical Passivation Treatments for Stainless Steel Parts*. Another useful document is ASTM Standard A380, *Practice for Cleaning, Descaling, and Depassivation of Stainless Steel Parts, Equipment, and Systems*. This document encompasses cleaning chemical solutions as well as nitric acid passivating solutions and processes. While materials such as citric acid and ammonium citrate are detailed in the standard, it is important to note that ASTM Standard A380 refers to these chemicals as cleaning acids and not passivating acids. This distinction is probably due to the fact that citric acid is not an oxidizing acid, whereas nitric acid is a true oxidizing acid.

It is again noted that the ability of nitric acid, when properly used, to effectively remove free iron on the surface of stainless steel causes nitric acid to effectively passivate stainless steel. It is also important to reiterate that the passivation process is intended to be a onetime operation, preferably performed prior to or at the beginning of use.

Passivation is an operation that is intended to be performed only one time during the useful life of the stainless steel. The passivation procedure should be performed prior to or at the beginning of the use period. Nonnitric acid passivation procedures performed at scheduled intervals during the equipment lifetime may indicate that the operation is actually a cleaning operation, as opposed to a true passivation procedure.

8.5.2   Rouging

On occasion, a residue with a reddish hue is observed on equipment components such as rotary pump gears and stainless steel piping. Often this residue occurs in USP and WFI systems, although it can occur in equipment intended for applications such as pumps used for tablet coating solutions. The observed action causing this residue is known as rouging. Rouging can occur in DIW operations, WFI systems, clean steam units, and plant or municipal feedwaters. Rouge is categorized in three classes:

1) Rouge generated by an external source – The residue is deposited on stainless steel surfaces throughout the system and can easily be removed by wiping.
2) This type of rouge appears as blisters on the surface and is the result of stainless steel corrosion and forms in the presence of chlorides, other halides, chloramines, and some organic acids such as oxalic and formic. This type of rouging is more difficult to remove than class 1 and requires surface grinding of affected components.
3) This rouge appears as a black or blue surface coloring. Subsequent to grinding, it will reappear over time. Since passivation most likely was not performed prior to use, little can be done at this point, and continued operation with the rouging is recommended unless it is interfering with product quality [7].

The main constituent of this residue or film is ferric oxide (Fe_2O_3), which is formed by the following reactions:

$$2Fe + 4H_2O \rightarrow 2FeO(OH) + 3H_2 \uparrow$$
$$2FeO(OH) \rightarrow Fe_2O_3 + H_2O$$

The constituents can also contain small amounts of nickel and chromium.

While the reddish film often can be removed easily, some forms require mechanical action, such as using a wire brush, to remove the film.

The residue in appearance may be aesthetically undesirable, but there is no indication that the presence of rouge in pharmaceutical water systems has a deleterious effect on water quality or clean steam. Thus, the FDA has no position on the presence of rouge on pharmaceutical water systems or the water itself. As long as the product water conforms to criteria such as conductivity, no special action is required at this time. It should be noted that rouging can affect conductivity readings and can possibly interfere with certain microbial reactions employed in biotech operations.

8.6   General Corrosion Protective Measures

Table 8.5 provides design information and techniques for reducing or eliminating many corrosion scenarios [4]. These methods should be evaluated with a focus on facility maintenance activities and manufacturing operations.

8.6.1   General Design Considerations for Corrosion Prevention

New or existing facility assets such as equipment and installed systems can also be evaluated in terms of adequate corrosion protection. This evaluation program can be initiated at the beginning of the design phase or as part of the maintenance planning program. Ideally, vendors involved with new installations should be able to demonstrate that corrosion protection measures have been incorporated in the design and/or development phase of the system installation and/or equipment they are responsible for implementing.

In addition to Table 8.5, some corrosion prevention design elements are presented as follows:

- Component shapes should be simple employing rounded corners and edges wherever possible; sharp corners and recesses should be excluded from these components. Sharp edges present the possibility of forming a discontinuity as opposed to maintaining a continuous surface. Recesses, of course, present the possibility of moisture buildup and the potential of forming a localized corrosion cell. Simple continuous forms are also advantageous when dealing with maintenance activities such as coating, painting, and equipment cleaning.
- Construction components such as angle iron, channels, I beams, etc. should be configured such that catchment areas are eliminated. If such configurations are deemed to be impractical, drainage holes with approved discharge elements should be incorporated. These considerations should be given special consideration if components are hidden behind walls, ceilings, and other locations where routine inspection and maintenance is not incorporated in the maintenance program.
- Piping installations that undergo temperature variations as a result of introducing materials at different temperatures (e.g. a crystallization/fermentation system that is used to transport raw finished product at high fermentation temperature and exposed to a lower crystallization cooling temperature and also used for steam cleaning of the piping), installation of expansion joints should be considered if a reasonable probability of formation of a stress corrosion cell exists. Even stainless steel piping is susceptible, particularly if chlorides and a pH below 7 are present as part of the processing operations.
- Tanks, containers, and pressure vessels and other similar structures should be free of direct contact with the ground or floor without an electrically insulating material separating

Table 8.5 Techniques for corrosion elimination.

Item	Suggestion
Elimination of "sump" areas, surfaces, and volumes where trapped moisture is in contact with metal	Analyze structural details of the equipment or component
	Provide properly located drain holes, specifying a minimum ⅛ in. drain hole to prevent plugging
Avoid nonabsorbent nonwicking materials (particularly relevant for packaging operations)	Determine water absorption qualities of materials to be used, including excipients and active ingredients
	Use approved epoxy and vinyl tapes and coatings, wax, and rubberlike materials for protective barriers
	Avoid use of wood, paper, cardboard, open cell foams, and spongelike materials, without protective hydrophilic coatings
Protect faying surfaces (the surface of an object to which it is fastened, e.g. plates, angle iron, etc.)	Employ approved (USDA, FDA) sealing materials (tapes, films, sealing compounds on all faying surfaces); consider using primers
	Lengthen possible liquid pathways to prevent forming an electrolytic cell
Use compatible metals or stainless steel	If magnesium and aluminum are present (for non-cGMP components), 5000 and 6000 series aluminum alloys are most compatible
	For magnesium steel couples, use tin or cadmium plated steel
	For bimetallic metals or alloys in the same group in Table 8.3, employ materials as close as possible in the table. Use tapes or primers on faying surfaces to prevent metallic or electrical contact
Select proper finishing systems	Select approved chemical treatments, paints, and plating on basis of service requirements
	Use past experience in similar applications as a selection guide
	Service test the selected system prior to formal use
Continuous water circulation	Avoid stagnation of water in tanks, heat exchangers, pumps, and piping. Even when not in a manufacturing mode, water that is used in manufacturing should be continuously recirculated, thereby minimizing allowable time for a galvanic or corrosion cell to form

the components from the process component. This is particularly relevant if the tanks or containers are stationary and not normally moved to various locations.

- Avoid joining or coupling of dissimilar metals (i.e. metals that are widely separated in the galvanic series, as shown in Table 8.3). While for most cGMP applications employing metals, stainless steels (types 304, 304L, 316, and 316L) are the most widely used. However, corrosion dynamics are not specific to cGMP components. For instance, facility utilities such as heating steam and plant compressed air for instrument operation are also susceptible to potentially costly corrosive activity as a result of galvanic action. For instance, if dissimilar metals are mechanically joined, adequate protection, such as insulated gasket material, should be used. For nonstainless metals, protective measures should definitely be considered. These measures include chemical or anodic coatings, approved organic coatings including epoxies, and urethane coatings. For non-cGMP components, surfaces should be painted or coated when possible. Where applicable, cathodic protection should also be evaluated.
- During assembly operations, caution should be used to avoid damaging metal surfaces causing denting, cracking, and unintentional grooves, particularly if separate metal parts are being joined. Also, welding and bolting operations should be in accordance with approved procedures for the operation and metal being used.

If atmospheric corrosion presents a potential corrosion problem for certain equipment or components (e.g. outside tanks, pressure vessels and fractionating columns, etc.), the design and materials of construction should be considered as part of the design phase, if possible. Such an evaluation would typically include potentially corrosive agents that will be used in the process. Also, organic and metallic coatings should be evaluated as a technique to reduce corrosion potential wherever applicable. Another important corrosion prevention action would to use butt welded joints rather than lap joints. If lap joints are unavoidable, exposed edges should be filled with caulking compound, soldered or welded to preclude the entrance stagnation of liquids (primarily water).

8.6.1.1 Design Considerations for Controlling Galvanic Corrosion

Galvanic corrosion is most common when dissimilar metals and alloys come in contact. The severity of corrosion is commonly a function of the degree of dissimilarity of electrode potentials of the alloys or metals having common contact (Table 8.1). Also, metals and alloys that are widely separated in the galvanic series (Table 8.3) will exhibit a greater proclivity for corrosion via a galvanic cell. As a result, if dissimilar metals must be used, appropriate actions should be implemented to prevent contact. The most common method to achieve this is by applying a sacrificial metal coating to the cathodic component that contains material similar to or identical to the anodic component. Also effective for galvanic protection are previously noted design approaches such as painting or coating surfaces to increase the resistance of a galvanic circuit.

When dealing with small anodic areas (relative to the cathodic area), the same metal or a more noble metal should be used for small fastenings, bolts, and screws. A common example is a rusting bolt and an accompanying nut or a visa versa configuration. This situation occurs most often when nonidentical nuts and bolts are used for an application. While in many cases this situation may be harmless and have little effect, if not corrected, significant weakening of the assembly can occur over time.

Cathodic protection, if feasible, is a preferred technique for prevention and control of galvanic corrosion. Corrosion inhibitors are also used in lieu of sacrificial cathodic protection. If used together, the combined use of cathodic protection and an inhibitor is not practical unless the inhibitor has no passivating effect on the sacrificial anode.

8.6.1.2 Design Considerations for Preventing Concentration Cell Corrosion

Just as previously noted as a general design consideration, sharp corners, angles, pockets, and other conditions where solids or liquids can accumulate should be avoided. On occasion, maintenance actions can inadvertently create such configurations.

Equipment involving liquid transport (piping, transfer tanks, storage tanks, etc.), should be designed such that turbulent flow (a Reynolds number >5000) and gas or air entrainment are minimal.

Insulating materials or other substances that absorb or retain water should not be in contact with metallic surfaces. If possible, place a protective coating or paint on the surfaces prior to installing installation. This is particularly important if the component is not in an enclosed environment or in an area where chlorides are present or an environmental concern, particularly if 300 series stainless steel is being used.

Butt welds are preferable to spot weld, lap joints, or mechanical fasteners. If lap joints or sport welds must be used, the joints should be sealed with solder, weld metal, or caulking compound with the intent of preventing trapping of corrosive chemicals, including water and water vapor.

8.6.1.3 Avoiding Stress Corrosion Failures

While it is difficult to confirm that best practices have been employed for purchased equipment and components, there are some design elements that should be standard for vendor-installed equipment and system components. While it might be difficult to verify vendor fabrication processes, incorporating best practices for facility maintenance operations is less of a difficulty. For instance, annealing should be an operation used to relieve internal stresses, while rolling operations or swaging should be used to induce compressive stresses at the surface. If these tasks are part of the facility maintenance and repair operations, these activities can be identified in the maintenance procedures. Also, alloys should be heat-treated to minimize the effect of intergranular corrosion. Annealing is also important to minimize the formation of grain boundaries in specific pharmaceutical equipment. An example would be the failure of punches used in tableting operations. Normally fabricated of stainless steel, unless the grain boundaries are properly annealed, the punch life can be significantly shortened because of a stress failure at the weakened grain boundary. This failure is, in addition to being expensive to replace, a significant loss of time, and consequently cost, because of the equipment downtime.

Appropriate protective coatings should, again, be utilized to protect the equipment, component, or structure from corrosive agents. On occasion, it may be necessary to protect a large continuous metal surface. In this case, electroplating of the component may be worth evaluating as a corrosion prevention technique.

8.6.1.4 Prevention of Fretting Corrosion

The rapid corrosion that occurs between highly loaded metal surfaces when subjected to vibratory motions is identified as fretting corrosion. Fretting corrosion is found on bearing surfaces, connecting rods, and structural trusses using highly loaded bolts often located near or part of HVAC components such as ducts, compressors, and fans. This can occur in metal, including stainless steel in certain operating environments.

Fretting corrosion prevention actions include cold working of susceptible surfaces and lubrication of the surfaces and critical components.

8.6.1.5 Preventing Corrosion at Joints and Faying Surfaces

A fraying surface can be defined as a contact surface between two, three, or more tightly joined objects. Commonly, two surfaces are normally in contact and typically joined by bolts. For older installations, rivets might be the mechanical joining method. For this situation, it would be beneficial to provide a bit more detail, specifically for corrosion prevention of at joints and faying surfaces involving metal parts and joined assemblies. At these locations, there can be contact between similar metals, dissimilar metals, or nonmaterials such as foam or another type of insulation. Corrosion can be initiated at these locations. Bolted or riveted joints, spot welds, brazed and soldered joints, and threaded fittings are also common contacting surfaces where corrosion can develop. The use of gaskets, shims, washers, clips, springs, inserts, and bushings can provide areas susceptible to corrosion. It is not uncommon that, during maintenance activities, a replacement component, such as a replacement washer or bolt, is used with little concern for the type of metal or alloy composition of the replacement part. Another concern involves welded structures or repair activities. Here, contact between dissimilar metals may start galvanic activity. Further, similar metals may be welded with a rod of the same or different composition, while the rods used in welding dissimilar metals may be the same as either one of the metals or that of an entirely different composition.

As is the case with other types of corrosion detailed thus far, the principal types of corrosion encountered with joints and associated contacts are galvanic and concentration cell corrosion.

Galvanic corrosion is manifested when dissimilar metals are joined and most pronounced when the metals of concern are distant in the galvanic series (Table 8.3). Concentration cell corrosion can be observed at washers, under applied protective coatings (e.g. under pipes that are in contacted with pipe bridge supports), on or adjacent to threaded joints and at surface defects in welds. Concentration cell action can start with a small amount of liquid trapped in crevices. The fault can be a result of faulty design, improper maintenance plans, or a lack of corrosion prevention knowledge. The phenomena can also be exacerbated by outdoor environments. For adverse outdoor environments, protection of joints and faying surfaces should undergo more frequent scheduled maintenance procedures during the critical weather periods.

The extent of corrosion protection needed for joints and faying surfaces is dependent to a large extent on intended service, the types of metal or alloys that are joined, and how dissimilar the metals are. As a minimum, reiterating, it is desirable to keep contact surfaces as dry as possible with coatings, including paints, insulation, sealing tape, polymer coverings, and maintaining tight joints to minimize corrosion of faying surfaces. These elements coupled with the measures described in Table 8.5 can significantly reduce the probability of this form of corrosion.

8.7 Pourbaix Diagrams

The McGraw-Hill Science and Technology Dictionary defines a Pourbaix (poor bay) diagram as a plot of standard electrode potential (Table 8.1) versus pH that is used to predict the thermodynamic tendency of a metal to corrode. The Pourbaix diagram also presents a range where metals in aqueous solutions are stable or exhibit passivity in specific pH ranges and standard electrode potential ranges. While the Pourbaix diagram is reasonably useful in forecasting corrosion scenarios, it should be recalled from the Thermodynamics Interlude of Chapter 2 that the Pourbaix cannot determine a rate of corrosion. Further, many of the concepts introduced in Section 8.1 (redox reactions, cathode, and anode activity, etc.) should be helpful when covering Pourbaix diagrams.

While the Pourbaix diagram is used to evaluate metal, an understanding of the system can best be explained by starting with a simple system, water. A relationship known as the Nernst equation serves as the basis for the development of Pourbaix diagrams. The Nernst equation is given as

$$E = E^0 - 2.303 \frac{RT}{2F} \log \frac{1}{\left(H^+\right)^2} \tag{8.3}$$

where

E = standard electrode potential at a given condition, volts
E^0 = standard electrode potential at 298.2°K
R = ideal gas constant, 8.314 J/°K-mol
T = temperature, °K
F = 96 500 coulombs
(H^+) = hydrogen concentration in moles

Recalling the definition of pH previously defined in Section 2.1 and employing the half cell reaction for hydrogen ($2H^+ + 2e^- = H_2\uparrow$) given in Table 8.1 and the Nernst Equation

(Equation 8.3), a linear relationship correlating the pH of a metal in a dilute aqueous solution and the electromotive potential of the metal can be derived. This relationship is given as

$$E = E^0 - 0.0592\,\text{pH} \tag{8.4}$$

As will be shown, Equation (8.4) is a fundamental element of the Pourbaix diagram.

Example 8.3 Determine the electrode potential of the half cell reaction $Al^{+++} + 3e^- = Al$ in an aqueous solution at $25\,°C$ and 1 atmosphere pressure atmosphere and a pH of 4.

Solution
Referring to Table 8.1, the standard electrode potential for the half cell reaction is $-1.63\,\text{V}$ (E^0). Inserting the values for E^0 and the pH (4) in Equation (8.4) yields

$$E = E^0 - 0.059\,\text{pH} = -1.63 - 0.059(4) = -1.866\,\text{V} \leftarrow \textbf{Solution}$$

This increased standard electrode potential increases with pH, signifying oxidation.

Using Equation (8.4) as a basis, a simple Pourbaix diagram for water has been constructed [8]. Figure 8.5 displays the Pourbaix diagram for water, as conceived by Marcel Pourbaix.

By examining the diagram, it is observed that the pH is plotted on the abscissa (x-axis or horizontal axis), while the electrode potential shown on the ordinate (y-axis or vertical axis) represents the pH. The Pourbaix diagram also has two diagonal lines labeled a and b. The horizontal lines separate the diagram into three zones marked "Oxidation," "Domain of water stability," and "Reduction," respectively.

- The region below line a is water that is unstable and decomposes (reduction) to hydrogen gas, H_2, through the decomposition of water, a lower pH, by means of the reaction: $2H^+ + 2e^- \rightarrow H_2\uparrow$.
- The region above line b is water that is in a stable state undergoes no significant change.
- The region above line b contains water that because of its thermodynamic properties and pH is unstable and oxidizes to give off oxygen (oxidation) O_2 from the reaction: $O_2 + 2H_2O + 4e^- \rightarrow 4(OH)^-$. As shown in Figure 8.5, another important factor is that metals having a standard electrode potential above hydrogen (Table 8.1) will undergo oxidation and thus exhibit corrosion potential.

The region below line b is composed of water that is stable, and if there is dissolved oxygen in the water, the oxygen will be reduced to water, H_2O, with no change in stability.

The elementary Pourbaix diagram for water illustrated in Figure 8.5 clearly identifies three regions, separated by the dashed diagonal lines, which are applicable to all Pourbaix diagrams. The three regions and their functions are as follows: the upper region where water is anodically oxidized to give off oxygen gas, the lower region where water undergoes a cathode reduction reaction yielding hydrogen gas, and the middle region where the water is stable and not subject to decomposition.

This fundamental water diagram is the basis for all metal (including alloys) composition Pourbaix diagrams. In fact, the Pourbaix diagram for alloys and metals is composed of superimposing the aqueous phase diagram, shown in Figure 8.5, superimposed on a metal/metal alloy Pourbaix diagram.

Figure 8.6 is an example of the water diagram superimposed on a metal; in this case the metal is aluminum. Figure 8.6 is a simplified Pourbaix diagram for a metal (aluminum) and an aqueous medium.

It must be emphasized that Pourbaix are for metals and alloys in an aqueous medium. Figure 8.6 has some additional factors not shown in Figure 8.5. In addition to the two dashed

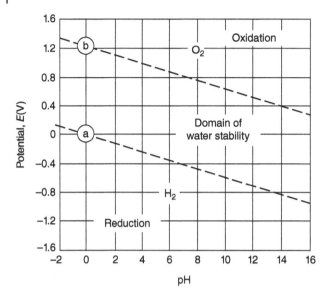

Figure 8.5 Pourbaix diagram of water.

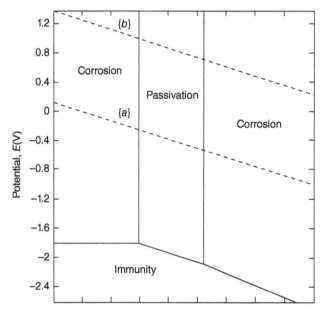

Figure 8.6 Pourbaix diagram for aluminum.

lines defining the water oxidation region (the area below line a where elemental oxygen is released), the aqueous stability region (the area between dashed lines a and b) and the reduction zone, below dashed line b, where elemental hydrogen is released from the water due to the reduction reaction occurring, there are two additional vertical lines shown in Figure 8.6. As can be seen in the diagram, the vertical lines differentiate corrosion regions from the passive, less reactive zone. Note that the less reactive or passive region occupies the neutral pH range (i.e. pH 4 to pH 9). The lines that are not vertical, exhibited in Pourbaix diagrams, differentiate species in aqueous solution that do not involve redox reactions and represent the region that is immune to corrosion, as the diagram notes. Typical of this region is the lower region shown in Figure 8.6 identified as "Immunity." This region is the lower part of the diagram where the

Figure 8.7 Pourbaix diagram for chromium.

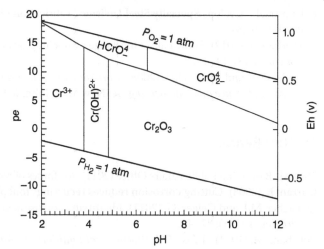

upper boundary is the nonhorizontal lines exhibiting decreasing immunity as the pH increases and the standard electrode potential decreases relative to hydrogen. Again, it is most important to reiterate that as the standard electrode potential becomes more negative, higher energy, in the form of voltage, is required to affect the region in the Pourbaix diagram. Consequently, as one might expect, the upper, more positive sections of the Pourbaix diagram are the regions most likely to undergo a transformation, which is typically the oxidation or corrosion reaction.

A more detailed Pourbaix diagram for chromium is shown in Figure 8.7. Although the diagram for chromium is more detailed due to various configurations, chromium can assume that the basic regions (oxidation, reduction, passivity, and immunity) can be identified as well as the changes occurring with pH variation and electrode potential.

Another mode of Pourbaix diagram analysis is by observing the classification of related reactions that can be identified by following comments:

1) Equilibrium in reactions that involve neither H^+ ions and not electrons are independent of pH and potential.
2) Equilibrium in reactions that involve H^+ ions but not involve electrons depends on pH but not on potential.
3) Equilibrium in reactions that do not involve H^+ ions but do involve electrons are independent of pH but depends on potential.
4) Equilibrium in reactions that involve both H^+ ions and electrons depends on both pH and potential [5].

References

1 Federal Highway Administration (1999). *Corrosion Costs and Preventative Strategies in the United States.*
2 Hutchinson, R. (2012). Microbiologically induced corrosion. *Flow Control* 7 (18): 24–40.
3 Brady, R.J. (2009). The other war. *Products Finishing* 73 (3): 10–15.
4 Corrosion and Corrosion Protection of Metals (1965), MIL-HDBK-721. Washington, DC: Department of Defense (November).
5 Pourbaix, M. (1971). *Thermodynamics of Dilute Aqueous Solutions.* London: Edward Arnold & Company (Translated by J.N. Agar).

6 Universal – Cyclops Specialty Steel Division "Fabrication of Stainless Steels" (1966), pp. 4, 5, Pittsburgh, PA.

7 Tverberg, J. (2012). Spray balls: One rouge source in pharmaceutical water systems. *Flow Control* 18 (3): 14–17.

8 Revie, R.W. and Uhlig, H.H. (2008). *Corrosion and Corrosion Control: An Introduction to Corrosion Science and Engineering*, 4e. New York: John Wiley & Sons, Inc.

Further Reading

78th Air Base Wing Public Affairs Office (2011). Air forces of nature. *Products Finishing* 75 (8): 14, 15.

Curran, E. (2008). Cutting corrosion reduces recurring "foul play." *Flow Control* (March): 30–38.

Esmacher, M.J. and Geiger, G. (2011). Identifying corrosion and its causes. *Chemical Engineering Progress* 107 (4): 37–41.

Pourbaix, M. (1974). *Atlas of Electrochemical Equilibria in Aqueous Solutions*, Second English Edition. Houston: National Association of Corrosion Engineers.

Smith, R.N. and Pierce, C. (1955). *Solving General Chemistry Problems*, 5e. San Francisco: Freeman and Company.

Sutton, B. and Larkin, C. (2010). Passivation basics. *Products Finishing* 74 (7): 18–21.

Tverberg, J. (2000a). A stainless steel primer, Part 1: Types of stainless steel. *Flow Control* 8 (6): 30–42.

Tverberg, J. (2000b). A stainless steel primer, Part 2: Corrosion mechanisms. *Flow Control* 9 (6): 34–39.

Tverberg, J. (2012). A stainless steel primer, Part 3: Selection of the proper alloy. *Flow Control* 10 (6): 28–36.

9

Pharmaceutical Materials of Construction

9.1 Introduction

Other than corrosion prevention and corrosion control, little time is spent on materials of construction and their applicability to pharmaceutical operations. Typically, types 304, 304L, 316, 316L, and various plastics (generally supplied as OEM components of a purchased as-built system) are the only materials pharmaceutical personnel deal with as part of common operations with little or no understanding of why specific materials are used. Consequently, it is desirable for pharmaceutical operating personnel to understand the materials selection activity, often to implement cost savings and reliability improvements.

9.2 Materials Selection and Performance Requirements

Pharmaceutical manufacturing operations have specific regulations that require an extra level of detail not commonly encountered in manufacturing operations These regulations are defined in 21 CFR 210 and 21 CFR 211. These specifications are referred to as current good manufacturing practice or cGMP. The need to conform to standard construction materials and design criteria are set forth in 21 CFR Subpart D, specifically 211.65. The relevant regulatory criteria for 211.65 (a) notes that "Equipment shall be constructed so that surfaces that contact components, in-process materials, or drug products shall not be reactive, additive, or absorptive so as to alter the safety, identity, strength, quality, or purity of the drug product beyond the official or other established requirements."

Another related section of cGMP regulations are 21 CFR 211.94 (a), drug product containers and closures:

Practical Pharmaceutical Engineering, First Edition. Gary Prager.
© 2019 John Wiley & Sons, Inc. Published 2019 by John Wiley & Sons, Inc.

> Drug product containers and closures shall not be reactive, additive, or absorptive so as to alter the safety, identity, strength, quality, or purity of the drug beyond the official or established requirements.

The above regulations are applicable to drug containers and manufacturing process components such as disposable single use manufacturing parts and equipments in contact with the constituents of the drug product or intermediate. Two terms often used when dealing with single use components in contact with drug and bioprocess components are leachables and extractables.

For a cGMP configuration, contact components include processing items such as pipes, valves and fittings, seals, instruments, tanks (including atmospheric tanks, pressure vessels, reactors, bioreactors, and fermentors), pumps, containers, and packaging. These relevant elements have also been detailed in other chapters (e.g. Chapters 1, 2, and 4).

The phrase, "...contact components, in-process materials or drug products..." located in 21 CFR 211.65 and identified as the basic definition of critical components for cGMP applications, also defines the need for including design elements as cGMP considerations. Consequently, the pharmaceutical process equipment identified above, when used in drug processing, can be considered critical components when coming in contact with pharmaceuticals or pharmaceutical manufacturing ingredients. In this vein, it is relevant to note some issues encompassed in materials of construction that may appear to be unrelated to the subject. However, as pharmaceutical engineering is a very broad subject, many details appearing to be unrelated, do, in fact have common themes when addressed more closely. An example of this commonality is 21 CFR 211.65.

9.2.1 Introduction of Polymeric Materials for Single Use Systems

Two important terms that are relevant to operations where single use components interact with cGMP constituents and products are extractables and leachables [1]. These terms can be defined, respectively, as:

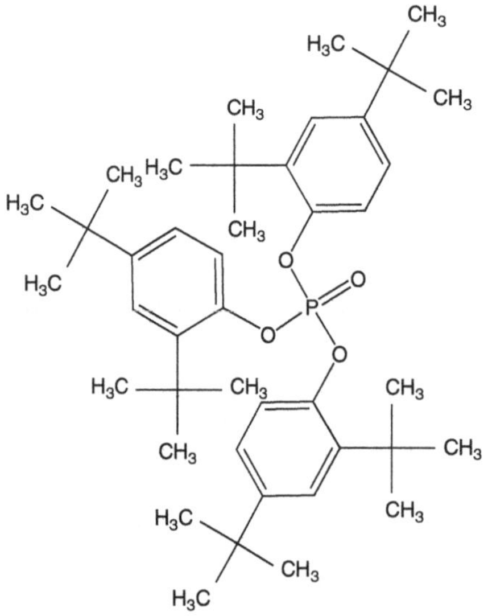

Figure 9.1 Molecular structure of bis(2,4-di-*tert*-butlylphenyl)phosphate.

- Extractables – Chemical compounds that migrate from any product contact material when exposed to appropriate solvent under exaggerated conditions of time and temperature [2].
- Leachables – Chemical compounds, typically a subset of extractables that migrate into the drug formulation from any product contact material, including elastomeric, plastic, glass, glass, stainless steel, or coating components as a result of direct contact with the drug formulation under normal process conditions and are found in their final drug product [2].

An extractable compound of increasing concern is bis(2,4-di-*tert*-butylphenyl)phosphate (bDtBPP), the molecular structure of which is shown in Figure 9.1. This compound is used often as a stabilizer for single use bag containers. It is known to inhibit cell growth and, thus, can affect biotech manufacturing. It is incorporated as an antioxidant additive in plastics, which is often for single use containers to prevent/minimize oxidation of the polymer containers such as polyethylene formulations [3].

The original intent of polyethylene piping was to be a replacement for stainless steel piping.

Studies have indicated that bDtBPP is formed from the breakdown of tris(2,4-di-*tert*-butyl-phenyl phosphate), known as Irgafac®, the molecular structure of which is given in Figure 9.1.

For single use disposable applications, polymer constituents are increasingly being used in packaging and patient/end user applications for items such as bags, hypodermics, and containers. These polymeric materials have demonstrated durability, strength, and adaptability as a viable storage and containment media.

In addition to incorporating single use components for manufacturing and development, leachables are very important elements since these polymeric constituents cannot significantly affect the bioprocess operation by exhibiting leaching of constituents prior to and subsequent to manufacture. Section 9.5.4 identifies the hazards of using polymeric materials with a proclivity for leaching as a potential source of contamination of product.

Several studies have found that most of these polymeric materials can induce leachables into the manufacturing process [4].

Conforming to provisions of 21 CFR is critical to successful pharmaceutical manufacturing. However, significant components of pharmaceutical manufacturing do not involve cGMP operations; these non-cGMP operations and equipment are often referred to as balance of plant or BOP. Typically, such BOP operations include facility utilities such as drinking water and HVAC supplied to administrative operations such as accounting, sales, and central engineering offices that are isolated from cGMP facility locations. While these installations are non-critical operations, they also require maintenance, repair, and upgrades. Generally, construction materials, such as 304 and 316 stainless steels, are used because they provide a resolution of CFR 21 Parts 210 and 211 criteria. Whether or not these materials are best suited for the required application is often overlooked during the selection process. If an as-built system is selected, often standard materials are used due to success in previous vendor supplied installations. The use of alternate materials is seldom evaluated for a specific facility due to cost and time constraints.

9.3 Advantages and Disadvantages of Stainless Steels and Polymers for cGMP and Non-cGMP Pharmaceutical Applications

With the increasing deployment of disposable polymeric single use systems (SUS) and disposable process components fabricated of a variety of high performance polymers, it is reasonable to wonder if SUS outperform equipment and components based on conventional stainless steel items.

Single use technology is a technology frequently used in biotech and, to a lesser extent, pharmaceutical laboratory operations for several years. Several vendors provide components such as single use disposable tubing, sanitary connectors, disposable holding containers for both products and reactants, shipping containers, disposable bioreactors for smaller batches, and disposable single use chromatography columns and filtration units.

Recent concepts, such as the WAVE™ bioreactor, are employed for biotech operations in lieu of standard agitator equipped bioreactors and reporting significant cost reductions. One source reportedly notes the Wave bioreactor can reduce a single cost run by about $400 or $7000/year [5], exclusive of disposal costs.

While the use of disposable SUS are gaining acceptance for several reasons, current applications of SUS technology is primarily for use in process development and clinical scale manufacturing (i.e. typically limited manufacturing of product to support new drug development trials). The reliance on developmental applications, as opposed to production scale manufacturing, is an impediment to further growth in SUS technology.

While single use bioreactors are, as noted, generally accepted at the research and development level for screening, optimization, scale-up, and pilot-scale operations, these single use components are being studied for incorporation in cGMP manufacturing operations [6].

While single use reactors and fermentor are envisioned for manufacturing applications, current technology limits scale-up applications to devices ranging from 1 to 7 ml and from 20 up to 10 l capacity [6].

The primary impetus for greater implementation of single use disposable technology is the possibility of achieving significant cost reductions in design, installation, and operation of units containing disposable components compared with stainless steel configurations. For instance, disposable bioreactor units, including polymeric reactors and reactor liners, eliminate much of the requirements to clean-in-place (CIP) and or sanitize-in-place (SIP) components of the system subsequent to batch manufacturing, which is standard for stainless steel system components. Elimination of these operations minimizes the use of water and steam with a corresponding reduction in costs (a cost analysis should also include the cost of disposition of the hard piped CIP and SIP system components performing the same function).

While single use technology offers many advantages, such as those described, SUS implementation does have certain limits [7]. For SUS, the working volume for single use disposable agitated bioreactors is approximately 396 gal ($1.5 \, m^3$) and 660 gal ($2.5 \, m^3$) for storage tanks. One important limiting parameter is largely due to the limits of the materials of construction, i.e. low density polyethylene, which has a limited bonding and sealing options. The bonding method most commonly employed for biotechnology and pharmaceutical operations is heat sealing. This technique minimizes the probability of foreign matter contaminating the feed, media, or product. Other limiting materials of construction factors for single use applications include yield strength, elongation, shear stress, and operating temperature ranges; all of which have values significantly lower than stainless steel. As a result, technicians and operators often require special skills set compared with personnel trained in stainless steel hard piping operations. An example of the different requirements needed would be a more frequent installation and maintenance operation with an emphasis on leak prevention and repair for the single use components.

Another consideration when evaluating SUS is materials compatibility. Careful evaluation is very important when specific solvents are components of a biotech or pharmaceutical composition that is in contact with disposable components. Unlike stainless components, polymeric SUS parts must be analyzed to verify that leaching of the material will not occur. Such an analysis would be an added cost for single use disposable operations. Normally, a compatibility analysis would not be a consideration for stainless steel installations.

The actual savings derived from implementation of disposables will depend on the specific process, geographic location, local costs, and technology mix.

9.4 Disposal of Single Use Components

Disposal of single use components is a topic requiring an in-depth analysis one the SUS components have been expended. Depending on the process, the used, disposable single use components must be evaluated using regulated classification criteria for the spent SUS components when dealing with disposal of these spent disposable components as a hazardous or nonhazardous waste. This procedure involves regulations set forth by the US Environmental Protection Agency (EPA, the agency responsible for the provisions of 40 CFR, et. seq.), the US Occupational Safety and Health Administration (OSHA, responsible for enforcement of regulations contained in 29 CFR, et. seq.), and the US Department of Transportation (DOT, the agency responsible for enforcement of the regulations contained in 49 CFR, et. seq.).

While most spent SUS components may not contain hazardous substances subsequent to processing, it is important to evaluate the potential for the presence of harmful constituents (hazardous wastes) prior to disposal of SUS material. The classification of hazardous wastes and the regulations involving treatment and removal of hazardous wastes are also most important as requirements for lifecycle analysis of wastes comes under greater scrutiny.

The US EPA, until recently, has not focused, in depth, on the generation of hazardous pharmaceutical wastes as a specific category. Often, these wastes are classified as medical wastes or industrial wastes. Currently, an effort is underway to more closely monitor the specific hazards and classification of hazardous materials generated from pharmaceutical manufacturing operations. As the impetus to further investigate the hazards posed by pharmaceutical wastes, the role of single use disposable constituents will come under further scrutiny. This more detailed investigation could significantly alter the existing criteria for component disposal and the related disposal costs.

Perhaps the most attractive disposal method for nonhazardous disposable components is recycling. Recycling, although initially appealing is impractical for many polymeric materials, particularly if component is composed of different polymers. Some of the constituents may contain nonrecyclable thermosetting plastics (cannot be remolded to another form). An example would be a widely used single use material, silicone rubber, which, by definition, is not recyclable. Another detriment to recycling is the requirement to process this material at an economic rate. With an annual processing rate of 1–3 million pounds, few sites would generate the quantity required to make recycling a viable option [8].

The origin of hazardous waste identification and management is contained in legislation initially passed by Congress in 1976 and identified as the Resource Conservation and Recovery Act (RCRA). Shortly after passage of RCRA (1980), the US EPA introduced the initial regulations for dealing with hazardous wastes, the provisions of which are identified in 40 CFR 261 Subpart C. The major elements of this subpart are the identification and classification of what constitutes a hazardous waste (identified by the letter and followed by a series of numbers). The letters characterizing the type of waste identified are D, U, P, K, and F. Wastes that have not specifically identified by the letters U, P, K, or F are considered hazardous waste if they exhibit one of four properties defined in 40 CFR Part 261 Subpart C. These four characteristics are subject to the RCRA provisions for hazardous wastes. These four properties are:

1) Ignitability (D001) – Ignitable wastes can create fires under conditions, are spontaneously combustible, or have a flash point less than 60 °C (140 °F). Included in this category are used solvents such as those used in biotech compositions.
2) Corrosivity (D002) – Corrosive wastes are acids (pH ≤2 or ≥12.5) that are capable of corroding metal containers such as storage tanks, drums, and barrels. Includes substances capable of corroding steel at a rate greater than 0.25 in./year.
3) Reactivity (D003) – Reactive wastes are unstable under "normal" conditions. They can cause explosions, toxic fumes, gases, or vapors when heated compressed or mixed with water.
4) Toxicity (D004–D043) – Wastes that are harmful or fatal when ingested or absorbed. If toxic wastes are land disposed, there exists the possibility contaminated liquid may leach from the waste and pollute ground water.

In addition to the four characteristics identified, hazardous wastes can also be identified by employing hazardous wastes using the previously letter identifications U, P, K, and F. Each letter also identifies a waste category and associated regulation. These letters and associated regulations are:

- U – Regulated by 40 CFR 261.33f – this category is composed of discarded commercial products, such as out-of-specification products. If SUS components are used, they should also be evaluated if in critical contact with the discarded commercial product.
- P – Regulated by 40 CFR 261.33e – this category regulates acutely hazardous discarded commercial chemical products, out-of-specification substances, container residues (including disposable bioreactor bags, if applicable), and spill residues.
- K – Regulated by 40CFR 261.32 – this species of hazardous waste is derived from known sources.
- F – Regulated by 40 CFR 261.31 – this category of hazardous waste consists of substances that have nonspecific source of origination.

There are also regulations for mixtures of hazardous wastes mixed with nonhazardous wastes. This may be the situation where single use components are involved in the processing of hazardous solvents such as a flammable liquid such as hexane, a D004 waste, or arsenic-containing pharmaceutical waste compounds, D084. Mixtures such as these are regulated by the provisions of 40 CFR 261.3(b) (2), commonly referred to as the "mixture" rule.

The amount of hazardous wastes accumulated and the time the wastes can be stored on site is also regulated by the EPA. Most hazardous wastes generators are limited to producing and storing between 100 kg (221 lb) and 2200 kg per month. For acutely hazardous wastes (category P wastes), the regulatory limit begins at 1 kg per month.

Hazardous waste generators also have a timeline, depending on the quantity of hazardous wastes generated, to dispose of stored hazardous wastes. Hazardous waste generators that generate between 100 kg (220 lb) and 1000 kg (2200 lb) can store waste for 180 days prior to removal. Generators of more than 1000 kg (2200 lb) must be dispose of the stored wastes within 90 days (it is important to reemphasize that SUS hazardous wastes include the disposable components used in generating hazardous wastes as well as the species classified as a waste). Hazardous wastes stored in excess of 90 days are classified as storage facilities under RCRA regulations. This classification, storage facility, involves significantly greater regulatory and permitting criteria. The regulations governing a storage facility are provided in 40 CFR 260.10, which define storage as: "The holding of hazardous waste for a temporary period at the end of which the hazardous waste is treated, disposed, or stored elsewhere."

Since there may be several categories of hazardous wastes stored in the facility, the regulations are complex and significantly more detailed than hazardous waste storage regulations that are stored for less than 90 days. Consequently, it is desirable to remove generated hazardous wastes as quickly as possible.

The EPA has exempted several categories of waste from the RCRA regulations. These exemptions are located in 40 CFR 261.4 and are divided into four general categories containing about 18 specific exempt wastes. The four specific exclusion categories are:

1) Materials that are not classified as solid wastes.
2) Solid wastes that are not classified as hazardous wastes.
3) Solid wastes that are exempt from certain regulations.
4) Laboratory samples.

If discarded, SUS components are disposed of as nonhazardous wastes, it is most important that the waste is carefully analyzed prior to off-site removal. With the increased use of disposable components with an ever-increasing use of products and processes, single use parts disposal will come under much greater scrutiny as the technology becomes more widely incorporated in biotech processes.

Defining whether or not a waste is hazardous or nonhazardous is determined by performing a various analyses to determine the concentration of the hazardous waste of concern. These test methods for evaluating and analyzing solid hazardous wastes, if not identified as D001, D002, or D003, are obtained in a government manual titled *Test Methods for Evaluating Solid Wastes, Physical Chemical Methods,* SW-846. The document, commonly referred to as SW 846, describes the techniques, sample size, sample holding time, and equipment needed to analyze for specific hazardous wastes.

The procedures used in SW 846 toxicity tests rely on an extraction procedure to determine toxicity (F listed wastes) characteristics for the specific waste.

For instance, a waste composition containing arsenic has a regulatory limit of 5.0 mg/l (5.0 ppm). Any waste composition containing greater than 5.0 ppm arsenic is classified as a D004 hazardous waste. Arsenic concentration is obtained by using SW846 test methods 760 and 761. In addition to being a D004 hazardous waste, certain arsenic-containing wastes that are used in the manufacture of veterinary products are classified as K 101, K 102, or K 84 wastes pharmaceutical manufacturing wastes (i.e. wastes from a specific source, as defined by 40 CFR 261.32).

Regulatory agencies are increasingly requiring a more detailed analysis of potential contaminants based on biopharmaceutical product leachables generated from disposable components such as reactor bags, tubing, storage containers, and other critical components of the manufacturing operation. This increased scrutiny has led to some novel approaches to ascertain the content and concentration of potentially harmful organic constituents. With the growth of biotechnology and SUS, quicker and more accurate analytical methods will be required to preclude bottlenecks in manufacturing operations due to analytical delays. New procedures such as relying on total organic carbon (TOC) analysis are being evaluated. While this evolving methodology exhibits potential, there is a difficulty encountered. The existing online TOC analyzers are effective only when the process solvent is an aqueous solution. The online analyzers are ineffective if the feed or product uses organic solvents.

This example represents just one of the difficulties encountered when more effective analytical tools are sought to address leachables analysis for pharmaceutical and biotech manufacturing [9].

Waste generators, including pharmaceutical manufacturers, in addition to determining the quantity of hazardous wastes generated, must also classify the category of waste produced, prior to off-site disposal. Disposal of hazardous material may include single use components employed in the processing of the hazardous wastes.

The EPA reference to solid wastes can be misleading. If a solid waste is blended with a liquid and subsequently determined to be a hazardous material, the original material was a solid; hence the waste, although in liquid form, is classified as a solid waste.

In addition to conforming to the EPA provisions noted, state environmental agencies may have other applicable regulations that must also be evaluated. Further, there is not one uniform single use disposal criteria. Pharmaceutical and biotech SUS wastes can contain both chemical and biological wastes. As an example, discarded disposable components involved in the production of monoclonal antibodies (MAb) may be classified a hazardous waste if intermingled with a solvent such as hexane that is used as part of the manufacturing operation. Currently, the FDA and EPA are investigating the risks associated with genetically modified organisms. It is conceivable that, as more data becomes available, closer scrutiny will require more detailed disposal requirements.

A growing concern dealing with disposal of drugs and biotech products is the presence of drug products in drinking water systems. Consumers often discard unused drug products in domestic disposal systems (e.g. sink, toilets) that are treated via municipal waste treatment system that are not equipped to treat products such as antibiotics and diuretics. As a result, the wastes cannot be treated and are discharged, often without proper waste treatment. In fact, in 1980, the EPA identified approximately 30 chemicals used as pharmaceuticals that met the criteria of RCRA hazardous waste [10]. As of 2013, the list has yet to be modified or updated. Further, a review process has yet to be established for providing a format to identify additional compounds used in biotech and pharmaceutical manufacturing that would classified as RCRA wastes. To preclude this type of contamination occurring from SUS components, it is critical the SUS waste stream be properly classified. Usually, if the components present a hazard, the primary concern is being a hazardous waste with medical wastes being secondary. Also, pharmaceutical and biotech operations cannot reduce the waste volume or treat the waste in any manner since drug and biotech operations typically are not licensed as a waste treatment facility, thereby increasing the expense of waste removal. Here, the treatment of discarded SUS components requires attention prior to disposal. If a waste component, such as tubing or connections contaminated species as a residue, the components can be rinsed several times to remove the accumulated hazardous residue. However, the rinse solvent cannot be discharged in standard facility treatment operations unless the facility discharge permit allows for discharge at the waste concentration specified by the permit. If the waste is below the permit level specified, it is usually removed by the site waste contractor. Discharge permits are reviewed at intervals and compared with the actual facility discharge rates to verify disposal procedures are performed in accordance to the permit regulations.

Neutralization of the waste to a sanitary sewer is generally permissible if the waste has no other hazardous or dangerous waste characteristics (P listed wastes) and the neutralization procedure does not alter the characteristics of the waste.

With the increasing use of disposable SUS items, it is reasonable to assume the rate of hazardous waste associated with SUS will increase. Removal of the contaminated material may eliminate the waste from the SUS component, but the waste is collected in the solvent (typically water) used for washing the SUS component. Some SUS constituents such as disposable chromatography columns and disposable filters may not be free of contamination after the three allowable washes, and consequently, the entire assembly is classified as a hazardous waste and subject to off-site removal to a licensed transportation, storage, and disposal facility (TSDF) for approved treatment such as incineration or hazardous waste burial in a hazardous waste landfill. A significant disposal task involves chemical analysis of the SUS components, subsequent to use. The analysis may be more detailed if the process of concern produces side products during, or subsequent to, manufacturing. For an in-place stainless steel configuration, cleaning or sanitization of the installation prior to additional manufacturing is normally satisfactory. However, if SUS components disposal is required, additional analytical testing may be necessary to determine if the unwanted side reaction products are present. This, in turn, could significantly alter the waste classification and disposal costs. This is often time consuming and cost intensive.

Perhaps just as important as the hazardous wastes generated through SUS operations are the single use components themselves. Disposal of these items, subsequent to use, is also a concern. Landfilling of these polymers and resins, even if free of hazardous materials, presents a further difficulty. Most of the polymers and resins will not biodegrade and are designed to be UV stable; the materials can remain stable for many years. Incineration, cogeneration, and recycling similarly have limitations [11].

Another incineration concern is the formation of hazardous compounds, including dioxins, during the incineration operation [12]. While dioxin formation occurs at higher temperatures in excess of 700 °C or 1292 °F, these operating conditions are not uncommon for incineration.

Another potential alternative to incineration is pyrolysis. Pyrolysis is essentially the decomposition of plastics (e.g. polyethylene and polypropylene) at high temperature (around 800 °C or 1472 °F) in the absence of oxygen. The process is attracting notice since a product of the pyrolysis operation is liquid fuels. Conversion rates for pyrolysis-produced fuels are claimed to be about 7 lb of plastic consumed for each gallon of fuel produced [13] with a supposed energy value in the range of 15 000–20 000 Btu/lb and a post-pyrolysis residue ranging from 0.15 to 5.82%, depending on the plastic [14].

While the extracted oil is very pure and burns cleaner than diesel fuel, pyrolysis is not very efficient and has a high capital cost, so a high input volume would be needed [8].

One important consideration that must be evaluated in the development of pyrolysis processes specific to single use plastic components is the formation of hazardous compounds generated during the operation. Several studies have indicated the formation of dioxins and dioxin precursors during pyrolysis operations [12]. Another study indicates dioxin formation occurs when the feed composition contains certain metals that act as catalysts in the presence of chlorides [12].

While the quantities of dioxins produced are very low, many studies are performed using small feed quantities and may not be representative of actual industrial operations. For disposal of plastic single use components undergoing pyrolysis, a careful, and costly, segregation pretreatment operation may be required prior to actual pyrolysis.

Another alternative to incineration is an autoclave. An autoclave is a device normally used to sterilize (a comprehensive definition of sterilization is located in Chapter 2, Section 2.1) equipment commonly using high pressure saturated steam for a specified processing time. Single use disposable components must be sterilized such that all species of bacteria, viruses, fungi, and spores are rendered harmless. For medical applications, autoclave typically operates at 121 °C (250 °F) for 15 min. However, recent infectious agents have been found that are capable of survival at temperatures above 121 °C (250 °F).

The autoclave temperature, coupled with the steam, can be problematic if the components are fabricated or contain plastics. Unless the components are fabricated of high performance polymers such as PVDF (the published melting point of PVDF is 177 °C [350 °C]), there is a strong chance the materials will melt during the sterilization procedure. Many single use components are fabricated of polyethylene with a melting point range of 120–130 °C (248–266 °F) or polypropylene 121 °C (250 °F).

While autoclaving is used to destroy hospital wastes, establishing a large operation is very costly and, thus far, not an operable option.

If a technique other than incineration is considered (i.e. pyrolysis), the situation becomes even more problematic. In this case, the operating temperature is 420 °C (788 °F) for polyethylene and 400 °C (752 °F) for polypropylene, indicating significantly higher energy requirements for this operation (pyrolysis).

Another issue requiring resolution if autoclave sterilization is to be evaluated as an option is an additional processing step prior to final disposition of single use components is required. While using an autoclave is often employed to sterilize wastes, the presence of single use components can complicate disposal. Processing details such as post-sterilization of single use component integrity and assurance of operating profiles with respect to temperature and time for the plastic components can complicate quality criteria and require significant qualification or validation, an additional cost factor. As noted in Section 2.1 (sterilization), autoclave operation requires assuring that the F_0 value is achieved in the operating cycle to achieve successful sterilization. The critical parameter, in addition to proper operating temperature (121 °C or 250 °F), is the processing pressure (30 psig at atmospheric pressure) and processing time of 15 min. For a large mass of plastic components, a larger autoclave design may be required for economic implementation, requiring even additional validation and possible design enhancement. Further, successful autoclave processing is preemptory to ultimate disposal such as a landfill that, as noted, could entail further transportation and processing costs.

While landfill disposal is a potential option, a thorough analysis must be performed to confirm biohazards such as those identified and virus species have to be satisfactorily removed or sterilized from the single use component. Again, this activity often involves transportation to an off-site treatment facility for a significant removal and disposal fee. Typically, in-house treatment such as autoclaving is not effective enough for landfill disposal, primarily due to economics and biosafety considerations.

Regardless of the approved disposal methods, disposal of SUS components represents a cost that must be included as capital or maintenance expenditures. This cost evaluation may be difficult to evaluate for several reasons. Often, a large quantity of SUS waste is generated by contract manufacturing organizations (CMO) that perform smaller-scale processing for the product owner. Common CMO operations using SUS technology are small clinical batches; small quantities of product intended for absorption, distribution, metabolism, excretion, and toxicology (ADMET) studies; and scale-up pilot studies.

Disposal of the wastes is typically the responsibility of the CMO. As a result, the owner may have little knowledge of the actual disposal costs of SUS spent components such as bioreactor bags, disposable filter units, single use chromatography columns, and disposable tubing and fittings. Commonly, the CMO forwards the costs to the owner as part of a contract deliverable, often generically classified as waste disposal, either hazardous or nonhazardous. As a result, actual SUS disposal costs may not be accurately accounted for when classified as solid or liquid wastes. Similarly, when in-house SUS waste disposal costs are considered, the expense may be imbedded in a broad waste disposal cost category. As a result, the actual SUS waste disposal expenses may not be representative of the true SUS components disposal costs. Further, the SUS disposal classification may also be difficult to ascertain since it can be combined with other waste streams that may be classified generically as industrial wastes or hazardous wastes with the same category, thus rendering an actual expense more complicated than often assumed.

A generalized checklist has been prepared to assist in single use component disposal options [13]. The relevant aspects of the checklist are:

- Is the component classified as a biohazard or a chemical hazard (e.g. D001–D0043)?
- If hazardous, are the disposal methods restricted?
- How are the wastes to be handled, packaged, and transported to the disposal facility (the disposal technique must be in accordance with the provisions of 29 CFR, 40 CFR, 49 CFR, and applicable state and local regulations)?
- Does the preferred disposal technique present the possibility of harmful or gaseous vapor discharges (e.g. dioxins, carbon dioxide, methane, or carbon monoxide)?
- Has an economic analysis been performed to justify the desired disposal method?

If a CMO is operating the SUS, these concerns should be carefully addressed during the stakeholder's vendor audit procedures.

Another limiting parameter for single use is the allowable size of hosing and tubing. The maximum tube diameter employed for single use applications is nominally 1 in. (It is important to note polymeric flexible tubing is specified by the outside diameter of the tubing and not the inside diameter; consequently, an exact inside diameter is somewhat problematic since tubing sizing is based on the outside tube diameter, OD.) The tubing diameter restricts the flow rate and, consequently, the transfer time. For instance, using the maximum inside tube diameter of 1 in. and a flow rate of 31 l/min (8.2 gal/min), the time required to transfer 100 l (926.4 gal) of material would be 3 min. However, if a production batch of 1000 l (264.2 gal) required transfer, the transfer time would be 39 min. If the tubing is ¼ in., in lieu of a 1 in. line, the transfer time from a 100 l vessel would be 167 and 59 min for a 100 l vessel. Depending on product characteristics, the longer transfer time could be a relevant scale-up factor when compared with the 100 l pilot size batch transfer time. This timing consideration implies the tubing transfer length may play an important role in the process and the tubing transfer length (i.e. the distance between production components such as rectors and transfer tanks) may need to be evaluated also due to time considerations. Additionally, the tubing wall thickness may need to be evaluated for ruggedness of the wall thickness and flexibility of the wall thickness. An associated evaluation element is the type and integrity of the hosing and tubing connections installed for single use applications. In addition to properly designed connections, hose and tubing integrity is also a critical connection. Whereas stainless steel components such as piping have high yield strengths of about 70 000 lb/in^2 for 300 series stainless steels, polymeric single use tubing and hosing has yield strength in the range of 3200–3800 lb/in^2 for FDA grade tubing such as polypropylene, LDPE, and HDPE, as presented in Table 9.2. The concern when using these disposable single use components is tubing material mechanical degradation during pumping operations when transferring media in various stages of manufacturing. Because of the lower yield strength, standard centrifugal pumps are often not considered because of the higher pumping pressures employed. As a general guideline, single use tubing is limited to pressures less than 8 psig. However, if reinforced tubing is used, the tubing withstands pressure to 90 psig [16]. Consequently, if transfer operations do not rely on gravitational flow, the alternative is often a much lower flow device such as a peristaltic pump. Stainless steel piping does not require such a detailed evaluation for pumping transfer operations. The major drawback of peristaltic pumps, in addition to expense, is the proclivity for the tubing material to degrade with time, often with limited use. The degradation most commonly occurs at the pump rollers, where the tubing is continuously compressed and uncompressed during the pulsing operation of the pump rollers. Consequently, frequent preventive maintenance and tubing replacement can be encountered. However, one advantage to polymeric tubing is, if the tubing is clear, it is possible to see constriction and blockage of the tubing during operations. The most common tubing employed for peristaltic pumps is polyvinyl chloride (PVC), silicone rubber, and fluoropolymers such as PVDF. Whatever type of tubing used, it is critical the material is an approved FDA grade polymeric to assure that product contamination and leaching will not occur under operating and storage conditions.

Sterile connectors are also very widely used SUS items that require special consideration. Presterilized single use connectors perform the critical operation of connecting two sterile single use components in an environment that may or may not meet the specified requirements of a sterile environment yet maintain a sterile fluid flow path. A sterile fluid path should have a sterility assurance level (SAL) of 10^{-6} (the SAL is the probability of reducing a microbial population of a million to a value very close to zero). Ideally, a sterile connector should protect the fluid path from exposure to the environment. Installing a sterile connector to a SUS and possibly to another sterile connector can compromise the sterility of the SUS configuration.

In addition to design factors such as transfer tubing and hosing, temperature is also a critical variable. Disposable components are used for biopharmaceutical operating temperatures ranging between 4 °C (39.2 °F) and 40 °C (104 °F). Also, some bags, tubing, and connectors may not meet low temperature requirements for storage. Most biopharmaceutical products are stored at −30 °C (−25.6 °F), but some products, such as frozen biological cells, require storage at −80 °C (−112 °F). These considerations are typically not evaluation concerns when dealing with stainless steel piping and stainless steel components due, primarily, to the higher strength and other physical and thermal properties.

Another detail that should be considered with single use applications is the vendor selection. Standard stainless steel components such as piping, sterile connections, and holding tanks, with few exceptions, are generally available by a large compliment of vendors at competitive pricing. Single use disposable components often have limited vendors supplying validated components such as sterile connectors and bags, a factor that should be considered when preparing capital budget maintenance and support requirements beyond the initial operating period. This period could require several years of providing for replacements and maintenance. Depending on the amortization period, this annual expense for system maintenance is often at least 7% of the initial installed cost, exclusive of single use components replacement such as bags, connectors, and tubing. This expense, in addition to the disposal costs of the SUS waste components, can create significant additional expenses that are direct charges to the system operating costs.

Capital expenditures are a very relevant component when considering the change from conventional stainless steel to single use disposable configurations. Cost savings achieved by installing a single use technology suite can be estimated as [7]:

- 25–45% reduction in process equipment.
- 37–40% reduction for piping and equipment installation.
- 36–52% reduction in instrumentation and automation.

Of course, a significant cost variable is the degree of disposable technologies employed in the system. For instance, if a system consists of three 2000 l disposable bioreactors intended for monoclonal body (Mab) processing (assuming 90% utilization) and exclusive of disposal costs, possible savings of about 5.2% of the capital cost may be achieved. A mixer that uses a disposable media prep bag can achieve a possible 1.8% capital cost savings, also exclusive of disposal costs. A system using mixer buffer prep liner, also exclusive of disposal costs, can result in an additional 4.8% savings. If a third mixed buffer prep bag is also part of the system, the additional savings, again, exclusive of disposal costs, can be about 2.2% of the capital costs. With the addition of more single use components such as disposable holding units, products units, and membranes, a total system savings of 22.8% of capital costs can possibly be realized, exclusive of disposal costs.

These capital expenditure savings are primarily predicated on the deletion or reduction of installations such as detailed piping configurations, including the elimination of CIP and SIP systems and associated installations such as integrated instrumentation and automation equipment. Another capital expenditure reduction is a minimization in the requirement for qualification and cleaning of single use components. Most single use components are presterilized, primarily by gamma irradiation, and followed by subsequent sterile packaging of disposable components. This method of sterilization is less costly than cleaning, sanitization, and sterilization methods used for stainless steel installations.

While implementation of single use technology does offer significant opportunities for reduction of capital expenditures, there are other significant cost considerations that must also be addressed. For instance, with the introduction of SUS technology, the need may arise to install additional clean room facilities since there will be a more frequent replacement requirement for the disposable components than stainless steel installations that are not subject to replacement, subsequent to completion of a manufactured batch. In addition to clean room considerations, the existing HVAC requirements for the SUS will need upgrading. Typically, the upgrade could require more frequent air change out, higher air filtration requirements, ceiling modifications, flooring modifications, and other regulatory and design features and requirements, such as those addressed in Section 3.3. These additional requirements will usually require additional capital expenditures not often required in standard stainless steel installations.

A single use disposable system will, normally, require additional single use components as the number of production batches is increased.

As noted, single use disposable components require replacement after each production batch. This also affects SUS component storage requirements and supply logistics. Single use disposable components and spares must be available in adequate quantities to ensure production would not be affected by shortage of available components due to delivery time lapses. Lead times for disposable reactor bags and holding bags are typically in the range of 9–15 weeks for custom designs [17]. Also, poor planning, resulting in excess disposable stored components can also be detrimental to capital expenditures since it involves excess inventory, a cost consideration. While the surplus disposable items will eventually be consumed, the excess capacity of single use items represents an expenditure that could have been applied to more critical needs. Consequently, it is important to provide an inventory that can accommodate the single use units at an economic and logistically supportable level. Applying relationships such as Equation (5.10) can be helpful in determining an adequate single-use components inventory.

Occasionally, SUS are cited as having an improved reliability when compared with stainless steel systems. Such a claim may be problematic. While single use components have a high reliability, they are only employed for a single batch prior to disposal. Disposable items, such as single use filters and chromatography columns, are usually effective for the one-time application. However, the reliability of single use components should be compared with other SUS components serving the same function. Just as "replacement in kind" is a fundamental criterion for maintenance activities, the same principle should apply to SUS parts.

If an outside vendor is responsible for maintaining a single use component inventory, the spares stock should match the actual production requirements of the SUS. Often, component inventory is not intended to meet the specific single use replacement requirements of the installed system, but rather rely on a standard or generic single use components inventory. This inventory may or may not be specific to accommodate the supplier's requirements, as opposed to the user's requirements. The spare parts float, or inventory. Without this specificity, the user does not benefit by outside parts/spares inventory maintenance.

This lack of standardization creates a difficulty for the supplier and the user. The user desires the capability of interchangeability, and the supplier intends to introduce value adding component differentiation and maintain dependence on the supplier's products [18].

Another consideration involving single use components that should also be evaluated is the reliance on items that are specific to the SUS. Reliance on disposable components that are specific to the system can render the user to be dependent on one supplier for the SUS components. Much of this predicament is because there is no standardization or interchangeability program for single use disposable parts or components such as valves and connectors and disposable single use reactor components such as bags. Stakeholders are, once again, largely dependent on sole source suppliers for system support. Such single sourcing, in addition to relying on a single, dependent on one supplier, can obviate the possibility of other vendors offering less expensive and better designed products.

Alternatively, sole source approved vendors are desirable, in some situations, if they can effectively control the economics of single use components by maintaining costs and components inventory controls or assuring "just in time" delivery of the single use consumables. This can present difficulties if long lead time items cannot always be integrated in the logistics cycle.

The relationship between engineering and economics is identified by the American Engineer's Council for Professional Development (ECPD) and their definition of engineering:

> The creative application of scientific principles to design or develop structures, machines, apparatus, or manufacturing processes, or works utilizing them singly or in combination; or to construct or operate the same with full cognizance of their design; or to forecast their behavior under specific operating conditions; all, as respects an intended function, economics of operation or safety to life and property.

This rather detailed definition clearly itemizes the relevance of economics to engineering operations. Clearly, the success of many processes, such as disposable technology, is predicated on cost effectiveness as a measure of project success (other projects, such as Department of Defense programs, public works projects, etc., are important because of protection of life and property to citizens).

Often overlooked in the process of implementing new technologies such as single disposables are the overall economic benefits and drawbacks of the proposed technology. Often an objective evaluation is problematic because there is a paucity of available "real-life" information. A great deal of material exists regarding the efficacy of single use technology, but there is a significant lack of verifiable data regarding end-of-life disposition of contaminated single use disposable components. Indeed, while a great deal of published material has been directed at SUS components, disposal techniques are not readily forthcoming, Further, some information regarding disposal that has been presented is often provided by SUS vendors whose experience in single use component disposal may not be adequate to complete the SUS lifecycle.

To adequately achieve the objective of economic single use disposal, the responsible owners/operators of the SUS must be involved in virtually all aspects of the single use technology; failure to be proactive can be a serious detriment to this evolving technology as well as the "bottom line" of the manufacturing operation.

9.5 Performance Considerations for Pharmaceutical Materials of Construction

Chapters 2 and 8 have also addressed factors that should be considered when evaluating stainless steel and other metallic compositions for use in pharmaceutical and biotech operations, particularly those requiring adherence to cGMP regulations. One particularly relevant aspect of materials of construction for pharmaceutical applications is the provisions of 12 CFR 820 (which is specific to the design, manufacture, packaging, labeling, storage, installation, and servicing of all finished devices for human use). To achieve this goal, it is imperative that devices such as knee implants, spine implants, and hip implants are fabricated of materials that pose no significant risk to users of the devices. Here alone, materials of construction and their performance can affect the health and safety of users as well as significant penalties to manufacturers of faulty devices.

For pharmaceutical applications, an understanding of the behavior of materials used in pharmaceutical and biotech manufacturing can be a valuable asset when evaluating quality system criteria such as corrective and preventive actions (CAPA) for both cGMP and non-cGMP defects where materials of construction may be involved (e.g. pipe failures, seal failures, and equipment failures).

9.5.1 Stainless Steels

This section addresses stainless steels, their specific properties, and their applicability for use with other emerging materials, with a focus on pharmaceutical piping systems. The advantages and disadvantages of stainless systems and components are also addressed.

While significant information relating to stainless steels is located in Chapter 8, it is specific to one important aspect of stainless steel performance (i.e. corrosion). It is also important to understand the design and operational aspects of stainless steel equipment when used for pharmaceutical cGMP and non-cGMP applications, as well as a background in stainless steel corrosion prevention, as presented in Chapter 8.

Decisions on the types of stainless steel to use to use are generally based on experience, specifications, and, ideally, corrosion tests employing the procedures of the National Association of Corrosion Engineers (NACE) or the American Society for Testing Materials (ASTM). However, such testing and evaluation is rarely employed, due primarily to costs, time, and other feasibility concerns. As a result, it is a common practice to rely on specifications and standards when determining an applicable stainless for pharmaceutical and chemical applications. While Table 8.4 is specific to corrosion applications as criteria, there are types of stainless that have selected applications for specific uses in cGMP and non-cGMP environments. However, Types 316 and 316L are usually selected for all parts in contact with the product (critical components) because of its inherent corrosion protection and greater assurance of product purity. For older installations, Types 304 and 304L are used in lieu of 316 and 316L. Table 9.1 identifies some specialty stainless steels and their applications [19].

Table 8.4 identifies common stainless steels used in pharmaceutical applications with an emphasis on corrosion characteristics of these more common stainless alloys and compositions. Table 9.1 identifies less common stainless steels and potential pharmaceutical applications.

For pharmaceutical applications, 300 series stainless steels is the preferred choice since the 400 series ferritic steels do not possess the corrosion resistance of the 300 series austenitic stainless steels, primarily because of the higher chromium content (15–30%) than 400 series ferritic (10.5–20%) that, as noted, imparts a greater corrosion resistance. The corrosion resistance of 300 series stainless steels is also attributable, in large part to the higher concentration of nickel. Typically, 300 series austenitic stainless steels contain between 7 and 10.8% nickel, whereas 400 series ferritic stainless steels range from 0.4 to 0.8% nickel.

A third classification of stainless is martensitic stainless steel. This type of stainless is not employed in pharmaceutical applications since its lower chromium content (11.5–18%) does not exhibit the corrosion resistance of austenitic stainless steels.

Pharmaceutical manufacturing uses many categories of constituents in manufacturing processes. While each ingredient should be evaluated independently, there are some reasonably

Table 9.1 Some nonstandard stainless steels and their application.

Stainless steel type	Principal application(s)
301	Bins and containers
303, 303 Se	Nuts, bolts, and machined components
305	Used for fasteners corrugated siding and roofing in cGMP and non-cGMP applications
312	Used for welding dissimilar steels
317	For applications where somewhat better corrosion resistance than can be obtained from Types 316 and 316L is needed
405	Has better weldability properties than other 400 series stainless types
410	When heat treated, Type 410 can be used for high pressure process tubing. A martensitic stainless intended for non-severe corrosive applications
416, 416 Se	Nuts and bolts and components where precision machining is required
420	Valve stems
442, 446	High temperature applications in high sulfur-containing materials preclude the use of nickel containing stainless alloys

accurate general guidelines for evaluating the applicability of certain stainless steels concerning their relative corrosion resistance when in contact with common pharmaceutical ingredients. While Chapter 8 is a more comprehensive source, it would be helpful to have a brief overview of stainless steel piping corrosion considerations from the perspective of piping design guidelines when dealing with active pharmaceutical ingredients (API) and intermediates common to drug manufacturing.

While cGMP operations predominantly employ 300 series stainless steels, several operations relevant to pharmaceutical operations use a variety of metals, metal alloys, and high performance polymers for both cGMP and non-cGMP applications, as is described below.

9.5.2 Copper and Copper Alloys

Copper is available in various compositions including pure copper (99.88%, minimum), alloys for casting and forging, and alloys for the manufacture of wrought iron, and they are commercially available

While copper components are typically not employed for cGMP applications, they are used extensively in pharmaceutical facilities as the preferential material for applications that are integral to manufacturing operations. This is attributable in part because copper is a hygienic metal with intrinsic antimicrobial properties and slows the growth of harmful organisms including, algae, fungi, *Escherichia coli*, and MRSA legionella. Some of the antimicrobial applications of copper include antifouling paints, antifouling medicines, oral hygiene products, hygienic medical devices, and antiseptics.

Most copper alloys are primarily composed of copper with additives including zinc, tin, and lead. Other alloying constituents can include aluminum, beryllium, iron, manganese, nickel, phosphorus, and silicon. Most of these alloys are heat treatable (i.e. altering the properties of an alloy by heating or cooling) with the exceptions of aluminum bronzes, copper–beryllium, copper–nickel–tin, and copper–nickel–phosphorus alloys. Casting alloys, such as copper tubing, typically contain two or three alloying elements. Total alloying additives can be about 42% with the copper content varying from 58 to 92% copper [15].

As is the case with several metals, copper is classified in two categories regarding its mode of fabrication. The two classes are wrought copper and cast copper. Wrought copper, and other metals are formed by mechanical operations such as hammering or pressing, whereas cast metals are formed by the liquid metal being poured in a cast or mold. Wrought metals exhibit greater strength and withstand extreme weather conditions (atmospheric corrosion) better than cast metals. Table 9.2 compares some mechanical properties of cast copper alloys with wrought copper alloys.

Also, copper alloys provide high electrical and thermal conductivity, good workability and ease of joining, good corrosion characteristics, and a useful range of mechanical properties. A wide range of mechanical properties may be achieved by the proper use of alloying. Typically, tensile strength, yield strength, and hardness of copper alloys decrease with an increase in temperature, while elongation exhibits erratic behavior at elevated temperatures. Further, most

Table 9.2 Mechanical property ranges of copper alloys [15].

Property	Cast	Wrought
Tensile strength, min, ksi	21–155	30–230
Yield strength, min, ksi	8–130	11–165
Elongation, %	1–25	1–60

copper alloys are significantly impaired when the temperature exceeds 500 °F (260 °C). Conversely, at subzero temperatures, the tensile strength increases with the yield strength and impact resistance also increase, albeit in an erratic manner. The elongation is also slightly altered.

In addition, copper and its alloys are readily finished by plating or lacquering and can be joined by conventional methods. Table 9.3 summarizes some important thermal and mechanical properties of copper.

For most pharmaceutical operations, copper tubing is the most common application of the metal. Common copper tubing applications, service, and operating limits are displayed in Table 9.4. The term NPS is an abbreviation for nominal pipe size, denoting the pipe outside diameter for identification purposes. The NPS is not intended for detailed calculations.

Table 9.3 Mechanical and thermal properties of copper at ambient conditions.

Material – copper/copper alloy	
Property	Pure form or alloy
Specific gravity	8.96 (pure)
Elongation (%)	15 (pure)
Tensile strength (lb/in^2)	32 000–45 000 (pure copper and alloy)
Flexural strength (lb/in^2)	17 000 (alloy)
Compressive strength (lb/in^2)	25 000 (pure)
Tensile elastic modulus (Young's modulus) (lb/in^2)	17 000 000 (pure)
Hardness, rockwell	B 40 (pure)
Melting point (°F)	1981.4 °F (alloy)
Thermal conductivity (btu/ft °f h)	231 (pure)
Specific heat	0.095 (pure)
Coefficient of thermal expansion ×10^{-4}/F	9.4 (pure)

Table 9.4 Copper tubing service data.

Material – copper tubing			
Service	NPS (in.)	Maximum pressure/ vacuum (lb/in^2)	Construction
Plant air (dew point – 39 °F [−39.4 °C], CO$_2$, nitrogen)	⅜–2⅛	150	Solder joint/flareless tube fittings
Instrument air	¼–2⅛	100	Solder joint/flareless tube fittings
Refrigerants	⅜–4⅛	150	Solder joints/flareless tube fittings
Chilled water supply and return – above ground	⅜–2⅛	150	Solder joint
Well water – above ground	⅜–2⅛	150	Solder joint
Process water (untreated) above ground (prior to treatment)	⅜–2⅛	150	Solder joint

Similarly, copper tubing is also identified by the outside diameter. For copper tubing, the minimum service temperature is 40 °F (4 °C), and the maximum service temperature is 180 °F (82 °C).

9.5.3 Carbon Steels and Alloy Steels

Carbon steels are, perhaps, the most widely used type of piping in pharmaceutical manufacturing facilities. Steel is identified as carbon steel when a minimum content is identified or specified for the elements aluminum, boron, chromium, cobalt, columbium, molybdenum, nickel, titanium, tungsten, vanadium, zirconium, or other elements that displays a desirable alloying property, provided the minimum copper content does not exceed 0.40%. Other criteria are the maximum content for manganese does not exceed 1.65%; silicon, 0.60%; and copper, 0.60% [20].

Carbon steel is differentiated from alloy steel due to the content of the composition range of elements identified above. Alloy steels contain constituents exceeding one or more of the following percentage compositions: manganese, 1.65%; silicon, 0.60%; and copper, 0.60%. Also defined as alloy steels are steels containing a maximum of 3.99% of cobalt, columbium, molybdenum, nickel, titanium, tungsten, vanadium, zirconium, or other alloying constituents that exhibit a preferred alloying property.

While stainless steel is an alloy steel, there is one element with a specific content range that differentiates the two steels. Stainless steel contains 10–11% minimum concentration of chromium, while alloy steels contain a maximum of 4% chromium. As noted previously, the chromium creates the oxide surface layer that imparts the corrosion resistance property of stainless steel.

9.5.3.1 Primary Constituents and Performance Properties of Carbon Steels

Carbon, the major hardening element of steel, increases the hardness and tensile strength of steel. The optimum carbon concentration is about 0.60%. Above that range, the hardness and tensile strength increases become very small. In fact, a carbon content of 0.85% results in a hardness and tensile strength that is less than the lower carbon ranges. Additionally, the ductility and weld ability decreases with increasing carbon content (recalling that 304L and 316L stainless steels are specific to welding these materials due to the lower carbon content).

Manganese also improves the strength and hardness, but not to the extent of carbon addition. The addition of manganese also affects the ductility and welding capability, but to a lesser extent than the addition of carbon. Manganese is also effective in minimizing the formation of undesirable iron sulfide since it more readily reacts with sulfur during the smelting process.

Phosphorus, while often classified as an impurity, imparts improved machinability due to increased strength and hardness, phosphorus also somewhat diminishes the ductility and impact strength. For this reason, the phosphorous concentration is maintained below a maximum concentration of 0.040% by weight.

Sulfur, a nonmetallic constituent of carbon steel, is considered an impurity and undesirable element with the exception when machinability is consideration. Sulfur-containing carbon steels reduce the occurrence of "chipping" or slivers that form while cutting carbon steel during machining operations. Depending on the carbon content, the sulfur content of carbon steel is generally in the range of 0.08–0.33%. If the carbon content ranges between 0.15 and 0.25%, the sulfur content rarely exceeds 0.13%, unless sulfur addition is required for enhanced machinability. In this case, the sulfur content can be as high as 0.35%.

Silicon-containing compounds (silicon carbide is commonly used) are added to carbon steel during manufacture to limit the oxygen content of the carbon steel. Depending on the type of carbon steel required, silicon content varies from 0.10 to 0.30% by weight.

Copper, whose concentration should, as noted, not exceed 0.4%, is added to enhance the atmospheric corrosion resistance of carbon steel. While the small addition of copper has no significant negative effect on mechanical properties, the copper tends to cause brittleness when hot, thereby adversely affecting the surface brittleness [20].

9.5.3.2 Primary Constituents and Performance Properties of Alloy Steels

While not used as often as carbon steels or stainless steels in pharmaceutical applications, they have noteworthy benefits and should be considered for specialized applications a particular alloy steel property can be considered.

As a rule, production of alloy steels are more difficult to produce than carbon steels for several reasons. The primary reason is chemical constituents must be present in a very precise and accurate range. Also, most alloy steels require specific procedures and specific cooling procedures as well as detailed testing and inspection to produce an approved product (i.e. billet)

As is the case with carbon steels, common alloying elements have distinctive characteristics. The alloying elements and their effects on alloy steels are described below.

The addition of nickel as an alloying agent dates to the early twentieth century. When nickel composition is 8–10% composition, the resulting alloy steel exhibits increased roughness, particularly at lower temperatures. As a result, this type of austenitic alloy steel (nonmagnetic steel having a face-centered cubic [FCC] crystal structure) is used for cryogenic applications such as liquid nitrogen storage tanks.

Nickel containing alloy steels, depending on the desired property, has a composition that varies from 0.35 to 0.75% to as high a content range of 1.65–2.00%.

Chromium, when alloyed with carbon steel, imparts an increased hardenability and improved abrasion resistance. This type of alloy steel is also easily machined and does not require special welding techniques such as tungsten inert gas (TIG) or magnesium inert gas (MIG) welding. Chromium is the most stable alloying elements and is a preferred alloy when high temperature applications are required.

When chromium is alloyed in excess of 10%, the corrosion resistance of the steel is significantly enhanced.

While vanadium is not an additive to alloys for piping applications, it is used in pharmaceutical projects. Alloy steels containing vanadium are widely used in structural steels for its unique properties. Vanadium, when alloyed with carbon steel (the composition ranges from 0.03 to 0.25% vanadium), produces an alloy of strength and lighter weight. For this reason, vanadium alloy steels should be evaluated for structural steel components when evaluating pharmaceutical structural projects.

Since there is a wide selection of carbon steel and carbon steel alloys that are used for pharmaceutical piping applications, it would be useful to identify where various carbon steels and alloy steels are employed in pharmaceutical manufacturing operations. Table 9.4 identifies the various designations for types of piping used in pharmaceutical applications. The piping designations contain the composition of the piping. For instance, ASTM A335 (seamless ferritic alloy steel pipe for high temperature service) has a higher amount of chromium (8–10%) and may be specified if a degree of corrosion resistance is desired. In this case, Grade P9 should be considered. Each ASTM designation identifies the composition and grade of the pipe types designated. Table 9.5 also identifies the appropriate American National Standards Institute (ANSI) designation for standard piping.

For non-cGMP applications, the most common type of carbon steel is ASTM A106, Grade B. Type A106 has three grades, Grade A, Grade B, and Grade C, distinguished by their respective compositions. Table 9.6 identifies the alloying elements and composition of ASTM A106, Grade B carbon steel.

Table 9.5 Standard piping designations with pharmaceutical applications.

ASTM designation	ANSI designation	Title
ASTM A53 (A53 is outdated and replaced by A106)	B36.1	Welded and seamless steel pipe
ASTM A72	B36.2	Welded wrought iron pipe
ASTM A106	B36.3	Stainless-carbon steel pipe for high temperature service
ASTM A120	B36.20	Black and hot-dipped zinc-coated (galvanized) welded and seamless steel pipe for ordinary uses
ASTM A134	B36.4	Electric-fusion (arc)-welded steel plate pipe sizes 16 in. and over
ASTM A135	B36.5	Electric-resistance-welded steel pipe
ASTM A139	B36.9	Electric-fusion (arc)-welded steel pipe, sizes 16 in. and over
ASTM A155	B36.11	Electric-fusion-welded steel pipe for high temperature service
ASTM A211	B36.16	Spiral – welded steel or iron pipe
ASTM A312	B36.26 (B125.16)	Seamless and welded austenitic stainless steel pipe
ASTM A333	B36.40 (B26.27)	Seamless and welded steel pipe for low temperature service
ASTM A335	B36.42	Seamless ferritic alloy steel pipe for high temperature service
ASTM A358	B36.47	Electric-fusion-welded austenitic chromium nickel alloy steel pipe for high temperature service
ASTM A369	B36.48	Ferritic alloy steel forged and bored pipe for high temperature service
ASTM A376	B36.43	Seamless austenitic steel pipe for high temperature central station service
ASTM A381	B36.49	Metal-arc welded steel pipe for high pressure transmission service
ASTM A405	B36.44	Seamless seamless ferritic alloy steel pipe specially treated for high-temperature service
ASTM A419	N/A	Electric-fusion (arc)-welded wrought iron plate for pipe
ASTM A523	N/A	Plain end seamless and electric-resistance-welded steel pipe for high pipe type cable circuits
ASTM A530	B36.57	General requirements for specialized carbon steel and alloy line pipe
ASTM A589	N/A	Seamless and welded water pipe

The primary differences between the various grades of carbon steel are primarily due to two elements in the composition, carbon and manganese. For ASTM A106, Grade A, the carbon composition is 0.25%, whereas for Grade C, the carbon concentration is 0.35%. For manganese, the Grade A concentration ranges from 0.27 to 0.93%. The Grade C composition of manganese varies from 0.29 to 1.06%. The low alloying elements of ASTM A106 (Section 9.5.3.2) precludes it from being classified as an alloy steel.

Table 9.6 The composition of ASTM A106, Grade B carbon steel.

Element	Composition (%)
Carbon, C	0.30, max
Manganese, Mn	0.29–1.06
Phosphorus, P	0.035, max
Sulfur, S	0.035
Silicon, Si	0.010, min
Chrome,Cr	0.40
Copper, Cu	0.40, max
Molybdenum, Mo	0.08

Table 9.5 identifies ASTM A312 as the designation for austenitic stainless steels. This standard addresses pharmaceutical stainless steels including the most common 304, 304L, 316, and 316L.

9.5.4 Polymeric Materials: Overview

While it is generally thought that most critical pharmaceutical applications employ stainless steel and glass as the standard cGMP processing construction material, there is another class of materials rapidly gaining acceptance as a biotech/pharmaceutical material for a wide range of application with the rapid growth of single use components. Commonly used, but often overlooked, is the incorporation of polymeric materials in pharmaceutical applications. One ubiquitous application, in addition to single use applications, is employing polymers for reverse osmosis (RO) system components such as membranes, tubing, and gasket material. Additionally, in the past few years, many other components and finished equipment have been fabricated of polymers. As noted, the increasing application of single use disposable cGMP components has significantly affected the rapid expansion of polymers, particularly, high performance thermoplastic piping, in cGMP applications.

Polymers are natural and synthetic compounds often of high molecular weight consisting of millions of repeated linked units, each of which is a relatively light, simple molecule referred to as a monomer. The process of linking the monomers to form polymers is referred to as polymerization. Polymerization most often occurs by exposing monomers to heat, pressure, or radiation with the addition of specific compounds often referred to as catalysts and cross-linking agents. It is the cross-linking and similar configurations that give polymers, commonly referred to as plastics, their strength and integrity.

For many years, plastics had limited applications in pharmaceutical operations due to undesirable physical and chemical properties. Older plastics such as PVC were not good candidates because of an inability to operate at temperatures much above 65.6 °F (50.6 °C). PVC also emits harmful gases when exposed to higher temperatures such as a fire scenario. For this reason, PVC piping is still rarely used in pharmaceutical applications except for possible underground drains. In fact, it is generally prohibited from being employed in most commercial applications by many fire codes.

With the acknowledgment that cost-effective solutions to pharmaceutical operations are necessary to remain competitive for biotech and pharmaceutical manufacturing, the role of plastics is, as noted, a significant factor. This is due to the fact that pharmaceutical and biotech are now seeking ways to leave behind the currently used troublesome materials of construction and to accelerate conversion to problem-free improved materials. For this reason, it is germane for designers, users, manufacturing operations, and maintenance organizations to stay current with new data and changes occurring in the area of materials, as applicable to pharmaceutical/biotech operations. In addition to health and safety concerns, efficiency and profits are also major considerations.

The materials of processing equipment used in these industries, stainless steel and glass, imposed constant and increasing problems: rouging, stress corrosion, pitting, metallic poisoning, aggravated compliance issues, costly, and often environmentally adverse cleaning protocols an inadvertent fracture, e.g. glass-lined reactor vessels, plus costly biofilm issues. Further, in some cases, increasingly complex pharmaceutical and biochemical products and processes can be limited by what can be synthesized and manufactured in glass-lined or stainless steel components.

In this vein, the piping system should also be a concern. Here, particular attention should be given to cGMP design criteria, such as those identified in Chapter 2. When evaluating plastic piping, the design criteria identified should be evaluated and incorporated where feasible. While these criteria are intended for pharmaceutical water systems, these design elements are also germane to other pharmaceutical and biotech processes where transfer via piping is employed.

As plastics applications became more widespread in the pharmaceutical and biotech industries, specific requirements for selection of piping candidates have evolved. Among the factors that must be evaluated when considering plastic piping, as well as stainless steel, installations are:

1) The piping should not contaminate the media being transported (leachability).
2) The inner surface should have a nondegrading smooth finish (180–400 grit or electropolish finish that remains constant over time and use).
3) Ease of installation – Installation, repair, and maintenance downtime should be minimal.
4) Sterilizable and sanitizable for standard media and temperatures – Sterilization consists basically of exposing, for this application, internal piping surfaces to a temperature of 270–285 °F (132–141 °C) for a specified exposure time. Chemical sterilization includes liquid sterilants such as chlorinated liquids as well as gaseous agents. Certain types of radiation emissions can also be used in sterilization. A common method of evaluating sterilization is the SAL. To achieve a sterile classification level, the SAL level should be less than or equal to 10^{-6}. This value indicates that there is one chance in a million that a microorganism will survive the time and temperature required for the sterilization process. On the other hand, the SAL for a sanitization operation is less than or equal to 10^{-3}, significantly lower than the sterilization criteria.
5) Cost competitive – The piping material should be competitive with stainless steel and similar performing polymeric piping materials. The total installation should be evaluated on the basis of the lifecycle and sustainability rather than the initial capital expense. Such an estimate should include factors including spares provisioning of parts such as valves, tees, branches and pipe hangers (if the pipe runs are significantly long), sanitization/sterilization costs, lead time to obtain replacement parts (identified in the mean time between repair [MTBR] portion of Chapter 5), disposal options, and piping preparation costs (e.g. passivation costs and inside surface finish costs).

Another important factor is the operating temperature the piping must withstand. For most pharmaceutical and biotech operations, the maximum temperature is usually the sterilization temperature (270–285 °F). When determining an applicable piping material, it would be advisable to specify piping that can operate at the sterilization temperature. In fact, it would be desirable to have a safety margin of about 10% when evaluating temperature applicability.

It would be of value to be familiar with the properties of high performance plastics as well as stainless steel, a piping material often employed in pharmaceutical applications. For this purpose, Table 9.2 presents relevant properties of two high performance plastic piping materials, polyvinylidene fluoride (PVDF), polytetrafluorethylene (PTFE), and 304 stainless steel, a material often employed in the pharmaceutical and biotech operations. It is important to note that the polymers identified in the table, as is the case with 304 stainless steel, are approved FDA and USDA 3A piping materials.

Table 9.2 presents relevant properties of common high performance plastic piping materials as well as properties for Type 304 stainless steel.

Additionally, when dealing strictly with stainless steels, Table 4.1 is useful for specific calculations located in Section 9.6.

Table 9.7 presents a comparison of selected properties for a widely used pharmaceutical stainless steel and two high performance polymers.

Pipe hangers are also a very important consideration. While stainless steel piping should employ hangers that are corrosion resistant and present little or no hazard of galvanic activity potential, compatibility of hanger and piping composition is not a significant concern. Generally, galvanized or mild steel hangers are satisfactory. This is also an economic concern when evaluating piping stainless steel and polymeric materials relating to fabrication and installation costs. Welding of selected plastic piping does not require the use of orbital welders, TIG or MIG specialty operations and personnel, whereas stainless piping joining techniques

Table 9.7 Properties of selected high performance piping polymers and 304 stainless steel.

Property	PVDF	PTFE	304 Stainless
Mechanical properties			
Specific gravity	1.78	2.18	7.97
Elongation, %	50	200–450	40.0 (in 2 in.)
Tensile strength (lb/in^2)	6.300	3500	75 000
Flexural strength (lb/in^2)	9700	No break	31 200
Compressive strength (lb/in^2)	9000	3500	44 962
Tensile elastic modulus (Young's modulus) (lb/in^2)	290 000	57 000	28×10^6
Hardness, Shore D	D75	D50–75	Applicable only for plastics and rubber, for stainless steel, the rockwell B hardness is 70
Thermal properties			
Melting point (°F)	332	621	2550–2650
Thermal conductivity, Btu/ft °F h	1.39	1.67–28.2 (32–122 °F)	9.4 (32–212 °F)
Specific heat	0.287–0.382 (@ 32 °F)	0.239–0.251	0.12 (32–212 °F)
Coefficient of thermal expansion $\times 10^{-4}$/F	$12–14 \times 10^{-5}$ °F	7×10^{-5}	9.6

Table 9.8 Recommended pipe hanger spacing for stainless steel and selected high performance polymeric piping (Sch 40).

Pipe diameter, nom. (in.)	Steel (304, 316), maximum span (ft.)	Polymer (PVDF and PTFE), maximum span (ft.)
1	8	6
2	10	6
3	12	7
6	17	8½

often do. Another consideration that should be evaluated is the use of sanitary fittings in lieu of a welding procedure.

Another important design concern when considering plastic piping configurations is the support hanger location. For comparison purposes, Table 9.3 presents generally accepted hangar spacing for ambient piping systems for stainless steel and high performance plastic piping such as PVDF and PTFE (also referred to as PFA).

Additionally, pipe hanger selection should employ ASME Specification B31.1 "Power Piping" (piping originating at boilers) or consider ASME Specification B31.3 "Process Piping" (pharmaceutical and other process applications). Also important is ANSI Standard SP-58, "Standard Pipe Support Components" for design and installation information.

Table 9.8 summarizes spacing for thermoplastic piping and steel piping, which includes stainless steels, carbon steels (addressed in Section 9.5.3.1), and alloy steels (addressed in Section 9.5.3.2).

It was noted above that pharmaceutical piping should be non-contaminating when exposed to processing media. As noted in Chapter 8, rouging (Section 8.5.2) is a phenomena associated with stainless steel piping. While rouging is not seen as a major concern in pharmaceutical water operations, it can be detrimental in biotech operations such as protein synthesis. This is due to the iron and other metals that deposit in the piping as part of normal erosion that occurs within the inside of the pipe. While the impurity level may be in the ppb range, it may be sufficient to cause contamination of the product. Internal pipe erosion occurs even if the inner wall has an electropolish finish.

9.5.5 Preventing Pharmaceutical Materials Component Materials Failures

An increasing concern in materials science as applicable to medical devices and pharmaceutical processing materials is the risk of materials failure [21]. Medical device product recalls, specifically, have increased steadily in the past 15 years. While tracking pharmaceutical manufacturing components may not be tacked as rigorously, it is reasonable to assume a similar failure record, albeit not as critical for devices, this situation also exists. While process component failures may not be as critical an issue as the risk of a device failure, process component failures associated with a materials failure can impose a cost and time handicap on the pharmaceutical manufacturing process. While early involvement in the materials selection operation in the system design procedure is important, predicting materials behavior for a particular pharmaceutical operation must often be evaluated on a case-by-case situation depending on specific use, operating environment, and materials compatibility. Some of these evaluation elements typically include operating temperature ranges, thermal coefficients of expansion, electrical properties, and dimensional specifications. However, some practical considerations to aid in preventing common materials failures have been identified to, hopefully, preclude and/or reduce these mistakes [21] and oversights. These specific issues include:

- Improper materials selection – While, as noted, each specific application often requires an in-depth evaluation of the materials being used or planning to be used, Tables 9.1–9.4 and associated information in this chapter may be an initial basis for more detailed materials evaluation of failure mitigation and prevention. Again, it is most important to integrate materials selection in all phases of a pharmaceutical system starting from initial concept through start-up and operational monitoring.
- Materials degradation – The behavior of a material over time is most germane to product safety as well as operating costs. Material wear through continuous use of a component is a most common cause of degradation of materials. This phenomenon is often encountered when a preventive maintenance program is nonexistent or not addressed with proper actions (A more in-depth coverage of this topic is contained in Chapter 5). Two major passive contributors to polymeric materials degradation are exposure to ultraviolet (UV) and atmospheric oxygen. While cGMP pharmaceutical operations are performed in controlled environments with respect to temperature and humidity, the possibility of oxidation due to exposure of polymeric components to oxygen and UV light (component of the electromagnetic spectra) is a concern. Polymers can also undergo a reaction known as auto-oxidation. This phenomenon also degrades plastics over time.
- Manufacturing process incompatibility [1] – The manufacturing process can also be impacted by the materials in the process. For instance, plastic piping that has improperly spaced hangers (Table 9.3) can create unplanned stresses and ultimately create a break in the line. Similarly, abrasive materials such as slurries can create pitting, erosion, and ultimately, excessive wear in the lines.
- Procedures – Verify approved procedures are in place to analyze materials failures employing techniques such as CAPA and failure modes evaluation and analysis (FMEA) assessments.

9.6 Practical Piping Calculations

Chapter 2 presented information and methods intended to provide the basics of fluid mechanics and heat transfer concepts for biotech/pharmaceutical operations and processes. The following material is intended to provide sufficient information and techniques to allow for further preparation of piping specifications relating to transferring fluids in processes with a minimum of required input regarding important characteristics of specified pipe sizes. For instance, a background in determining how pipe sizing relates to internal pressure of the fluids being handled is critical to safe and economic operations. Also, it would be of value to understand how available data (i.e. pipe schedule) can relate to operating pressure and allowable stress using a straightforward and rapid calculation. Conversely, by having access to allowable stress and internal pressure data, the pipe schedule, a critical requirement for pipe specifications, can also be readily obtained.

When a fluid is moving through a piping configuration, the fluid exerts a pressure on the pipe. This situation is similar to a pressure vessel, the design elements of which were covered in Chapter 4, specifically, Section 4.1.2.

There is one significant similarity and one important difference between thin wall pressure vessels and piping (thin wall pressure vessels and piping are about one tenth the radius [22]). A pressure vessel is designed to contain fluids both circumferentially (the pressure is exerted radially outwardly) and axially (the resultant force component of the pressure is also exerted perpendicular to the resultant circumferential pressure component). When fluids flow in the direction of the radial component, the circumferential force is reduced because the fluid pressure against the pipe wall is transferred to the direction of the flowing fluid (the important difference between pressure vessels and piping). As a result, the pressure exerted against the pipe wall (circumferentially) is the critical variable for piping determinations specific to wall thickness and subsequent specification requiring sizing details. That is, the circumferential stress on a pipe is approximately twice the stress of the longitudinal or axial component. Section 4.1.2 presents a more detailed explanation of why the circumferential stress is the critical stress value.

Again, referring to Section 4.1.2, specifically, Equation (4.5) where $t = PR/(SE - 0.6P)$, and employing reasonable assumptions and a slight rearrangement, a valid relationship for obtaining an accurate relationship to evaluate pipe wall thickness can be obtained.

Pipe manufacturing is typically a continuous operation. As such, there is no need to consider weld efficiency since a continuous weld is 100% efficient due to the continuous welding operation. Also, the factor, $-0.6P$ is a safety factor specific to pressure vessels, and since piping configurations are not pressure vessels, the factor can be eliminated. Finally, by rearranging the modified equation, a relationship for determining the wall thickness of pipes can be expressed as

$$S_a = \frac{PR}{t} \tag{9.1}$$

where

S_a = allowable stress, lb/in^2;
P = internal pressure, lb/in^2 (circumferential pressure on the pipe wall)
t = pipe wall thickness, lb/in^2

For stainless steels, the allowable stress is determined by the method described in Section 4.1.2 or using the data provided in Table 4.1.

For polymeric piping, the International Standards Organization (ISO) provides a working relationship for determining the allowable stress for polymeric piping materials. This relationship, also referred to as Barlow's Formula, is given by

$$S_a = \frac{P(R-1)}{2} \tag{9.2}$$

where

S_a = allowable stress, lb/in^2
P = Internal pressure, lb/in^2 (circumferential pressure on the pipe wall)
$R = D_o / t$ (D_o = outside wall pipe diameter, in., t = pipe wall thickness, in.).

Example 9.1 A 316L stainless steel USP water pipe that has an inside diameter of 2.157 in. is subjected to an internal pressure of 190 lb/in^2. While the allowable stress for 316L stainless steel is 8572 lb/in^2 (Table 4.1), the corporate piping specifications have a service limit of 2000 lb/in^2 at ambient conditions (allowable stress, S_a). Calculate the minimum thickness required to keep the allowable stress at, or below, 275 lb/in^2.

Solution
Equation (9.1) serves as the basis of calculation for this example. Rearranging the equation to a workable form for this calculation yields

$$t = \frac{PR}{S_a}$$

Inserting the given values in the modified form of equation:

$$t = \frac{\left(190\text{lb}/\text{in}^2\right)\left(1.0785\text{in.}\right)}{2000\text{lb}/\text{in}^2} = 0.10.3\text{in.} \leftarrow \textbf{Solution}$$

The calculated required wall thickness for the configuration described is 0.103 in. However, this is not typical standard pharmaceutical piping. Most company standards rarely specify pipe stress values at high levels since there are few large-scale pharmaceutical production processes that operate at such high requirements. Most pharmaceutical pressures rarely exceed $300\,\text{lb}/\text{in}^2$ as a service limit. Operations requiring high service requirements (i.e. a hydrogenation reaction) are often performed by CMO.

Pipe specifications require a specific format for accurate identification. The standard identification format is described in Section 2.1.1 (specifically, Table 2.4). As noted, pipe wall thickness is identified by the pipe schedule. In this case, the appropriate nomenclature is 2 in. Sch. 10S (often, the S is deleted in the nomenclature), Type 316L stainless steel.

For most pharmaceutical water production, the common pipe diameter rarely exceeds a nominal diameter of 2 in. (2.375 in. O.D.).

This example is specific to stainless steel piping. When dealing with piping that is not stainless (e.g. carbon steel), it is recommended a corrosion allowance be added to the calculated diameter. A common corrosion allowance is 0.125 in. (⅛ in.).

PVDF and other high performance piping is often used in specific pharmaceutical applications such as USP pharmaceutical water and high purity water for injection (WFI) where the temperature of the operation is commonly below 107 °C (225 °F) and the pressure is not excessive (Conservatively, at or below $30\,\text{lb}/\text{in}^2$). In fact, pharmaceutical water production operations are often processed and stored at ambient temperatures since most common purification operations do not require elevated temperatures for production, with the exceptions being evaporation or distillation for WFI. It is important to point out for many resin systems parameters are not fixed values and can, and often do, vary among vendors. As a result, there is usually an operating range for polymeric piping of the same composition, although most operate within a range. Consequently, brand selection is often determined by pricing, specifications, and vendor preference.

As installation costs become more relevant, the role of high performance piping polymers will engender greater consideration due to several factors. One very significant positive consideration is the obviation of passivation considerations. Since performance polymers such as PVDF contain no free surface iron passivation, as detailed in Chapter 8 (Section 8.5.1), there is no need to be concerned with passivation options.

Rouging, another passivation related a phenomenon, is also not a concern when dealing with high performance polymers. Here again, because iron is not present, rouging is not a consideration.

Since stainless steel piping, in addition to iron, is alloyed with several constituents such as chrome, molybdenum, and nickel, these components appear in small quantities in the water during processing. Depending on pH, these trace compounds can appear as metal ions or small particles in the water. Normally, these trace elements are removed during the ion exchange or RO operation in the purification operation. However, these compounds, which occur as part of the pipe flow, add additional loading to the purification operations and may require a more frequent replacement or regeneration of the ion exchange resins over time. This consideration is not necessary when dealing with high performance resin piping since there are no ions that can leach or erode out of the piping.

Unlike steel piping, fungi and other bacterial growth is not a concern when dealing with polymeric piping, due in large part because polymers do not exhibit the high heat transfer properties of stainless steel. Also, high performance polymers have smoother inside wall finish than stainless steel. The smoother inner wall makes it more difficult for bacterial growth since there are fewer crevices for a colony to grow.

The fact that there is a lower load on ion exchange units (cations) and RO systems should reduce maintenance requirements relating to ion exchange and RO cartridge replacement over time.

Example 9.2   The use of PVDF is being considered for a USP water production system. The skid mounted purified water unit, operating at ambient conditions (68 °F), is designed to use 1½ in. (1.900 in. O.D.) Sch. 40 PVDF piping (0.145 in. wall thickness) with an internal pressure of 30 lb/in². When preparing the specifications for the PVDF pipe, a service limit is required. Determine the service limit (i.e. the allowable stress) for the pipe.

Solution
Equation (9.2) or Barlow's formula is the basis of calculation for this example. Identifying the required variables:

P, internal pressure $= 30\,\text{lb/in}^2$
t, wall thickness $= 0.145\,\text{in.}$
$D_o =$ pipe diameter, 0.1.900 in.

$$R = \frac{D_o}{t} = \frac{1.900\,\cancel{\text{in.}}}{0.145\,\cancel{\text{in.}}} = 13.10$$

Inserting the variables in Equation (9.2) yields

$$S_a = \frac{P(R-1)}{2} = \frac{\left(30\text{lb}/\text{in}^2\right)(13.10-1)}{2} = 181.5\text{lb}/\text{in}^2 \leftarrow \text{Solution}$$

The allowable stress for the PVDF piping assembly described is 181.5 lb/in². The service limit for 1½ Sch. 40 PVDF pipe should be 181.5 lb/in² for piping specifications at a temperature of 68 °F (20 °C), or ambient conditions.

High performance thermoplastics such as PVDF and PTFE are often preferred because of their relatively high melting points, as shown in Table 9.1. As Example 4.2 has shown, these materials perform well at ambient conditions. However, as the working temperature rises, some modifications of performance can be expected. For plastic piping, the important operating feature is the allowable stress. To correct for the reduction in allowable stress with increases in operating temperature, correction factors are available. Table 9.4 presents some temperature correction factors when dealing with PVDF and PTFE at elevated temperatures, based on vendor data. While the correction factors given at higher temperatures are useful, more detailed values, if required, should be obtained by actual materials testing of the selected piping.

Table 9.9 exhibits correction factors for two high performance plastics at selected operating temperatures.

Table 9.9 Allowable stress correction factors for PFTE and PVDF plastic piping.

Temperature		Correction factor
68 °F	20 °C	1.0
80 °F	27 °C	0.95
100 °F	38 °C	0.80
120 °F	49 °C	0.68
200 °F	115 °C	0.36

Example 9.2 had an allowable stress of 181.5 lb/in^2 at 68 °F. If the process temperature was altered to operate at 120 °F, the service limit would be 0.68×181.5 lb/in^2 or 123.42 lb/in^2 ($\approx$124 lb/in^2).

One aspect of piping specifications and calculation that have not been addressed, thus far, is determining the piping schedule required for pipe sizing identification (further details for pharmaceutical applications are located in Chapter 2, Section 2.1.1). A simplified technique has been developed to specify a required pipe schedule with reasonable accuracy when dealing with steel pipe. The relationship is given as

$$\text{Sch.} = \frac{1000P}{S_a} \tag{9.3}$$

Sch. is the pipe schedule, dimensionless;
P is the internal pressure, lb/in^2
S_a is the allowable stress, lb/in^2

When performing a calculation using Equation (9.3), the schedule number should be rounded to the higher value (i.e. a calculated schedule of 28.5 should be rounded to the next higher value that would be Sch. 40).

Example 9.3   Calculate the schedule number of the pipe selected for Example 9.2.

Solution
The basis of calculation for this problem is Equation (9.3). Inserting the required variables from Example 9.2 in Equation (9.3) yields

$$\text{Sch.} = \frac{1000P}{S_a} = \frac{1000\left(190\text{lb}/\text{in}^2\right)}{8572\text{lb}/\text{in}^2} = 22.17 \text{ Calculated pipe schedule}$$

The calculated value 22.17 rounded to the next highest standard value is a Schedule 40 pipe. The value required is **Sch. 40 ← Solution**

As noted, this calculation is valid for various types of steel piping, but if plastic piping is involved, a more rigorous analysis is required. For such configurations, vendor data and/or testing procedures in accordance with applicable ASTM standards are the preferable option.

On occasion, a situation may arise where a pipe wall thickness may be required and only the pipe schedule and internal diameter are known. Such a scenario occurs when an incomplete

P&ID is the only information source. This occurs when the diagram has an incomplete revision history or the diagram was set aside to address a higher priority project without being completed. Urgent field modifications lacking a complete bill of materials (BOM) can also be a reason for the omission of the required information. Of course, the piping required for replacement can be old and painted with identification markings no longer visible and corrosion is evident. An empirical calculation that provides a reasonable pipe wall thickness can be obtained if only the pipe internal diameter and schedule number are known. This solution can be obtained by the relationship:

$$t = \left(\frac{\text{Sch. No.}}{1000} \right)\left(\frac{d}{2} \right) + 0.156 \tag{9.4}$$

where

Sch. No. is the pipe schedule, a dimensionless value
t is wall thickness, in.; and
D is the pipe I.D., in.

Equation (9.4) is, as indicated, an empirical calculation and is best suited as a first approximation, particularly if the variables such as internal pressure or allowable stress cannot be readily determined.

Example 9.4   Compare the results of Example 9.1 with the results obtained by using equation for pipe wall thickness.

Solution
Recalling the calculated wall thickness for the schedule 10S NPS 2 in. is 0.103 in. (Table 2.4, Chapter 2); the values required for calculating a value for Equation (9.4) are inserted to yield

$$t = \left(\frac{\text{Sch. No.}}{1000} \right)\left(\frac{d}{2} \right) + 0.156 = \left(\frac{10}{1000} \right)\left(\frac{2.157\,\text{in.}}{2} \right) + 0.156\,\text{in.}$$
$$t = 0.1668\,\text{in.} \rightarrow \textbf{Solution}$$

The calculated wall thickness using Equation (9.4) is 0.1668. This represents a difference of 38.25%. If Equation (9.4) were used as the basis for sizing, instead of a schedule 10S pipe, schedule 40S or 80S would be considered. Since ASTM A106, schedule 40S is one of the most widely used pipe sizes for non-cGMP applications, it would be the preferred choice.

References

1 Drucker, C., Kolodziejski, M., Jacques, M., et al. (2014). Risks of single-use bioprocess containers. *American Pharmaceutical Review Supplement*.
2 Bestwick, D. and Colton, R. (2009). Extractables and leachables from single-use disposables. *BioProcess International* 7 (2): 88–94.
3 Hammond, M., Nunn, H., Rogers, G., et al. (2013). Identification of a leachable compound detrimental to cell growth in single-use bioprocess containers. *PDA Journal of Pharmaceutical Science and Technology* 67 (2): 123–134.

4 Drucker, C., Kolodziejski, M., Jacques, M., et al. (2014). Risks of single-use bioprocess containers. *American Pharmaceutical Review* 17 (6, Supplement): 1–4.

5 DePalma, A. (2012). GE healthcare improves single-use abilities. *Genetic Engineering & Biotechnology News* 32(20): 34.

6 DePalma, A. (2014). Single use bioreactors dare to scale. *Genetic Engineering & Biotechnology News* 34(14): 24–27.

7 Kraemer, P. and Eitel, T. (2012). Stainless steel nowhere near over the hill. *Genetic Engineering & Biotechnology News* 32 (15): 48–49.

8 BPSA (bio-process systems alliance) roundtable addresses disposal of disposables (2008). *BioPharm International* 21(7): 12.

9 Thibon, V. and Creasey, J. (2013). Measuring potential leachables in single-use manufacturing assemblies using total organic carbon (TOC) analysis. *American Pharmaceutical Review* 16 (3): 18–26.

10 Office of the Inspector General, U.S. Environmental Protection Agency (2012). EPA Inaction in Identifying Hazardous Waste Pharmaceuticals May Result in Unsafe Disposal, Report No. 12-P-0508 (26 May 2012). Washington, DC: Office of the Inspector General, U.S. Environmental Protection Agency.

11 Ziemlewski, J. (2009). The new disposable evolution. *Chemical Engineering Progress* 105 (7): 23–29.

12 Garcia-Perez, M. (2008). *The Formation of Polyaromatic Hydrocarbons and Dioxins During Pyrolysis*. Pullman, WA: Washington State University.

13 Pora, H. and Rawlings, B. (2009). Managing solid waste from single-use systems in biopharmaceutical manufacturing. *BioProcess International* 7(1): 18–25.

14 Wells, B., et al. (2007). Guide to disposal of single-use bioprocess systems. *BioProcess International* 11 (5): 22–28.

15 Department of Defense (1965). *Copper and Copper Alloys, Military Standardization Handbook*, MIL-HDBK-698 (MR). Washington, DC: Department of Defense.

16 Pora, H. (2007). Keeping pace with today's disposable processing applications. *BioPharm International* 20 (4): 50–56.

17 Shahidi, A.J. (2008). Major benefits of single use systems in the bio pharmaceutical industry: An evolving technology. Presented at INTERPHEX, New York, p. 12.

18 Mary Ann Liebert (2013). Single-use bioreactor roundup. *Genetic Engineering & Biotechnology News* 33 (8): 26–35.

19 Climax Molybdenum Company (1966). *Stainless Steel and the Chemical Industry*. New York: American Iron and Steel Institute.

20 Bethlehem Steel Corporation (1967). *Modern Steels and Their Properties*, 6e. Bethlehem, PA: Bethlehem Steel Corporation.

21 Heintz, A. (2016). Mitigating the risks of materials failure. *Medical Device and Diagnostics* 38 (1): 38, 39.

22 Roark, R.J. and Young, W.C. (1975). *Formulas for Stress and Strain*, 5e, pp. 445–466. New York: McGraw-Hill Book Company.

Further Reading

Archer, R., Cook, N.H., Crandall, S.H., et al. (1959). *An Introduction to the Mechanics of Solids*, 121–163. New York: McGraw-Hill Book Company.

Bodeker, B. (2012). Improving biologic manufacturing operations and plant design through single-use technologies application. *Pharmaceutical Outsourcing* 13(3): 12–17.

Committee of Stainless Steel Producers (1977). *Design Guidelines for the Selection and Use of Stainless Steel*. Washington, DC: American Iron and Steel Institute.

Committee of Stainless Steel Producers (1979). *Welding of Stainless Steels and Other Joining Methods*. Washington, DC: American Iron and Steel Institute.

Committee of Stainless Steel Producers (1980). *Design Guidelines for Stainless Steel in Piping Systems*. Washington, DC: American Iron and Steel Institute.

Depalma, A. (2008). Versatility concerns impact facility design. *Genetic Engineering News* 28 (9).

DePalma, A. (2013). Single-use systems entice multiple users. *Genetic Engineering & Biotechnology News* 33 (2): 1, 24, 25.

Flue gas cooling puts the damper on dioxins (1996). *Chemical Engineering* 103(7): 21.

Freeman, H.M. (1989). *Standard Handbook of Hazardous Waste Disposal*. New York: McGraw-Hill Book Company.

Langer, E.S. and Price, B. (2007). Biopharmaceutical disposables as a disruptive future technology. *BioPharm International* 20(6): 48–56.

MacLauchlan, A., Jenness, E., and Gupta, V. (2010). Selecting the appropriate sterile connector. *Genetic Engineering & Biotechnology News* 30 (1): 35.

Pinier, K. (2016). The relationship between materials selection and machining processes. *Chemical Engineering* 123 (2): 46–53.

Rader, R.A. and Langer, E.S. (2012). Upstream single-use bioprocessing systems. *BioProcess International* 12 (2): 12–18.

Sinclair, A. (2008). Disposable bioreactors: The next generation. *BioPharm International* 21 (4): 38–40.

Sinclair, A. and Manage, M. (2009). Disposables cost contributions: A sensitivity analysis. *BioPharm International* 22 (4): 28–32.

10

Commissioning and Validation

10.1 Introduction to Commissioning and Validation

Commissioning is a process that is common to virtually all construction programs. As such, commissioning is an important aspect of pharmaceutical projects since a commissioning program will, in most cases, allow for a more efficient installation, qualification (validation), and subsequent operation such as manufacturing and facility maintenance.

Few construction and installation projects can successfully be implemented without a comprehensive program containing preapproved construction standards or specifications. By starting with a background and commencing through the commissioning phase and terminating with validation operations, the complete scope of the documentation process for a pharmaceutical construction and installation project can be addressed. Further, a great deal of the information contained in this chapter is also applicable to general engineering and construction programs other than pharmaceutical projects.

10.1.1 Introduction to Construction Specifications

A key requirement of commissioning is the need for construction and installation procedures and guidelines to perform the required task. In order to effectively implement a project involving construction and installation activity, a set of standards has been prepared. These standards are identified as construction specifications. A specification is a set of requirements for a material, product, or service intended for construction activities. The primary guide is a document identified as the National MasterFormat®. The document was prepared by two organizations: the Construction Specifications Institute and Construction Canada. The MasterFormat is employed by architects, engineers, owners, contractors, and manufacturers to identify procedures and products required for the construction project.

Practical Pharmaceutical Engineering, First Edition. Gary Prager.
© 2019 John Wiley & Sons, Inc. Published 2019 by John Wiley & Sons, Inc.

For pharmaceutical projects, similar specifications, often unique to the industry, are also needed in the same type of organized format. However, specifications for pharmaceutical applications are also incorporated in the MasterFormat classification.

A widely employed system for classifying and identifying construction specifications is called MasterFormat. As stated, MasterFormat is a formalized system for organizing construction specifications, drawings, and related data. The MasterFormat is currently divided into 50 Divisions, with each Division representing an element of a specific project. Each Division is further divided into subdivisions that define the specific tasks for the project. These requirements are identified in the form of written specifications. These specifications are supplied as part of the contract to perform the construction. There are four types of specifications:

1) Prospective specifications – This is often a proprietary and the most rigorous type of specification. It is definitive in identifying products by brand name and does not allow for bidding on the product. The prospective specification does, however, allow the designer to perform installation activities as required to complete the task.

2) Base bid specifications – This is similar to a prospective specification with the exception of the phrase "equal to" or "equivalent to" followed by the brand of product to be used in the specification.

3) Performance specifications – Performance Specifications provide a more detailed breakdown of materials and installation procedures required to perform the installation task. A typical requirement for this type of specification would be a Division 23 (Heating, Ventilating, and Air Conditioning) specification defining the qualifications of welders performing tasks requiring stainless steel welding. Because performance specifications have a great deal of detail regarding the installation project, it is significantly more involved than the previous types of specifications. Consequently, this type of specification can be expected to be more difficult to prepare. Similarly, work performed using this type of specification should involve more construction oversight than specifications involving a great deal of subcontractor responsibility for installation.

4) Reference standard specifications – Perhaps requiring the least input and oversight, reference standard specifications rely a great deal on recognized test and evaluation standards. Often, a specification of this type defines the performance on defined results that are detailed in the specification. This result is often in the form of recorded test results. Typically, this type of specification uses test methods prepared by organizations such as the American Society for Testing and Materials (ASTM), the American National Standards Institute (ANSI), the National Association of Corrosion Engineers (NACE), the American Society of Heating, Refrigeration, and Air-Conditioning Engineers (ASHRAE), the National Electrical Code (NEC), the Occupational Safety and Health Administration (OSHA), and specific trade organizations.

To allow for a convenient indexing and reference availability, the MasterFormat Divisions are organized into subgroups as follows:

Procurement and Contracting Requirements

- Division 00 – Procurement and Contracting Requirements Group
- Division 01 – General Requirements

Facility Construction Subgroup

- Division 02 – Existing Conditions
- Division 03 – Concrete
- Division 04 – Masonry

- Division 05 – Metals
- Division 06 – Wood, Plastics, and Composites
- Division 07 – Thermal and Moisture Protection
- Division 08 – Openings (Doors, Windows, etc.)
- Division 09 – Finishes
- Division 10 – Specialties
- Division 11 – Equipment
- Division 12 – Finishing
- Division 13 – Special Construction
- Division 14 – Conveying Equipment
- Divisions 15–19 – Reserved for future expansion

Facilities Services Subgroup

- Division 20 – Reserved for future expansion
- Division 21 – Fire Suppression
- Division 22 – Plumbing
- Division 23 – Heating, Ventilating, and Air Conditioning
- Division 24 – Reserved for future expansion
- Division 25 – Integrated Automation
- Division 26 – Electrical
- Division 27 – Communications
- Division 28 – Electronic Safety and Security
- Division 29 – Reserved for future expansion

Site and Infrastructure Subgroup

- Division 30 – Reserved for future expansion
- Division 31 – Earthwork
- Division 32 – Exterior Improvements
- Division 33 – Utilities
- Division 34 – Transportation
- Division 35 – Waterway and Marine
- Division 36 Through Division 39 – Reserved for future expansion

Process Equipment Subgroup

- Division 40 – Reserved for future expansion
- Division 41 – Material Processing and Handling Equipment
- Division 42 – Process Heating, Cooling, and Drying Equipment
- Division 43 – Process Gas and Liquid Handling, Purification, and Storage Equipment
- Division 44 – Pollution Control Equipment
- Division 45 – Industry-Specific Manufacturing Equipment
- Division 46 Through Division 49 – Reserved for future expansion

Specification Sections are subgroups of Divisions, as briefly noted previously. For instance, if a project involves metal gratings and installed struts, applicable specification would be identified as specifications:

- Specification 05120 – Metal Gratings
- Specification 05430 – Slotted Channels (Framing)
- Specification 05450 – Metal Supports (Struts)

The specification starts with Division 05. This identifies it as a construction or installation involving metal working. The subsequent numbers are subdivisions that identify detailed aspects of the task.

Another example of a specification that is relevant to pharmaceutical operations is contained in Division 10 – Specialties. Division 10 contains a subgroup 8100. The complete specification is identified as 108100 – Pest Control Devices. Referring to 21 CFR Part 211.56 (c): "There shall be written procedures for use of suitable rodenticides, insecticides, fungicides, fumigating agents, and cleaning and sanitizing agents."

As will be seen and has been noted, without construction specifications, commissioning and validation, in accordance with cGMP criteria, would be a most difficult task to accomplish for pharmaceutical operations.

A typical large pharmaceutical construction or installation project could involve hundreds of specifications dealing with detailed system components of the project.

To provide for a complete specification that contains the required information to perform a specific task, a standard specification format is generally employed. This standard format is composed of three sections:

- Part 1 General – General Requirements. This part describes the administrative, procedural, and temporary requirements unique to the specification.
- Part 2 Products – A description of the materials, products, equipment, fabrications, systems, and assemblies required for project implementation is presented in this section.
- Part 3 Execution – The purpose of part 3 is to describe the installation or application, including preparatory actions and post installation protection required for proper installation. Part 3 also includes site-built assemblies and site manufactured that may be required.

Specifications are legal documents and must be accurate as well as complete. A document that is complete can also be a time-saving and cost-cutting document since it will require fewer "request for information (RFI) inquiries" by the contractors, which can cause needless delays.

Figure 10.1 is a specification for an electrical specification based on the Division 26 Specifications employed by the State of Wisconsin. Some typical subdivisions and associated tasks for a specific project include:

- 260500 – Common Work Results for Electrical
- 260523 – Control Voltage Electrical Power Cables
- 260529 – Hangers and Supports for Electrical Systems
- 260553 – Identification for Electrical Systems
- 261302 – Medium-Voltage Pad-Mounted Switchgear
- 262200 – Low-Voltage Transformers Detail
- 262702 – Equipment Wiring Systems
- 262813 – Fuses
- 265113 – Interior Lighting Fixtures, Lamps, and Ballasts
- 265629 – Site Lighting

Figure 10.1 presents an actual specification for a Division 26 project. In this case, Division 26, Subsection 2813, is a detailed description of fuses for equipment installations.

It is most important to note that in addition to approved construction specifications, the approved as built drawings must be included with the construction specifications. When dealing with as built drawings (owner-supplied P&IDs), it is suggested that Chapter 3 be reviewed

SECTION 26 28 13
FUSES
BASED ON DFD MASTER ELECTRICAL SPEC DATED 11/07/16

This section has been written to cover most (but not all) situations that you will encounter. Depending on the requirements of your specific project, you may have to add material, delete items, or modify what is currently written. The Division of Facilities Development expects changes and comments from you.

This section should not be used if fuses are covered in the other electrical specification sections.

PART 1 - GENERAL

SCOPE
The work under this section includes 250 and 600 volt fuses. Included are the following topics:

PART 1 - GENERAL
    Scope
    Related Work
    Submittals
    Regulatory Requirements
    Extra Materials
PART 2 - PRODUCTS
    Fuses
PART 3 - EXECUTION
    Installation

RELATED WORK
Applicable provisions of Division 1 govern work under this Section.

Section 01 91 01 or 01 91 02 – Commissioning Process

SUBMITTALS
Provide device dimensions, nameplate nomenclature, and electrical ratings.

Submit manufacturer's product data sheets with installation instructions.

REGULATORY REQUIREMENTS
Listed by Underwriter's Laboratories, Inc., and suitable for specific application.

EXTRA MATERIALS
Provide three (3) spares of each size and type fuse.

PART 2 - PRODUCTS

FUSES
Fuses 600 Amperes and Less: Dual element, time delay, [250] [600] volt, UL Class [RK 1.] [RK 5.] [J.] Interrupting Rating: 200,000 rms amperes.

Fuses 601 Amperes and Larger: Low Peak, time delay, 600 volt, UL Class L. Interrupting Rating: 200,000 rms amperes.

Fuses 30 Amperes and less: Time-Delay, 600 volt, UL Class CC. Interrupting rating: 200,000 rms amperes.

Provide storage enclosure for spare fuses. Enclosure shall be a hinged-cover junction box, minimum size of 12" x 12" x 6" D. Enclosure shall be labeled as "Spare Fuses". Install enclosure in main electrical room.

PART 3 - EXECUTION

INSTALLATION
Fuses shall not be installed until equipment is ready to be energized.

END OF SECTION

DFD Project No.
26 28 13 - 1

Figure 10.1 Specification 262813 (Division 26, Subsection 2813): Fuses.

to note the important elements of these project components. Particular attention should be given to the walk down as related to project preparation.

10.2 Commissioning

Commissioning is the process of ensuring through systematic challenging and verification employing testing and documentation that a facility, a system, and equipment are installed, started, and functionally operated together through normal modes and conditions in accordance with design, installation requirements, and documented acceptance criteria and design intent.

The commissioning process serves to verify facilities and process systems are also designed, installed, functionally tested, and capable of operating in accordance with design standards and specifications.

The commissioning process has many practical advantages for both the owner and the contractor. For the owner, commissioning, when properly planned and executed, offers the following opportunities:

- Reduction in change orders during the construction phase.
- Reduction in delays during the construction phase.
- Shorter system turnover period.
- Reduction in maintenance and equipment replacement.
- Lower energy costs.
- Better trained operations staff.
- Operation and maintenance (O&M) manuals are complete at completion of the project (turnover).

For the contractor, a well-organized and well-executed commissioning process can offer the following:

- Better planning – This should significantly improve the installation operations.
- Greater emphasis on quality control – The contractor has access to the same quality control criteria as the owner. This permits inspections and evaluations to be performed simultaneously, employing preapproved procedures and inspections.
- Quicker resolution of problems during the installation and quality control operations – Tasks such as modifying as built drawings and equipment changes can be accomplished in a timely manner, allowing for greater mutual cooperation for the owner, construction manager, and subcontractors.

Two significant differences between commissioning and validation are that commissioning is intended to confirm the equipment or process operates as intended; validation requires the equipment or process operates at specific performance levels for given time periods. Another difference is the fact that validation is required when the equipment, process, or facility is regulated by the provisions of 21 CFR 210 and 211 for pharmaceutical operations. This is more commonly referred to as conforming to cGMP regulations.

Commissioning procedures for pharmaceutical operations usually involves limited installations. The most common commissioning activities involve Mechanical, Division 23; Fire and Suppression, Division 21; Electronic Safety and Security, Division 28; and Electrical Installations, Division 28.

While it is common engineering practice to perform commissioning operations on several installations or subsystems, validation is required only when dealing with critical systems having direct or indirect product contact or where the operation can significantly affect the final product.

As is the case with the most formalized processes and operations, they are specific definitions common to commissioning operations. Terms most often employed in commissioning operations include:

Redline drawings – This set of drawings represents the final configuration of the completed installation. Redline drawings show the approved changes that were performed and required for successful operation of the completed system. The redline drawings are submitted upon completion and handover of the completed project to the owner. As expected, the redline drawings are submitted as part of the engineering turnover package (ETOP).

Checklists – Records that provide documentation, via signed offs by recognized commissioning personnel, that specific requirements of a commissioning plan have been completed. Check list forms include installation verification, initial operational verification, and system verification acceptance forms.

Critical (FDA definition) – A process step or process control (e.g. process condition, test requirement, or other relevant parameter or item) that must be controlled within predetermined criteria to ensure that the drug substance meets its specification.

Critical component – A component within a system where operation, contact, data, control, alarm, or failure will have a direct impact on product quality.

Critical punchlist items – Items that must be completed before the start-up of systems can proceed. This list is most often generated by input from the commissioning leader, commissioning team, and/or the system contractor.

Factory acceptance test (FAT) – Formal confirmation that specific components have been manufactured and function in accordance with specifications and related contract documents. Since the FAT is a formal contractual procedure, witnesses and signatures are required to confirm the completion of this task.

Installation verification – The element of the commissioning program that verifies the equipment and systems is installed and conforms to contract obligations. As is the case with the FAT, witnesses and signatures are part of the verification.

Operational staff training – Verification that required operator training has been completed. This is also most often a contract obligation and requires approved witnesses and signatures for completion.

Operational verification – The segment of commissioning that verifies components and systems are installed in accordance with design specifications and criteria. Since this activity is another contract requirement, witnesses and approved signatures are also part of the procedure.

Punchlisting – The procedure that documents, through a detailed listing, the defects, deficiencies, and remaining tasks to be completed for system acceptance by the owner. Punchlisting is most often initiated shortly before the system completion and hand over to the owner. Coordination and completion of punchlist tasks are to be the responsibility of the project construction manager.

Start-up – The confirming procedure that verifies the installed system, upon completion of the commissioning of the tasks, and operates in accordance with contract requirements and intent. Successful start-up further confirms that operating parameters such as proper alignment, calibration, balancing, controls, power supply, safety concerns, consumables, and spares have been adequately addressed to assure the installed system will function in accordance with accepted and approved criteria.

Test reports – Documentation confirming the measurements performed on critical aspects of the system, subsystem, or components of an installation. These reports are part of the contract deliverables required by the contractor.

The process of commissioning is most often performed by a commissioning team. The commissioning team personnel are generally composed of owner personnel. Typically, a commissioning team is composed of the following:

- Project manager
- Commissioning leader
- User group representative
- Quality assurance
- Validation
- Manufacturing
- Maintenance
- Engineering
- Subject matter experts (SME)
- Construction manager
- Equipment/system representative
- Other personnel as required (e.g. calibration, corrosion/passivation, automation consultants)
- Subcontractors

The role of the commissioning leader is particularly important. Understanding his role and responsibilities gives an understanding of the activities required to perform a commissioning activity. It is the responsibility of the commissioning leader to provide and implement commissioning, planning, and technical support, maintain a current overview of the commissioning process, and verify inspections and performance goals' objectives and criteria that are current with the commissioning milestone activity completion dates. The commissioning team leader either supports or is the project manager, depending on the length and complexity of the commissioning activity. The project manager represents the owner. The commissioning team leader reports to the construction manager, particularly relevant in the case when the construction manager represents an outside, non-owner operation.

The commissioning team leader is responsible for assessing the need for outside (non-owner assets) commissioning support such as SME consultants. External support requirements should be predicated on education, training, experience, and certification. The qualifications of outside commissioning personnel involved directly in commissioning operations should be retained and easily accessible.

The primary role and main tasks of the commissioning leader is to coordinate the myriad of tasks requiring completion for a successful commissioning program. The major responsibilities and tasks of the commissioning leader are:

- Commissioning plan preparation, including scheduled team reviews, updates, and distribution.
- Monitor the progress of the design and construction phases of the project to verify conformance to documented project schedule and requirements.
- Approval of commissioning documents for inspection, functional testing, acceptance criteria, and documentation of the results.
- Verify through completion and documentation that installations and materials in equipment and related project components are in conformance with specifications and related project

requirements, including "as built" drawings and "redline" diagrams. This responsibility includes maintaining required project documentation in an accessible format.

- Coordinate, verify, and document that spare parts requirements programs, walk downs and tagging, FAT, equipment functional testing, static testing, and systems testing conform to acceptance criteria and design intent. It is also the responsibility of the commissioning leader to confirm relevant documentation is distributed to needed assets such as validation organizations and involved project engineers in a timely manner. Any deficiencies noted should also be presented to the project engineer.
- Provide, verify, and document that required training for operating personnel, maintenance personnel, contractors and subcontractors, installers suppliers and associated personnel has been performed.
- Verify, approve, document, compile, and distribute commissioning results as required.
- Coordinate with contractor, resolve and complete, and approve and document the final punchlist items required for installation completion.
- Upon completion of the commissioning process, the commissioning leader shall compile and confirm required commissioning requirements documentation that conforms to the approved elements of the commissioning plan. The commissioning leader is also responsible for submission of the contractor-supplied ETOP containing O&M manuals, equipment manuals, and other related documentation, upon completion of the commissioning program.
- It is the responsibility of the project manager to provide distribution of all, or required, portions of the completed commissioning plan to responsible matrix elements.

The commissioning team is a formally organized entity that typically includes the project manager or other owner representative, commissioning team leader, construction manager, equipment manufacturers, as appropriate, user representative, subcontractors (contract installer) as appropriate, quality assurance, and in the case where HVAC installation is involved, a certified independent testing and balancing contractor.

10.2.1 Description of Tasks

Commissioning will commence subsequent to noncritical punchlist item completion. This task can be completed by corporate operations, but most commonly this is performed by contractors. In addition to having an organized commissioning plan, the plan should employ a formalized milestone schedule that can be easily tracked. The schedule allows for accurate and efficient tracking of required activities. While the format of the schedule can vary from a Gantt chart to a more formalized software package such as Microsoft® Project, having a user friendly milestone schedule is, as noted, most helpful.

The procedures employed for commissioning tasks are described as follows:

1) Verification of factory acceptance operations.
2) Compilation and documentation of contractor O&M manuals.
3) Installation verification checklist.
4) Completion of *critical punchlist items* checklist.
5) Start-up – This procedural step is usually performed with the vendor or manufacturer in attendance and generally performed by owner staff personnel trained in operating the system.
6) Operational verification – Documentation the system functions in accordance with specifications, procedures, and design intent.
7) Submit ETOP.
8) Distribution of commissioning deliverables.

9) User design review, as required – Issues relating to equipment and equipment information may arise during the construction or installation phase of the project. When this occurs, the construction manager will request information not previously clarified or addressed. This action is called a request for information (RFI). Documenting the RFI and the response to the request must be included as part of the ETOP.

10) Installation walk downs – Typically, the owner representative, the construction manager, the safety officer, and the subcontractor will walk the installation location. The purpose of the walk down is to review P&IDs while actually inspecting the installation for safety concerns, visible flaws, and observable deviations from the project drawings, documents, and specifications. The number of walk downs varies with the size of the project; smaller projects usually require only one or two walk downs, whereas larger programs may undergo three or four. Walk downs are generally performed when the project is within a couple of weeks of handover to the owner.

The commissioning plan is the document intended to detail the installation and operational verification of the intended equipment, system, or installation that has been constructed in accordance with checklists specifically prepared for the intended commissioning project. Included in the commissioning checklists are verified confirmations that the required functional tests have been performed.

As might be concluded from the general commissioning plan outline presented thus far, the formal commissioning plan involves little or no extensive operation of the completed system or installation of concern. Rather, the approved commissioning plan is composed of a series of checklists designed to confirm the system has been properly installed with a high probability of operation and successful start-up and qualification (validation). The actual construction and installation activities are most often performed by contractors (often referred to as subcontractors) under the supervision of the owner's construction manager or contracted construction manager. Most often, the construction manager is responsible for preparation of the commissioning plan checklist, and, as would be expected, a major portion of the commissioning plan checklist is derived from the construction and installation specifications, as well as the as built drawings that are an integral part of the contract.

Perhaps the best technique to understand the format of a commissioning plan is to view a typical section of an installed facility component. Table 10.1 is a template of a commissioning plan specific to fume hoods that are part of a pharmaceutical manufacturing operation. The parties responsible for approval are the owner representative (Owner), construction manager (Const. Mgr.), and the architecture and engineering firm hired to prepare drawings and documents required for design and implementation of the installation. While this procedure is specific to fume hoods, the Installation Guideline (Table 10.1) is applicable to other installations such as deionized water installations, electrical installations, and building management system (BMS).

For a particular installation, in this case, fume hoods, the commissioning portion will often start with a system description of the installation to be commissioned.

It is most important each system requiring commissioning be identified by a unique number or nomenclature. Often, the manufacturer's serial number is used. If more than one installation of the same design is installed, then each unit will have a unique identifier. For instance, if there is more than one fume hood system involved, such as the example below, then a commissioning protocol will be required for each individual fume hood. In the case of the example fume hood commissioning shown below, the only difference between the commissioning plans is the identification number, in this example, the serial number, employed.

While the commissioning plan shown below was prepared in a Microsoft® Word format, it is not uncommon to prepare a commissioning plan in the form of an Excel spreadsheet. Any format that can present the necessary commissioning elements in a clear and comprehensive manner is generally acceptable.

Example 10.1   Part of an extensive pharmaceutical manufacturing overhaul and equipment installation involves the installation of several new fume hood units. A commissioning plan is needed for the new units. Prepare a paradigm that can be used as a commissioning protocol for the fume hoods.

Solution

Laboratory Fume Hood System

System Description

Fume hood criteria shall be based on implementing and maintaining a <0.10 ppm containment level that has been tested as per the procedures and requirements set forth in ASHRAE Standard 110-1995, Method of Testing Performance of laboratory Fume Hoods.

Fume hood volumetric air flows are predicated on hoods maintaining an average face velocity of 90 ft/s at the hood face (sash). The airflow is to function with the sash 50% open. In all cases, the height is measured from the top of the airfoil.

All fume hoods must be equipped with visual and audible alarms that actuate when the exhaust falls below the design limit.

If so equipped, cabinets intended for solvent storage will be properly ventilated.

Table 10.1 Fume hood installation commissioning guideline.

Subsystem Location: Room123, Bldg 15	Fume Hood (Serial #FH1234567)

(Initial & Date where applicable)*

Task No.	Task	Const. Mgr	A/E	Owner Representative
1	Confirm manufacturer has submitted O & M manuals as well as required warranty information	*Construction Manager is responsible for maintaining current records for project personnel required to use signature or initial authority or approval	*Signature and initials are on record with Construction Manager	*Signature and initials are on record with Construction Manager

2	Verify required factory acceptance test results are documented and contained in the ETOP			
3	Verify fume hoods have been installed as documented in manufacturer's recommended procedures			
4	Verify duct connection is unimpeded to discharge point			
5	Verify utility connections have been installed and connected in accordance with approved P & ID related documentation			
6	Verify associated piping has been flushed, cleaned and, if required, sanitized as per SOP			
7	Verify BMS and fume hood are integrated and I/O commands operate as programmed with test results incorporated in the ETOP			
8	Confirm fume hood balancing has been performed with results documented in the ETOP			

9	Confirm "Problem and Resolution" issues have been resolved & documented			
10	Fume hood control, testing and calibration has been completed as per specifications and SOP's			
11	Confirm fume hood operation has been simulated with the sash at various positions with no operating problems			
12	Confirm that for a simulated failure of the BMS installation, the fume hood installation will operate based on last control command			
13	Confirm RFI's have been resolved an documented as part of the ETOP			
14	Confirm shop drawings and required submittals are complete and contractor Red Line Drawings are complete, documented and submitted in the ETOP			
15	Confirm required O&M manuals have been delivered and submitted as part of the ETOP			

16	Owner representative training has been completed and documented			
17	Commissioning, identified start-up activities have been performed and documented			
18	All Punchlist items have been resolved by all parties.			
19	All equipment warranty periods have been identified and documented in the warrantee manual			

The Construction Manager hereby asserts the work associated with this system is complete except for deficiencies listed below, if any, that are not crucial to the operation of the system and confirms a good faith effort will be made to complete remaining deficiencies in a timely manner.

Construction Manager   Date

The owner hereby accepts the tasks associated with this system, Laboratory Fume Hood, Serial Number FH 1234567, as complete, except for deficiencies, if any, listed below and assumes responsibility and possession of the installed system.

Owner Representative   Date

The A/E (e.g. ABC Architects and Engineers Inc.) hereby asserts the proper documentation has been submitted and review as per contract requirements and work associated with this system has been completed except for deficiencies, if any, listed below.

ABC Architects and Engineers Inc.   Date

Corrective List: See Attached Corrective List

The commissioning documentation is part of the facility engineering organization. The completed documentation is to assist in engineering and maintenance operations. Documentation related to commissioning is not required to be submitted to the FDA; its function is for in-house use. Unlike validation programs, which do require FDA involvement, there are no mandatory storage periods or formalized system identification procedures. Nevertheless, commissioning programs should be available through the facility document control organization since, when properly completed and documented, the commissioning documentation is most useful for O&M tasks and future system modifications.

10.2.2 Commissioning Costs

Commissioning costs are, to a certain degree, contingent on what element of the installation or construction is responsible for preparing and performing the commissioning operations. If the bulk of the commissioning task is assigned to the project construction manager, as is often the case, the total commissioning cost is greater than having the task performed by the owner with final approval. This is a common method of performing commissioning operations. It is convenient since the commissioning costs are generally included in the contract cost. This method is desirable since the construction manager is usually well versed in performing this activity, whereas the owner and owner representative are probably not as experienced in this portion of the project. Also, the construction manager is most likely more familiar with the subcontractors responsible for installation and construction of the various tasks required.

But regardless of how the commissioning plan is prepared, there should be a template or cost estimate that can provide a rationale on the costs for commissioning. In general, the total commissioning cost should not exceed 3–5% of the construction costs, including change orders. Another reliable estimate is to use a range of $0.23–0.30/ft^2 as a more quantitative cost. For more specific installation costs, the following values are often employed as commissioning cost estimates:

- Commissioning should be 1–1.5% of the electrical costs (Division 26).
- Commissioning should be 1.5–2% of the mechanical costs (Primarily Division 23 but could involve tasks in Divisions 42 and 43).
- Commissioning should be in the range of 0.75–2.5% of the HVAC and BMS costs. The BMS is the automated control center that continuously monitors and controls facility variables such as temperature, humidity, and, in certain scenarios, pressure and/or pressure differentials for certain controlled areas. These tasks are often regulated through use of programmable logic controllers (PLC).

10.3 Validation

Because of the nature of pharmaceutical manufacturing, it is most important to have a system or process that is capable of collecting, analyzing, and documenting data specific to a product's performance. Such a system or process is referred to as validation in the pharmaceutical and medical device industries. The FDA broadly defines validation as "establishing documented evidence which provides a high degree of assurance that a specific process will consistently produce a product meeting its predetermined specifications and quality attributes."

While it may appear that the validation process is merely a technique intended to create paperwork and regulatory morass, this is certainly not the case.

More specifically, 21 CFR 211.100 (Current Good Manufacturing Practice or cGMP) specifically notes:

a) There shall be written procedures for production and process control designed to assure that the drug products have the identity, strength, quality, and purity they purport or are represented to possess. Such procedures shall include all requirements in this subpart. The East written procedures, including any changes, shall be drafted, reviewed, and approved by the appropriate organizational units and approved by the quality control unit.
b) Written production and process control procedures shall be followed in the execution of the various production and process control functions and shall be documented at the time of performance. Any deviations from the written procedures shall be recorded and justified.

The validation process is often referred to as the Qualification Process. Essentially, qualification (Validation) is a study intended to demonstrate that equipment, instruments and certain utilities meet requirements specified prior to purchase (Design Qualification), have been installed correctly (Installation Qualification), operate within the supplier and/or user specifications (Operational Qualification) and can be used for intended operation. The Qualification of a process, instrument or equipment installation is a term used to encompass the entire validation process and is often used interchangeably with validation.

These statements define the regulated need for a validation program for drug, biological, and medical devices and supporting electronic systems.

While cGMP regulations 21 CFR 210 and 21 CFR 211 are specific to drugs, 21 CFR 820 serves similar regulatory criteria for medical devices. Similarly, 21 CFR Part 11 is specific to electronic signatures and electronic records.

The validation process employs several terms that may be unfamiliar to many people. Before an understanding of the validation process is undertaken, some common definitions often used in the validation process should be defined:

Design qualification (DQ) – A document that defines the functional and operational specifications of equipment or installation that is prepared and approved prior to purchase. The DQ can also apply to an existing equipment or installation that is intended for an application not previously specified. Of course, a significant component of the DQ is confirming the equipment selected will meet the user's needs. Typically, the DQ protocol contains the following details [1]:

- User requirement specification (URS)
- Vendor documents and specifications
- Facility layout
- Purchase orders
- Design documentation
- FATs
- As built drawings
- Data sheets

The output of the DQ phase is a report and a standard documentation list (SDL) that contains the following:

- Design requirements
- Bidding requirements
- Purchasing and order documentation
- Vendor supplied documents list
- As built drawings (drawings and related information defining the project in the current status, prior to construction or upgrade)
- Component lists
- Inspection lists
- FATs

Facility/utility qualification – An approved procedure that involves IQ, operation qualification (OQ), and performance qualification (PQ) of the building and equipment. It covers several utility systems that are required to operate a manufacturing facility, but only a few have direct or minimal contact with the product and thus does not require qualification (validation).

GMP instrument – Instruments used to determine and record product quality; all critical and noncritical instruments are GMP instruments.

Installation qualification (IQ) – An approved procedure that has been prepared to demonstrate that an equipment/instrument or equipment/instrument system is constructed and installed in accordance with design requirements and established specifications and that the manufacturer's recommendations, specifications, and requirements are adequately addressed. Simply stated, the IQ is intended to verify that the system, installation, instrument, or equipment is properly installed. IQ criteria and considerations include:

- Equipment design features (e.g. materials of construction, cleanability, etc.).
- Installation conditions (wiring, utilities such as steam supply and plant air).
- Calibration, maintenance (including SOPs), cleaning procedures, and cleaning schedules.
- Safety features.
- Supplier documentation, prints, drawings, and manuals.
- Software documentation.
- Spare parts list, including location of spares.
- Environmental criteria such as clean rooms, temperature, and humidity (Chapter 3).
- Installed instruments must have current calibration documentation as well as a calibration schedule noting when recalibration of the instrument is required. Additionally, calibration tags, indicating the date of the most recent calibration, as well as the date for the next calibration must be clearly marked and visible.

Output IQ documentation and SDL file should include [1]:

- Project changes
- Tests performed
- Calibration
- Vendor-supplied documents
- Equipment information
- IQ deviations
- Consumables
- Spare parts list
- Facility SOPs and vendor supplied documentation such as manuals, P&IDs, and instructional documents (SOPs)

It may be observed the IQ information appears similar to the elements of a commissioning plan. While many of the elements of a commissioning plan are indeed similar to the IQ format, the commissioning plan does not have the rigor most often found in an IQ protocol. However, in recent years there has been a significant effort to combine or merge commissioning with the IQ process.

Operational qualification (OQ) – An approved procedure that has been prepared to confirm, through objective testing, that an equipment/instrument or equipment/instrument system is operating in accordance with established specifications and produces a result that functions within predetermined control limits. OQ considerations include:

- Process control limits (time, temperature, pressure, line speed, etc.).
- Software parameters (as differentiated from software documentation, software documentation does not require testing the software, as is the case with the OQ).
- Raw material specifications.
- Process operating procedures (SOPs).
- Material handling requirements.
- Process change control procedures.
- Training plans and documentation.

- Identification of process failure modes, worst-case conditions, and required action levels.
- Electronic operators such as critical alarms, interlock alarms, and alerts are set at the correct levels and function as intended.

Performance qualification (PQ) – Establishing through an approved procedure, and subsequent operation, installed equipment can operate reliably as intended for the process under routine, minimum, and maximum operating ranges. PQ evaluation variables include:

Actual product and process parameters employing procedures (SOPs) identified in the OQ. A common method of product evaluation is to evaluate several powder production batches and combine the batches, commonly three production batches, into one evaluation lot for evaluation of the product using the criteria defined previously. The major objective of the PQ portion of the validation program consists of manufacturing the product under actual conditions and confirming acceptable product can be produced consistently during normal operating conditions. An important element of the PQ operations is a series of in-process tests identified as a challenge. Challenges to the process should be performed under conditions encountered during manufacturing operations. While operating variables may be tested at, or near, the operating limits, they should not exceed the approved operating limits. The purpose of a challenge of the PQ is not to operate until the approved manufacturing limits are operated until a failure occurs, but rather to verify operations will perform under normal conditions and testing the approved process limits. A typical challenge test would be to maintain the temperature of a PQ operation at or near the maximum operating temperature for an extended period while continuously manufacturing product. The intent of the challenge, as noted, is to submit the operation to various operating limit tests. Among the elements that should be addressed in the PQ are:

- Acceptability of the manufactured product – The material produced should meet the approved quality control standards established for the product.
- Process repeatability – The product analysis is within the limits of the approved product specifications. The process must also exhibit acceptable production capability through an extended (long-term) period.
- Performance of defined operating parameters (e.g. temperature, differential pressure, relative humidity, flow rates).

Confirm process parameters and procedures (SOPs) are properly prepared and adequately address the process requiring validation. Process parameters and procedures (SOPs) should be established prior to or during the OQ phase of the qualification operation.

The intent of process validation is by establishing, through testing, operation and documentation that a manufacturing process consistently results in a product that conforms to the predetermined requirements for acceptance.

As is the case with the PQ, it is standard practice to blend three powder production lots into one batch for evaluation. Testing and analysis is thus performed using the one product blended batch. The requirements for liquids and parenterals require a protocol format that must be evaluated by more in-depth analysis. Parenterals, if biological in nature, are also regulated by the provisions of 21 CFR Part 600: Biological Products: General, 21 CFR Part 601: Licensing and Part 21 CFR 610: General Biological Products Standards.

Testing techniques to verify product conformance can vary from sophisticated analytical techniques such as spectroscopy to simple melting point determinations, depending on what the analytical methods development organization requires.

In 2011, the process validation was modified to reflect a greater conformity to concepts put forth by both the FDA and the International Council on Harmonization (ICH).

Upon completion of validation activities, a final report is prepared. The final report is part of the validation document. The final report should indicate the validation was successfully completed. Deviations from the protocols should also be recorded. The final report should also describe how the deviation was resolved. The final report should be reviewed and approved by responsible personnel and delegated management representatives.

It is most important to verify and date every activity of the protocol as it is being performed. The individual or individuals responsible for performing or witnessing the validation procedure must initial and date the validation operation upon completion of the particular test. When a particular activity of the protocol is completed, the results are verified, reviewed, signed, and dated by authorized individuals responsible for oversight of the qualification protocol.

There is a specific order in which the IQ, OQ, and PQ is to be performed. While it is not uncommon to perform the IQ and OQ simultaneously, it is more common to perform the IQ first, followed by the OQ. However, whether the IQ and OQ are performed simultaneously or sequentially, under no circumstances can the PQ be performed until both the IQ and OQ are completed.

An important output of the OQ and PQ is the identification of continuous operating variables.

It is not uncommon to have equipment vendors prepare and observe validation operations. If outside consultants (e.g. consulting firms or engineering companies specializing in preparation of validation procedures and protocol implementation) are to be considered, it is important to maintain continuous communications between the vendor and the client when outside protocol preparation is undertaken. It is also advisable to ensure both vendor and client personnel assigned to the tasks are knowledgeable in preparation, implementation, and monitoring of validation activities.

While it is anticipated, protocol execution will be successful, and it is most critical to have a prepared procedure to rely upon if all or parts of the protocol fail. Procedures must be identified, either in an approved SOP or documented as part of the protocol, what type of additional action is required should protocol test and or evaluation segments require retest of the deviations or if the entire protocol must be redone.

Often, a test or evaluation may produce a deviation in protocol testing. If the deviation is minor and easily resolved, the remedial action may require only a brief explanation be documented in the deviations section of the final report. However, if it is determined the deviation is significant, a risk assessment should be performed to verify whether the failure is an equipment or procedural failure. Again, failure to carefully document follow-up procedures in the event of a protocol malfunction should be clearly defined and transmitted to all parties involved in the validation operations. All parties, particularly the quality assurance operation, should be fully aware of the approved steps required should protocol implementation fail to perform as intended. Having an SOP in place that clearly identifies the steps to be employed, in the case of a malfunction, is of little value if key personnel are not familiar with the approved procedure that should be followed.

The validation procedural protocol must also clearly distinguish between minor failures and catastrophic failures. In such cases, the protocol procedure must clearly identify when validation testing and evaluation operations must be halted and when operations can continue with detailed documentation of the deviation being documented in the deviation report. It is not uncommon to bypass a catastrophic validation failure and continue with the qualification. This erroneous misstep can occur if proper oversight is not maintained. Even if a backup operation automatically replaces the failed primary operators (electrical, mechanical, or human), it is imperative these malfunctions be corrected prior to continuing the protocols. Unless corrective measures are employed prior to continuing or restarting the protocol, there is a high probability the malfunction will arise during the manufacturing operations. If such a scenario does occur, the costs are significantly greater than would be the case if the defects were not remediated during the OQ or PQ phase.

While a detailed analysis of a protocol operational failure is the most effective technique for finding the root cause of a failure, two common failure modes are the failure to confirm the new equipment that can compatibly operate with existing manufacturing software such as interlocks and the failure to adequately verify calibration standards for new and existing equipment.

When preparing validation protocols, it is important to understand there are no formal procedures required by the FDA. As explained previously, the only formal requirement the FDA has imposed is the need to provide documented procedures and results confirming the capability that an equipment, instrument, or system will perform as intended. With this in mind, it would be beneficial to know what actual qualification protocols contain. Thus, the material below presents the basic input required in an actual IQ, OQ, and PQ is prepared. The material below is a basic template and should be adjusted, as needed, to fulfill the requirements of the equipment or component requiring validation. The DQ document is not included since this information is most often prepared prior to the actual validation operations and serves primarily to identify the equipment, instrument, and/or utility requiring qualification activity.

Example 10.2   A generic pharmaceutical firm is installing some pharmaceutical equipment and laboratory instruments for the quality control operation. Rather than seeking outside consultants for the preliminary qualification work, it was decided to prepare in-house drafts of qualification protocols and associated material.

Prepare a typical validation protocol and explanation of the sections required for a comprehensive validation template for the hardware requiring validation.

Solution
Preparing a table of contents that identifies the validation activities and associated information is the primary requirement. The protocol page numbering should be a form such that each page number is identified and the total number of pages in the protocol is also visible on the page. For instance, a protocol may contain a total of 40 pages; the cover page would be identified as page 1 of 40, the protocol approval page, usually the second page, would be numbered as page 2 of 40, while the table of contents would be page 3 of 23, and so on. Thus, the "purpose" section would be on page 3 of the protocol. Also, every protocol page should have a header that clearly identifies the name and identity of the equipment to be validated.

The original issue of a qualification protocol is generally identified as revision 0. Since the protocol is reviewed by several organizations prior to final approval (e.g. manufacturing, engineering, quality assurance), several versions of the protocol may be prepared and modified prior to issuing a protocol that is prepared to undergo formal validation or revalidation.

While individual validation protocols may vary, it is critical that the information required be included in protocols to enable efficient tracking of items and approved procedures. It is through the availability of documented data that corrective and preventive actions can be employed should difficulties arise during the validation implementation as well as post-validation operation.

Company Name	Protocol Number
Validation Protocol for (Insert Device or Equipment Name)	**Page 1 of 26**

Prepared By:_____   Date:_____
                   (Name & Organization)

PROTOCOL EXECUTION APPROVAL

_____   _____   Date:_____
Name & Title   **Signature**

_____   _____   Date:_____
Name & Title   **Signature**

PROTOCOL COMPLETION APPROVAL

_____   _____   Date:_____
Name & Title   **Signature**

_____   _____   Date:_____
Name & Title   **Signature**

_____   _____   Date:_____
Name & Title   **Signature**

Company Name	Protocol Number
Validation Protocol for (Insert Device or Equipment Name)	**Page 2 of 26**

Table of Contents

Company Name	Protocol Number
Validation Protocol for (Insert Device or Equipment Name)	Page 3 of 26

Purpose
This section contains a brief description of the equipment or device that is to be validated. It typically contains the company and location of the equipment. The purpose should be a brief statement of a few sentences noting that the intent of the validation is to confirm that the validation protocol, upon successful execution, will verify that the equipment operates within the levels of reliability, reproducibility, and accuracy specified.

Scope
This section is a brief statement explaining that the protocol is specific to the equipment identified in Tables 10.2–10.7. This section should also note any elements that are not included in the validation protocol. An example would be related to software that would be validated separately under the provisions of 29 CFR Part 11.

Company Name	Protocol Number
Validation Protocol for (Insert Device or Equipment Name)	Page 4 of 26

INSTALLATION QUALIFICATION, IQ

Table 10.2 Installation qualification identification equipment identification template, IQ.

Equipment Description	Manufacturer	Model Number	Serial Number	Equipment Location (Bldg./Rm.)	Verified □ Yes □ No	Initial/Date	Comments

Verified By: _____   Date: _____

Reviewed By: _____   Date: _____

Table 10.2 is a typical template that identifies the information normally required for an equipment survey. As can be surmised, the size of the table is contingent upon the size of each particular data entry. Consequently, Table 10.2, as is the case with most of the required material, would be larger than the example shown.

Table 10.3 identifies the components of the installed equipment. The components are the replaceable or repairable parts that comprise the total installation. For instance, if the installed equipment to be validated is a chart recorder intended to determine and record the pressure and temperature of a large pharmaceutical powder blender, the installed unit would be the described chart recorder. Table 10.2 should be completed using the information required for the chart recorder information.

Table 10.3 further details the components of the equipment installed in Table 10.2. The purpose of Table 10.3 is to identify the replaceable parts that comprise the equipment installation identified in Table 10.2.

Company Name	**Protocol Number**
Validation Protocol for (Insert Device or Equipment Name)	**Page 5 of 26**

Table 10.3 Equipment components template, IQ.

Equipment Description (Table 10-2, Column 1)			
Item	**Specification** (Insert data for Items 1, 2, 3)	**Verified** (Initial/Date)	**Comments:**
1. Component Name		☐ Yes ☐ No	
2. Manufacturer		☐ Yes ☐ No	
3. Part Number		☐ Yes ☐ No	

Verified By: _____   Date: _____

Reviewed By: _____   Date: _____

For Table 10.3, components include details such as relief devices, regulators, control units, and other spare parts required for maintenance of the installation. Items such as manufacturer and part number should be entered in column two (specification).

Table 10.3 is intended to identify hardware components for the equipment installation of concern. Many equipment installations also have computer-controlled control units that include operating hardware and imbedded software operations. Tables 10.4 and 10.5 present the IQ templates specific to hardware and software identification and documentation.

Company Name	Protocol Number
Validation Protocol for (Insert Device or Equipment Name)	**Page 6 of 26**

Table 10.4 Hardware identification template, IQ.

Hardware			
Item	**Specification** (Insert Data for Items 1, 2, 3)	**Verified** (Initial/Date)	**Comments**
1. Manufacturer		☐ Yes ☐ No	
2. Model Number		☐ Yes ☐ No	
3. Part Number		☐ Yes ☐ No	

Verified By: _____   **Date:** _____

Reviewed By: _____   **Date:** _____

Company Name	Protocol Number
Validation Protocol for (Insert Device or Equipment Name)	Page 7 of 26

Table 10.5 Software identification template, IQ.

Software			
Item	Specification (Insert Data for Items 1,2,3,4)	Verified (Initial and Date)	Comments
1. Software Name		☐ Yes ☐ No	
2. Developer		☐ Yes ☐ No	
3. Development Tool (Programming Language)		☐ Yes ☐ No	
4. Version		☐ Yes ☐ No	
5. Software	Validated ☐ Yes ☐ No ☐ N/A	☐ Yes ☐ No	
Validation Status			

Verified By: _____   Date: _____

Reviewed By: _____   Date: _____

The most common item for utilities validation is probably electric power. Other utilities often requiring validation are plant air for pneumatic instruments, nitrogen and inert gases for blanket gases. In most cases, nitrogen and inert gases do not affect the product quality if in contact with pharmaceutical ingredients. Electrical power, on the other hand, can affect product quality if variables such as voltage, amperage, and backup power are not operating at intended levels. The capability of the equipment to generate electromagnetic pulses/fields is also important since spurious pulses can significantly affect equipment and instrument performance. Consequently, it is desirable to evaluate for electromagnetic interference (EMI) potential of the equipment to be validated. If such a potential exists, Table 10.6 presents a typical utility template for electric supply.

By observation, it can be seen that a good deal of the information required in a typical IQ protocol can be gleaned from previously documented commissioning data. However, care should be taken to verify that commissioning data is accurate and applicable to the IQ. For instance, confirm that motors, pumps, fans, and agitators are all identical. In an installation where several motors are installed, for instance, it is critical that the commissioning motors from which data is taken are identical to the motors that are subject to validation (i.e. they are identical down to the serial numbers).

Company Name	Protocol Number
Validation Protocol for (Insert Device or Equipment Name)	Page 8 of 26

Table 10.6 Utility identification template, electrical, IQ.

Item	Specified	Actual	Verified (Initial & Date)	Comments
1. Voltage, VAC/VDC			☐ Yes ☐ No	
2. Current, amps			☐ Yes ☐ No	
3. Frequency, Hz			☐ Yes ☐ No	
4. Phases, φ (1,2, or 3)			☐ Yes ☐ No	
5. Source			☐ Yes ☐ No	
6. Grounding			☐ Yes ☐ No	
7. Breaker Rating			☐ Yes ☐ No	
8. Breaker Number			☐ Yes ☐ No	
9. Breaker Location Bldg/Room Number			☐ Yes ☐ No	
10. Backup Power Source (UPS, Battery, etc.)			☐ Yes ☐ No	
11. EMI Potential Evaluated(Yes/No/N/A)			☐ Yes ☐ No ☐ N/A	

All calibrated test and measurement equipment or instruments used during the protocol execution must indicate the original calibration certification location, date of calibration, and recalibration date. The test and measurement equipment must be traceable to NIST or an equivalent standard. Table 10.7 identifies the relevant information required to verify the test and measurement equipment required for validation operations are adequately identified.

Verified By: _____   Date: _____

Reviewed By: _____   Date: _____

Company Name	Protocol Number
Validation Protocol for (Insert Device or Equipment Name)	**Page 9 of 26**

Table 10.7 Validation test equipment template, IQ.

Instrument Description	Serial Number	Calibration Certificate Location	Calibration Date	Recalibration Date	Verified (Initial & Date)	Comments
					☐ Yes ☐ No	
					☐ Yes ☐ No	
					☐ Yes ☐ No	

Verified By: _____   **Date:** _____

Reviewed By: _____   **Date:** _____

Company Name	Protocol Number
Validation Protocol for (Insert Device or Equipment Name)	**Page 10 of 26**

Table 10.8 A–documentation template, IQ.

Manuals

The information below is a list of manuals on file for the (insert name of equipment to be validated) installation, operation, and maintenance. Additionally, the manuals may contain drawings for the installed equipment.

Manual Title	Manual Number	Revision	Date	Location Bldg./Rm.	Initials/Date
Comments_____					

Drawings

The information below documents the drawings of record and on file for the (insert name of equipment to be validated). The documented information includes installation drawings, specifications, and utilities, including electrical and associated P&IDs.

Company Name	Protocol Number
Validation Protocol for (Insert Device or Equipment Name)	**Page 11 of 26**

Table 10.8 B–drawings template, IQ.

Drawing Title	Drawing Number	Revision	Date	Location (Bldg./Rm.)	Initial & Date
Comments_____

Verified By: _____   Date: _____

Reviewed By: _____   Date: _____

Company Name	Protocol Number
Validation Protocol for (Insert Device or Equipment Name)	**Page 12 of 26**

Table 10.9 Maintenance template, IQ.

Maintenance Program in place	☐ Yes	☐ No
Written Procedure in place	☐ Yes	☐ No
Performed by an outside contractor	☐ Yes	☐ No
Performed on an "as needed basis only"	☐ Yes	☐ No
(Some equipment do not require scheduled maintenance due to monitoring devices that may be integral to the equipment)		
Procedure Number _____		
Location of Procedure (Bldg. & Room No.) Bldg._____ **Room No.**_____		
A copy is attached to this Protocol	☐ Yes	☐ No
Maintenance Manual is Available	☐ Yes	☐ No
A copy is attached to this Protocol	☐ Yes	☐ No
Location of Maintenance Manual (Bldg. & Room No.) Bldg._____ **Room**_____		
A Spare Parts List is Available	☐ Yes	☐ No
Location of List (Bldg. & Room No.) Bldg._____ **Room No.**_____		
A copy is attached to this Protocol	☐ Yes	☐ No
Comments:_____ _____		

Verified By: _____   **Date:** _____

Reviewed By: _____   **Date:** _____

Company Name	Protocol Number
Validation Protocol for (Insert Device or Equipment Name)	**Page 13 of 26**

Acceptance Criteria

Verification for the IQ has been completed with deviations identified and reviewed for acceptance. All data sheets contained in this IQ are dated and verified where indicated.

1. Acceptance Criteria have been met and conform to the approved specifications and

installed procedures. ☐ Yes ☐ No

2. Utilities are installed in accordance with manufacturer and/or company specifications,

where applicable. ☐ Yes ☐ No

3. Any deviations observed during the execution of the IQ section of this protocol are

listed in the Deviation Log attached to this protocol ☐ Yes ☐ No

If No, present details below:

Comments:_____

_____

_____

_____

Verified By: _____ **Date:** _____

Reviewed By: _____ **Date:** _____

Company Name	Protocol Number
Validation Protocol for (Insert Device or Equipment Name)	**Page 14 of 26**

OPERATIONAL QUALIFICATION, OQ

Table 10.10 Template for operator training, OQ.

Acceptance Criteria	Results Acceptable (Yes/No)		Initials & Date
An approved training program for this equipment has been developed	☐ Yes	☐ No	
Operators have been trained in the operation of this equipment	☐ Yes	☐ No	
Operators have been instructed in the safety requirements and emergency shutdown procedures for this equipment	☐ Yes	☐ No	
The operator training has been properly documented	☐ Yes	☐ No	

Verified By: _____   **Date:** _____

Reviewed By: _____   **Date:** _____

Company Name	Protocol Number
Validation Protocol for (Insert Device or Equipment Name)	**Page 15 of 26**

Table 10.11A Standard operating procedures (SOP's) template.

SOP Title	SOP Number	Revision	Effective Date

Comments:_____

_____

_____

_____

Verified By: _____   **Date:** _____

Reviewed By: _____   **Date:** _____

Company Name	Protocol Number
Validation Protocol for (Insert Device or Equipment Name)	**Page 16 of 26**

Table 10.11B Criticaland non critical instruments template*, OQ.

Type_____ Critical Instrument ☐ Yes ☐ No

Manufacturer_____

Model_____

Serial Number_____

Company Identification Number_____

Instrument Location Bldg _____   Room No._____

Range_____

Scale Division_____

Range_____

Next Calibration Date_____

Calibration Frequency_____

Initials & Date_____

* The purpose this Table is to confirm critical instruments requiring calibration conform to design and/or manufacturing of specifications. Non critical instruments do not require calibration.

Verified By: _____   Date: _____

Reviewed By: _____   Date: _____

Company Name	Protocol Number
Validation Protocol for (Insert Device or Equipment Name)	**Page 17 of 26**

Table 10.12 Critical and non critical instruments template* (continued).

Type_____ Critical Instrument ☐ Yes ☐ No

Manufacturer_____

Model_____

Serial Number_____

Company Identification Number_____

Instrument Location Bldg._____   Room No._____

Range_____

Scale Division_____

Range_____

Next Calibration Date_____

Calibration Frequency_____

Initials & Date_____

*The purpose this Table is to confirm critical instruments requiring calibration conform to design and/or manufacturing of specifications. Non critical instruments do not require calibration. <u>One form for each instrument is required.</u>

Verified By: _____   Date: _____

Reviewed By: _____   Date: _____

Table 10.2 is intended to identify by means of operating the equipment to be validated using the approved SOPs and recording the actual results. Acceptance criteria are predicated on the manufacturer's specifications, regulatory requirements, and company standards. The results of the OQ testing are recorded and evaluated using the expected results as the evaluation standard. Based on the recorded results, the individual test procedures, which comprise the bulk of the OQ Procedures, are then documented as Acceptable or Nonacceptable (Yes or No). It is most important that procedures are performed and witnessed by authorized personnel, as confirmed by the Signature Log contained in this Protocol (page 22 of 23). Similarly, any deviations from the OQ results or procedures are recorded on the Deviation Log (Table 10.17). The severity of deviations and effect on the OQ, if not specified in the SOP, are normally evaluated by facility operations, including Manufacturing, Engineering, and Quality Assurance and/or health/safety personnel on an as needed basis.

The objective of the OQ tables identified below is to identify and document requirements and criteria in for equipment identified as critical. It is again most important to note that the OQ Protocol does not include software validation that it is regulated, for the most part, by 21 CFR Part 11 and performed under a separate auspice. Also, the information identified in the OQ does not necessarily contain actual sequences, information, or detailed results. The OQ presents what challenges (tests) are required. In all cases, the owner is responsible for performing OQ tests. However, the protocol can be prepared by a consultant, A/E firm or the owner (manufacturer). When successfully completed, the OQ confirms the equipment operates within the defined limits and tolerances.

The operating limits and tolerances are the challenges (tests) defined in the approved protocol. The expected test results are identified as the Acceptance Criteria. The actual text results are then compared with the approved Acceptance Criteria. If the test procedures are within the tolerances and specifications required, the OQ results are then approved. The challenge tests should be performed on the system over the full operating range as defined by the specification.

The limits of capability do not mean challenging or testing until destruction, but rather the OQ challenge tests are performed within the limits of quality that are defined. If the criteria are not met, an investigation and corrective action must be undertaken prior to retest.

While Table 10.12 is a single sheet, in practice more sheets will be employed so that the procedures identified in the SOP and/or manufacturer's standards are fulfilled.

Company Name	Protocol Number
Validation Protocol for (Insert Device or Equipment Name)	**Page 18 of 26**

Table 10.13 Test procedure, OQ.

Test Procedure	Expected Result	Actual Result is Acceptable ☐ Yes ☐ No	Initial & Date
1.			
2.			
3.			
4.			
6.			

Verified By: _____   **Date:** _____

Reviewed By: _____   **Date:** _____

Company Name	Protocol Number
Validation Protocol for (Insert Device or Equipment Name)	**Page 19 of 26**

Table 10.14 Alarm and interlock verification*.

Test Procedure	Expected Results	Actual Results	Results Acceptable	Initial & Date
	☐ Audible ☐ Visual	☐ Audible ☐ Visual	☐ Yes ☐ No	
	☐ Audible ☐ Visual	☐ Audible ☐ Visual	☐ Yes ☐ No	
	☐ Audible ☐ Visual	☐ Audible ☐ Visual	☐ Yes ☐ No	
	☐ Audible ☐ Visual	☐ Audible ☐ Visual	☐ Yes ☐ No	
	☐ Audible ☐ Visual	☐ Audible ☐ Visual	☐ Yes ☐ No	

*To be completed, if applicable

Verified By: _____   Date: _____

Reviewed By: _____   Date: _____

Company Name	Protocol Number
Validation Protocol for (Insert Device or Equipment Name)	**Page 20 of 26**

Table 10.15 Operational qualification summary template, OQ.

Item	Acceptable		Initial & Date
	☐ Yes	☐ No	
1. The OQ conforms to the company and manufacturer specifications.	☐ Yes	☐ No	
2. All validation equipment requiring calibration has been calibrated and a copy of the certificate is attached to this protocol.	☐ Yes ☐ N/A	☐ No	
3. Training procedures are in place for the safe operation of the equipment and personnel operating the equipment have been properly trained	☐ Yes	☐ No	
4. All required information is being maintained in appropriate equipment logs.	☐ Yes	☐ No	
5. All applicable data sheets have been completed in accordance with cGMP regulations.	☐ Yes	☐ No	
6. All deviations observed during the execution of the OQ portion of this protocol and are documented in the Deviation Log	☐ Yes ☐ N/A	☐ No	
7. The OQ has successfully met the criteria established in the protocol	☐ Yes	☐ No	

Verified By: _____   Date: _____

Reviewed By: _____   Date: _____

PERFORMANCE QUALIFICATION, PQ

The objective of the Performance Testing is to verify that the (insert name of equipment) operates satisfactorily and consistently performs in accordance with the manufacturer's and the company's specifications during manufacturing operations. The PQ must also demonstrate that in addition to performing in accordance with approved specifications and standards, the end products must also meet the quality standards defined by the finished product. Of equal importance, the results must demonstrate a level of reliability that results in a consistently acceptable product over time.

The PQ procedure is usually prepared by implementing the manufacturer's operational procedures, which are presented in the supplier manuals. In the rare situation where manufacturer's operating procedures are unavailable, in-house documentation such as SOPs, product specifications, and batch records can be used to create an acceptable PQ testing procedure. But, again, it is a very unusual scenario wherein an equipment does not have an accepted procedure upon which a PQ test protocol is not available.

The actual OQ test procedure generally consists of producing a few batches of product under normal operating conditions. The completed batch or batches are then evaluated as normal manufactured product employing the standard quality control procedures for required for approval.

Company Name	Protocol Number
Validation Protocol for (Insert Device or Equipment Name)	Page 21 of 26

PERFORMANCE QUALIFICATION

Table 10.16 Test procedure, PQ.

Test Procedure	Expected Result	Actual Result is Acceptable ☐ Yes ☐ No	Initial & Date
1.			
2.			
3.			
4.			
6.			
7.			

Verified By: _____   **Date:** _____

Reviewed By: _____   **Date:** _____

The Acceptance Criteria for the PQ must be completed and deviations recorded and reviewed for acceptability as specified in the company SOP and other relevant documentation. All data test forms are to be completed initialed, dated, and signed in locations identified in the PQ test procedure.

The PQ test results will be evaluated and compared with the approved acceptance criteria. For this reason, it is most important that approved acceptance criteria are in accordance with manufacturer's specifications and company product specifications. Consequently, a comprehensive review is required prior to acceptance or rejection of a PQ protocol.

It is most desirable to have PQ acceptance and failure criteria clearly defined prior to commencing the PQ test phase. Failure to do so can create unneeded angst among stakeholders, evaluators, manufacturer, and approvers (i.e. Quality Assurance) and other organizations involved in the PQ.

Company Name	Protocol Number
Validation Protocol for (Insert Device or Equipment Name)	**Page 22 of 26**

ACCEPTANCE CRITERIA

All tests executed in this protocol conform to the predetermined specifications, procedures, and deviations from the approved protocol that have been determined as ACCEPTABLE and deviations are documented in the deviation log. Completed data sheets have been reviewed, verified, and dated, indicating they are in compliance with the approved protocol and conform to cGMP regulations.

Verified By: _____   **Date**_____

Reviewed By: _____   **Date**_____

Company Name	Protocol Number
Validation Protocol for (Insert Device or Equipment Name)	**Page 23 of 26**

Table 10.17 Deviation log template.

Deviation Number	Deviation Description	Deviation Closed (Initial & Date)*
1.		
2.		
3.		
4.		
5.		
6.		
7.		

*Initial & Date the "Deviation Closed" column when the deviation has been resolved

Verified By: _____   **Date:** _____

Reviewed By: _____   **Date:** _____

Company Name	Protocol Number
Validation Protocol for (Insert Device or Equipment Name)	Page 24 of 26

Table 10.18 Deviation report template.

Deviation Report Number:	Deviation Report Date:

Deviation Report Description: _____
_____
_____
_____

Verified By: _____ **Date:** _____
Reviewed By: _____ **Date:** _____

Root Cause: _____
_____
_____

Verified By: _____ **Date:** _____
Reviewed By: _____ **Date:** _____

Corrective Action: _____
_____
_____
_____

Verified By: _____ **Date:** _____

Reviewed By: _____ **Date:** _____

Company Name	Protocol Number
Validation Protocol for (Insert Device or Equipment Name)	**Page 25 of 26**

Table 10.19 Signature log template*,

Name (Print) Company/Title	Signature	Initials	Date
1.			
2.			
3.			
4.			
5.			

*EACH INDIVIDUAL INVOLVED IN THE EXECUTIONOF THIS PROTOCOL MUST COMPLETE A LINEIN THIS TABLE.

Verified By: _____ Date: _____

Reviewed By: _____ Date: _____

Company Name	Protocol Number
Validation Protocol for (Insert Device or Equipment Name)	Page 26 of 26

APPENDIX[*]

ATTACHMENTS FOR VALIDATION PROTOCOLS

[*] Attachments are not part of the protocol. Attachments may contain information such as copies of the raw data sheets, applicable SOP's and specifications as well as other material which may be helpful.

Example 10.2 describes the information required to complete a typical validation for equipment or installation composed of several individual components. In such cases, each individual component will be validated. For instance, if an installed system is intended to perform mixing for a pharmaceutical blend, the agitator, motor, mixing tank, and control accessories will be evaluated. Thus, while the total blending operation is to be validated, each critical component must be evaluated.

When a facility is constructed (i.e. vaccine manufacturing facility, clinical supply facility), a unique document must be prepared to, among other details, define the extent of validation required for the new facility. This document is known as the validation master plan (VMP). The master validation plan defines requirements for validation of a new or significantly revised cGMP facility. More specifically, the VMP is intended to be a management tool that describes the validation requirements for cGMP facilities, manufacturing and control systems that exist, or will exist, within the facility. The general contents of a VMP include:

- A description of the facility including the site plan, individual floor plans, relevant structural features, process areas, and the required utilities.
- A description of the production process detailing the rooms, equipment, process procedures, air quality, and personnel movements.
- A description of the validation program including qualification procedures and prerequisites, the equipment to be qualified, and the documentary requirements for validation.
- A list of departments responsible for the components of the validation program.
- Computer system validation plan (usually the information technology {IT} department).
- QC instruments and equipment.
- Instrument and equipment cleaning validation plans (with the advent of disposable instruments and components, the requirements for cleaning validation is beginning to have less emphasis. Chapter 9 reviews elements of disposable technology).
- Descriptions of the requirements for maintenance, revalidation, and change control procedures.
- Applicable SOPs and related systems manuals.
- Approved vendors listing.
- Milestones and schedules for task implementation.

10.4 Process Validation

The validation procedures described in the previous section are specific to components, equipment, and/or systems. The performance validation and associated qualifications (DQ, IQ, OQ, and PQ) are a procedure that is specific to the manufacturing system the process validation program is intended to support.

Formerly, the FDA definition of the performance validation (qualification) was defined as "establishing confidence that the process is effective and reproducible."

However, with the introduction of the modified process validation guidance, the FDA definition of process validation is now "Process Validation is establishing documented evidence which provides a high degree of assurance that a specified process will consistently produce a product meeting its predetermined specifications and quality characteristics."

While there may be some slight modifications in the definition, the intent of the process qualification scope and presentation, the basic procedures, and format for the task remain reasonably consistent. Similarly, the procedures, which should be validated, are also constant.

In addition to product manufacturing, among the common operations that typically require process qualification are operations such as [2]:

- Sterilization processes.
- Clean room operations.
- Aseptic filling processes.
- Sterile packaging sealing processes.
- Lyophilization process.
- Heat treating process.
- Plastic injection molding operations (most likely supplied by an outside manufacturing firm or Original Equipment Manufacturer [OEM]).
- Instruments used in manufacturing (exclusive of quality control laboratory functions) such as colored testing, turbidity, pH, and HPLC.
- Certain filling operations such as tablet bottling and filling operations.

With the processes requiring validation identified, a qualification procedure should be prepared. Typical topics addressed in the protocol development should include, as a minimum:

- Identification and description of the process to be validated.
- Identification of the material to be validated (this could be an active pharmaceutical ingredient (API), finished pharmaceutical product, or a medical device regulated by 21 CFR 820).
- Objective of the protocol.
- Duration of the validation process.
- Equipment and operators involved in the qualification.
- Description of the process to be validated.
- Relevant SOPs that relate to the product, components, and manufacturing materials germane to the process qualification.
- Process parameters to be monitored during the validation operation.
- Product characteristics requiring monitoring during protocol execution.
- Characteristics of the product that are to be monitored during the qualification process.
- Methods employed for data collection analysis.
- Modification, change control, and revalidation documentation.

A process qualification format will generally address the following elements for evaluation:

    I. Objective
   II. Process Description
  III. Responsibilities
  IV. Prequalification Requirements
   V. Performance Qualification
  VI. Acceptance Criteria
 VII. Documentation
VIII. Modification/Change Control and Revalidation
  IX. Index of Attachments

A process qualification, though quite detailed, is not designed to create excess documentation (i.e. paperwork). Similarly, the process qualification is clearly not intended to create an additional superfluous FDA requirement.

Process validation is a task predicated on detailed and documented testing and monitoring. One significant detail of the operation is that sample sizes are larger and often undergo more frequent sampling. Another important constituent of the process validation format is that the actual sample batch is composed of at least three consecutive manufactured batches. Further, the product must be uniform within the lots used for the process validation by conforming to the specified criteria.

Normally, a committee or team is assigned the responsibilities of administering, preparing, evaluating, and approving the process qualification. Salient elements and responsibilities of the team members include [3]:

- Manufacturing operations – This element is responsible for preparing validation batches and assuring the batches are processed as routine manufacturing lots.
- Quality assurance – Team member of this organization verifies documentation and compliance specifications and procedures are identified evaluated and documented.
- Quality control – In addition to reviewing relevant protocols and other germane documents, this operation is responsible for testing and approving raw materials, in-process sampling, and finished product.

Other organizations involved in the process qualification include (but not limited to) training, facilities engineering/maintenance engineering, site engineering, quality control, and quality assurance.

Example 10.3 illustrates how a typical process validation is prepared. While the actual manufacturing operation may be more detailed than the process described in Example 10.3, it is a reasonable representation of process qualification documentation.

Example 10.3 Small API generic manufacturing firm, ABC Generics, plans to produce a generic product that is, once manufactured, to be shipped in bulk to a tableting and finishing operation. The API product, isocarboxazid (an antidepressant), has successfully performed the equipment validation (IQ, OQ, PQ). Once manufactured, the API isocarboxazid is purified by filtration and recrystallization, with the purity confirmed by chromatography analysis and melting point determination. Product purity, in addition to the manufacturing process, requires a process validation for the product, isocarboxazid recrystallization. The components required for manufacture of isocarboxazid manufacture are:

1) Isoxazole ester
2) Benzylhydrazine
3) Isopropanol

A basic process qualification template is described below. While the process qualification format may vary between manufacturers, contract manufacturing organizations (CMO), and various consultants, the content of the protocol prepared below is representative of the requirements and content of most process qualifications.

Solution
The documentation below serves as a reference format for a process qualification. The information presented below is for a generic operation. Depending on the process to be validated, the qualification requirements may vary. For instance, a heat sealing packaging operation will have different procedures and process variables than the process described below. However, the basic protocol format is applicable for most process qualifications, even though the actual process is unique to a specific product, isocarboxazid in this case.

Note that the process qualification (PQ) is designed only to validate manufacture of an API compound, additional processing such as tableting, and fill and finish operations that are also often performed by other manufacturing operations specializing in these areas.

ABC, Inc. Process Qualification Protocol				
<u>Protocol Number</u> ABC- 001	<u>Title</u>: Isocarboxizad Process Qualification	<u>Revision Number</u> 0	<u>Date</u>	Page 1 of 10

PROTOCOL DOCUMENT APPROVAL

_____   _____

Validation/QA   Date

_____   _____

Production   Date

_____   _____

Production   Date

_____   _____

ABC, Inc. Management   Date

_____   _____

ABC, Inc. Management   Date

ABC, Inc. Process Qualification Protocol				
Protocol Number ABC-001	Title: Isocarboxazid Process Qualification	Revision Number 0	Date:	Page 2 of 10

TABLE OF CONTENTS

ABC, Inc. Process Qualification Protocol				
Protocol Number ABC-001	Title: Isocarboxazid Process Qualification	Revision Number 0	Date:	Page 3 of 10

I. OBJECTIVE

The objective of this protocol is to define the requirements and acceptance criteria for the manufacture of Isocarboxazid at the ABC, Inc facility located at (Insert **address of facility**). Successful completion of the qualification requirements will verify the process employed conforms to predetermined specifications and procedures.

ABC, Inc. Process Qualification Protocol				
Protocol Number ABC-001	Title: Isocarboxazid Process Qualification	Revision Number 0	Date:	Page 4 of 10

II. DESCRIPTION

(This section should provide a brief description of the steps involved in synthesizing the final product. The description does not require how intermediate products are manufactured; the process description need only includes the manufacturing steps normally required to isolate the final product as well as any reaction steps incorporating intermediates to obtain the final product. Typically, isolation of product would include filtration, distillation, centrifuging, drying, HPLC, and/or other separation operations. Relevant process temperatures should also be included since temperature control is a critical process parameter.)

A more detailed process description is found in the Attached SOP (Attachment Number 2) identifying process procedures.

ABC, Inc.				
Process Qualification Protocol				
Protocol Number ABC-001	Title: Isocarboxazid Process Qualification	Revision Number 0	Date:	Page 5 of 10

III. RESPONSIBILITIES

This protocol has been developed by the ABC, Inc. Validation Group. The validation team will assign pertinent responsibilities to validation team members, ABC, Inc. employees, consultants, and suppliers to verify the process validation is successfully completed. Responsibilities include:

1) Development of the isocarboxazid protocol.
2) Coordinating overall adherence to and implementation of this protocol.
3) Coordinating and monitoring the validation with the appropriate personnel to assure that the installation qualification (IQ) and operational qualification (OQ) are completed on schedule.
4) Verifying the protocol test requirements are completed and assembled in the summary report for management review and approval.
5) Coordinating the validation with designated individuals.
6) Review and approval of this protocol and summary report.
7) Supplying all procedures, data, manuals, drawings, and documentation necessary for the generation and execution of protocols and completion of the summary report.
8) Calibration of appropriate instrumentation.

ABC, Inc. Process Qualification Protocol				
Protocol Number ABC-001	Title: Isocarboxazid Process Qualification	Revision Number 0	Date:	Page 6 of 10

IV. PREQUALIFICATION REQUIREMENTS

1) The equipment qualification must be completed and documented. The first batch (es) will serve as development batch (es) that will qualify the equipment. The development batch will be executed using approved standard operating procedures (SOPs).

V. PERFORMANCE QUALIFICATION

The performance qualification is executed for the key processing steps to verify the consistency and effectiveness of the operation. Three consecutive approved batches of the product are evaluated to confirm product assurance.

The performance qualification testing procedures and requirements are:

A. Validation Batches

The validation batches will be executed using approved SOPs as regular production batches. All batch records, associated documents and log books, should be identified as "isocarboxazid validation batch" and the batch number. Additionally, the standard identification number must also be included. All approved SOPs applicable to the process are listed on Attachment Number 2 (SOP List).

ABC, Inc. Process Qualification Protocol				
Protocol Number ABC-001	Title: Isocarboxazid Process Qualification	Revision Number 0	Date:	Page 7 of 10

B. Raw Materials

All materials must conform to ABC, Inc. specifications prior to release. Raw materials employed in the process, including suppliers and manufacturers, are identified on Attachment Number 3 (Raw Materials List). Additionally, the release status of the raw materials must be verified and documented.

C. Monitoring of Key Process Parameters and Testing Summary

The effectiveness of the key processing steps will be monitored, and critical data collected will be reviewed and compared with the specified parameters. Key parameter trends and variations must be analyzed to assure proper process and equipment performance. The specific process parameters and evaluation criteria are documented in Attachment Number 4 (Monitoring of Key Parameters).

ABC, Inc. Process Qualification Protocol				
Protocol Number ABC-001	Title: Isocarboxazid Process Qualification	Revision Number 0	Date:	Page 8 of 10

VI. ACCEPTANCE CRITERIA

1) All monitoring and testing of the key process steps will be completed and approved. The process qualification is performed on the key processing steps to verify and document the effectiveness and consistency of the operation.
2) All materials used in these production runs must conform to specifications and be released by API, Inc. Quality Operations.
3) All in-process and final testing must conform to approved specifications and approved SOPs.

VII. DOCUMENTATION

All validation preparation will be performed and defined in the approved protocol, and all documentation will be completed concurrently with the execution of the protocol. Data entries must be completed during the validation testing and when samples were taken during the trial period. Any deviation or abnormality encountered during protocol execution must be documented in the summary report (Attachment Number 7).

ABC, Inc.				
Process Qualification Protocol				
Protocol Number ABC-001	Title: Isocarboxazid Process Qualification	Revision Number 0	Date:	Page 9 of 10

Any changes required, subsequent to validation team member approval, must be documented and justified in a Protocol Addendum and reapproved by the same validation team members or their designees. All required documentation (approved protocol, completed data sheets, and copies of laboratory reports) will be filed in the validation project file.

Upon completion of the protocol execution, a summary report will be prepared, which documents the validation efforts. The summary report will include information and data from the installation qualification (IQ) and the operational qualification (OQ).

The summary report will be reviewed and approved by the validation team as well as ABC, Inc. management. Approval will be documented on the summary report approval page (Attachment Number 6) of this protocol.

ABC, Inc.				
Process Qualification Protocol				
Protocol Number ABC-001	Title: Isocarboxazid Process Qualification	Revision Number 0	Date:	Page 10 of 10

VIII. MODIFICATION/CHANGE CONTROL AND REVALIDATION

Any modification or changes to the system will be documented in compliance with modification/change control of validated systems (SOP, ABC CC 012).

IX. INDEX OF ATTACHMENTS

1) Prequalification Checklist
2) SOP Review
3) Raw Material List
4) Monitoring of Key Process Parameters
5) Process Qualification Summary
6) Summary Report Approval Page

Attachment Number 1	PRE-QUALIFICATION CHECKLIST	Page 1 of 6
Protocol ABC-001	ISOCARBOXIZAD QUALIFICATION	REVISION NUMBER 0

DESCRIPTION	DATE	ACCEPTABLE (Y/N)
Development Batch(es) Completion		
Analytical Procedures for final testing an in process testing identified and approved		
Required equipment, facilities and maintenance documentation is approved and identified		
Quality Assurance procedures including deviations, change control, and final review are identified and approved		
Key process variable operating ranges have been identified and approved		
Relevant SOPs are identified and approved		
Raw materials and product specifications are identified and approved		
At least three approved batches are involved on the process qualification		

Completed By_____ Date_____

Reviewed By_____ Date_____

Attachment Number 2	STANDARD OPERATING PROCEDURES (SOPs)	Page 2 of 6
Protocol ABC-001	ISOCARBOXIZAD QUALIFICATION	REVISION NUMBER 0

SOP TITLE	SOP Number	Revision 0	Date	Verified By
Isocarboxizad, pure	N/A			
Change Control of Validated Systems	EQ012			
Preparation and Approval of Permanent Product/ Process Changes	MFO16			
Completing Batch Records	Q013			
Gowning Procedure	EM016			
Instrument/Equipment Calibration	EMF001			
Operation, Revision and Distribution	MFQ001			
Standard Production Procedures	M001			
Revision of Standard Production Procedures	EMF002			
Evaluation and Documentation of Production Variations	EMQQ007			
Assignment of Lot Numbers	QMF014			
Training of API Personnel	MFTQ019			
Labeling	QP021			
Purchase/Release of Raw Materials	MQC001			
Processing of Samples and Lots	QCMF002			
Assignment of Validation Code #	VMFQ001			

Prepared By _____   Date_____

Reviewed By _____   Date_____

Attachment Number 3	RAW MATERIALS LIST	Page 3 of 6
Protocol ABC-001	ISOCARBOXIZAD QUALIFICATION	REVISION NUMBER 0

Raw Material/Code Number	Supplier/Manufacturer	Released (Y/N)
Isocarboxizad		
Isopropanol		
Carbon		
Filtration Media		

Document any discrepancies in the Process Qualification Summary _____

_____

_____

_____

_____

Prepared By_____ Date_____

Reviewed By_____ Date _____

Attachment Number 4	MONITORING OF KEY PROCESS PARAMETERS	Page 4 of 6
Protocol ABC-001	ISOCARBOXIZAD QUALIFICATION	REVISION NUMBER 0

Key Processing Parameters	Validation Batch	Batch/Lot	
SOP Operation/Step	Specified	Observed	Acceptable (Y/N)
Drying Temperature (SOP QCMF002)	Dry to a maximum temperature of 35± 5° C*		
Analytical Results for isolated isocarboxizad (SOP QCMF002)	≤ 0.8% Isoxazole ester*		
Melting Point (MP) of isolated isocarboxizad (SOP QCMF002)	100-103 °C*		

*The values recorded are not actual specification criteria but are typical values

Document any discrepancies in the Process Qualification Summary Re port_____

_____

_____

Performed By_____ Date_____

Reviewed By_____ _____ Date_____

Approved By_____ Date_____

Attachment Number 5	Process Qualification Summary	Page 5 of 6
Protocol ABC-001	ISOCARBOXIZAD QUALIFICATION	REVISION NUMBER 0

Reference:_____

Discrepancy/ Variation_____
_____
_____   _____

Resolution_____
_____
_____   _____

Satisfactorily Completed? (Y/N)_____ _____   _____
                                          Signature   Date

Reference:_____

Discrepancy/ Variation_____
_____   _____
_____

Resolution_____
_____   _____
_____

Satisfactorily completed ? (Y/N)_____ _____   _____
                                          Signature   Date

Reference:_____

Discrepancy/ Variation_____
_____   _____
_____

Resolution_____
_____
_____ _____

Satisfactorily completed? (Y/N)_____ _____   _____
                                          Signature   Date

Reviewed By:_____   Date: _____

Attachment Number 6	SUMMARY REPORT APPROVAL PAGE	Page 6 of 6
Protocol ABC-001	ISOCARBOXIZAD QUALIFICATION	REVISION NUMBER 0

VALIDATION DATA APPROVAL

_____   _____
Validation Team   Date

_____   _____
Manufacturing   Date

_____   _____
Maintenance   Date

_____   _____
Facility Engineering   Date

_____   _____
Quality Assurance   Date

_____   _____
Quality Control   Date

_____   _____
Management Approval   Date

ABC, Inc. SUMMARY REPORT		
Protocol ABC-001	Revision Number 0	Page 1 of 6
Isocarboxizad Process		

TABLE OF CONTENTS

Page

ABC, Inc. SUMMARY REPORT		
Protocol ABC-001	Revision Number 0	Page 1 of 6
Isocarboxizad Process		

I. SUMMARY

Three batches of isocarboxazid, 0031, 0041, and 0051, were produced in (month, day, and year). Protocol Number ABC-001 was followed. The validation batches met all specifications.

Protocol ABC-001 is one part of a project validating the purification of the isocarboxazid process, as described in the isocarboxazid validation master plan, Revision Number 0. Based on the acceptable execution of Protocol ABC-001, the isocarboxazid's final purification is considered to be successfully validated.

II. OBJECTIVE

The objective of this report is to summarize and document execution of protocol ABC-001, Isocarboxazid Process.

ABC, Inc. SUMMARY REPORT		
Protocol ABC-001	Revision Number 0	Page 2 of 6
Isocarboxizad Process		

III. DESCRIPTION

The process qualification is performed at ABC, Inc., facilities at (insert, address, city, state, and country, if not the United States of America) as follows:

- Isocarboxazid is dissolved in isopropyl alcohol (isopropanol) heated to boiling point and filtered.
- Subsequent to cooling, the product is further filtered and given a final wash with isopropanol. The Isocarboxazid is dried at $35 \pm 5\,°C$ ($95 \pm 9\,°F$).

ABC, Inc. SUMMARY REPORT		
Protocol ABC-001	Revision Number 0	Page 3 of 6
	Isocarboxizad Process	

IV. RESULTS

Shown below are the results of the process qualification execution:

A. <u>Prequalification Requirements (Attachment Number 1)</u>

Completion of development batch 159 using the approved SOPs pre qualified the equipment. There were no significant discrepancies or variations observed.

B. <u>Standard Operating Procedures (SOPs) List (Attachment Number 2)</u>

All applicable SOPs were reviewed and listed in Attachment Number 2.

C. <u>Raw Materials (Attachment Number 3)</u>

Raw materials used to manufacture validation batches are listed on the batch records and in Attachment Number 3 and were released prior to use.

The isocarboxazid recrystallization batches were produced from validated and regular production batches as shown in the following table.

ABC, Inc. SUMMARY REPORT		
Protocol ABC-001	Revision Number 0	Page 4 of 6
Isocarboxizad Process		

Isocarboxizad Batch Number (Development Batch)	Isocarboxizad Recrystallization Batch Number (Validation Batch)
159 259 359	0031
259 459 559	0041
D 559 659R	0051

D. Monitoring of Key Process Parameters (Attachment Number 4)

All critical operating parameters were monitored and documented as specified in the protocol. The critical parameters were maintained as specified in the protocol. The critical parameters were maintained within the specified range during the processing of the validation batches. The results of the operations are shown in the following table.

SOP Operational Step	Specified Results (Validation Batch)[a]	Batch 0031 Recorded Value[a]	Batch 0041 Recorded Value[a]	Batch 0051 Recorded Value[a]
Heat Isopropanol	50°C (122°F) ±3°C (5.4°F)	50°C (122°F)	50°C (122°F)	50–52°C (122–126°F)
Drying Temperature	35±5°C (95±9°F)	35±5°C (95±9°F)	35±5°C (95±9°F)	34.5°C (94.1°F)
Impurities Analysis	Less than 0.5% impurities	<0.5%	<0.5%	<0.5%

[a] The values shown are for example purposes only. These values are not representative of actual physical properties of the products.

ABC, Inc.		
	SUMMARY REPORT	
Protocol ABC-001	Revision Number 0	Page 5 of 6
	Isocarboxizad Process	

E. Process Qualification Summary (Attachment Number 5)

There were no significant discrepancies or variations observed during the isocarboxazid process validation batches.

Development Batch 659R is a rework of Batch 659. Batch 659 required a rework because the product quality was lower than previous development batches. The lower quality of the original batch is attributed to:

1) Failure to employ an inert gas blanket during the cooling phase.
2) Improper cleaning of the reactor.

The rework of Batch 659 produced an acceptable batch that, when reworked, was reidentified as Batch 659R.

The manufacturing and documentation changes were performed in accordance with the procedures defined in Protocol ABC-001, Attachment Number 2, SOP MF016.

All of the key processing performed within the specified ranges and all results for the validation batches were acceptable.

ABC, Inc. SUMMARY REPORT		
Protocol ABC-001	Revision Number 0	Page 6 of 6
Isocarboxizad Process		

V. CONCLUSIONS

The process qualification protocol, ABC-001, isocarboxazid process, has been found to be acceptable. The acceptable execution of the protocol has demonstrated that the process will perform as planned. Based on the demonstrated results, the isocarboxazid process is considered validated. ← **Solution**

It may be worth noting the following comments concerning a process validation:

- The content of this protocol appears to be repetitious. It is important to note several elements are responsible for implementing and approving a protocol such as this (e.g. Quality Control, Quality Assurance, etc). It would be needlessly time consuming for every involved facility operation to review a protocol since many of the elements are not germane to all operations. Typically, quality control is only concerned with results and procedures dealing with monitoring, raw materials, and analysis; the technology of the process is, for the most part, of passing relevance. Similarly, quality is not deeply involved in calibration, whereas maintenance and manufacturing operations are responsible for items such as instrument calibration status.
- There is usually a timeline for completion of validation operations. This requires a great deal of information to be processed in a finite time period. This assumes relevant information can be found in the protocol with a minimum of lost time. Having relevant information handily available in several locations within the protocol can be valuable when protocol evaluation and approval times are at a minimum. For operations involved in a process validation that are generally not on overhead (e.g. manufacturing), this should be an important aspect time management for process validation evaluation.

10.5   Electronic Records and Electronic Signatures

21 CFR, Part 11 is a regulation specific to electronic signatures and electronic records. In recent years, the emphasis on automated record keeping and tracking of data has become an important part of pharmaceutical operations. Systems such as TrackWise® and SAP® software have been employed to ease the record keeping tasks for many operations. As these types of software gain importance as facility monitoring operations, the provisions of 21 CFR, Part 11 take on a greater importance.

Integral to electronic records is the implementation of microprocessor-based devices currently being employed in manufacturing, laboratory operations, inspection procedures, and environmental monitoring such as temperature control, humidity, and electronic storing of the data collected. While an understanding of computer hardware and software is important, it is often not enough background when dealing with microprocessor-based equipment. In this situation, successful validation requires an in-depth knowledge of the equipment of which the microprocessor is a component. Consequently, Part 11 validation procedures can be more complex than standard equipment validation. While a Part 11 validation can require more tasking, it is not uncommon for owner and vendor organizations to make an electronic records/ electronic signature system validation more complex than required. Thus, by understanding the key contents of 21 CFR Part 11 and the methods that could simplify the task, significant time and, in many cases, cost savings can be realized. The discussion below is intended to provide basic considerations intended to minimize time and costs. The criteria identified are intended to be tested and validated in the IQ, OQ, and PQ format as previously detailed.

Prior to initiating a Part 11 type of validation, a preliminary study should be performed to determine if the installed electronic system requires validation. The criteria for a computer system validation need are predicated on a cGMP applicability determination (it should be noted that many computerized records management and control systems do not necessarily require validation because they are installed in a pharmaceutical facility).

By evaluating some important elements used to determine if a computer system is regulated by cGMP, it is possible to perform an evaluation before committing time and assets to a system that may not require qualification activity. Table 10.20 is a checklist that will, for most systems, indicate if the installed computer is a cGMP system. As the table indicates, a "Yes" response to any of the items indicates the probable need for validation. In general, if the computer system directly or indirectly affects the safety, identity, strength, quality, purity, or efficacy of a product, it is very likely a cGMP system exists and validation is required.

While the aforementioned table is very useful in determining the status of an electronic system regarding the need for validation, it would be very useful to apply a more quantitative evaluation technique that allows for a more definitive result. That is, it would be very valuable to employ a numerical scale to determine the value or criticality of a computer system. While such a technique would be very useful, this type of evaluation is best performed through the use of knowledgeable personnel with extensive experience in hardware and software applications, particularly for the cGMP Part 11 system under consideration. So, when evaluating a system based on Part 11 standards, it is most important that a consensus exists among participants or selecting numerical values.

Since the risk analysis is not as comprehensive as the technique identified for performing a formal failure mode, effects, and analysis (FMEA) described in Chapter 5, it is, nevertheless, an effective measure of risk and associated system criticality.

Table 10.21 presents a template for evaluating the risks of a Part 11 (cGMP) computer system.

With numerical values now identified for the relative risk for a Part 11 cGMP computer system, it is now possible to determine the level of risk, or criticality, presented by the computer system of concern. In performing the assessment, the numerical values should identify the minimal criticality of a computerized voiced mail system or a payroll operation when compared with a system that records and documents production batch records. When the risk evaluation is completed in accordance with the numerical factors identified in Table 10.21, it is then possible to perform a criticality assessment on the designated computer system. The criticality rating is a measure of the impact a failure of the kind identified in Table 10.21 will have on the computer system. Table 10.22 describes a common method employed to determine a criticality assessment rating.

Once values are determined from Table 10.21, the relationship shown below will be employed to obtain a criticality assessment rating (low, medium, high).

Multiply assessment rating number obtained from Table 10.21, by category, and by weighting factor and then enter the product in the total column. Then record the sum in the score row.

Upon completion of the scoring phase of Table 10.22, the system criticality rating can now be employed by using the information in Table 10.23.

Table 10.20 Computer system validation evaluation.

Preliminary cGMP Evaluation		
Item	☐ No	☐ Yes → cGMP System
1. Does the system control the manufacturing process?	☐ No	☐ Yes → cGMP System
2. Does the system process product release or stability information?	☐ No	☐ Yes → cGMP System
3. Does the system process required regulatory information?	☐ No	☐ Yes → cGMP System
4. Does the system support cGMP activity?	☐ No	☐ Yes → cGMP System
Preliminary cGMP Manufacturing Evaluation		
5. Does the computer system perform functions or calculations that are used to affect the bill of materials for a component or pharmaceutical product?	☐ No	☐ Yes → cGMP System
6. Does the computer system perform functions or calculations that affect any critical parameters of product manufacturing such as temperature, low rate, and reaction time?	☐ No	☐ Yes → cGMP System
7. Does the computer system generate and/or store alarms, events, or transactions that are generated during the manufacturing or packaging operation?	☐ No	☐ Yes → cGMP System
8. Does the computer system store data concerning actual manufacturing or packaging variables for a given component or a product (e.g. low material weight, blending time, process temperature and pressure, and process equipment)?	☐ No	☐ Yes → cGMP System
9. Does the computer system perform manufacturing process control?	☐ No	☐ Yes → cGMP System
10. Does the computer system generate labels packaging components or identify a drug product or API?	☐ No	☐ Yes → cGMP System
Product Release: cGMP Considerations		
11. Does the computer system control or generate data from equipment, instrumentation, or identification used in testing release products or samples	☐ No	☐ Yes → cGMP System
12. Does the computer system perform functions or calculations that have an impact in the expiration dating of a component or drug or finished product? (A finished product includes components of the product.)	☐ No	☐ Yes → cGMP System
13. Does the computer system collect data that impacts calculations employed to determine quality variables affecting the finished product?	☐ No	☐ Yes → cGMP System
14. Does the computer system store data used to determine the stability of a finished product	☐ No	☐ Yes → cGMP System
15. Does the computer system store data relating to the release status of a finished product?	☐ No	☐ Yes → cGMP System
16. Does the computer system store data identifying the location of a finished product or component?	☐ No	☐ Yes → cGMP System
17. Does the computer system store environmental conditions (e.g. temperature, humidity) for the manufacturing and storage of the finished product?	☐ No	☐ Yes → cGMP System

Table 10.20 (Continued)

Preliminary cGMP Evaluation	

Preliminary cGMP Regulatory Assessment

18. Does the computer system manufacturing equipment data over time or for single batches?   ☐ No   ☐ Yes → cGMP System

19. Does the computer system generate and/or equipment data?   ☐ No   ☐ Yes → cGMP System

20. Does the computer system collect or store data involving cleaning or sterilization of manufacturing equipment?   ☐ No   ☐ Yes → cGMP System

21. Does the computer system create, store, or maintain approved procedures (e.g. SOPs, directives, and batch records) for the manufacture of a finished product?   ☐ No   ☐ Yes → cGMP System

22. Does the computer system perform calculations that will support a regulatory submission or registration?   ☐ No   ☐ Yes → cGMP System

23. Does the computer system have impact on regulatory submissions?   ☐ No   ☐ Yes → cGMP System

24. Does the computer system record lot or batch data for specific equipment or procedures?   ☐ No   ☐ Yes → cGMP System

25. Does the computer system store corrective and preventive actions (CAPA) investigations for cGMP equipment and/or products?   ☐ No   ☐ Yes → cGMP System

26. Is any output of the computer system subject to audit or archived?   ☐ No   ☐ Yes → cGMP System

27. Does the computer system generate data that will interface with a regulated system?   ☐ No   ☐ Yes → cGMP System

Miscellaneous cGMP Activity Evaluation

28. Does the computer system store, generate, control, record, or monitor laboratory or clinical data?   ☐ No   ☐ Yes → cGMP System

29. Does the computer system generate process or approve computer maintenance management system (e.g. MAXIMO) work orders involving cGMP equipment or activities?   ☐ No   ☐ Yes → cGMP System

30. Does the computer store or generate cGMP-related training records?   ☐ No   ☐ Yes → cGMP System

31. Does the computer system interfaces or support networks or servers that process cGMP data?   ☐ No   ☐ Yes → cGMP System

Table 10.21 Part 11 electronic records: Electronic signatures risk evaluation.

Risk Identification	Numerical Value	System Numerical Rating
1. Quality impact (select one numerical value)		
Direct impact on product quality and/or safety	6	
Indirect impact on quality/and or safety	3	
No impact on product quality and/or safety	0	
2. Regulatory risk (select one numerical value)		
System supports data required by FDA as per regulations	6	Selected value=
System could be audited by FDA	3	Selected value=
Minimum risk of FDA audit	1	Selected value=
3. Data impact risk (select all that apply)		
Regulatory and business impact	6	Selected value=
Financial and Business impact	4	Selected value=
No impact	0	Selected value=
4. cGMP system failure will affect regulatory criteria		
Nonmanufacturing system, if inactivated, will create a regulatory compliance issue if backup or manual mode not available	3	Selected value=
Productivity affected but regulatory or compliance risk is minimal	2	Selected value=
No impact	0	Selected value=
5. Software maturity (select one numerical value)		
Software is new	4	Selected value=
Software in production less than 2 years	2	Selected value=
Software in production more than 2 years	1	Selected value=

Table 10.22 Criticality assessment rating.

Risk Assessment Category	System Numerical Rating (Table 10.21)	Weighting Factor	Total
		5	
		5	
		3	
		2	
		1	
SCORE =			

Table 10.23 System criticality rating.

Criticality	Score
Low	6–50
Medium	51–75
High	76–100

Example 10.4   A pharmaceutical firm has installed a widely used 10-year-old software system intended to record, track, and document approvals and, via electronic signatures, facility SOPs. The IT operation is the manager of the software and is responsible for effective administration. The system has been successfully validated and it was determined a criticality evaluation is needed. In addition to the IT operations (composed of the IT manager, IT maintenance, and an on-site software support vendor who installed the system), the assembled team included personnel from quality assurance, validation, facility training, and records management/document control personnel. The team has determined the following numerical values for the system risk determination (Table 10.21):

1. Quality impact	Numerical value = 3
2. Regulatory risk	Numerical value = 6
3. Data impact risk	Numerical value = 6
4. cGMP failure will affect regulatory criteria	Numerical value = 3
5. Software maturity	Numerical value = 1

Solution
Using the selected values from Table 10.21 and multiplying by the weighting factors defined, the results are recorded in the total and score column to yield:

Risk Assessment Category	System Numerical Rating (Table 10.21)	Weighting Factor	Total
	3	5	15
	6	5	30
	6	3	18
	3	2	6
	1	1	1
SCORE =			70

Referring to Table 10.23, the score for the criticality rating is 70. This score, 70, indicates the system is a Medium Risk ← **ANSWER.**

21 CFR 11 subpart B contains the relevant validation issues that must be addressed when dealing with this type of validation. The format for closed systems (i.e. a system operating in an environment controlled by persons who are responsible for the content of electronic records that are in the system) is the type most often used for validation since these systems are typically cGMP regulated. The essential elements that must be tested for a comprehensive Part 11 protocol are:

1) Validation of systems to ensure accuracy, reliability, consistent intended performance, and the ability to discern invalid or altered records. This statement defines the need for system validation.
2) The ability to generate accurate and complete copies of records in both human readable and electronic form suitable for inspection, review, and copying by the agency (i.e. FDA).
3) Protection of records to enable their accurate and ready retrieval throughout the record retention period. 21 CFR 211.180 identifies a minimum record retention period of one year after the product expiration date and longer for other products as specified in 21 CFR 211.137.

4) Limiting system access to authorized individuals. In effect, this statement defines the need to have a secure system with limited access.

5) Use of secure, computer-generated, time-stamped audit trails to independently record to date and time of operator entries and actions that create, modify, or delete electronic records. Record changes shall not obscure previously recorded information. Such audit trail documentation shall be retained for as long as that required for the subject of electronic records and shall be available for agency review and copying.

6) Use of operational system checks to enforce permitted sequencing of steps and events, as appropriate. Essentially, this statement reinforces the need for system protection.

7) Use of authority checks to ensure that only authorized individuals can use the system, electronically sign a record, access the operation or computer system input or output device, alter a record, or perform the operation at hand. This statement confirms the need to verify system use by individuals.

8) Use of device (e.g. terminal) checks to determine as appropriate, the validity of the source of data input or operational instruction. Data input and data collection techniques must be verified.

9) Determination that persons who develop, maintain, or use electronic record/electronic signature systems has the education, training, and experience to perform their assigned tasks. In this instance, it is required to have trained users with documented training records.

10) The establishment of, and adherence to, written policies that hold individuals accountable and responsible for actions initiated under their electronic signatures, in order to deter record and signature falsification. This item notes approved policies (i.e. SOPs) must be employed for system use.

11) Use of appropriate controls over systems documentation including:
   - Adequate controls over the distribution of, access to, and use of documentation for system O&M.
   - Revision and change control procedures to maintain an audit trail that documents time-sequenced development and modification of systems documentation. Here, the importance of approved change control procedures is emphasized.

When dealing with equipment and/or system validations, it is very important to employ reputable firms with a proven record of successful validation activities. The same comments apply when dealing with electronic signature/records qualifications. Generally, the FDA does not have sufficient personnel to carefully evaluate every Part 11 validation that is prepared. Thus, to a large degree, companies are assumed to perform comprehensive validation programs using "best practice" for all qualifications. In addition to consultants that may be involved in Part 11 validations, it is equally important that owner and staff representatives maintain the highest level integrity when preparing, implementing, and evaluating these protocols. Also, it is important that careful planning and understanding of the requirements for a Part 11 protocols be employed. A longer time frame to comprehensively detail the planned qualification is more desirable than a shorter less detailed plan. Again, it is most important that reputable personnel are selected for the task to confirm that the owner is purchasing a cost-effective, comprehensive, and compliant product. To achieve these goals, critical details such as careful analysis of recording instruments and devices are properly selected and, as is the case with cGMP equipment, currently validated.

For electronic systems, it is also desirable to confirm, through detailed analysis, that cost-effective and non-overly complex data collection devices are adequately evaluated and installed. If feasible, less complex data collection tools may be preferable due to lower cost and higher reliability.

10.5.1 Application of Risk Assessment Methods to Outsourcing

It is becoming more common to outsource validation operations while commissioning has most often been the responsibility of the construction engineering contractor. In fact, the trend is to outsource entire IT operations of pharmaceutical and biopharmaceutical firms, and within a few years much of this work will be outsourced. As validation outsourcing becomes more common, particularly when dealing with IT functions, a risk assessment format would be advisable for evaluating contractor capabilities and performance. Preparing a risk assessment format for validation contractors can best be achieved when prepared by owner staff personnel. An in-house effort can often obviate unintended bias that can be included in contractor prepared assessments.

There are several factors that should be evaluated when considering preparing an outsourced validation protocol. A very important consideration is timing. It is desirable to have a risk assessment plan in place prior to contractor implementation of the validation operations. Both the contractor and the user should have common agreement on the scope of the validation. Both the user and the consultant must understand the scope of the task regarding what work is to be completed and what efforts are not parts of the assignment. A complete and detailed description of the statement of work should be incorporated in the contract deliverables along with a project timeline. By employing a team approach composed of involved stakeholders, identification of all germane risks is conceivable.

Special attention should be given to elements that are scored as high risk factors. Optimally, the contractor should be apprised of the high risk elements, via the completed risk assessment, of the project. This is most important to maintain cost control and project harmony.

It is most important that contractor/stakeholder communications be clearly delineated and recognized as a risk factor. There should be no margin for error because of miscommunication due to erroneous assumptions by the contractor or misinterpretation of the contractor's actual work scope. Also, it is imperative that a clear defined communications chain be incorporated in the assessment. This risk element should include personnel responsible for specific tasks as well as a clear description of each individual's specific task. Personnel contact information should also be clearly identified.

While owner participation is a very important factor, it should be realized that the value of outsourcing is the "freeing up" of user staff and assets. As such, the importance of a risk assessment matrix cannot be overemphasized. It is expected that, as a minimum, the contractor should have a standard risk assessment format in place when addressing oversight operations. The contractor oversight risk assessment procedures should include identified risks associated with the work scope as well as a quantitative evaluation system. Identification of this aspect of the outsourced task should be included as an evaluation factor for the user's risk assessment.

As is the case with any outsourced task, performance monitoring is very relevant. Here, scheduled contractor and owner staff should establish a schedule for performance monitoring as well as employing the responsibility matrix to address unplanned performance monitoring related events.

To assure a comprehensive, effective, and successful program, these procedures, at a minimum, should be considered when considering outside contracting assistance. Above all, it is most important to note that regardless of the product produced by a contractor or consultant, the ultimate responsibility for the out sourced task is the owner.

10.5.2 Validation Costs

Cost estimates for a validation procedure are not as clearly defined, as is the case with estimating commissioning cost estimates. While some factors can be used to approximate a validation expense, actual costs are often difficult to ascertain due to a lack of incorporation of overhead costs with actual project costs. For instance, qualification protocols must be approved by the quality assurance (QA) operation prior to implementation. In actuality, many QA operations are devoid of the type of technical personnel that can actually investigate the content of a protocol. Personnel not familiar with the technical aspects of complex systems such as motor control centers, tablet coaters, or tableting operations often have little recourse but to focus on non-germane factors such as "dotting the i and crossing the t." Continuous changes in superfluous validation protocol details often challenge schedules and manpower allocation. Since most QA operations involved in protocols are charged as overhead, rather than a direct project expense, it is most important technical personnel involved in protocols clearly transmit the intent and technical details to QA personnel involved in protocol approval. Similarly, a QA operation with a knowledgeable engineering presence can significantly improve protocol efficiency.

Another element that significantly affects validation costs is the team selected to perform the protocol preparation. An increasingly common protocol preparation technique is to have the manufacturer prepare the protocol (i.e. IQ, OQ, and PQ) with the owner's representative performing the actual testing. For complex systems, this is a viable option, and the protocol preparation is typically a small fraction of the total equipment and installation cost and generally included in the total project cost, if a validation plan is included.

Employing a consulting or engineering firm for protocol preparation is also a technique used to prepare validation protocols. While this is a common method, care is advised when using a consultant or engineering operation. For certain protocols, this is a viable option. For utilities such as electrical installations, water, and HVAC, this may be an acceptable qualification mode, but when, as identified above, the protocol involves a detailed equipment installation, a detailed evaluation may be required. Even if the same engineering firm selected to install the equipment to be validated, their knowledge may be specific to the installation procedures and not necessarily the details of the equipment operation. Additionally, it is important to verify that individuals assigned to the program are actually the staff performing the task. For this reason, it would be valuable to review qualifications, including resumes, of all personnel and not just senior project personnel. Also, labor expenses and direct and overhead costs should be closely and continuously monitored by the project cost center. This may prove to be valuable in preventing unplanned for cost overruns.

A typical validation protocol consisting of an IQ, OQ, and PQ typically involves two or three people, a senior validation manager and one or two associates. For smaller programs such as a chart recorder that records temperatures during a manufacturing operation, the time frame from start of protocol preparation to approval of the complete protocol should be between 4 and 6 weeks. This would be between 120 and 180 man-hours. With a margin of error, 200 man-hours of protocol preparation, testing, and follow-up is a reasonable expectation. Of course, it must be recalled that some operations such, as QA, are generally overhead expenses and not chargeable as direct costs.

A typical validation operation would normally compose 2–4% of the installed cost of the equipment or facility, depending on the total budget.

For large programs employing several phases involving requirements (including user requirements specification), design, procurement and construction, and a validation phase, it is not uncommon for the program to require 24 months of individual effort. For such large projects, the validation could be about 6 months in length.

Table 10.24 Comparison Between Commissioning Plans and Validation Protocols.

Commissioning	Validation
• Supplier responsibility until system handover	• User responsibility
• Objective is to identify and rectify problems	• Demonstrate the process operates as specified and under control
• Approved protocol is not required; likely to follow "accepted practice/standards"	• Must follow an approved protocol
• Owned and operated by supplier until handover	• Owned and operated by user
• Not all data and adjustments are recorded and reviewed	• All data and adjustments are recorded, reviewed, and corrected
• No written report unless specified	• Written report is required
• Reviewed for acceptance by engineering/project team	• Reviewed and approved by quality assurance

10.6 Comparison Between Commissioning and Validation

Table 10.24 briefly compares the commissioning process and the validation process.

References

1 Muchemu, D. (2011). Principles of cleanroom validation. *Controlled Environments* 14 (10): 13–15.
2 Hojo, T. (2004). Quality Management Systems – Process Validation Guidance, 2e. The Global Harmonization Task Force (GHTF), Study Group #3 (January 2004).
3 Buzzone, S. (2001). Process validation of solid oral dosage forms, Part I. *IKEV Meeting*, Turkish Pharmaceutical Society (31 May–1 June 2001).

Further Reading

American Institute of Chemical Engineers (1985). *Guidelines for Hazard Evaluation Procedures.* New York: American Institute of Chemical Engineers.
Blanchard, B.S. (1974). *Logistics Engineering and Management*, 3e. Englewood Cliffs, NJ: Prentice-Hall, Inc.
Fetterolf, D. (2007). Developing a sound process validation strategy. *BioPharm International* 20 (12): 38–46.
Flanagan, N. (2003). Bioprocess validation still a complex process. *Genetic Engineering News* 23 (2): 39, 40, 47, 55, 62.
Prager, G. (2004). *Equipment and Operational Qualification.* Clifton, NJ: Source Tech Design Inc.
Schofield, D. (2010). How to apply risk assessment techniques to outsourcing. *BioPharm International* 23 (7).
Shaw, A. (1993). *Guideline on General Principles of Process Validation.* Silver Spring, MD: U.S. FDA.
Signore, A. (2005). Integrating Commissioning and Validation on Pharmaceutical Manufacturing Facility Projects – Opportunities for Better Results. Integrated Project Services (16 June 2005).

Taylor, G.L. (2005). *Inspecting Commercial, Industrial, and Residential Construction.* New York: McGraw-Hill.

Thomas, P. (2010). The unintended consequences of software validation. *Pharmaceutical Manufacturing* 9 (1).

U.S. FDA (1994). *Guide to Inspections of Oral Solid Dosage Forms Pre/Post Approval Issues for Validation.* Silver Spring, MD: U.S. FDA.

U.S. FDA (1999). *Guide to Inspections of Quality Systems.* Silver Spring, MD: U.S. FDA.

U.S. FDA (2000). *21 CFR Part 210, Current Good Manufacturing Practice in Manufacturing, Processing, Packing or Holding of Drugs.* Silver Spring, MD: U.S. FDA.

U.S. FDA (2000). *21 CFR Part 211, Current Good Manufacturing Practice for Finished Pharmaceuticals.* Silver Spring, MD: U.S. FDA.

U.S. FDA (2011). *Guidance for Industry Process Validation: General Principles and Practices,* Revision 1. Silver Spring, MD: U.S. FDA.

Wong, W. (2011). FDA process validation. *Controlled Environments* 14 (5): 13–15.

11

Topics and Concepts Relating to Pharmaceutical Engineering

In an effort to provide additional product reliability for pharmaceutical and biomedical devices, the FDA has, in recent years, implemented several initiatives intended to achieve this goal (enhanced reliability, maintainability, and product safety). This chapter describes several programs currently being employed to enhance pharmaceutical and biotechnology operations. This chapter commences with basic statistical techniques and applications and culminates with tools and program elements designed to streamline and enhance manufacturing operations by applying the methods described. Several of these tools employ common statistical methods used in this chapter.

11.1 Preliminary Concepts

This section is intended to provide a background for dealing with material often employed in pharmaceutical engineering operations. While much of the material covered may be a review for many individuals, it nevertheless serves a basis for topics addressed in this chapter. The statistics and associated mathematical details covered are intended to provide users the basic tools to perform typical analyses relating to several elements addressed in this section and the remainder of the chapter. As such, the basic statistical and associated information will, hopefully, provide users with tools to sufficiently perform several tasks common to pharmaceutical engineering projects such as such as design of experiments (DOE), basic statistical process control, process capability, and other Six Sigma applications introduced in this chapter.

11.1.1 Basic Statistical Concepts and Computational Techniques

Simply defined, statistics deals with the collection, analysis, interpretation, and presentation of numerical data in a cogent presentation format. By presenting large arrays of data and reducing this numerical information to a single or smaller grouping, it is often possible to obtain valid and accurate results in a more efficient approach by minimizing the time required to achieve valid conclusions when dealing with a large amount of numerical data.

Practical Pharmaceutical Engineering, First Edition. Gary Prager.
© 2019 John Wiley & Sons, Inc. Published 2019 by John Wiley & Sons, Inc.

A rudimentary statistical technique is known as the median. Here, the median is merely the middle value of a set of numbers that are arranged by increasing the order of magnitude. For example, the median for the following set of numbers – 3, 5, 7, 7, 9, 10, 14, 16, 17, 17, and 21 – is 10.

In the case where the number set is an even amount of values, the median is determined by dividing the sum of the two middle values by 2. For instance, the median value for the following set of numbers – 2, 4, 8, 10, 12, and 12 – is $(8 + 10)/2 = 9$.

Often, the median is used in lieu of the mean or the average when seeking basic statistical trends. This approach is often reasonably valid if the data sets are closely grouped together and the population is large enough to be statistically valid (typically 10 values as a minimum). Additionally, it is important to consider if the sample group has several values that differ widely in value from other values in the set. Often, just one value in a sample of 10 can significantly alter the mean in the final result. Consequently, a bit more scrutiny should be exhibited when considering the median technique.

The mode is another technique that can be used for obtaining an elementary technique to describe a set of numbers with one value. The mode is merely identifying values from a group of numbers that occur with the greatest frequency, for instance, for the numbers 8, 2, 5, 9, 9, 10, and 15 having the value of 9 as the mode. If the number group has more than one value repeating in the numbers, the number group is then identified as a bimodal. An example would be that the number group 4, 7, 7, 8, 8, 8, 10, 13, and 17 is bimodal since the numbers 7 and 8 repeat as part of the number set.

Perhaps the most basic detailed computational statistical tool is the mean, also referred to as the average or arithmetic mean and commonly noted by the symbols $\bar{X}$ and mathematically defined as

$$\bar{X} = X_1 + X_2 + X_3 + X_4 + \cdots + X_n = \sum \frac{X_i}{N} \tag{11.1}$$

where

X_i is a given set of numbers $X_1, X_2, X_3, X_4, \ldots, X_n$
N is the quantity of numbers in the set.

While $\bar{X}$ is the standard symbol for the average or arithmetic mean, when dealing with more complex statistics, the Greek symbol μ is also denoted to identify the mean or average value.

Example 11.1 A pharmaceutical facility uses the municipal water system as the feed for its water purification operations. The facility is located near a large farming operation that uses quantities of nitrate fertilizer. The feedwater monitoring and analysis program specified by the facility quality control unit and approved by the state regulatory agency requires the average nitrate concentration during a 24 h period not to exceed 10 mg/kg (ppm) of nitrates. The sampling calls for a feedwater sample to be taken every 4 h. The readings can be rounded off to the nearest whole number.

Calculate the mean (average) for the following number set: 7, 4, 9, 13, 10, and 11 ppm.

Solution
Using Equation (11.1) and inserting the given values yields

$$\bar{x} = \frac{7\text{ppm} + 4\text{ppm} + 9\text{ppm} + 13\text{ppm} + 10\text{ppm} + 11\text{ppm}}{6} = \textbf{8.16} \approx \textbf{8 ppm} \leftarrow \textbf{Solution}$$

As the analytical results indicate, the feedwater concentration for nitrates conforms to the established feedwater standards.

The mean, or average, is often used in situations where a reasonable number is needed to quickly estimate a value that can be used to represent a typical value of a given group of values. A typical application would be obtaining a value for particle counts in a classified manufacturing area as a check to assure the expected particle counts are within the specification tolerances. The mean value is also employed in preventive maintenance operations where the mean time between failures (MTBF) and mean time to repair (MTTR) are important, details of which are located in Chapter 5.

A variation of the arithmetic mean is the root mean square (RMS), which is typically used in electric circuit applications and other physical events (e.g. calculating a gas velocity) related to pharmaceutical manufacturing. The RMS is given by

$$X_{RMS} = \sqrt{\frac{X_1^2 + X_2^2 + X_3^2 + \cdots + X_N^2}{(N)^{1/2}}} \tag{11.2}$$

where

X_{RMS} = root mean square
N = quantity of numbers in a given set
X_i = the given numbers in the set

Example 11.2 Calculate the RMS and the mean of the following set of numbers and the average of the numbers: 2, 4, 6, 8, and 10.

Solution
A. The RMS
Inserting the given number set in Equation (11.2) yields

$$X_{RMS} = \sqrt{\frac{\left(2^2 + 4^2 + 6^2 + 8^2 + 10^2\right)}{5}} = 6.63 \leftarrow \textbf{Solution}$$

B. The mean

$$\overline{\overline{X}} = \frac{2 + 4 + 6 + 8 + 10}{5} = 6 \leftarrow \textbf{Solution}$$

Another method in which a data set can be represented as a single value is the geometric mean, G. The geometric mean, G, is composed of a group of positive numbers (X_1, X_2, X_3, X_4, ..., X_N). Here, X_N is the Nth root of the total product of the group of the positive numbers specified. Mathematically, the geometric mean is given as

$$G = \sqrt[N]{(X_1)(X_2)(X_3)(X_N)} \tag{11.3}$$

Example 11.3 Determine the geometric mean of the number set given in Example 11.2, which is 2, 4, 6, 8, and 10.

Solution
Inserting the given values in Equation (11.3) yields

$$G = \sqrt[5]{(2)(4)(6)(8)(10)} = 5.21 \leftarrow \textbf{Solution}$$

It appears the calculated geometric mean is similar to the root mean square obtained using the same values obtained in Example 11.2 (6.63) and lower than the average value or mean value of 6 as calculated in Example 11.2, B.

Since the root mean square is not often used in general pharmaceutical applications, it would be worthwhile to be aware of this method for applications described previously. However, for data interpretation requiring a quick generally accurate value, the mean (average) is the value most often employed.

The mean is a value that is independent of values used in the calculation. The geometric mean, on the other hand, is generally used for evaluating data that is a function of the values selected. For instance, if it was desired to calculate average values for fungus growth in a fermentor (Chapter 4), the individual values are dependent on the previous readings and are not independent of the values in the data group for calculating the mean (average) since biological growth occurs in an exponential growth mode. The calculation of values for the average is singular (i.e. finding the average of a number of tablets produced in a tablet press or the number of labels affixed to bottles in a given time period) and not dependent on previous tablet counts or number of labels processed. For this reason, the geometric mean is not used as a technique for representing a group of values as a single value. And, as noted, the basic statistical tool employed to represent a group of independent values is the mean or average.

While the mean can serve as a method to describe a collection of numerical values in the form of a single term, there is a shortcoming in the calculated mean. The average does not identify the dispersion of the data set. That is, the mean does not indicate how closely the calculated values actually represent the mean. Two values, 46 and 100, have an average of $(46 + 100)/2$, yielding a mean of 73. Similarly, two values, 2 and 144, also have an average value of $(2 + 144)/2 = 73$. While the calculated mean is the same in both cases, the data sets selected are quite different from one another. Consequently, it would be very useful if a relatively simple and quick statistical technique was available to accurately represent numerical data by taking into account the range of data groups as well as an accurate representative or sample value. Such a straightforward technique has been developed and is referred to as the standard deviation.

Before delving into the details of the sample and population standard deviation, it might be best to present the details of the calculation and the resulting mathematical equation. For pharmaceutical engineering purposes, the emphasis herein is on determining values for a sample, as opposed to a total population. For instance, if it were desired to ascertain if increasing the pressure of a tablet press by using additional force (kilo pounds) would result in a significantly greater dissolution time, a sample of the manufacturing batch would be selected for testing rather than the entire quantity of tablets produced (i.e. the population). Hence, a random number of tablets would be selected for testing and subsequently used as the basis for the sample standard deviation, s, and analysis.

Calculating the standard deviation for a sample involves performing of five calculations:

1) Compute the mean of samples, X, in this case the dissolution time at the modified tablet press force.
2) Compute the deviation. This step involves subtraction the mean from each value or $X - \bar{X}$.
3) Square each deviation or $(X - \bar{X})^2$.
4) Sum the squared deviations, $\sum_{i=1}^{N}(X - \bar{X})^2$.
5) Divide the result of step 4 by 1 less than the sample size $(N - 1)$.

Calculate the square root of the values obtained from steps 1 to 5 to obtain the term identified as the sample standard deviation, which is given by

$$s = \frac{\left(\sum_{i=1}^{N}(X-\bar{X})^2\right)^{1/2}}{(N-1)^{1/2}} \tag{11.4}$$

The sample deviation, s, differs from the population standard deviation, denoted by the Greek symbol σ, because the population standard deviation denominator, σ, is represented by N in the denominator and, as noted, s, for the sample deviation with an $N-1$ in the denominator. The primary reason the $N-1$ term is used when determining the sample deviation is that with a smaller sample, it is possible a bias could exist. By using $N-1$ in the denominator, any bias should be eliminated. As the sample denominator gets larger, $N-1$ approaches N, thereby rendering the 1 less and less significant. Consequently, where $N \geq 30$, the $N-1$ value is virtually identical to the N value (a difference of 3.3%).

Although this chapter is primarily concerned with the sample standard deviation, s, it would be worthwhile to also define the population standard deviation, σ, to avoid possible confusion:

$$\sigma = \frac{\left(\sum_{i=1}^{N}(X-\bar{X})^2\right)^{1/2}}{(N)^{1/2}} \tag{11.5}$$

A function that will be used as this section proceeds is the variance. The variance is the square of the population deviation and given by σ^2. Just as the term implies, the variance is merely the square of the population deviation (here the variance is the population standard deviation without a square root as the final calculation step).

Equation (11.4) is, as noted, the equation for determining the standard deviation for a sample selected from a total population. For instance, if, over the course of a year, a pharmaceutical operation manufacturing continuously (24 h and 7 days a week) sampled the incoming municipal water quality twice a day, a total of 730 samples would have been collected and analyzed. To obtain an accurate profile of the incoming water quality, the standard and the mean are employed for a statistical value that is valid. Once the average value $(\bar{X})$ has been determined, a random number of samples are selected from the 730 values recorded. Typically, a satisfactory number of values would be based on the square root of the population plus one and rounded to the nearest whole number. This calculation would be $(740)^{1/2} + 1 = 28$. Thus, selecting 28 random values from the population of 730 analyses would be the sample size. The 28 values would next be inserted in Equation (11.1) to obtain the mean. Subsequent to calculating the mean, the standard deviation would then be determined by employing Equation (11.4).

Actually, determining the average and the standard deviation is, today, a relatively straightforward operation allowing for a quick solution without actually performing detailed calculations. Currently, there are many online programs that provide nearly instantaneous solutions once the input values have been entered. Most handheld calculators also provide this calculating option by merely entering values and pressing a function key. And, of course, Microsoft® Excel for Windows® also has the capability of performing the operations with additional statistical options available. Nevertheless, an understanding of how these calculations are performed (i.e. mean and standard deviation) is most relevant, particularly in pharmaceutical engineering operations where, often, the ongoing project may require the scrutiny of a facility quality assurance operation that does not possess an in-depth knowledge of statistical techniques.

Subsequent to a basic example problem presented later, the purpose of the standard deviation will be explained in some detail, and, hopefully, the intent of the need for this statistic will be understood.

It is important to remember the values employed in the following example are intended only to become accustomed to calculating the standard deviation. The numbers selected are random numbers and have no significance, other than being numbers.

Example 11.4   The following six numbers were randomly selected from a total of 36 numbers ranging from –5 to 31. The numbers randomly selected were 5, 8, 24, 3, 20, and –4.

Calculate the average and sample standard deviation of the selected values.

Solution

Calculation of the mean

$$\bar{X} = \frac{5+8+24+3+20+(-4)}{6} = 9.33 \leftarrow \text{the mean value}$$

Calculation of the standard deviation
1) The mean $= 9.33 \leftarrow$ The mean
2) Compute the deviation

$$(5-9.33)=-4.33$$
$$(8-9.33)=-1.33$$
$$(24-9.33)=14.67$$
$$(3-9.33)=-6.33$$
$$(20-9.33)=10.67$$
$$(-4-9.33)=-13.33$$

3) Square the deviation

$$(-4.33)^2 = 18.75$$
$$(-1.33)^2 = 1.77$$
$$(14.67)^2 = 215.21$$
$$(-6.33)^2 = 40.01$$
$$(10.67)^2 = 113.84$$
$$(-13.33)^2 = 177.69$$

4) Sum the squared deviations

$$\Sigma(18.75)+(1.77)+(215.21)+(40.01)+(113.84)+(177.69)=565.50 \leftarrow \text{Sum of the deviations}$$

5) Divide the sum of the deviations by $N-1$ (since there are six values, division is $N-1=5$)

$$\frac{665.50}{5} = \mathbf{133.10} \leftarrow \text{Sum of the deviations with } N-1 \text{ in the denominator}$$

6) Calculate the square root of step 5

$$s = \sqrt{133.10} = \mathbf{10.63} \leftarrow \text{The sample standard deviation}$$

We now have a value for a standard deviation of a random group of 30 or less values, which is identified as a sample standard deviation. For a comparison, if the population standard deviation were used (i.e. N in the denominator vs. $N-1$ in the denominator), the population standard deviation, σ, would be 10.53 in lieu of 10.63 (difference of 0.94%). Again, the sample standard deviation is intended for samples as opposed to populations, a detail that should not be overlooked.

It is also important to note the units for the standard deviation are the same as those identified for the mean. The units for the average in Example 11.1 were given in ppm; these units are also the same as the standard deviation, were it to be calculated.

Hopefully, the calculation steps described have been mastered and present no significant difficulty. The objective is now to understand the intent of the standard deviation and how to effectively use this statistical tool. Since pharmaceutical engineering, as is the case with most engineering operations, is concerned with useful and valid applications that enhance productivity tools (e.g. statistics, fluid mechanics, and heat transfer), an extensive emphasis on a theoretical treatise, while of interest, may not be particularly germane to the ultimate objective at hand, which is to efficiently apply useful tools, such as mean and the sample standard deviation, and apply them to pharmaceutical engineering solutions that may be encountered.

Example 11.4 resulted in a sample standard deviation of 10.63. When presenting the standard deviation, whether the value is for a population, σ, or a sample, s, the standard deviation is often associated with the average of the values of interest. Thus, from Example 11.4, the result is described as 9.33 ± 10.63. The sample standard deviation of 10.63 states that for one standard deviation (± 10.63), 68.3% of the six random numbers used in Example 11.4 (about four values) are within one standard deviation of the calculated mean (9.33). As a result, four of the random numbers will be higher than the mean (9.33), and four values will be lower than the mean. Similarly, two standard deviations from the mean would indicate that 95.45% of the values greater than the mean (5.97 values) and 95.45% of the six values would be below the mean (also 5.97 of the six values). Practically, two standard deviations from the mean, either above or below the mean, encompass the entire six-number sample. Further, if three standard deviations from the mean were relevant, this would result in 99.7% of the six values being above the mean and below the mean of 5.99.

Another interpretation of the standard deviation can be found by observing a lower standard deviation indicating that the bulk of the data points are closer to the mean whereas the greater the standard deviation, the further the individual data points are from the mean.

Another statistical mode often encountered in pharmaceutical quality control operations is the relative standard deviation (RSD). The relative standard is also referred to as coefficient of variation (COV). The RSD is mathematically defined as

$$\text{RSD} = \frac{100s}{\bar{X}} \tag{11.6}$$

where

RSD = relative standard deviation, %
s = population standard deviation
$\bar{X}$ = mean

When covering topics such as statistics and probability, it is worth noting that while a basic understanding and working knowledge of these topics will not guaranty a perfect solution to a specific problem, this basic knowledge will provide for a more reliable result than merely guessing what the solution is. By becoming familiar with the basic statistical material presented and the material to be presented, a greater reliance and confidence on these methods should lead to a higher successful result and reliability by employing the topics encompassed.

Example 11.5 Determine the relative standard deviation (RSD) of Example 11.4.

Solution
From Example 11.4, the value for the standard deviation is $s = 10.63$ and the mean $\bar{X} = 9.333$. Inserting the required values in Equation (11.6) yields

$$\frac{10.63(100)}{9.33} = 113.9\% \leftarrow \textbf{Solution}$$

The RSD is a measure of precision of the values. Recall that the values selected for Example 11.4 were random numbers. Because the values are random, a particular precision is not expected. For instance, an analytical specification states a particular compound should be 2.453 g/ml ± 0.002 g, and the analytical results would result in an RSD of 0.002(100)/2.453 = 0.0815%. Since Example 11.5 is predicated on randomly selected values, it can be expected that the RSD calculation would have little precision. Typically, in a pharmaceutical manufacturing environment, the RSD value would be in the range of 1.5–5%. It is important to note the FDA is responsible for defining the need for pharmaceutical statistical specifications. As a guidance tool, the FDA uses an RSD value of 5% or less for several operations such as blending reproducibility and analytical results. The value is based on a decision commonly referred to as the Barr Decision. The operator or owner of the manufacturing process is responsible for meeting the specifications defined by the manufacturer of the product, whereas the FDA is concerned that the standards and specifications, for the most part, conform to the results approved by the manufacturer. In this case, the operator (owner) is responsible for defining the RSD values for a product, while the FDA must assure the manufactured item conforms to the RSD value specified by the operator. The operator (owner) quality assurance operation is responsible for verifying the results (acting as the FDA on-site representative).

The statistical material covered thus far is intended to introduce a most important statistical technique known as a normal distribution (also referred to as a Gaussian distribution).

At this point, a brief diversion concerning the origin and meaning of the normal distribution would be helpful. The normal distribution, coupled with the standard deviation, is a powerful tool that can be used when evaluating and resolving many engineering problems where an analysis of data can resolve design, research, and manufacturing issues where valid data is or can be made available. A good starting point is the function curve known as a Gaussian distribution curve or familiar term often referred to as a "bell-shaped curve," shown in Figure 11.1. Note that the bell-shaped curve is asymptotic since both tails are asymptotic to the x axis as it approaches 100% or what is also referred to as a probability of 1. However, the curve is never equivalent to a value of 100%, or a probability of 1. Additionally, the area under the bell-shaped curve is equal to 1. In fact, Figure 11.1 is also referred to as a normal distribution curve or a probability distribution curve. Also, a less common identifier is the term "continuous distribution." Regardless of the nomenclature used, it is again emphasized that the area under the curve is equal to 1.

Figure 11.1 Gaussian distribution (bell-shaped curve).

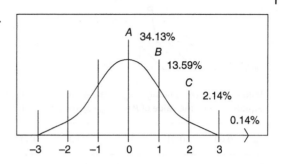

As will be seen, the normal distribution or Gaussian distribution curve is of particular importance when addressing the Six Sigma methodology addressed in the section.

The Gaussian distribution is represented by the function

$$f(x) = \frac{1}{\left(2\pi\sigma^2\right)^{1/2}} e^{-\frac{(x-\mu)^2}{2\sigma^2}} \tag{11.7}$$

where

σ^2 = the variance (the square of the population standard deviation)

μ = the mean of the population or sample, denoted by the Greek symbol μ

e = the natural logarithm base, 2.718

x = an independent variable value of the population data to be evaluated (for this chapter, an uppercase X is interchangeable with a lowercase x)

π = 3.14159

By observation, it can be seen that Equation (11.7) can be quite cumbersome when attempting to perform a calculation. As a result, the equation has been digitized to visually present the continuous function (Figure 11.1) in a workable format. Table 11.1 is known as a cumulative normal distribution or a table of normal distribution (the term normal defines the midpoint mean, μ, {also point A in Figure 11.1}, as the zero value and the variance, $\sigma^2 = 1$).

Table 11.1 will be employed shortly, but Equation (11.7) is, for the moment, the focus. As noted, Table 11.1 is, basically, a digitized version of Equation (11.7).

A cursory examination of Equation (11.7) indicates that the relation is exponential with a natural base, e. The natural logarithmic base implies that this is a naturally occurring relationship. In fact, the Gaussian distribution is found to be a basis for many natural phenomena occurring in biological science, chemistry, physics, engineering, and other sciences. For instance, if a large quantity of sand is dumped from a significant height, the mound of sand formed subsequent to the drop appears to be in the form of a bell-shaped curve (for this example, the sand pile will be defined in two dimensions only, length, the x axis and height, the y axis). A closer examination of the sand pile shows that the pile is basically symmetric in that the amount of sand to the left of the centerline is about equal to the quantity of sand to the right of the centerline. By mentally comparing the sand pile with the Gaussian or bell-shaped curve by superimposing the curve over the sand pile, the sand pile and the curve are identical in shape.

With Figure 11.1 superimposed on the sand pile, we can see that point A on the curve would correspond to the highest level or greatest amount of sand; point A is actually the mean point or average value of Figure 11.1 (designated by $\bar{X}$, μ, or in this case A). It follows that point A is the average amount of sand on the pile, since Figure 11.1 has been superimposed on the sand pile.

Table 11.1 Cumulative normal distribution.

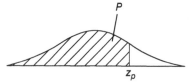

Values of P corresponding to z_p for the normal curve. z is the standard normal variable. The value of P for $-z_p$ equals 1 minus the value of P for $+z_p$, e.g. the P for -1.62 equals $1-0.9474=0.0526$

z_p	0.00	0.01	0.02	0.03	0.04	0.05	0.06	0.07	0.08	0.09
0.0	0.5000	0.5040	0.5080	0.5120	0.5160	0.5199	0.5239	0.5279	0.5319	0.5359
0.1	0.5398	0.5438	0.5478	0.5517	0.5557	0.5596	0.5636	0.5675	0.5714	0.5753
0.2	0.5793	0.5832	0.5871	0.5910	0.5948	0.5987	0.6026	0.6064	0.6103	0.6141
0.3	0.6179	0.6217	0.6255	0.6293	0.6331	0.6368	0.6406	0.6443	0.6480	0.6517
0.4	0.6554	0.6591	0.6628	0.6664	0.6700	0.6736	0.6772	0.6808	0.6844	0.6879
0.5	0.6915	0.6950	0.6985	0.7019	0.7054	0.7088	0.7123	0.7157	0.7190	0.7224
0.6	0.7257	0.7291	0.7324	0.7357	0.7389	0.7422	0.7454	0.7486	0.7517	0.7549
0.7	0.7580	0.7611	0.7642	0.7673	0.7704	0.7734	0.7764	0.7794	0.7823	0.7852
0.8	0.7881	0.7910	0.7939	0.7967	0.7995	0.8023	0.8051	0.8078	0.8106	0.8133
0.9	0.8159	0.8186	0.8212	0.8238	0.8264	0.8289	0.8315	0.8340	0.8365	0.8389
1.0	0.8413	0.8438	0.8461	0.8485	0.8508	0.8531	0.8554	0.8577	0.8599	0.8621
1.1	0.8643	0.8665	0.8686	0.8708	0.8729	0.8749	0.8770	0.8790	0.8810	0.8830
1.2	0.8849	0.8869	0.8888	0.8907	0.8925	0.8944	0.8962	0.8980	0.8997	0.9015
1.3	0.9032	0.9049	0.9066	0.9082	0.9099	0.9115	0.9131	0.9147	0.9162	0.9177
1.4	0.9192	0.9207	0.9222	0.9236	0.9251	0.9265	0.9279	0.9292	0.9306	0.9319
1.5	0.9332	0.9345	0.9357	0.9370	0.9382	0.9394	0.9406	0.9418	0.9429	0.9441
1.6	0.9452	0.9463	0.9474	0.9484	0.9495	0.9505	0.9515	0.9525	0.9535	0.9545
1.7	0.9554	0.9564	0.9573	0.9582	0.9591	0.9599	0.9608	0.9616	0.9625	0.9633
1.8	0.9641	0.9649	0.9656	0.9664	0.9671	0.9678	0.9686	0.9693	0.9699	0.9706
1.9	0.9713	0.9719	0.9726	0.9732	0.9738	0.9744	0.9750	0.9756	0.9761	0.9767
2.0	0.9772	0.9778	0.9783	0.9788	0.9793	0.9798	0.9803	0.9808	0.9812	0.9817
2.1	0.9821	0.9826	0.9830	0.9834	0.9838	0.9842	0.9846	0.9850	0.9854	0.9857
2.2	0.9861	0.9864	0.9868	0.9871	0.9875	0.9878	0.9881	0.9884	0.9887	0.9890
2.3	0.9893	0.9896	0.9898	0.9901	0.9904	0.9906	0.9909	0.9911	0.9913	0.9916
2.4	0.9918	0.9920	0.9922	0.9925	0.9927	0.9929	0.9931	0.9932	0.9934	0.9936
2.5	0.9938	0.9940	0.9941	0.9943	0.9945	0.9946	0.9948	0.9949	0.9951	0.9952
2.6	0.9953	0.9955	0.9956	0.9957	0.9959	0.9960	0.9961	0.9962	0.9963	0.9964
2.7	0.9965	0.9966	0.9967	0.9968	0.9969	0.9970	0.9971	0.9972	0.9973	0.9974
2.8	0.9974	0.9975	0.9976	0.9977	0.9977	0.9978	0.9979	0.9979	0.9980	0.9981
2.9	0.9981	0.9982	0.9982	0.9983	0.9984	0.9984	0.9985	0.9985	0.9986	0.9986
3.0	0.9987	0.9987	0.9987	0.9988	0.9988	0.9989	0.9989	0.9989	0.9990	0.9990
3.1	0.9990	0.9991	0.9991	0.9992	0.9992	0.9992	0.9992	0.9992	0.9993	0.9993
3.2	0.9993	0.9993	0.9994	0.9994	0.9994	0.9994	0.9994	0.9995	0.9995	0.9995
3.3	0.9995	0.9995	0.9995	0.9996	0.9996	0.9996	0.9996	0.9996	0.9996	0.9997
3.4	0.9997	0.9997	0.9997	0.9997	0.9997	0.9997	0.9997	0.9997	0.9997	0.9998

Further, Figure 11.1 identifies the standard deviations from the mean, which in this case is denoted by the abscissa (horizontal line) or x value of the curve. As can be observed, values to the right of the mean, point A, represent 1 (34.13%), 2 (13.59%), and 3 (0.14%) standard deviations, respectively. And also by observation, values to the left of the mean represent standard deviations of a decreasing value. These three standard deviations are commonly referred to as the 68–95–99.7 rule. That is, rather than being detailed about specifics, 1 standard deviation is rounded to 68%, 2 standard deviations are 95%, and a 3 σ value is 99.7%.

Since the values on the abscissa represent all of the sand in the pile, in effect, we are looking at a distribution of 100% of the sand pile. This implies that from −3 standard deviations to +3 standard deviations, 100% of the sand (the values are so close to 100%; it is safe to consider that the few grains that blew off the pile are irrelevant) is accounted for in the distribution represented in Figure 11.1. Continuing with the description, we can now state that the values on the abscissa represent a probability distribution of 0–100% (going from left to right on the horizontal axis). Now, by taking our visualization one step further, it can now be stated that 68.2% of the sand is within −1 standard deviation of the mean (the mean value being point A on Figure 11.1) and +1 standard deviation of the mean or point B (the values noted on the abscissa [the horizontal axis]).

The area under the curve in Figure 11.1 is also the probability curve. That is, the area under the curve represents the probability of an event or component meeting a required or desired value. It can also be noted that the total area under the normal distribution curve is 1.00. Again, by observation, the area to the right of the midpoint of Figure 11.1 (point A) represents 50% of the area under the curve, or a probability of 0.5. Since Figure 11.1 is symmetrical, the area to the left of point A is also 50%, or a probability of 0.5. From calculus, the area under the curve represented by Equation (11.7) can be determined by inserting specific values in the equation. Since integration of a somewhat complex function, Equation (11.7), can be cumbersome, Table 11.1 allows for a convenient method of obtaining relevant information for probabilistic and statistical solutions to a variety of situations by using a convenient technique.

While Figure 11.1 is helpful in dealing with a few locations (i.e. 1, 2, and 3 standard deviations from the mean), it would be of real value if a full spectrum of values could be evaluated. Fortunately, there is a technique available to predict, with reasonable accuracy, a method to determine probabilities and standard deviations when a value for the mean is known. The information given in Table 11.1 is how values other than the 1, 2, or 3 standard deviation probabilities can be ascertained.

In order to effectively use the data contained in Table 11.1, some relationships and definitions are required. The fundamental relationship needed for using Table 11.1, the cumulative normal distribution, is given by

$$z = \frac{x - \mu}{\sigma} \tag{11.8}$$

where

z = the standard normal distribution
x = a random variable
μ = the mean
σ = the population standard deviation

Equation (11.8) is, as indicated, a very helpful statistical tool with many valuable applications, both for Six Sigma projects and for other pharmaceutical engineering applications. By calculating a value for z, it will be possible to determine probabilities and standard deviations for many real-time operations. Additionally, it is important to note that the z value represents the number of standard deviations of a selected value from the mean value, μ.

While the calculation methods used to determine the solutions in this section are basic methods to become familiar with the calculation steps that are required, the results can easily be obtained by using standardized techniques available with Excel software, Minitab, or a programmable calculator such as the TI-83 or higher.

Before delving into the applications of the normal distribution curve (Figure 11.1 and Table 11.1, an ability to work with Table 11.1 is required; specifically, working with the z value is key.

Given the following values, $x = 500$ (randomly selected number), $\mu = 450$, and $\sigma = 18$, the z value is

$$\frac{500 - 450}{18} = 2.777 \approx 2.78$$

(When determining a z value, two decimal places are required.)

Referring to Table 11.1 and referring to the value of 2.7 in the left column and going across the column until the value for 0.08 is found (2.78), a value of 0.9973 is found. The value 0.9973 represents the area under the normal distribution curve. As a percent, the value 0.9973 states that 99.73% of the area under the normal curve (recall that the area under the curve starts at zero and progresses to a probability of 1.00) is less than the calculated z. The probability is less than the value selected, but not equal to that value. Because of the almost infinite number of values that can be selected, finding one single value is virtually impossible; thus, the answer is given as less than a selected value. Consequently, it can be stated that any X value selected will have a probability of being less than the value (500) selected. Recall that 3 standard deviations comprise 99.7% of the area under the normal curve presented in Figure 11.1. It is also seen that virtually all values of X are included. Again, it is important to note that the z value, in this case 2.78, represents the number of standard deviations, μ, and the chosen value (500) varies from the mean value, $\mu = 450$.

Delving a bit further, by selecting a random X to the left of the mean value of A in Figure 11.1, a random value of 410 is selected. The mean, μ, is still 450 and the population standard deviation remains as 18. Inserting the values in Equation (11.8) yields

$$z = \frac{410 - 450}{18} = -2.22$$

The calculated probability for a z value of -2.22 (the number of standard deviations less than the mean value) is determined by referring to the technique described in the header of Table 11.1. For a z value of -2.22, the result is 0.9868. Further, referring to the header in Table 11.1, the result is $1 - 0.9868 = 0.0132$ or 1.32%. The result, 1.32%, states that the probability that the value selected is less than the mean value, 450, is only 1.32%. This seems reasonable since the z value, -2.22, represents over 2 standard deviations, which is greater than 95.5%, thus leaving a small area under the curve to the left of the value.

For many applications, it is necessary to know the probability over a given range. By referring to Table 11.1, a typical situation can be identified when it is recalled that the value A in the mean value is 450. The example above ($X = 500$, $\mu = 450$, $\sigma = 18$) is a convenient subject. The probability of obtaining a value less than the selected value (500) is 99.73% or about 3 standard deviations from the mean value.

Addressing the opposite side of the probability curve, the probability that a selected value was less than the mean value (450) is 1.32% or slightly more than 2 standard deviations (2.22) from the mean value. But what if it was desired to know the probability between the two values obtained (e.g. 99.73 and 1.32%), in effect obtaining the area under the normal distribution curve (Table 11.1) within the two values identified (99.73 and 1.32%)?

Situations may arise when the probability of a specific region of the normal distribution is required. For instance, it might be necessary to know the probability of a value being between the higher probability (99.73%) and the lower probability (1.32%). Finding the value of the interval is relatively straightforward. The value is determined by subtracting the larger probability from the smaller probability. For this example, the solution is 99.73 – 1.32% = 98.41%. Thus, it can be stated that the probability of a value being between 0.9973 and 0.0132 is 0.9841 or 98.41%.

Example 11.6 A batch of aspirin tablets have been manufactured and bottled to distribute as a free sample for an income tax preparation service. The tablets were manually bottled in a sanitary filling and counting area. The average number of tablets in the bottle is 60 ($\mu = 60$) with a standard deviation of 10 ($\sigma = 10$). How many tablets should a bottle contain to be confident that the bottle will contain at least 60 tablets? The confidence level is a probability of about 95%. What is the probability a bottle will contain less than 60 tablets?

Solution
Referring to Table 11.1, a probability close to 95% is 0.9505. Referring to Table 11.1, this corresponds to a z value of 1.65. Equation (**11.8**) can be applied to obtain the desired value. Inserting the given values in the equation yields

$$1.65 = \frac{X - 60}{10}$$
$$X = 78.5 \approx 77 \leftarrow \textbf{Solution}$$

For the bottles to contain 60 or more tablets with a 95% probability, 77 tablets should be in the bottle. The probability of a bottle containing less than 60 tablets is a straightforward calculation. Since the maximum probability is 1.00 and the probability of a bottle containing in excess of 60 tablets is 95.05% (95%), the result is 1.00 – 0.9505 = 0.0495 ≈ **0.05%.** ← **Solution**

The result indicates that about 5% of the bottles will contain less than 60 tablets.

Since the aspirin tablets are complementary, the average amount of tablets in the bottle should not be a serious consideration, unless labeling is a concern.

The normal distribution curve is applicable to many areas. If a mean and standard deviation can be calculated, many important probabilistic scenarios can be solved with a reasonable degree of accuracy, as noted. Example 11.7 involves a previously encountered problem and describes an additional application of the probability distribution curve.

Example 11.7 Example 5.2 determined the mean time between failures (MTBF) of an agitator to be 1580 h with a standard deviation, σ, of 200 h (rounded off to the nearest hour). The maintenance schedule provides for a preventive maintenance inspection at 2000 h of operation. Determine the probability that the agitator will fail prior to the 2000 h inspection. All recorded values represent a normal distribution.

Solution
Equation (**11.8**) is the basis of calculation for this example. The variables required are

$\mu = \text{MTBF} = 1580 \, \text{h}$
$\sigma = 200 \, \text{h}$
$X = 2000 \, \text{h}$

Inserting the known variables in Equation (**11.8**) and solving for z yields

$$z = \frac{2000 - 1580}{200} = 2.10$$

Referring to Table 11.1, the z value for 2.10 is **0.9821 or 98.21%.** ← **Solution**

The probability that the agitator will fail prior to the 2000 h inspection is 98.21%. Based on the result, an inspection and preventive activity should be rescheduled employing a shorter time interval.

11.2 Introduction to Six Sigma

Many manufacturing processes are not run as efficiently as they could be. The Six Sigma methodology provides a foundation for continuous improvement and for solving quality, operational, and other problems [2].

Six Sigma is a methodology that has become an important part of business, and to a lesser extent government, operations. The methodology, when properly implemented, has shown to be an effective tool for cost savings and improved customer satisfaction. The Six Sigma concept has been widely embraced by many major corporations such as Motorola, the originator, and General Electric, the major implementers of Six Sigma.

11.2.1 Six Sigma Organization and Background

While there are several applicable definitions for the concept of Six Sigma, a convenient one could be that "Six Sigma is a program whose goal is to detect and eliminate defects and errors in manufacturing and service operations."

Six Sigma is, as the name implies, the term for six standard deviations from a mean value, assuming a normal distribution represents a quality goal.

The goal of Six Sigma is to implement a process that will reduce defects and errors to a value of 3.4 defects per million opportunities (DPMO). The value of 3.4 DPMO is the equivalent of six standard deviations from the mean of a population or sample (Section 11.1.1). Table 11.2 indicates the approximate defect (DPMO) level over a range of standard deviations. When reviewing the values given in Table 11.2, it would be helpful to recall that 10 000 parts per million (ppm) is equivalent to 1%.

Table 11.2 Standard deviation versus defects per million opportunities (DPMO).

Standard deviation, sigma	DPMO
1.5	500 000
2	308 537
3	66 807
4	6 210
5	233
6	3.4

Source: From May [1].

Some details regarding the value of Six Sigma (99.99966%) opportunities compared with 3.8 sigma (99.95%) opportunities are as follows:

- On average, at a 99.7% (3 standard deviations) success rate, nationally, 20 000 articles of mail are lost in 1 h. At the Six Sigma level the number of lost articles is reduced to 7 mail items per hour.
- At the 99.7% level, 5000 surgical procedures are incorrect in a week. At the Six Sigma level, incorrect surgical procedures would be reduced to 1.7 in a week.
- There are 200 000 wrong drug prescriptions per year. The Six Sigma level would be reduced to 68 incorrect prescriptions annually.

To achieve this rigid quality objective, a program has been developed to identify, quantify, and implement relevant variables that can implement, in a planned approach, a project that successfully achieves the desired goal (i.e. a product or service that conforms to the 3.4 DPMO goals).

The word "opportunity" can be interpreted as a task that is performed with a measurable outcome. For instance, a worker may have the task to visually inspect prepackaged syringes as a final inspection step. Subsequent to syringe inspection, quality assurance checks indicate that the inspection procedure is revealing a failure rate of 2 DPMO; the operation conforms to the Six Sigma criteria. Another example of a Six Sigma application is the document controls administrator whose task is to properly file and record the visual inspection results of the syringe inspection.

Another example of an opportunity is the correct diameter of a syringe needle. If the needle diameter is larger than the specification requirement, it is not acceptable and cannot be incorporated in the final syringe assembly. If the inspection detects 3.2 DPMO, the production batch is acceptable, even though the difference between acceptance and rejection is quite small.

While one would consider the word opportunity to apply to manufacturing, Six Sigma programs are also applicable to service organizations. An opportunity can be defined as a property or characteristic that is evaluated or compared with a defined specification or procedure. If the comparison meets the criteria of the specification or procedure, the result is a success. If, on the other hand, the evaluation does not conform to the defined criteria, the part or operation is defined as a defect or contributing to a defective component or operation. A typical service application would be keeping a record of how often a worker incorrectly copies documents in the reproduction center of a large pharmaceutical firm.

Since calculating DPMO is a mathematical concept, it would be convenient to quantify DPMO in a workable format that incorporates the number of defects detected, the number of units, and the calculation basis of one million (10^6). Starting with a basic relationship,

$$\text{DPO} = \frac{\text{No. of defects}}{\left(\text{No. of units}\right)\left(\text{No. of opportunities / unit}\right)} \tag{11.9}$$

where

DPO = DPU = defects per operation or opportunity or unit (sample)
No. of defects = total number of defects observed
No. of units = U = total number of units evaluated
No. of opportunities = O/U = number of product inspections or number of procedural checks (opportunities) performed per unit

Once the DPO is known, the DPMO can be determined. DPMO is defined as

$$\text{DPMO} = \text{DPO} \times 10^6 \tag{11.10}$$

Example 11.8   A generic pharmaceutical manufacturer is contracted to produce an experimental tablet for a phase I clinical trial from an active pharmaceutical ingredient (API) supplied by the customer, a large drug company. The contract manufacturer has agreed to produce a product (finished tablet) that conforms to the Six Sigma criteria (3.4 DPMO) for specific parameters of the tablet. Rather than employing small-scale laboratory tablet press, the generic manufacturer will use a large quantity, double-sided high speed tablet press (7000 tablets/min operating at 4 kP (kilo pounds) and one tablet coater intended for full-scale manufacturing. For an initial pilot test, 10 finished tablets were randomly selected from a preliminary batch for an initial evaluation of the specified tablet parameters. The physical tablet parameters to be initially evaluated are:

1) $\text{Mass} = 200 \pm 10 \text{ mg}$
2) Dissolution time $= 150 \text{ seconds} - \text{minimum}$
$ 300 \text{ seconds} - \text{maximum}$
3) Friability – less than 5% mass loss (10 mg) when tested with friabilator at 25 RPM for 4 min

Based on the tablet information recorded, calculate the DPMO for the required parameters.

Solution
To perform the calculation, a parameter matrix or table identifying the variables and outcomes is prepared such as the summary table that follows.

Finished tablet parameters.

Tablet	Tablet mass (mg)	Dissolution time (s)	Friability (≤10 mg mass loss), Yes/No	Defective tablet, Yes/No	Number of defects per tablet
1	205	273	203	No	0
2	208	209	207	No	0
3	203	328	197	Yes	1
4	207	333	195	Yes	2
5	212	321	197	Yes	2
6	204	231	201	No	0
7	206	310	195	Yes	2
8	209	313	194	Yes	2
9	201	291	185	Yes	1
10	268	317	252	Yes	2

From the preceding above,

$D = $ total number of defects $= 12$
$O = $ number of opportunities (mass, dissolution time, friability) $= 3$
$U = $ number of tablets (units) $= 10$
$\text{TOP} = $ total opportunities $= U \times O = 10 \times 3 = 30$
$\text{DPO} = $ defects per unit (tablets) $= D/\text{TOP} = 12/30 = 0.40$

Applying Equation (**11.10**) yields

$$\text{DPMO} = \text{DPO} \times 10^6 = (0.4) \times (1.0 \times 10^6) = 400\,000\,\text{DPMO} \rightarrow \textbf{Solution}$$

The significant deviation from 3.4 DPMO indicates that a very small sample should not be relied upon for a reliable value. Typically, for a phase I study, a sample size of a few hundred people (~300) would be used as study subjects. For instance, with subjects ingesting two tablets a day for 1 year, a batch in excess of about 240 000 tablets would be manufactured. A conservative sample size would be about 500 tablets. This small sample would not warrant initiation of a Six Sigma program. Such programs are usually implemented during the production phase of a manufacturing operation.

The small sample size indicated a lengthy dissolution time and a friable tablet. This could be an indication that the addition of an organic binder excipient might reduce the friability. Since the sample tablets tended to be of lower mass, the inert excipient, might not adversely affect the final product. However, further studies involving excipient addition would be required if future batches reveal a problem.

Commonly, these additional studies would focus on specific interactions in the body of the additive. These studies focus on five specific effects on the body. These effects are identified in the pharmaceutical industry as ADMET, where A (absorption) is the method of administration (i.e. oral or parenteral), D (distribution) is how the drug is distributed in the body, M (metabolism) is the time required to digest in the system or organ of intent, E (excretion) is the mechanism by which the waste by-products exit the body, and T (toxicology) determines if the product or by-products are harmful to the body or specific organs of the body.

With the basic intent and primary quantitative aspects of Six Sigma defined, a more detailed examination of the Six Sigma process can proceed.

As noted briefly, a Six Sigma program is designed to be implemented in an ongoing production operation and is applicable to, as noted, both a product and a service. The program can be implemented on a corporate, division, departmental, or group basis. The basic managerial, technical, and administrative principles apply to virtually any organizational entity, regardless of size, although Six Sigma programs are generally used in larger organizations due to costs such as training and consultants. The basic Six Sigma approach is valid in most instances by recognizing the need and correctly employing the Six Sigma format to correct or greatly minimize a recognized manufacturing defect that can benefit by Six Sigma.

While implementing a Six Sigma program often indicates an organization is "cutting-edge" managerial and quality oriented, the objective of a Six Sigma initiative is to enhance operational efficiency. In essence, operational efficiency enhancement translates to improving the profitability of the company. The basic measure of the profitability of an operation is often determined by the net sum available to an organization subsequent to the investment required. The net sum, profit, is normally obtained by determining the profit over a period of time. Accounting criteria of the organization determine the net profit of the investment. This sum is known as the return on investment (ROI) and can be expressed as

$$ ROI = \frac{Net\,income}{Investment}(100\%)\pi \tag{11.11} $$

where

ROI = return on investment
Net Income, \$
Investment, \$

Equation (11.11) appears to be straightforward. However, several variables should be evaluated prior to using Equation (11.10) as a tool to evaluate the worth of a Six Sigma program initiative. The source of funding a program is a major factor for evaluation. One important consideration is if the assembled team charge project time as a direct cost or will overhead expenses be the source. There also exists the possibility of both sources underwriting the cost. For instance, quality assurance representatives if assigned to a Six Sigma project are overhead costs, as they typically are in pharmaceutical-related quality assurance functions. Similarly, engineering activities are often overhead costs. Conversely, production is usually a direct cost. Funding source is crucial since the "bean counters" of organizations depend on acceptable inputs and results; there is very little margin of error when analyzing the inputs and results when determining the ROI of the project. Yet, many Six Sigma programs fail to achieve the anticipated ROI. A major reason for the unanticipated results shortfall is due to using the "best-case" scenario. The resulting inaccuracies commonly occur because each facility has unique properties, process parameters, and data properties. Consequently, it may be problematic to accurately estimate the actual benefits, such as the ROI, before actual completion of the Six Sigma undertaking and analyzing the results.

When performing a Six Sigma study, it is necessary to clearly understand what procedures are required to successfully enhance the balance sheet (i.e. ROI). Using a high level analysis that fails to include elements such as maintenance and support requirements can lead to an incorrect final ROI. These factors can often be manifested when vendors do not have a foundation of the individual operations involved. Thus, the customer must be cognizant of vendor capabilities if vendor support is used as part of the project [4].

Another consideration is whether or not Equation (11.10) is the most effective measurement tool for evaluating a Six Sigma program. The ROI is a common accounting tool. However, it is most commonly used when determining the return on an investment such as new equipment or a new technology. In the case of new equipment, the item cost can be depreciated over time, thus allowing the owner to write off a portion of the cost on an annual basis. In the case of a Six Sigma initiative, the benefit gained is often an enhanced productivity or operating efficiency, a cost that is not as quantifiable as amortizing equipment as a fraction of the total cost at given time intervals (e.g. annually). Also, it is common to ignore several items not addressed in Six Sigma programs but very germane to the overall process. Two examples that may be overlooked are:

- Has the possibility of downstream bottlenecks, subsequent to Six Sigma project implementation, been considered in the overall operation?
- Will the project create limitations both upstream and downstream of the project?

While Six Sigma projects are often most effective when dealing with manufacturing programs, pharmaceutical operations primarily use the Six Sigma initiatives to enhance nonmanufacturing functions such as reducing inventory, reducing fixed costs, or improving product and raw material logistics factors and finished product considerations such as delivery times and distribution operations. Pharmaceutical distribution is important since pharmaceutical products have expiration dates and quicker distribution to end users is quite important. Also, in addition to expiration dates, many pharmaceutical products require that products be stored and transported under strict controls such as temperature control, security, chain of custody, and regulatory compliance criteria (e.g. 21 CFR, 49 CFR, etc.) during storage and transit. Storage and transportation and related logistics operations can involve several internal corporate operations as well as noncorporate operations such as land, air, and third-party logistics operations. Here, a Six Sigma program could result in significant efficiencies and significant cost savings [5]. Such a large operation often requires a large Six Sigma program that can adequately address the myriad of organizations, tasks, and inputs required to complete the assigned mission in an efficient and cost-effective manner.

The role of Six Sigma in pharmaceutical engineering, specifically, is a nascent area that, when properly implemented, can present significant benefits with a minimum of expenditures by addressing the engineering aspects of Six Sigma tools.

Two concepts where Six Sigma programs can significantly affect pharmaceutical manufacturing operations are process analytical technology (PAT) and quality by design (QbD), items that will be addressed further in Sections 11.3 and 11.4.

Recognizing a product or service operation that can significantly benefit from a Six Sigma program is, perhaps, the most important item of a proposed initiative. A comprehensive analysis of the benefits of implementing a Six Sigma project must be thoroughly investigated. The best project is where there is a gap in performance, the reason for the gap is not known, and closing the gap is also problematic. An approved Six Sigma program based on a poorly selected project can result in a needless waste of time and assets (e.g. wasted man-hours, computer time, etc.). Often failure to recognize a problem that could benefit from a Six Sigma approach also results in a loss by failing to mitigate or remediate the existing impediment.

In addition to problem recognition, a recognized Six Sigma presence in the organization is required. The central individual in the organization is typically a managerial level, or higher, individual possessing the credentials to qualify as the leader or sponsor. This individual may be certified by an accredited organization such as the American Society for Quality (ASQ).

A certified Black Belt is an individual who has demonstrated knowledge of the principles of Six Sigma by successfully passing an examination and exhibiting proficiency by successfully completing assigned Six Sigma projects. Individuals identified as Black Belts or master Black Belts are generally located in larger operations where Six Sigma initiatives are an ongoing enterprise. As a result, the Black Belt is considered a full-time operation whose time is totally dedicated to ongoing Six Sigma improvement projects throughout the organization (i.e. the time is chargeable to Six Sigma program). While not specific, a Black Belt expert can be involved in as many as four projects a year with an individual project ROI in excess of $150 000 per project [6].

Another Six Sigma professional level is the Green Belt, usually obtained after a 2-week training regimen, usually in the range of 40–80 h. The primary difference between the two levels is the degree of experience between the two levels. While both levels may exhibit the same skill sets and skill levels, a Black Belt is more experienced than the Green Belt counterpart. Regarding project involvement, a typical Green Belt would contribute an ROI of about $35 000 per project and participate in approximately two projects per year. For pharmaceutical engineering applications, programs of this scope (~$35 000) are well suited for Six Sigma programs. As a result, there should be a minimal need to use an outside consultant to perform the task. Selected staff personnel with an overview to Six Sigma and a basic knowledge of the program fundamentals, and directed by a certified Green Belt, should adequately master the Six Sigma project. Projects of this scope often involve a single manufacturing operation where personnel are most familiar with the operation and can involve a minimum of personnel assigned to the Six Sigma tasking.

While larger Six Sigma projects can involve up to 20 individuals, the Green Belt initiatives can be successfully implemented with as few as 5 or 6 participants for a minimum time frame, often about 3 or 4 weeks of chargeable time or overhead.

While the Black Belt and the Green Belt represent the major knowledge base for Six Sigma, there are other categories whose input is an effective component of Six Sigma projects. An individual with a Yellow Belt is an example. Yellow Belts are typically staff employees who have been trained in the basic Six Sigma tools and are familiar with Six Sigma methodology. Another Six Sigma team member is the White Belt. White Belt individuals are personnel who have completed a cursory overview of Six Sigma concepts.

Depending on organization size, the number of individuals possessing a Green Belt can vary from 5 to 10% of the organization. As is the case with a Black Belt, a Green Belt has demonstrated skills in diverse areas such as leadership, statistics, business analysis, and problem-solving capabilities, yet operates under the guidance and leadership of a Black Belt or master Black Belt. In addition to Green Belt certification, it is important for the Green Belt assigned to a program to be familiar with the process and operations involved in Six Sigma analysis. For Six Sigma programs specific to pharmaceutical engineering applications, a successful outcome can be anticipated since many of the key elements of a program involve a knowledge base familiar to engineers dealing with pharmaceutical operations (e.g. statistics, FMEA, process design, etc.).

11.2.2 DMAIC: The Basic Six Sigma Acronym

DMAIC is a term many have been introduced to and many individuals may know the origin of the acronym. It represents the working tool of Six Sigma. It also represents the problem-solving methodology employed to perform successful Six Sigma programs. The DMAIC acronym is:

- Define – Identify the objective and products expected upon completion of the program, including a project charter.
- Measure – Identify and record the input and output parameters of the project (e.g. if the Six Sigma program involves improving the yield of a bioreactor, historic variables such as yield, temperature, and operating pressures will be documented, as well as previously recorded data).
- Analyze – Determine through techniques such as statistical analysis and other techniques (e.g. root cause analysis) the defects responsible for the operational difficulties.
- Improve – Implement techniques that eliminate defects causing the operational impediments or problems.
- Control – Incorporate and implement the results of the Six Sigma program (DMAIC investigation results) required to continuously monitor, record, and control activities required to maintain the operational improvements gleaned from the Six Sigma program. The key deliverable of the control stage is the control plan.

While the brief explanations provided above present a basic definition of the DMAIC factors, the following material is intended to provide a significant background and knowledge base in the DMAIC aspects of Six Sigma projects.

11.2.3 Define

This element, Define, is essentially a statement of the Six Sigma program objective. Briefly, this area specifies the beneficiary of the program. Normally, this would involve external (e.g. improving the warehousing and distribution of finished pharmaceuticals) or internal customers (e.g. increasing the yield of a bioreactor). For pharmaceutical engineering-oriented Six Sigma programs in the range of $35 000, in addition to a Green Belt acting as the titular leader, while under the guidance of a Black Belt, and one or more White Belts and Yellow Belts, it would be beneficial to select team members from various functional areas to support specific Six Sigma programs.

With the problem identified and the objectives defined, including the initial problem and a defined completion objective (commonly referred to as the Charter in Six Sigma terminology), the next element of Define is another acronym, the SIPOC, which is basically a subset of the Define element of DMAIC. The SIPOC – short for Supplier, Input, Process, Output, and Customer – is a process map or diagram that provides the envelope or boundaries the project involves. The terms map and diagram can also be a detailed written description of the SIPOC elements. Suppliers are the sources of the materials (also referred to as consumables) that are used to produce the desired product or end item. Inputs are the materials and resources required to perform the project to be improved. Inputs include not only the consumables needed but also assets such as computer resources, process procedures, and other outside organizational support such as maintenance and purchasing operations. The process represents the procedures that create the final product by transforming the supplier input to a finished item. The output is, as can be surmised, the completed product that results from implementing process procedures. The last SIPOC item, customer, is the end item user of the finished item, which can be a product or a service. For an internal product, the customer is the organization responsible for the next phase of the process. For instance, if a Six Sigma program involves improving the yield on a tablet coater by increasing the rotational speed and temperature, the next step of the process would normally be the fill and finishing operations. In this case, the customer would be the manager of the packaging operation, assuming the coated tablets conform to quality control specifications.

For large Six Sigma programs, SIPOC is a detailed tool requiring an extensive and detailed qualitative and quantitative effort. However, pharmaceutical engineering Six Sigma projects are geared for small projects that can be effectively directed by a Green Belt (~$35 000); many of the SIPOC inputs can be minimized. A major simplification involves the actual SIPOC preparation task. For instance, if the project involves a pharmaceutical manufacturing operation, the input SIPOC element is, in most cases, available in existing formats. These existing items are the piping and instrumentation diagram (P&ID) and the process flow diagram (PFD). The PFD includes the mass and energy balances for the process, while the P&ID identifies the steps of the process.

While other organizations would be involved in the program, the task of the Six Sigma team is to confirm and verify that the project conforms to the first element of the DMAIC program elements. And since other organizational divisions and departments must be involved to confirm that the Six Sigma project conforms to regulatory, corporate, and facility specifications and regulations (e.g. change control procedures), a high degree of coordination is needed, even for relatively small projects. Consequently, it is most important that the Define phase of DMAIC be implemented with the germane elements of the facility operations (e.g. quality assurance and quality control) to minimize or address undesirable consequences as a result of a planning oversight.

A clearly defined Six Sigma program should also address elements that will not be evaluated. For instance, the program intended to increase bioreactor yield, unless specified, would not address possible need for greater product container capacity once the yield increases. The boundary, or limits, of the project scope must be clearly presented in the Define phase to ensure the suppliers, customers, and process operations have an understanding of what to expect once the program is complete.

The final step employed for each element of DMAIC is the Tollgate. The Tollgate is basically a review in which all the Six Sigma team members review the results of each phase and, ultimately, concur that the DMAIC program is complete as per requirements (e.g. the Define phase of the project has conformed to the criteria set forth in the Charter). For pharmaceutical engineering applications, the Tollgate is essentially called the milestone review; in this case, it would be the milestone review for phase one of the DMAIC.

11.2.4 Measure

In addition to identifying the key input and output parameters, as identified in the SIPOC, the other major elements of Measure for a Six Sigma program include:

- Operational definitions – In addition to defining the Six Sigma project, key program deliverables must also be clearly identified. This includes a specification for identifiable defects, individual units (e.g. tablets, mixing times, and bioreactor volumes), approved baseline performance including Six Sigma performance results, calculated results including plots and graphs, etc.
- Data collection plan – A clear presentation of the measurement system to be used is critical to success. Typically, for pharmaceutical applications, measurement variables might include reaction rate dimensions and product acceptability (upper specification limits [USL] mean values, lower specification limits [LSL], standard deviations, temperatures in consistent units, pressure, in consistent units and other variables, depending on the program, the process capability, C_p). Also critical to data collection is the need to analyze and interpret the results.

11.2.4.1 Process Capability Indices

A term noted at the end of the conclusion of the data collection explanation was the process capability, also referred to as the process capability index. The process capability, C_p, is a technique developed to measure ability, or capability, of a process to meet specifications of a process. The process capability, C_p, is defined as

$$C_p = \frac{\text{USL} - \text{LSL}}{6s} \tag{11.12}$$

where

USL = upper specification limit
LSL = lower specification limit
s = the population standard deviation (Equation **11.4**)

C_p is, on closer examination, the ratio of the allowable spread to the measured spread. Table 11.3 compares some C_p values with corresponding failure in parts per million (ppm).

Example 11.9, in addition to performing a C_p calculation, describes how the process capability index can be employed to predict the capability of a process prior to actual implementation.

In essence, perhaps the best definition of a capability index (C_p and C_{pk}) is a measure of the ability of a process or operation to conform to approved specifications required by the customer or the process or operation owner.

Example 11.9   Prior to shipment, a small volume, manually packaged specialty pharmaceutical product is boxed and temporarily stored in a warehouse prior to off-site transportation and distribution. While in temporary storage, packaged products are randomly selected to verify that the net weight of the packages conforms to the approved specification. This is done to assure the shipment weight conforms to the Bill of Lading requirements, which is required by the US Department of Transportation for interstate shipment. The average weight of the packaged product (one box of bottled tablets) is 10.1 lb. The finished product specification requires a maximum weight per box of 10.5 lb, the upper specification limit (USL). Similarly, the lower specification limit for a box is 9.8 lb. Subsequent to sampling, the standard deviation was

Table 11.3 Comparison of capability index with failures, ppm.

Capability index, C_p	Failures (ppm)
0.33	317 311
0.67	45 500
1.00	2 700
1.10	967
1.20	318
1.30	96
1.33	63
1.40	27
1.50	6.8
1.60	1.6
1.67	0.57
1.80	0.067
2.00	0.002

Source: From Williams et al. [3].

calculated as 0.08. Determine the process capability of the sample lot and specify if employing the specification levels and standard deviation obtained is acceptable.

Solution
Equation (**11.12**) will serve as the basis of calculation with the following variables:
$\mu = 10.1\,\text{lb}$, $\text{USL} = 10.5\,\text{lb}$, $\text{LSL} = 9.8\,\text{lb}$, $\sigma = 0.08$. Inserting the given values in Equation (11.12) yields

$$C_p = \frac{10.5 - 9.8}{6(0.08)} = 1.45 \approx 1.5 \leftarrow \textbf{Solution}$$

When specifying a C_p value for the Measure component of DMAIC, the following general guidelines should be considered [3]:

- $C_p > 2$. The process is capable of being successfully operated.
- $C_p = 1.00$–1.33. The process is capable of being performed with tight control of the process parameter (i.e. USL, LSL, σ).
- $C_p < 1.00$. The process is incapable of achieving a successful outcome.

For pharmaceutical engineering applications, C_p and the related functions covered below are actually the germane activity for the Measure aspect of DMAIC. Most of the other Measure components are introduced to provide a basis for the capability indices. Consequently, for an identified pharmaceutical engineering project, many individuals involved in the program may, or easily can, possess the skills to perform and evaluate the Measure component in a judicious and expeditious manner. As such, the Measure aspect task should be completed rather quickly.

Based on the C_p calculation and the general guidelines presented, the process described is capable of yielding the results desired for full operations.

Another important function employed in process capability is the C_{pk} index. Unlike C_p, which measures the potential capability of the process to meet the specifications, the C_{pk} measures the process capability, taking into account process centering, and not the process potential. C_{pk} is defined as the smallest ratio between the following values:

$$C_{pk} = \frac{(USL - \mu)}{3\sigma}, \quad \frac{\mu - LSL}{3\sigma} \tag{11.13}$$

where

USL = upper specification limit
μ = average value
σ = standard deviation
LSL = lower specification limit

Equation (11.13) can be slightly modified to be more specific to USL and LSL. When considering performance of the USL, Equation (11.12) is

$$C_{PU} = \frac{USL - \mu}{3\sigma} \tag{11.14}$$

When considering the LSL, the relationship is given as

$$C_{pl} = \frac{\mu - LSL}{3\sigma} \tag{11.15}$$

When evaluating C_{PU} and C_{pl}, it is important to note, when evaluating these values, that the results for C_p must be greater than the values obtained for C_{pk} (i.e. $C_p > C_{pk}$).

While C_p is useful to evaluate a process for potential success, C_{pk} is an analytical tool that is most effective when applied to real-time process control, a topic that will be furthered addressed in Section 11.2.7. Also, the issue of short-term capability will be further detailed.

When evaluating process capability, an ideal configuration is the situation where $C_p = C_{pk} = C_{PU} = C_{pl}$. Such equality occurs when two statistical measures are equal: the USL and the LSL must be separated by the same standard deviation from the mean. As a general guide, a C_{pk} in excess of 1.33 indicates that the process is capable or acceptable. When the C_{pk} value for a process or operation is greater than 1.33, it is said to have a short-term capability. For a value greater than 1.33, for instance, in Example 11.9, the mean, μ, is 10.1 and the USL is 10.5. If the LSL were 10.5 instead of 9.8 and both values were 4 standard deviations from the mean value ($\mu = 10.1$), the equality $C_p = C_{pk} = C_{PU} = C_{pl}$ would be an acceptable short-term capability. However, for pharmaceutical products, with few exceptions, the specification limits are established and approved by several internal organizations such as research and development (R&D), quality assurance, pharmacology, clinical supplies, and other relevant operations prior to FDA approval. The role of manufacturing is to produce a product that conforms to the specifications in a quantity that produces a net income. Thus, it would be very difficult to change a specification limit solely for improving a manufacturing objective. Consequently, a large Six Sigma program geared for pharmaceutical manufacturing would be problematic. But, in lieu of large Black Belt managed programs, the use of a Green Belt program with a specified objective at a nominal program cost can be effective. An example would be a Six Sigma Green Belt program that could improve mixing homogeneity by altering the reactor impeller diameter, which in turn could shorten the mixing time, thereby possibly allowing additional manufacturing runs because of the shortened mixing time.

11.2.5 Analyze

The Analyze phase is highlighted by the completion of several deliverables that include:

- Defining and analyzing the relationship between dependent variables (commonly designated as the y value, or ordinate) and the x value (the independent variable, or abscissa). The standard analysis technique is to plot the x coordinate horizontally, while the y value is plotted vertically. Constructing a line that passes through the most plotted data points is the objective. The line offers the capability of defining an equation that describes the behavior of the data. There are several techniques available for defining the best line, or curve, for the data (e.g. visual and manual plot regression analysis, least squares fit). Most of the analytical tools employed for the techniques for the Analysis phase are readily available online or by using software such as Excel and ANOVA (analysis of variance). For small pharmaceutical engineering Six Sigma projects, a linear plot and the capability indices (C_p) are usually satisfactory since the ultimate goal is to demonstrate an improvement in the Six Sigma project objective rather than achieving a predetermined goal. Also, if the data covers a small range, a linear equation is usually satisfactory. Cost containment is also a critical part of a Six Sigma project. For smaller projects, it is important to consider the most effective analysis for the project. Using a data reduction mode such as ANOVA may not be required for the task; merely using the technique because it is available may not enhance the value of the project.
- Another important deliverable is the need to compare the current performance with the performance objective or goal. In pharmaceutical operations, this comparison is referred to as gap analysis. For large corporate Six Sigma Black Belt lead programs, a gap analysis is specific to a service or product. For a product, the program typically involves several elements of the company and can involve several departments and/or divisions. As emphasized previously, a pharmaceutical engineering Six Sigma initiative involves one, or two, at most, organizations, usually engineering and/or manufacturing. There are commonly five components involved in preparing a gap analysis:
  1) Identify the existing process – This element is essentially the material identified in the SIPOC as the PFD and the P&ID that describes the process.
  2) Identify existing operation – In addition to a process description, the current operating results should be detailed. This includes the data analysis performed (e.g. the critical x and y parameters, the reduced data in graph form, and other analytical tools that were used).
  3) Identify the desired performance goal of the Six Sigma program – For pharmaceutical engineering applications, this objective should be a quantitative value. For instance, referring to the agitator initiative where the impeller change is to be studied, an actual value of the number of extra runs that can be performed should be provided. While the desired goal may not be ultimately achieved, the target value will assist in determining the procedures and improvements for the existing process to achieve the desired goal. For this agitator project, measurements and data analysis should include data collection for the agitated Reynolds number (information and design material is found in Chapter 4, Section 4.1.5), viscosity, density, and reaction rate, if applicable (Chapter 4, Section 4.1.4), as well as operating temperature and pressure data. The operating variables required should be ascertained by the organization(s) responsible for performing the Six Sigma pharmaceutical engineering project. The services of an in-house or outside subject matter expert (SME) may also be helpful.

Prepare and document a comprehensive gap analysis – As a result of the data collection and analysis, the deficiencies or shortfalls of the program should be identified. For example, the agitator studies may reveal that the agitator blade diameter is too small to achieve proper mixing or the agitation speed (rpm) is too low. A gap analysis would involve identifying the symptoms (e.g. low yield from agitation operation) and causes (e.g. agitator diameter, improper Reynolds number, agitator speed). When it is agreed to by the Six Sigma team that a comprehensive root cause analysis has been completed, this generally occurs during the Tollgate, and the next DMAIC activity, Improve, can be performed.

11.2.6 Improve

The important product expected from this activity, Improve, is presenting the results of the Six Sigma project with a defined procedure describing how the results will enhance the performance of the program. Subsequent to developing the best solution, this phase also involves the evaluation criteria selected to achieve the best solution.

As part of the Measure phase and the Analyze phase, data was collected and analyzed to determine what variables of the Six Sigma program were critical to achieving the defined final product. The Improve phase is designed to present the parameters that will provide for a successful program result. The parameters, or variables, should be organized such that they can be readily evaluated by ascertaining the effect that planned changes have on the selected Six Sigma process. A methodology that can measure these effects is available and is identified by the term design of experiments [7]. The intent of a designed experiment is to generate sufficient data involving a product or a process that will allow for an effective improvement by employing the results of the data obtained through a series of planned experiments referred to as treatments.

Before delving into the procedures typically used in a DOE program, it would be helpful to define terms common to DOE operations:

- Factor – A planned activity that may affect the responses (results) at a given level.
- Level – The number of values of a factor (variable) to be evaluated in the experiment.
- Response variable – The factor that has been evaluated and produces recordable data.
- Treatment – An experiment using a predetermined number of factors at a specified number of levels (e.g. evaluating a chemical reaction at two temperatures and two reaction times). Treatments can be understood to be the combination of levels and factors required to evaluate the experimental variables selected for the DOE study.

Prior to implementing the DOE program, several program characteristics should be identified and defined. At a minimum, the plan should include [8]:

1) Objective – This element differs from Section 11.2.3 since it is specific to a procedure within Six Sigma and not a program explanation. For this element, the intent of the DOE Objective is to detail the experimental procedures.
2) Process variables – Specific variables of the DOE are fully identified.
3) Experimental design – The DOE elements should be specific to the variables that are selected for evaluation. For example, if the DOE selects temperature and time as variables, installing a strobe to measure agitator speed (rpm) would be unnecessary.
4) Execute the design – The DOE plan should be performed in accordance with approved procedures which are specific to the approved design plan.
5) Verify and analyze – Results should be consistent with expected results. Significant deviations should be analyzed to address the deviation(s).
6) Additional action – Based on experimental results, the need for additional data collection should be determined.

For Six Sigma programs, employing DOE, the design plan employed is using a matrix-like format. The variables for this matrix-like configuration are the factors that are defined above. The operation consists of identifying the factors that are critical to process performance. These factors, or variables that have been addressed in the Measure and Analyze phases as critical components of the process, can be identified as the factors. Some typical factors in a pharmaceutical Six Sigma process could involve temperature, concentration, and time. For Green Belt Six Sigma pharmaceutical programs (~\$35 000), the number of factors considered critical will typically range from two to no more than six factors. The next DOE requirement is the number of levels required. For instance, if a DOE study wishes to obtain reaction information for pilot plant reactions at two operating temperatures, 100 and 120 °C, there would be two experimental levels that data will be collected for as the reaction proceeds. The higher temperature, 120 °C, is identified with a + notation, while the lower temperature, 100°C, is noted with a − notation. In addition to studying, via experiments, the effect of temperature on the reaction rate, the effects of varying the agitator speed (rpm) on the program is also to be studied. Two levels were identified: a low speed, 90 rpm, designated by a "−," and a higher speed, 150 rpm, designated by a "+." For Green Belt projects such as the temperature of the reaction and pilot plant experiments focused on agitator speed, it is desirable to limit the number of levels to levels to two levels: high, +, and low, −. Two or three levels are useful since they will minimize the number of bench-scale or pilot plant-scale experimental runs. This particular DOE configuration would consist of two factors, reaction temperature and agitator speed. Similarly, the experimental design would have two levels: + for the higher reaction temperature and − for the lower temperature and a − for the higher and lower agitator speeds, respectively.

The term that identifies the number of factors and the number of levels evaluated in a DOE study can be integrated into a single term; the term for the combined variables, number of levels, and number of factors is, as noted earlier, treatments. By knowing the number of treatments, the number of experiments, or runs, the data input required for a DOE study can be calculated. A relationship that correlates the number of levels, the number of factors, and the number of treatments needed can be expressed by the relationship [7]

$$NOTR = 2^n \tag{11.16}$$

where

NOTR = the number of treatments required
2 = number of levels to be evaluated
n = number of factors, or variables to be evaluated

For pharmaceutical engineering Six Sigma projects, where the funding is in the \$35 000 range, the number of levels should be limited to two, if possible. The planned improvement should be based on a high level evaluation (+) and a low level evaluation (−). Since the major purpose of these smaller Six Sigma initiatives is to verify that a process or manufacturing modification can favorably affect an existing operation, exhibiting a favorable outcome in a DOE procedure is often significant since it can be viewed as a "proof of principle" demonstration and can act as the basis for implementing a change control action.

Example 11.10 A project employing Six Sigma principles has been approved to ascertain the feasibility of shortening the production run time for a reaction that produces an intermediate compound used in a drug product. The production staff believes that by increasing the reactor agitator speed from 60 to 110 rpm and increasing the reaction temperature from 110 °C (230 °F) to 120 °C (248 °F), a significant decrease in the processing time can be achieved. Prior to obtaining approval for the changes, an indication of success is required. Based on a limited number of runs, design a DOE matrix employing levels and factorials that will satisfy the experimental requirements. The reactor is designed and equipped to accommodate the increased temperature of the proposed study.

Solution
The matrix is composed of two levels (high [+] and low [−]) and two factors (temperature and agitator speed). This is a 2^2 full factorial design. The basic experiment matrix would consist of performing the following treatments (experiments):

Factor: temperature, +/− (level)	Factor: agitator speed, (level) −/+
−	−
+	−
−	+
+	+

Inserting the values into the above table for a 2^2 full factorial yields the following:

Temperature (°C)	Agitator speed (rpm)
110	60
120	60
110	120
120	120

The above matrix presents the procedural test outline for a full evaluation of the DOE required to implement the program.

For projects that require higher levels and more factors, Table 11.4 identifies the appropriate full factors, levels, and treatments for DOE common to typical pharmaceutical engineering projects.

The results of the DOE studies are intended to provide a solution or solutions that offer the best solution or solutions. When the optimal solution has been selected, an implementation

Table 11.4 Factors, levels, and treatments for smaller DOE programs.

Number of factors	Number of levels per factor	Number of runs (treatments)
2	2	4
2	3	9
3	2	8
3	3	27
4	2	16
4	3	81

plan is required. This plan identifies the critical elements that should be modified or prepared to assure successful operation. Typical items include:

1) Preparation of evaluation criteria.
2) Prepare a contingency plan that addresses problems that can arise.
3) Preparation of a risk analysis addressing the selected solution. This is often prepared in the form of a FMEA, the details of which is documented in Chapter 10.
4) Revision of process documents such as P&IDs and PFD, if applicable.
5) Develop an Evolutionary Operation (EVOP). This item clearly describes how through small changes in the operation allows for integration of the Six Sigma solution into the existing process with minimal operational disruption (e.g. for the temperature and agitator speed modification example, the plan would describe how the temperature is first adjusted, and then after several batches, the agitator speed could be increased, providing a record to observe the improved results).

When the relevant aspects of the "Improve phase" have been addressed, the Tollgate submission for this phase is then submitted.

11.2.7 Control

The Control phase, in addition to being the last element of DMAIC, has several important deliverables associated with this phase. For a pharmaceutical engineering DMAIC program, deliverable should include:

- An approved monitoring plan – This submission includes the activities that will provide activities that allow for real-time data collection and storage of the performance of the modified process.
- Approved SOPs addressing the modifications of the process as well as the modified procedures.
- An approved response plan – This submission details the actions required to correct, maintain, or adjust excursions that may occur during operation. Additionally, the response plan includes a troubleshooting plan identifying defects in operations and equipment that can occur. Corrective actions should also be identified.
- Transfer of ownership – This submittal, often referred to as the engineering turnover package (ETOP), is the body of documentation produced during the DMAIC. The ETOP, once submitted, is now the responsibility of the process owner to implement and control.

Recalling earlier, it was noted that Equations (11.14) and (11.15) were primarily applicable for real-time process control, whereas Equation (11.12) was more of a proof of principle relationship, which will indicate the potential success of a program when a specific C_P value range is obtained. When C_P is calculated, the value is based on the specification limits of the process or operation. That is, the USL and the LSL are the salient values. The problem is that the mean value for C_P does not enter into Equation (11.12). For instance, the mean value can be skewed to the left or right with no indication of where the mean value for the specification the range lies. Theoretically, a single C_P value can have many mean values yet maintain the same C_P value. As noted, other than exhibiting a capability of a successful outcome, based on a calculated C_p value, the C_P value, or short-term capability, provides little information concerning a timely evaluation of the process or operation performance in a given operating period. Without a reference value, it would be difficult to correlate capability with an ongoing process or operation.

In order to obtain useful data and employ the concept of capability indices contained in Equations (11.14) ($(USL - \mu)/3\sigma$) and (11.15) ($(\mu - LSL)/3\sigma$) can be incorporated. Note that these two equations use the mean value as a focus. That is, rather than having the entire range of values between USL and LSL, as is the case with Equation (11.12) ($C_P = (USL - LSL)/6\sigma$), the focus is placed on the USL and its distance from the mean, $USL - \mu$, and the LSL and its distance from the mean, $LSL - \mu$. Because the entire range of values is not used (i.e. USL − LSL), as is the case with Equation (11.12), only the 3σ range on either side of the mean is considered when dealing with C_{pk}. However, it can be seen that the mean is a key variable of C_{pk} evaluations. Referring to Figure 11.1, and recalling that point A is the mean value, with a reference value, point A, it is possible to compare the performance of a process or procedure over time, a task that cannot easily be performed when evaluating C_P. For instance, by measuring C_{pk} for a full production cycle, specifically C_{pl} and C_{PU}, determining these values subsequent to production runs allows for tracking the reliability of the process for a given production time. The C_{pk} values compare actual performance with the distance from the mean. By recording the results over time, trends in the process can be observed (i.e. over time, an increase, decrease, or no change will be recorded) based on the recorded data (this data can be recorded in the form of charts or a more detailed statistical analysis such as online curve fit routines, Excel routines, TI-83 calculator programs, etc.).

11.2.8 Lean Six Sigma

Simply stated, the term Lean Six Sigma is a concept brought to America in 1990 by James P. Womack, with the goal of eliminating "waste" through a planned, organized, and documented approach.

While the term Lean Six Sigma is recent in origin, Henry Ford, for one, can be credited with contributing the concept of Lean in 1926:

> One of the most noteworthy accomplishments in keeping the price of Ford products low is the gradual shortening of the production cycle. The elapsed time between the receipt of raw material and its appearance as finished merchandise bears strongly on the retail prices. The longer an article is in the process of manufacture and the more it is moved about, the greater its ultimate cost.

In this context, Lean means eliminating waste by reducing cost, making the organization (company) more competitive, and making the organization more profitable [1].

Two terms employed frequently in Lean thinking are accuracy and precision. Accuracy is commonly defined as the degree to which the average results of a measurement, calculation, or specification conform to the correct value, a standard, or a specification.

The second term, precision, is described as the closeness of agreement between results. Precision can also be interpreted as the ability to repeat the same measurement in identical environments. Figure 11.2 illustrates the general relationship between accuracy and precision.

Figure 11.3 identifies a target or reference value as the key reference point with accuracy being the deviation from the target value and the mean value of a normal probability distribution (Figure 11.1).

Figure 11.3 is another useful illustration. This illustration visually describes the difference between accuracy and precision.

The term waste, in terms of "Lean thinking," is composed of several aspects, as compared with the standard concept of useless or careless consumption.

Figure 11.2 Relationships between accuracy and precision. *Source:* Courtesy of All Sensors Corporation.

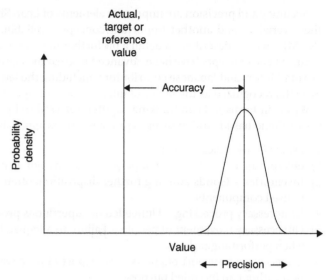

Figure 11.3 Differences between accuracy and precision. *Source:* Courtesy of All Sensors Corporation.

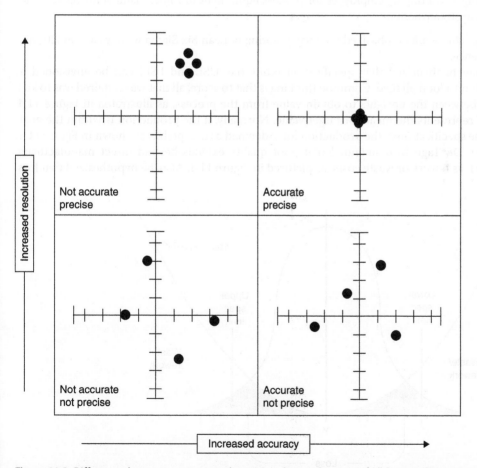

Accuracy and precision are important elements of Lean Six Sigma. The relationship between these variables and another important concept – reliability, availability, and maintainability (RAM) – will be described in detail a bit further in this section.

Since Lean concepts have been advanced in great measure by Japanese organizations such as Toyota Motors and Japanese contributors, including the late Dr. Ohno and the late Dr. Genichi Taguchi, recognized as the father of quality engineering.

Waste, in terms of Lean thinking, is often referred to by the Japanese term Muda, meaning waste. Ohno has identified seven types of waste, as applicable to Lean concepts [1]:

1) Defects – Mistakes to be fixed
2) Overproduction – Production of goods that are not wanted
3) Inventories – Goods awaiting further disposition, often resulting in inventory backlog and unused components
4) Unnecessary processing – Unneeded or superfluous processing operations
5) Unnecessary movement of people – Failure to adequately analyze movement of employees when performing tasks
6) Unnecessary transport of goods – Transport of processing material from location to location without an intended purpose
7) Waiting – Waiting by employees for process equipment to finish manufacturing activities or an upstream activity

Dr. Taguchi's work emphasized the incorporating of Lean Six Sigma with waste, quality, and maintenance.

Dr. Taguchi theorized that specification limits (i.e. USL and LSL) can be envisioned as goalposts on a football field. Common thinking is that to score, all that was required was to kick the ball between the uprights to obtain value from the process, as illustrated in Figure 11.4, where *Y* represents the mean, or target value. Normally, if the production is within the goalposts (the specifications), the production lot is deemed as acceptable, as shown in Figure 11.4.

Further, Dr. Taguchi maintained that poor quality extends beyond direct manufacturing costs such as rework or waste costs, as pictured in Figure 11.4. Also, he hypothesized that loss,

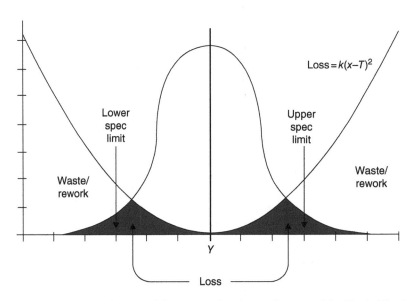

Figure 11.4 Process losses and the target value. *Source:* Courtesy of the Marshall Institute.

rather than being a linear function, is proportional to the square of the distance, x, from the target value, T, as shown in Figure 11.4.

Referring again to Figure 11.4, T represents the target value (in this figure, $Y = T$ and product loss is given by the parabolic relationship, Loss $= k(x - T)^2$, where k is the proportionality constant) [12]. The value $x - T$ also represents the accuracy of the value. Another important detail shown in Figure 11.4 is the difference in values between USL and LSL. This difference in values represents the accuracy. Interestingly, Dr. Taguchi hypothesized that additional losses are produced as a result of failure to achieve greater accuracy and precision relative to the mean target value, Y. In Figure 11.4, identifiable losses are identified in the areas bounded by the specification limits, the x axis (the abscissa), which in this case represents the distance from the target value, Y or T, and the loss curve, and below the loss function curve, identified as loss in Figure 11.4. Additionally, as Figure 11.4 indicates, Dr. Taguchi is implying that any deviation from the target value is identified as a loss.

Figure 11.4 serves to further define these important elements when action is initiated to reduce waste generation. As Figure 11.5 illustrates, to achieve the least loss, the target value must consistently be "hit" every time. "Once in a while" is not acceptable as common quality operations viewed as the prevailing thinking.

Figure 11.5 portrays a greater precision since the USL and LSL are closer to the target value. Similarly, since the target value, Y, is achieved more frequently, the accuracy is improved. These results produce reduced loss in the form of waste, scrap, and rework, also referred to as quality losses. Assuming manufacturing perfection, there would be no deviation from the target value and, consequently, little or no losses due to poor accuracy. Greater precision would also result in lower losses since the defined operating specifications are also tightened. In actuality, the distribution curve shown in Figure 11.4 is what actually occurs, a significant improvement but not perfection [12].

Further, narrowing of the operating specification values, tighter operating specifications, would reduce the excess production quantities resulting from out of specification (OOS) product or, as identified earlier, product loss. The resulting reject material would require a rework.

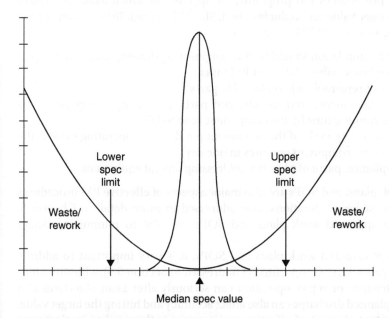

Figure 11.5 The effect of consistently hitting the target and the goalpost analogy to production specifications. *Source:* Courtesy of the Marshall Institute.

If the reject material is not destined for rework, the waste products are most often classified as nonhazardous or hazardous waste – an option subject costly to off-site disposal.

With waste reduction identified as a significant factor in Lean Six Sigma programs, it is logical to question what techniques can be applied to actually effect a significant improvement in many pharmaceutical operations as well as nonpharmaceutical manufacturing operations. An effective answer has actually been detailed previously in Chapter 5. Chapter 5 focuses on system maintainability, maintenance, and reliability and the relationship of these elements. By focusing RAM procedures, it is also possible to significantly reduce process wastes and rework of rejected batches. Such a reduction would also reduce lost time due to the disposition of reject material.

Recalling Equation (5.8) of Chapter 5, $A_i = \text{MTBF}/(\text{MTBF} + \text{MTTR})$, this relationship when integrated with the Lean elements defined previously can be used to design a program to affect losses in many pharmaceutical manufacturing operations. Further, these improvements can be implemented by using elements of the material previously presented in Chapter 5.

The salient content of Chapter 5 deals with the goal to achieving an availability as close to unity as possible. To achieve this goal, topics such as reliability, human factors, maintainability, preventive maintenance (PM) procedures, and maintainability were introduced. These elements are primarily focused on improving the performance (i.e. failure and repair times) of equipment and systems. Little or no coverage was given to the relationship between RAM and the target value. Chapter 5 also deals with improving the performance and safety of systems and equipment and how to extend the working life of systems by shortening the mean time to repair (MTTR) and the mean time between failures (MTBF).

An effective maintenance activity is a key element that can result in "hitting the target value" with increased frequency as a result of less downtime and shorter repair time. These factors, when effectively implemented, result in increased reliability and reduced costs.

An effective maintenance activity uses a well-documented proactive PM program as a foundation. Section 5.4 details the activities and tasks required for a program that is designed to improve accuracy and precision of PM programs, the specifics of which allow for greater frequency of hitting the target value and reducing the USL – LSL spread. Briefly, some of the characteristics identifying a successful PM program are:

- Compile data – The data compilation should include engineering drawings, parts list, repair manuals, approved procedures, failure data, and PM tasks.
- Determine the criticality of repairable and replaceable parts.
- MTBF data for critical equipment, components, and parts average operating period, is obtained by dividing the running time by the component time to fail.
- Determine the average operating cycle of the equipment per day. The operating cycle is the amount of time, in hours; the equipment operates in one day.
- Maintain schedule compliance, particularly when addressing critical equipment.

Effective task orders, job plans, and SOPs are also major aspects of effective PM procedures and, consequently, enhanced Lean operations, also addressed in more detail in Chapter 5, Section 5.4. Well-written approved work plans and SOPs are vital to minimizing delays and waste.

In addition to equipment-centered work plans and SOPs, it is very important to address human factors and safety procedures (a topic further addressed in Section 5.5); a life-threatening poorly defined maintenance or repair operation can seriously alter Lean objectives at a minimum. Further, an unplanned discharge can also affect accuracy and hitting the target value.

Perhaps the most important element of effective maintenance is the need to implement a conscious awareness of the interrelationship between effective maintenance, accuracy, precision, and safe operations. By allowing delays and interruptions to be the norm, rather than the exception, even the best Lean operations will be affected [12]. To achieve this goal, more than effective SOPs, specifications, and items such as availability of spares and tools are required; a conscious Lean thinking mindset should be encouraged. Fortunately, since effective maintenance operations coincide with Lean objectives, a significant synergy can be achieved by effective RAM program implementation.

For pharmaceutical engineering applications as well as other operations involved in Lean thinking, the ability to translate definitions and concepts to practical calculations is most important to further implement Lean (Lean Six Sigma) programs. Consequently, some additional Lean terms, as defined by the US government, should be introduced. These significant terms include:

Compile data – The data compilation should include engineering drawings, parts list, repair manuals, approved procedures, failure data, and PM tasks.

Determine the criticality of repairable and replaceable parts.

MTBF data for critical equipment – This information, if covered by cGMP criteria, should be stored in an easily retrievable format that conforms to the requirements of 21 CFR Part 11, Electronic Signatures.

Determine the average operating cycle. The operating cycle is the amount of time, typically in hours; the equipment operates in one day. The average operating period, h_{av}, is obtained by dividing the number of operating days into the total hours the equipment is operated while in service (defined by Equation 5.1 of Chapter 5):

$$h_{av} = \frac{h_1 + h_2 + h_3 + \cdots + h_n}{d_t}$$

where

h_{av} = average operating cycle per day, h/day

$h_1 + h_2 + h_3 + \cdots + h_n$ = the sum of the individual operating hours per day, hours

d_t = total number of operating days

Completion rate – The number of tasks completed in a specified time period, commonly expressed in tasks/day.

Work in process (WIP) – The amount of products or transactions that are being processed or waiting to be processed. WIP is given in the time (minutes, hours, days) required to produce a given number of units.

Total lead time (TLT) – Total time to develop a product or deliver a service to a customer, typically expressed in days and can be given as

$$TLT = \frac{\text{Number of items in the process}}{\text{Average completion rate}} \qquad (11.17)$$

The TLT can also be expressed as a function of the work in process (WIP) by the relationship

$$TLT = \frac{WIP}{\text{Completion rate}} \qquad (11.18)$$

With the terms work in process and total lead time defined, the completion rate can be expressed in a more useful form as

$$\text{Completion rate} = \frac{WIP}{TLT} \qquad (11.19)$$

The relationships defined earlier can also be categorized as measurement variables as identified in Section 11.2.4. But these items are also part of Lean metrics since they can identify potential sources of waste in manufacturing operations. This is accomplished by comparing actual results with actual manufacturing operations. Example 11.11 presents a typical Lean metrics problem employing some of the definitions previously presented.

Example 11.11 A generic tableting operation is contracted to produce 9 000 000 tablets in a day (1 day of manufacturing is composed of three 8 h manufacturing shifts, with 2 h of each shift reserved for maintenance and other downtime operations). The tablet press is configured to produce 8000 tablets/min. What is the total required lead time (TLT) required, and what is the planned completion rate?

Solution
Equations (11.18) and (11.19) are the basis of calculation for this example:

WIP = 9 000 000 tablets
Completion rate = (8000 tablets/min) (60 min/h) (18 h/day) = **8 640 000 tablets/day**

Inserting the given variables in Equation (11.18) yields

$$\text{TLT} = \frac{\text{WIP}}{\text{Completion rate}} = \frac{9000000\,\text{tablets}}{8640000\,\text{tablets/day}} = \textbf{1.04 days} \leftarrow \textbf{Solution}$$

The lead time required to manufacture 9 000 000 tablets is 1.04 days (based on an 8 h shift) or about 8.25 h.

In reality, the actual completion rate would be somewhat longer. One reason for the longer completion time would be the need to remove finished tablets from the manufacturing area. This task is often a manual process since the tableting press is in a confined area and space is at a minimum. Consequently, during product removal, as well as rejects, this may require a brief cessation of operations. Other potential delays are the need to replace punches during tableting operations. Similarly, the buildup of tablet dust during operations impedes operations and may require unanticipated maintenance.

The values determined in Example 11.11 are planned production values. Actual results are usually less than the planned number of tablets actually produced. Similarly, the actual time required is often longer than planned. Tablet press punch failure (MTBF), spare punch delivery time, installation time, restore time (MTTR), and the time required to commence operations are all factors that waste, in this example, both time and lost product.

11.3 Process Analytical Technology

While the term process analytical technology may be fairly recent nomenclature, the basic principle has been employed since the publication of USP 23 with the requirement to monitor water purity continuously monitored in µS/cm rather than the conductivity units. The objective of the modification was to eliminate the time-consuming analytical delays and sample contamination often caused by a lack of timely water quality analysis and to replace this existing technique with a continuous analytical method that analyzes, monitors, and records pharmaceutical water properties such as total organic carbon (TOC) concentration and conductivity.

Additionally, recordkeeping created difficulties due to a lack of uniform records management and, in some cases, a paucity of traceable "chain of custody" procedures for samples and documentation. Real-time processing and records management initiated a need for a new approach to pharmaceutical manufacturing and quality assurance. As a result, a PAT initiative was created to address the aforementioned and associated manufacturing impediments typical of pharmaceutical operations. Further, with the publication of PAT guidance, the FDA has taken the position that quality cannot tested into drug products, but rather quality must be inherent to the manufacturing process. Additionally, 21 CFR 211.165 (a) requires comprehensive testing prior to release of a drug product.

The FDA views PAT as a tool "...intended to support innovation and efficiency in pharmaceutical development, manufacturing, and quality assurance. The framework is founded on process understanding to facilitate innovation and risk based regulatory decisions and the Agency (FDA). The framework has two components: (1) a set of scientific principles and tools supporting innovation and (2) a strategy for regulatory implementation that will accommodate innovation."

A working definition of PAT is given as follows [10]:

> PAT is a system for designing, analyzing, and controlling manufacturing through timely measurement (that is, during processing) of critical quality performance attributes of raw and in-process materials and processes, with the goal of ensuring final product quality. Although the term analytical in PAT is broadly defined to include chemical, microbiological, mathematical, and risk analysis, it is performed in a planned and logical sequence.

In addition to PAT being an official position of the FDA, there are several reasons for pharmaceutical manufacturers to implement PAT. These rationales include:

- Reduction of production cycle times
- Increasing yield on scarce drugs
- Shorter time to market new and existing products to consumer
- Improved efficiency by reducing product batch variability
- Reduce rejects, scrap, and rework (addressed in more detail in Section 11.2.7)
- Significantly reduce disposal costs by elimination of chemical reagents
- Replace traditional wet chemistry techniques with faster methods
- Ability to analyze multiple components in a real-time format
- Improve operator safety (addressed in Chapter 5, Section 5.5)

As part of the PAT applications, several methods have been identified as effective items and techniques to use as a basis for PAT implementation. The FDA further states these tools and concepts can be categorized as follows [13]:

- Multivariate tools for design, acquisition, and analysis – Mathematical approaches such as statistical DOE (Section 11.2.6) and statistical evaluations employing the material in Section 11.1.1 as a basis are encouraged in PAT applications.
- Process analyzers – These tools are implemented in PAT programs as a replacement for time-consuming holdup time laboratory analysis prior to batch release. The thrust of process analysis is to provide nondestructive techniques that minimize or eliminate the need for quality control laboratory analysis. Process analyzers emphasize the use of automated systems that can measure, control, and store data in real-time without interfering with the

continuing manufacturing operations. The analytical tools used to achieve these results are commonly of three categories:

1) At-line – This type of installation is most similar to the standard analysis where samples are removed from the process and analyzed, typically in a laboratory.
2) Online – This technique involves measurements where samples are diverted from the manufacturing process, often by means of a loop in the operation, and, subsequent to automated instrumental analysis, reintroduced to the manufacturing process stream.
3) In-line – This design involves analysis while the process is operating without isolating the samples for laboratory analysis or bypassed in a loop. Table 11.5 identifies some of the instruments employed for this PAT application. Since elements 2 and 3 are quite similar, the term online is often used to denote both types of installations.

Table 11.5 Selected process analytical technology (PAT) evaluation parameters and current tools.

Parameter	PAT tools	For consideration
Mathematics and statistics	• Design of experiments (DOE) (Section 11.2.6) • Capability indices (Section 11.2.4.1) • ANOVA • Minitab • Excel • Multivariate tools for design, data acquisition, and analysis	• Minimize levels and factors (Equation 11.14) • Focus on C_{pk} values • For small projects, TI-83 or higher, calculator, and Excel are satisfactory tools in lieu of Minitab for a particular application
Analytical instruments applicable to PAT	• Optical sensors for real-time monitoring of oxygen levels and pH for bioreactors [9] • High performance liquid chromatography (HPLC) and ultrahigh performance liquid chromatography (UHPLC) • Raman and near-infrared (IR) spectroscopy [10] • Fourier transform/near-IR spectroscopy • Online temperature monitoring and data acquisition • Particle size analysis by laser diffraction [11]	• Single-use and/or disposable parts • Disposable HPLC columns • A robust, reliable technology widely used for laboratory studies and real-time analysis of particle analysis for food, powder coatings, and drugs
PAT information technology (IT) applications	• Real-time monitoring and data collection • Process control • Alarms and notification if process excursions occur • Data collection can indicate areas where process modifications can enhance efficiency and quality	• Data collection techniques should be compliant with the criteria defined in 21 CFR Part 11 (Electronic Signatures, Tracking, etc.) • Time-dependent PAT initiatives should be evaluated over a representative and defined time frame

- Process control tools – Process control tools are specific to two categories, namely, enhanced real-time analysis and greater use of statistical tools such as DOE, capability indices (C_P, C_{pk}), and risk assessment. Enhanced use of statistical tools allows for a more definition when evaluating specifications for manufacturing operations that must operate within these limits. Incorporating these mathematical formats and online analysis will allow for a tighter, speedier, and more accurate online process control, an important objective of PAT applications.
- Continuous improvement and knowledge management – An appropriate combination of some, or all, of these tools may be applicable to a single-unit operation or to an entire manufacturing process and its quality assurance. With the implementation of PAT analytical techniques, statistical methods, and risk assessment, manufacturing improvements such as equipment cleaning and maintenance changes can be implemented more rapidly and with a greater accuracy and reliability than traditional operations.

PAT applications are, for the most part, the responsibility of the user (i.e. the owner). The FDA Industry Guidance document (Ref. [13]) states:

> FDA does not intend to inspect research data collected on an existing product for the purpose of evaluating the suitability of an experimental process analyzer or other PAT tool. FDA's routine inspection of a firm's manufacturing process that incorporates a PAT tool for research purposes will be based on current regulatory standards (e.g. test results from currently approved or acceptable regulatory methods).

This guidance permits manufacturing firms to define the PAT systems that are most effective for their particular operations. The guidance document also indicates that PAT applications are reliant on the user's decision, and since most PAT tools are vendor supplied, the user is responsible for decisions to implement specific PAT applications. To effectively perform the task, the user should have an in-depth knowledge of the PAT requirements and the vendor solutions recommended. Also, PAT is intended to provide analytical tools that perform, ideally, fast online analysis and accurate data acquisition integrated as a complete operating system. The objective of PAT is to augment existing processes with enhanced analysis and not replace the existing process, an initiative that will be addressed in Section 11.4. PAT is also applicable to operations that support the manufacturing process. Operations such as acceptance of incoming raw materials, validation, and cleaning procedures are examples.

An example of the configuration of a PAT evaluation format and categories is presented in Table 11.5.

Table 11.5 identifies several PAT tools and concepts currently being used. It would be helpful to further define some of the items in the table to permit a more detailed overview of the table as well as highlight evolving PAT techniques and applications.

Table 11.5 also identifies various instruments and techniques used in PAT applications; there are also many online tools currently employed to measure, record, monitor, analyze (in real time), and control various operations of pharmaceutical manufacturing. PAT tools used frequently are online real-time monitoring and data acquisition and data storage, documentation of cGMP installations such as WFI and USP water operations, and temperature monitoring. Chapter 2 details the relevant features of these water manufacturing systems.

Temperature monitoring and control systems are often critical to many pharmaceutical operations, many of which are employed as PAT applications. In addition to water systems, common pharmaceutical operations such as HVAC, reactors, and fermentors (Chapters 3 and 4) are commonly used for temperature applications and requirements. Since temperature information monitoring, temperature control, and temperature data acquisition and storage are so important, it warrants attention to certain details not generally considered in cGMP installations. For instance, the tools used for temperature monitoring are often vendor supplied with little concern for the device employed. With the advent of QbD, detailed in Section 11.4, there is an increased requirement for more detailed and intricate analysis than previous vendor-installed and vendor-supported systems. One example of the increased need for detail is the selection of temperature probes. With the advent of PAT and QbD, the two most commonly employed temperature probes, namely, thermocouples and resistance temperature devices (RTD), require more scrutiny than previous non-PAT applications. While both thermocouples and RTDs are used in PAT and QbD applications, an RTD is considered for operations due to:

- The temperature deviation being less than 2 °C (3.6 °F)
- The operating temperature being in the –200 to 500 °C (–328 to 932 °F) range
- High accuracy
- Repeatability
- Stability, particularly where stability for precision applications
- Low drift

Thermocouples also offer advantages over the RTD, including factors such as:

- Stability.
- Cost – A thermocouple configuration is typically 2.5–3 times less than RTD installations.
- Ruggedness – Because of a less complex design, thermocouples can withstand more system stress during operation (e.g. physical forces applied to the thermocouple).
- For pharmaceutical applications, the two most common types of thermocouples are type J with an operating range from –40 to 750 °C (–40 °F to 1380 °F) and type K with an operating range from –200 to 1350 °C (–328 to 2460 °F).

In addition to proper sensor selection, such as a thermocouple or an RTD, a monitoring device is also integral for an effective online process analyzer, and all sensors must conform to the range and accuracy stated. In many cases, the failure of instrument sensors, if undetected, can have a significant effect on online analysis.

Since the feedback time between the onset of the real-time analysis and the recorded value is often critical, it is worth noting that many of the PAT initiatives involve applications where such feedback is very important; this is often the case for inorganic API processes. Biotechnology analysis often allows for a less responsive feedback mechanism since, in general, bioreactions, because of their slower reaction rates, require longer processing periods. Consequently, the need for quicker real-time analysis should be carefully evaluated. If a PAT application involves temperature monitoring or control, the need for a RTD or thermocouple should be defined by the vendor prior to installation.

In addition to instrumentation sensors, such as thermocouples and RTDs, it is also important to obtain comprehensive details concerning vendor-installed and vendor-supported PAT installations such as hardware, control system software, data analysis software, and process control software. Without a clear understanding of the purpose of the PAT-related installation and support obligations of the system supplier, post-installation conflicts can result in significant delays and unintended costs. These issues should be clearly defined and understood prior to installation and operation.

Again, referring to Table 11.5 it is observed that Raman and near-infrared (IR) spectroscopy are also commonly employed PAT tools because of their ability to perform real-time analysis. Another advantage of these real-time analytical techniques is the reduced probability of contamination due to manual handling and external environments.

Another useful and evolving PAT application is the handheld/portable spectrometer, which allows for identity confirmation of raw materials on the loading dock, often without opening liners (minimization of exposure hazards) [15].

While instrumentation applicable to PAT applications has shorter analysis times and exhibit enhanced accuracy, most analytical methods rely on physical properties as analytical tools. Online analytical tools such as HPLC rely on physical measurements such as viscosity, refractive index, and spectroscopy to determine constituent concentration. By employing physical measurements, the sample preparation time is significantly reduced. Traditional procedures such as drying the analytical sample to constant weight are not normally required, thus allowing for quicker analysis. Also, instruments such as Raman spectroscopy are not affected by water concentration. This too shortens the analysis time and permits real-time online analysis.

An often used PAT analytical tool is Fourier transform infrared spectroscopy (FTIR), identified in Table 11.5. Basically, FTIR is an analytical method that uses the longer nonvisible wavelength IR spectrum (700 nm–28 µm) to measure the absorption of light, emitted by a lamp at a specific spectral range. Each organic compound has a certain absorption range, and by detecting and analyzing the absorption spectrum of the compound, the composition can quickly be determined. A Fourier transform is integrated into the analytical algorithm to assist in the analysis. FTIR is increasingly popular with the evolution of lightweight handheld units that provide quick and accurate analysis. As a result, pharmaceutical operations such as certification of incoming material can have significantly reduced processing times, which would permit a reduction of the holdup time for manufacturing operations employing the required organic material.

With the greater analytical capability and analysis, PAT will significantly produce more data requiring much larger data storage requirements. This increased requirement indicates that 21 CFR Part 11 requirements such as time stamping, document tracking, security, and increasing deployment of automated analysis will become more relevant.

As a result of the implementation of PAT initiatives, it is possible to more closely monitor and record manufacturing process control operations. For instance, until the implementation of PAT instrumentation, real-time data acquisition, and analysis, few techniques were available to determine the progress of a chemical reaction. Often, reaction rate progression was determined by parametric measurements such as reactant concentration, viscosity, oxygen consumption, pH changes, and carbon dioxide concentration for biochemical operations. However, with the advent of real-time data analysis and acquisition, it is possible to track the progress of bioreactions and inorganic reactions in an accurate direct measurement mode. Not only is the accuracy enhanced, but these in situ, online analytical instruments, and associated electronics offer a significantly faster analytical result for measuring critical quality attributes.

Chemical companies have long used continuous processes for production of large volumes of materials, in situations where the reaction conditions are best operated in a flow paradigm, and where process control can be achieved with in-line controls such as those employed for PAT applications. The legacy approach to process design in the pharmaceutical has leveraged batch operations. Batch manufacturing has provided a convenient method to scale up from laboratory to facility manufacturing. Also, batch processing served as an effective method to monitor operations since sampling and analysis prior to PAT was often time consuming and completed batches are often delayed until lot approval is obtained. However, the introduction of online continuous analysis presents the possibility of designing manufacturing operations that are more akin to chemical company operations employing continuous operations. With effective real-time analysis, API manufacturing operations need not be halted due to quality control analysis and associated delays; the product can be forwarded to the next processing step with little delay. Rather than labor-intensive transport of bulk product, continuous lines can be employed to move the API with no significant holdup time.

Another benefit derived from evaluation and implementation of PAT analysis is a significant reduction of equipment downtime due to cleaning operations. Since the same equipment is used for a continuous or, at a minimum, longer batch run, less time is required to prepare the next manufacturing operation.

Powder analysis for pharmaceutical powders is often problematic since there is a paucity of FDA or USP analytical tools specific to pharmaceutical powder operations. While several empirical and semiempirical methods are described in Chapter 7, the techniques described are not widely used, thus far, in widespread PAT applications. One technology that is being actively employed is particle size analysis by the application of laser diffraction, as identified in Table 11.4. Laser diffraction is recognized by the USP as a technology applicable to pharmaceutical particle sizing and addressed in USP Edition 30, Monograph 429 (USP <429>). As is the case with many current and evolving PAT applications, particle analysis by laser diffraction relies on physical measurements rather than chemical analysis. This permits for a rapid real-time measurement with the goal of having an online operation.

The basic principle of laser particle size analysis involves passing a laser beam through a specifically prepared carrier solution containing the powder particles to be analyzed (laser particle analysis is also applicable to atomized mixtures where the particles are intended for nasal administration. The technique is also applicable for particle analysis for an injectable solution). When the incident coherent laser beam interacts with the particles in the sample solution, the particles cause the laser beam to scatter at various angles. The smaller diameter particles scatter the light relatively weakly at wide angles, while the larger diameter particles scatter strongly at angles close to the incident beam. The scattered light distribution is detected and recorded in real-time via analyzers and sophisticated analytical software. Subsequent to detection, the detected angular data is converted, via the software, to a particle size distribution using statistical techniques.

This nondestructive laser diffraction method is fast due to its ability to rapidly measure complete particle distribution rather than cumulative measurements of individual particles that would require many test runs. The technique is also desirable since sample preparation is straightforward once the media characteristics are defined. Also, because the complete particle size distribution is detected in a single analysis, there is no need to recalibrate subsequent to single laser scattering analysis [11].

While laser particle diffraction is becoming an important tool for powder particle sizing, there are several elements that, as USP 30, Monograph <429>, states, should be considered and evaluated when considering this analytical technology:

1) The wavelength selected for particle analysis should be unaffected by other particles that are present.
2) The system constituents should be compatible with instrument materials of construction.
3) The particles to be evaluated cannot dissolve in the solution in which they are suspended.
4) The particles must be easily dispersed in the chosen solution.
5) The viscosity can be readily determined.
6) The particle refractive index must be different from other compounds that may be present.
7) Health and safety hazards.

Laser diffraction while currently a popular technique has some analytical concerns that should be evaluated prior to use in pharmaceutical powder and particle analysis. One key assumption of operation is that the particles being analyzed by laser diffraction are randomly distributed in the solution and are in a turbulent flow regime. If the flow is laminar, the procedure can cause nonspherical particles to orient in the direction of the flow, thereby often negating the assumption of random particles in solution.

Another questionable assumption relating to laser diffraction is that volumetric (mass) results can be correlated with spherical particles. Often, in the computational paradigm selected, the volumetric variable is determined by using the cube of the radius to approximate an analytic volumetric variable. As a result, the calculated volume may actually be a measure of the particle surface area and not a particle volume. Studies have been performed to confirm that one technique (Coulter laser diffraction) practically coincides with the projected surface area, rather than the volume (recalling the surface area of a sphere expands more rapidly than the volume of a spherical particle of the same radius).

While laser diffraction is a promising technology, much of its applicability is dependent on effective analysis through the use of reliable numeric algorithms and software that can be validated. Proper validation involves tests of accuracy and use of instruments that can effectively measure nonspherical powder particles.

11.4 Quality by Design

Since its introduction in 2003, QbD has allowed regulatory bodies and pharmaceutical manufacturing to evaluate process design and performance with an emphasis on risk understanding and mitigation. The design of processes with quality in mind allows the developer to achieve processes with much greater understanding and control; by providing well-designed operational reliability, QbD removes regulatory risk and lowers cost. The multiple perspectives on QbD in pharmaceutical and biotech manufacturing will shed light on this important tool for process performance conformance and product quality [16].

The formalized concept of QbD is arguably the consequence of reports such a 2006 Georgetown study concerning pharmaceutical manufacturing. The study noted [17] that "If the FDA could change the way it regulated...the industry could save 10% to 50% of the cost of goods sold." As a result of the study, and other similar reports, the FDA produced a new guidance documents incorporating guidance covering concepts such as product life cycle, QbD, risk-based approaches (addressed in Chapter 10), and statistical process control (specific details are addressed in Section 11.2.4).

QbD is a topic that has many similarities to Six Sigma program elements, most notably an increased requirement for secure documentation and data storage. Also, several process elements of QbD are identical to identified aspects of DMAIC, as will be seen.

The International Conference on Harmonization (ICH), an organization seeking to develop common pharmaceutical standards, defines QbD as "A systematic approach to development that begins with predefined objectives and emphasizes product and process understanding and process control, based on sound science and quality risk management."

The development of QbD can be viewed as a logical consequence of PAT since QbD employs many of the same analytical data capturing tools in real time that PAT emphasizes. The relevant difference in the philosophy behind PAT and QbD is that PAT emphasizes utilization of techniques and analytical tools to essentially improve processing times for production operations and, to a lesser extent, R&D schedules.

In addition to emphasizing online testing in preference to end product testing, QbD is intended to be evaluated and implemented by incorporating tests and planning involving:

- Development studies – Typically, this aspect involves employing preliminary results of laboratory studies, pilot data, statistical procedures including Six Sigma techniques, and analytical calculations to serve a basis for a completed process design.

- Risk assessments – For this topic, detailed evaluations of the elements of the process design are undertaken. Risk formats, such as described in Chapter 10, can be used as a basis for many processes.
- Evaluation of process consumables – Materials such as HPLC matrices, ion exchange resins, filtration, and reagents should be evaluated for costs, disposal, replacement frequency, and associated materials and procedures based on the life cycle of the system.
- Microbial control – While QbD integrates PAT, online analysis of microbial activity presents a challenge. Bioreactor operations must be capable of analyzing process progress and also detecting the presence of harmful microbes and endotoxins in the reaction broth. As an alternative, increased focus will be given to nonprocessing activities such as cleaning and cleaning validation [18].
- Process monitoring – This concept indicates that, unlike past practices, once a process has been qualified via process validation (Chapter 10, Section 10.4), continued process monitoring and sampling at the process validation level is "recommended." Specifically, the FDA notes:

  > We recommend continued monitoring and/or sampling at the level established during the process qualification stage until sufficient data is available to generate significant variability estimates. Once the variability is known, sampling and/or monitoring should be adjusted to a statistically appropriate and representative level.

  While the guidance indicates the monitoring level can be reduced to levels that confirm that the process operates continuously within the operating limits defined (i.e. USL and LSL), this additional requirement suggests the need for additional personnel to specify, monitor, and operate the additional statistical and analytical requirements imposed on the guidance. For smaller manufacturing operations, this guidance could significantly increase the costs required to obtain product approval. Should these enhanced programs continue to be implemented, smaller operations as well as larger manufacturers will require additional resources to assure costs, resulting from additional monitoring, sampling, and process controls, are incorporated into the process in an economically viable approach. As is the case with PAT, additional resources such as enhanced engineering knowledge, more sophisticated statistical technology, additional process control capabilities, and more sophisticated software and IT capabilities will be needed. Successful QbD will plan, in advance, to accommodate these additional requirements. The need for advance planning should be evaluated using the system life cycle as a basis for evaluation. That is, rather than planning and provisioning system support on an annual basis, planning needs such as spare parts, maintenance, scheduled inspections, and contractor support should be predicated on multiyear needs.
- Continuing process improvement – The new criteria for process operations, as outlined in part by QbD, is intended to make it easier for manufacturers to implement continuous process improvement. The FDA guidance notes, "Data gathered (during the continued Process Verification stage) might suggest ways to improve and/or optimize the process by altering operating ranges and set-points, process controls, or in-process materials." So, the FDA clearly expects manufacturers to study, learn, and improve their processes [17]. One obvious impact of this is the emphasis on a continuous evaluation of processes as opposed to reliance on past data. Consequently, areas such as continuous process control (including distributed control systems [DCS], continuous monitoring of equipment in real time, and greater reliance on built-in testing [BIT], and built-in test equipment [BITE]) will require additional skill sets to be introduced to pharmaceutical manufacturing. Again, these requirements may offer significant challenges to smaller pharmaceutical manufacturing operations.

The intent of QbD is to, in addition to wider use of PAT tools, implement manufacturing operations where many new procedures and processes are an inherent part of the design. Currently, PAT innovations are incorporated subsequent to existing system installation and deployment. QbD is intended to integrate many elements of PAT as a basic design consideration for new pharmaceutical/API processes. In addition to enhanced data acquisition, data storage, and process control, techniques such as risk assessment and FMEA are baseline activities in the QbD vision of pharmaceutical and many API projects.

Some of the contrasts between the existing pharmaceutical manufacturing paradigm and the QbD concept of pharmaceutical manufacturing are presented in Table 11.6 [12]. Several significant benefits can be achieved for pharmaceutical operations by adopting QbD concepts [14]. An analytical approach to manufacturing by incorporating QbD principles results in producing finished items with reduced manufacturing difficulties and less waste as a result of a reduction in rejected batches. One example of this is the "streamlining" of the review process, which, when applied significantly, reduces the review process, resulting in speedier approvals. Another effective QbD initiative allows for a continuing improvement in manufacturing and resulting product quality. With the acceleration of data analysis, manufacturing and product modifications can be implemented with a shorter lag time. Another significant improvement offered by QbD is the ability to interact with the FDA in a format predicated on a more scientific overall approach rather than a process specific approach. This "global" approach to manufacturing processes allows for less emphasis on stratified or fixed procedures and more emphasis on the overall effect of a more dynamic process or product alteration without excess FDA examination due, in part, to implementation of risk assessment and FMEA procedures.

Table 11.6 Overview of standard pharmaceutical operations and quality by design.

Manufacturing variable	Standard technique	QbD technique
Pharmaceutical development	Empirical evaluation – single-variable experiments	Systematic – the design is a total system and the effects are evaluated by a multivariable approach
Manufacturing	Fixed-product acceptance determined by existing specifications and procedures	Adjustable (variable) – with the advent of PAT, rapid processing changes can be implemented
Process control	In-process testing and offline analysis, often affecting a slow response time	Implementation of PAT tools allows for quicker operating feedback, process correction, and adjustment in real time
Product specification and quality	Post-production batch records are primary quality control method	Quality of product can be continuously monitored and adjusted in real time during manufacturing operations
Manufacturing process control	Post-production testing of finished and intermediate products	Real-time online analysis of finished and intermediate products, resulting in a significantly reduced holdup and storage period
Product life cycle considerations	Out of specification (OOS) and related manufacturing problems evaluated for acceptance (quality) subsequent to batch completion	PAT and related options allow for real-time analysis of potential OOS products, thus minimizing or eliminating potential quality problems

Source: Form Winkle [14].

Here, the goal is to shorten the time required to deliver a product to the ultimate customer without additional risks. While QbD appears to have wide application to liquid process, there is also a need to incorporate the QbD concept in solids processing, specifically tablet and capsule manufacturing.

References

1 May, E. (2012). Lean/Six Sigma Overview, Part 1. Presented at the New Jersey Institute of Technology, Newark, NJ (20 June 2012).
2 Lieberman, G. (2011). Apply Six Sigma for process improvement and problem solving. *Chemical Engineering Progress* 107 (3): 53–60.
3 Williams, M. (2007). *CSSBB Primer* (the Certified Six Sigma Black Belt Primer), 2e. West Terra Haute, IN: Quality Council of Indiana (1 August 2007).
4 Lininger, A. (2012). A better way to calculate equipment ROI. *DC VELOCITY* 10 (7): 91–96.
5 Douglas, M. (2012). Navigating pharma logistics. *Inbound Logistics* 32 (8): 45–62.
6 Gygi, C., DeCarlo, N., and Williams, B. (2005). *Six sigma for dummies*. Hoboken, NJ: John Wiley & Sons, Inc.
7 Stagliano, A.A. (2004). *Rath & Strong's Six Sigma Advanced Tools Pocket Guide*. New York: McGraw-Hill.
8 May, E. (2006). Six Sigma Seminar. Fairfield, NJ (9 February 2006).
9 Guenther, D. and Mattley, Y. (2012). Real time oxygen and pH monitoring. *Genetic Engineering and Biotechnology News* 32 (16): 44–46.
10 Mendhe, R., Rathore, S., and Krull, I.S. (2012). Tools for enabling process analytical technology applications in biotechnology. *BioPharm International* 25 (8): 27–35.
11 Kippax, P. (2013). Effective application of particle size analysis from development to manufacture. *Powder & Bulk Solids* 31 (10): 25–29.
12 Strawn, T.T. (2012). Adding value to society. *Maintenance Technology* 25 (9): 36–38.
13 U.S. FDA (2004). *Guidance for Industry, PAT – A Framework for Innovative Pharmaceutical Development, Manufacturing, and Quality Assurance*. Rockville, MD: U.S. FDA.
14 Winkle, H.N. (2007). Implementing Quality by Design. Rockville, MD: Food and Drug Administration (24 September 2007).
15 Reid, G.L., Ward, H.W., Palm, A.S., and Muteki, K.. (2012). Process analytical technology (PAT) in pharmaceutical development. *American Pharmaceutical Review* 13 (4): 49–55.
16 Kelly, R.N. and Etzler, F.M. (2006). *What Is Wrong with Laser Diffraction?*, Special Edition, pp. 1–7. Donner Technologies. donner-tech.com
17 Watler, P. (2012). Bringing science to the plant. *Pharmaceutical Manufacturing* 9 (3): 31–33.
18 Regulatory challenges in the QbD paradigm (2012). *BioPharm International* 25 (9): 44–53.

Further Reading

Bernstein, L.A. (1965). *Statistics for Decisions: A Tool for Everybody*. New York: Grosset & Dunlap.
DePalma, A. (2016). See the bioprocess, be the bioprocess. *Genetic Engineering & Biotechnology News* 36 (15): 24, 26.
Dutton, G. (2008). In-line analytics improve manufacturing. *Genetic Engineering & Biotechnology News* 28 (15): 52–55.

Eckrich, T.M. (2013). Emerging technologies create opportunities in the API outsourcing industry. *Pharmaceutical Outsourcing* 14 (5): 4, 5.

Evans, J.R. and Lindsay, W.M. (2005). *The Management and Control of Quality*, 6e. Mason, OH: Thomson South-Western.

Larson, R. and Farber, B. (2006). *Elementary Statistics.* Upper Saddle River, NJ: Pearson Prentice Hall.

Li, M. (2010). Process analytical technology-based in-line buffer dilution. *Pharmaceutical Technology* (Supplement to October 2010 Issue) 34 (10): s18–s22.

Moroney, M.J. (1967). *Facts from Figures.* Baltimore, MD: Penguin Books.

Robertson, I.A., Tiwari, S.B., and Cabelka, T.D. (2012). Applying quality by design for extended – release hydrophilic matrix tablets. *Pharmaceutical Technology* 36 (10): 106–116.

Shotter, T. (2012). Understanding accuracy and precision. *Design World* 7 (11): 80–84.

Spiegel, M.R. and Stephens, L.J. (1998). *Theory and Problems of Statistics*, 3e. Schaum's Outline Series. New York: McGraw-Hill.

U.S. Army Materiel Command (1963). *Experimental Statistics, Section 1, Basic Concepts and Analysis of Measurement Data, AMCP 706-110*, U.S. Government (Department of the Army). Washington, DC.

U.S. Army Materiel Command (1975). *Engineering Design Handbook, Maintenance Engineering Techniques, AMCP 706-132.* Alexandria, VA.

PAT expected to streamline pharma process (2005). *Flow Control* 21 (3): 10.

United States Pharmacopeia (2007), Edition 30, Chapter 429. Rockville, MD.

Optimizing the tableting process with a quality – by design approach (2012). *Pharmaceutical Technology* 36 (5): 48–64.

Index

Practical Pharmaceutical Engineering, First Edition. Gary Prager.
© 2019 John Wiley & Sons, Inc. Published 2019 by John Wiley & Sons, Inc.